Assignment 1

wk1   chpts 1 & 2
                    2
wk2   chpt 3
wk2   chpt 4

HANDBOOK OF
SEMICONDUCTOR SILICON TECHNOLOGY

MATERIALS SCIENCE AND PROCESS TECHNOLOGY SERIES

Editors

Rointan F. Bunshah, University of California, Los Angeles *(Materials Science and Process Technology)*

Gary E. McGuire, Microelectronics Center of North Carolina *(Electronic Materials and Process Technology)*

DEPOSITION TECHNOLOGIES FOR FILMS AND COATINGS: by *Rointan F. Bunshah et al*

CHEMICAL VAPOR DEPOSITION IN MICROELECTRONICS: by *Arthur Sherman*

SEMICONDUCTOR MATERIALS AND PROCESS TECHNOLOGY HANDBOOK: edited by *Gary E. McGuire*

SOL-GEL TECHNOLOGY FOR THIN FILMS, FIBERS, PREFORMS, ELECTRONICS AND SPECIALTY SHAPES: edited by *Lisa A. Klein*

HYBRID MICROCIRCUIT TECHNOLOGY HANDBOOK: by *James J. Licari* and *Leonard R. Enlow*

HANDBOOK OF THIN FILM DEPOSITION PROCESSES AND TECHNIQUES: edited by *Klaus K. Schuegraf*

IONIZED-CLUSTER BEAM DEPOSITION AND EPITAXY: by *Toshinori Takagi*

DIFFUSION PHENOMENA IN THIN FILMS AND MICROELECTRONIC MATERIALS: edited by *Devendra Gupta* and *Paul S. Ho*

SHOCK WAVES FOR INDUSTRIAL APPLICATIONS: edited by *Lawrence E. Murr*

HANDBOOK OF CONTAMINATION CONTROL IN MICROELECTRONICS: edited by *Donald L. Tolliver*

HANDBOOK OF ION BEAM PROCESSING TECHNOLOGY: edited by *Jerome J. Cuomo, Stephen M. Rossnagel,* and *Harold R. Kaufman*

FRICTION AND WEAR TRANSITIONS OF MATERIALS: by *Peter J. Blau*

CHARACTERIZATION OF SEMICONDUCTOR MATERIALS—Volume 1: edited by *Gary E. McGuire*

SPECIAL MELTING AND PROCESSING TECHNOLOGIES: edited by *G.K. Bhat*

HANDBOOK OF PLASMA PROCESSING TECHNOLOGY: edited by *Stephen M. Rossnagel, Jerome J. Cuomo,* and *William D. Westwood*

FIBER REINFORCED CERAMIC COMPOSITES: edited by *K.S. Mazdiyasni*

HANDBOOK OF SEMICONDUCTOR SILICON TECHNOLOGY: edited by *William C. O'Mara, Robert B. Herring,* and *Lee P. Hunt*

HANDBOOK OF POLYMER COATINGS FOR ELECTRONICS: by *James J. Licari* and *Laura A. Hughes*

Related Titles

ADHESIVES TECHNOLOGY HANDBOOK: by *Arthur H. Landrock*

HANDBOOK OF THERMOSET PLASTICS: edited by *Sidney H. Goodman*

SURFACE PREPARATION TECHNIQUES FOR ADHESIVE BONDING: by *Raymond F. Wegman*

HANDBOOK OF SEMICONDUCTOR SILICON TECHNOLOGY

Edited by

William C. O'Mara

Rose Associates
Los Altos, California

Robert B. Herring

Thermco Systems
San Jose, California

Lee P. Hunt

Hunt Associates
Edmond, Oklahoma

Copyright © 1990 by Noyes Publications
No part of this book may be reproduced or utilized in any form or by any means, electronic or mechanical, including photocopying, recording or by any information storage and retrieval system, without permission in writing from the Publisher.
Library of Congress Catalog Card Number: 89-77167
ISBN: 0-8155-1237-6
Printed in the United States

Published in the United States of America by
Noyes Publications
Mill Road, Park Ridge, New Jersey 07656

10 9 8 7 6 5 4 3

Library of Congress Cataloging-in-Publication Data

Handbook of semiconductor silicon technology / edited by William C. O'Mara, Robert B. Herring, Lee Philip Hunt.
    p.   cm.
  Includes bibliographical references.
  ISBN 0-8155-1237-6 :
    1. Semiconductors--Handbooks, manuals, etc. 2. Silicon crystals--Handbooks, manuals, etc. I. O'Mara, William C. II. Herring, Robert B. III. Hunt, Lee Philip.
TK7871.85.H334   1990
621.381'52--dc20   89-77167
                                          CIP

Preface

Semiconductor silicon has become the most important and characteristic material of our age—the silicon age. It has achieved this distinction with a rather modest volume of production as compared to that of other basic industrial materials. For example, in 1989, about 6,000 metric tons of polysilicon were produced worldwide for silicon crystal growth, resulting in 3,000 tons of crystal produced in the United States, Japan, and Europe. This silicon crystal was converted to approximately 1,500 million square inches of wafer, or about 90 million individual wafers used for integrated circuit and discrete device production. For comparison, the annual worldwide production of steel and aluminum amounts to hundreds of thousands of tons.

In spite of its relatively small volume, the impact of silicon production is multiplied manyfold by the devices and electronic systems that are based on its properties. There have been many attempts to find improved materials with 'better' properties than silicon, but candidates such as sapphire, silicon carbide, diamond and III-V materials all lack some essential ingredient for manufacturing in quantity. Examples of these missing ingredients include: ease of growing large perfect crystals, freedom from extended and point defects, existence of a native oxide, or other essential properties, many of which are discussed in this book.

Basic information about silicon—how it is made, and its important physical, chemical and mechanical properties—is hard to find, and one of the motives for this volume is to make fundamental information available in handbook form. This also absolves the authors from having to include the relevant papers in their field that were published in the last twenty-four hours.

Early work in silicon science and technology was excellent, as evidenced by the fact that the original crystal growth process is still used in manufacturing today. That process was developed at Bell Laboratories by Teal and Buehler, following the original crystal growth process developed for germanium by Teal and Little. This initial work was done in spite of device engineers who were convinced that polycrystalline material would be adequate for transistor manufacture.

In the past, there has been only one book on semiconductor silicon technology, by Walt Runyan, formerly with Texas Instruments. Following his lead and inspiration, we have undertaken to produce a work of similar utility, since his original volume has gone out of print. Today it requires ten people to do what he was able to do by himself. In part this is testimony to the development that has occurred in all of these areas, and each chapter of the original work is now a separate discipline. We are fortunate to have excellent contributors for each of the topics discussed here, but we wish to salute Dr. Runyan for his original and enduring contribution to the field.

Los Altos, California
January 1990

William C. O'Mara

Contributors

W. Murray Bullis
Siltec Corporation
Mountain View, CA

Robert B. Herring
Thermco Systems
San Jose, CA

Lee P. Hunt
Hunt Associates
Edmond, OK

Theodore D. Kamins
Hewlett Packard Company
Palo Alto, CA

Richard L. Lane
Rochester Institute of
　Technology
Rochester, NY

H. Ming Liaw
Motorola, Inc.
Phoenix, AZ

William C. O'Mara
Rose Associates
Los Altos, CA

Leo C. Rogers
Polycrystalline Silicon Technology
　Corporation
Mesa, AZ

Dieter K. Schroder
Arizona State University
Tempe, AZ

Richard A. Seilheimer
Thermco Systems
San Jose, CA

NOTICE

To the best of our knowledge the information in this publication is accurate; however the Publisher does not assume any responsibility or liability for the accuracy or completeness of, or consequences arising from, such information. Mention of trade names or commercial products does not constitute endorsement or recommendation for use by the Publisher.

Final determination of the suitability of any information or product for use contemplated by any user, and the manner of that use, is the sole responsibility of the user. We recommend that anyone intending to rely on any recommendation of materials or procedures for semiconductor silicon technology mentioned in this publication should satisfy himself as to such suitability, and that he can meet all applicable safety and health standards. We strongly recommend that users seek and adhere to the manufacturer's or supplier's current instructions for handling each material they use.

Contents

1. **SILICON PRECURSORS: THEIR MANUFACTURE AND PROPERTIES** 1
   Lee P. Hunt
   - 1.0 Introduction 1
   - 2.0 Precursor Manufacture 2
     - 2.1 Metallurgical-Grade Silicon (MG-Si) 2
     - 2.2 Trichlorosilane 3
     - 2.3 Silicon Tetrachloride 5
     - 2.4 Dichlorosilane 7
     - 2.5 Silane 7
   - 3.0 Physical Properties and Critical Constants 9
     - 3.1 Non-Temperature Sensitive Properties and Constants 9
     - 3.2 Vapor Pressure 9
     - 3.3 Density (Liquid) 10
     - 3.4 Heat Capacity (Liquid) 11
     - 3.5 Heat Capacity (Gas) 12
     - 3.6 Free Energy, Standard Enthalpy, and Standard Entropy of Formation 14
     - 3.7 Enthalpy of Vaporization 19
     - 3.8 Surface Tension 20
     - 3.9 Viscosity (Gas) 21
     - 3.10 Viscosity (Liquid) 21
     - 3.11 Thermal Conductivity (Gas) 22
     - 3.12 Thermal Conductivity (Liquid) 24
   - 4.0 Safety 24
     - 4.1 Health Hazards 24
     - 4.2 Fire and Explosion Hazards 25
     - 4.3 Materials of Construction 26

Appendix..28
References...28

2. POLYSILICON PREPARATION............................33
Leo C. Rogers
- 1.0 The Technical History of Polycrystalline Silicon..........33
  - 1.1 Early and Present Polysilicon Manufacturers.........33
  - 1.2 Semiconductor-Grade Polycrystalline Silicon Precursors..............................35
    - 1.2.1 Silica (SiO_2).........................36
    - 1.2.2 Silicones............................36
    - 1.2.3 MG-Si and Ferrosilicon..................36
    - 1.2.4 Silicon Purity Beyond MG-Si..............37
    - 1.2.5 Feedstocks for Semiconductor-Grade Polysilicon.........................37
  - 1.3 Semiconductor-Grade Polysilicon...............38
  - 1.4 Potpourri of Other Methods to Manufacture Polysilicon..............................39
- 2.0 Polysilicon Production Technology (Most Practiced)......39
  - 2.1 The Feedstock System......................41
  - 2.2 Polysilicon Reactors.......................41
    - 2.2.1 General Bell-Jar Reactor Design............42
    - 2.2.2 General Bell-Jar Reactor Operation..........45
  - 2.3 Reactor Operation Criteria...................50
  - 2.4 Deposition Objectives for Polysilicon............51
  - 2.5 Reactor Exhaust-Gas Recovery–General Practice.....52
  - 2.6 Reactor Exhaust-Gas Recovery–General Design.....53
  - 2.7 Specific Reactor Exhaust—Gas Recovery Designs.....54
    - 2.7.1 Level-One Recovery System..............55
    - 2.7.2 Level-Two Recovery System..............57
    - 2.7.3 Closed-Loop Recovery System............57
  - 2.8 Capital and Operating Costs..................59
- 3.0 Alternative Chlorine-Based, Semiconductor-Grade, Polysilicon Feedstocks............................60
  - 3.1 Silane As A Polysilicon Feedstock..............61
    - 3.1.1 Silane Purity—Grade 6 to 10 for Bell-Jar Reactor Polysilicon....................61
    - 3.1.2 Silane Manufacturing Costs—Grade 4........61
    - 3.1.3 Silane Safety—Grade 4 for Chlorosilane-to-Silane, and 5 for Hexafluorosilicic-Silane.....62
    - 3.1.4 Silane Alternative Sources—Grade 0.........63
    - 3.1.5 Silane Transportability—Grade 2..........63
    - 3.1.6 Silane Storage—Grade 4................63
    - 3.1.7 Silane By-Product Recovery—Grade 2.......63
    - 3.1.8 Silane By-Product Use—Grade 5 for Chlorosilane-to-Silane; Grade 3 for Hexafluorosilicic-Silane..............................63
    - 3.1.9 Silane Deposition Rate—Grade 3..........64

Contents xi

	3.1.10	Silane Construction—Grade 5 for Chlorosilane-to-Silane, and 3 for Hexafluorosilicic-Silane 65
	3.1.11	Silane Reactor Choices—Grade 5 Going to 7 ... 65
	3.1.12	Silane Electrical Energy Usage—Grade 5/10 ... 65
3.2	Dichlorosilane As A Polysilicon Feedstock 66	
	3.2.1	Dichlorosilane Purity—Grade 7. 66
	3.2.2	Dichlorosilane Manufacturing Costs—Grade 6 .. 66
	3.2.3	Dichlorosilane Safety—Grade 3 66
	3.2.4	Dichlorosilane Alternative Sources—Grade O .. 66
	3.2.5	Dichlorosilane Transportability—Grade 2 67
	3.2.6	Dichlorosilane Storage—Grade 3 67
	3.2.7	Dichlorosilane By-Product Recovery—Grade 3 67
	3.2.8	Dichlorosilane Use of By-Products—Grade 5 ... 68
	3.2.9	Dichlorosilane Deposition Rate—Grade 5 68
	3.2.10	Dichlorosilane Construction Methods—Grade 5 68
	3.2.11	Dichlorosilane Reactor Choices—Grade 3 68
	3.2.12	Dichlorosilane Electrical Energy Usage—Grade 5 69
3.3	Trichlorosilane As A Polysilicon Feedstock 69	
	3.3.1	Trichlorosilane Purity—Grade 7 for Bell-Jar Polysilicon; Grade 5 for FBR Polysilicon 69
	3.3.2	Trichlorosilane Manufacturing Cost—Grade 7 .. 70
	3.3.3	Trichlorosilane Safety—Grade 9 71
	3.3.4	Trichlorosilane Alternate Sources—Grade 10... 71
	3.3.5	Trichlorosilane Transportability—Grade 9 71
	3.3.6	Trichlorosilane Storage—Grade 9 71
	3.3.7	Trichlorosilane By-Product Recovery—Grade 9 72
	3.3.8	Trichlorosilane By-Product Use—Grade 8 72
	3.3.9	Trichlorosilane Deposition Rate—Grade 7 72
	3.3.10	Trichlorosilane Construction Methods—Grade 9 74
	3.3.11	Trichlorosilane Reactor Choices—Grade 9 74
	3.3.12	Trichlorosilane Electrical Energy Usage—Grade 4 75
3.4	Silicon Tetrachloride As A Polysilicon Feedstock 75	
	3.4.1	Silicon Tetrachloride Purity—Grade 10 75
	3.4.2	Silicon Tetrachloride Manufacturing Costs—Grade 7 75
	3.4.3	Silicon Tetrachloride Safety—Grade 10 75
	3.4.4	Silicon Tetrachloride Transportation—Grade 9 75
	3.4.5	Silicon Tetrachloride Alternative Sources—Grade 9 Going to 3 76
	3.4.6	Silicon Tetrachloride Storage—Grade 9 76

            3.4.7   Silicon Tetrachloride By-Product Recovery—
                    Grade 10 76
            3.4.8   Silicon Tetrachloride By-Product Use—
                    Grade 9 76
            3.4.9   Silicon Tetrachloride Deposition Rate—
                    Grade 4 76
            3.4.10 Silicon Tetrachloride Construction Methods—
                    Grade 10 76
            3.4.11 Silicon Tetrachloride Reactor Choices—
                    Grade 9 77
            3.4.12 Silicon Tetrachloride Electrical Energy
                    Usage—Grade 1 77
     4.0   Alternate Polysilicon Reactor Selections 78
            4.1   Free-Space Polysilicon Reactors 78
            4.2   Polysilicon FBRs 81
     5.0   Evaluation of Semiconductor-Grade Polysilicon 84
     6.0   Future of Polysilicon 86
     References 88

3. CRYSTAL GROWTH OF SILICON 94
   H. Ming Liaw
     1.0   Introduction 94
     2.0   Melt Growth Theory 97
            2.1   Thermodynamic Consideration 97
            2.2   Heat Balance In Crystal Growth 99
            2.3   Crystal Growth Mechanisms 105
            2.4   Mass Transport of Impurities 106
            2.5   Constitutional Supercooling 121
     3.0   Practical Aspect of Cz Crystal Growth 125
            3.1   Crystal Pullers 125
            3.2   Dislocation-Free Growth 128
            3.3   Growth Forms and Habits 133
            3.4   Automatic Diameter Control 136
            3.5   Doping Techniques 138
            3.6   Variations in Radial Resistivity 140
            3.7   Oxygen and Carbon in Silicon 147
            3.8   Techniques for Control of Oxygen 154
            3.9   Control of Carbon Content 157
     4.0   Novel Czochralski Crystal Growth 157
            4.1   Semicontinuous and Continuous Cz 157
            4.2   Magnetic Czochralski (MCz) Crystal Growth 165
            4.3   Square Ingot Growth 166
            4.4   Web and EFG Techniques 167
            4.5   The Float-Zone Technique 173
     5.0   Trends in Silicon Crystal Growth 177
     6.0   Summary and Conclusion 179
     References 181

4. SILICON WAFER PREPARATION 192
Richard L. Lane

- 1.0 Introduction. 192
  - 1.1 Wafer Preparation Processes. 193
  - 1.2 Silicon Removal Principles 196
    - 1.2.1 Mechanical Removal 197
    - 1.2.2 Chemical Removal. 198
    - 1.2.3 Chemical-Mechanical Removal (Polishing) . . . 201
- 2.0 Crystal Shaping. 201
  - 2.1 Cropping. 202
  - 2.2 Grinding 203
  - 2.3 Orientation/Identification Flats. 205
  - 2.4 Etching. 206
- 3.0 Wafering 207
  - 3.1 Historical. 207
  - 3.2 The ID Blade 207
  - 3.3 Blade Tensioning. 209
  - 3.4 Process 211
    - 3.4.1 Crystal Mounting. 211
    - 3.4.2 Orientation 212
    - 3.4.3 Blade Condition 212
    - 3.4.4 Wafering Variables. 214
  - 3.5 Equipment. 214
  - 3.6 Unconventional Wafering Methods 217
    - 3.6.1 Slurry Sawing. 217
    - 3.6.2 Other Methods 217
- 4.0 Edge Contouring. 217
  - 4.1 Background 218
  - 4.2 Reasons for Edge Contouring. 218
    - 4.2.1 Silicon Chips and Wafer Breakage 220
    - 4.2.2 Lattice Damage. 220
    - 4.2.3 Epitaxial Edge Crown. 220
    - 4.2.4 Photoresist Edge Bead 220
  - 4.3 Commercial Equipment 221
- 5.0 Lapping. 222
  - 5.1 Background 222
  - 5.2 Current Technology. 223
  - 5.3 Wafer Grinding 226
- 6.0 Polishing 226
  - 6.1 Description of Polishing 226
  - 6.2 Historical Background 227
  - 6.3 Current Polishing Practice. 230
    - 6.3.1 Polishing Variables. 231
    - 6.3.2 The Optimum Polishing Process 233
    - 6.3.3 Other Methods of Polishing. 233
- 7.0 Cleaning 234
  - 7.1 Mechanical Cleaning. 235
  - 7.2 Chemical Cleaning. 236

xiv Contents

	7.3	Other Cleaning Methods	236
	7.4	Equipment	237
8.0	Miscellaneous Operations		238
	8.1	Heat Treatment	238
	8.2	Backside Damage	239
	8.3	Wafer Marking	240
	8.4	Packaging	241
9.0	In-Process Measurements		241
	9.1	Wafer Specifications and Industry Standards	242
	9.2	Mechanical Measurements	244
		9.2.1 Diameter and Flat Length	244
		9.2.2 Crystallographic Orientation	245
		9.2.3 Thickness and Thickness Variation	245
		9.2.4 Flatness	246
		9.2.5 Bow and Warp	247
		9.2.6 Edge Contour	248
		9.2.7 Surface Inspection	248
10.0	Discussion		249
References			251

5. SILICON EPITAXY 258
Robert B. Herring

1.0	Introduction		258
	1.1	Homoepitaxy and Heteroepitaxy	259
	1.2	Applications of Epitaxial Layers	259
		1.2.1 Discrete and Power Devices	259
		1.2.2 Integrated Circuits	260
		1.2.3 Epitaxy for MOS Devices	262
	1.3	Epitaxy As the Complement to Ion Implantation	262
2.0	Techniques for Silicon Epitaxy		262
	2.1	Chemical Vapor Deposition	262
	2.2	Molecular Beam Epitaxy	263
		2.2.1 MBE—Process Description	264
		2.2.2 MBE Equipment	265
	2.3	Liquid Phase Epitaxy (LPE)	270
	2.4	Solid Phase Regrowth	273
		2.4.1 Regrowth of Amorphous Layers	273
		2.4.2 Recrystallization of Thin Films	273
3.0	Surface Preparation for Silicon Epitaxial Growth		275
	3.1	Surface Cleaning and Oxide Removal	275
		3.1.1 Surface Precleans	275
		3.1.2 Drying the Wafers	276
	3.2	Insitu Gas Phase Cleans	277
		3.2.1 Removal of the Surface Oxide	277
		3.2.2 Removal of Adsorbed Water Vapor	279
		3.2.3 Oxide Removal	283
		3.2.4 Carbon on the Surface	283
	3.3	Insitu Etching	284

- 4.0 Growth of Silicon Epitaxy by CVD 286
  - 4.1 Growth Chemistries 286
    - 4.1.1 Disproportionation 287
    - 4.1.2 Pyrolytic Decomposition 288
    - 4.1.3 Reduction of Chlorosilanes 290
  - 4.2 Growth Kinetics and Mechanisms 293
    - 4.2.1 Kinetics of Growth from Silane 294
    - 4.2.2 Kinetics of Growth from Dichlorosilane 295
    - 4.2.3 Kinetics of Growth from Trichlorosilane and Silicon Tetrachloride 296
  - 4.3 Nucleation 299
    - 4.3.1 Homogeneous Nucleation 299
    - 4.3.2 Heterogeneous Nucleation 299
- 5.0 Dopant Incorporation 301
  - 5.1 Intentional Dopant Incorporation 301
    - 5.1.1 Measurements of Dopant Incorporation 302
    - 5.1.2 Effect of Temperature 303
    - 5.1.3 Effect of Growth Rate 303
    - 5.1.4 Effect of Pressure 305
  - 5.2 Unintentionally Added Dopants (Autodoping) 305
    - 5.2.1 Sources of Autodoping 305
    - 5.2.2 Lateral Autodoping 308
    - 5.2.3 Suppression of Autodoping 309
- 6.0 Surface Morphology and Epitaxial Defects 313
  - 6.1 Substrate Orientation Effects 313
  - 6.2 Spikes and Epitaxial Stacking Faults 313
    - 6.2.1 Growth Spikes 313
    - 6.2.2 Epitaxial Stacking Faults 314
  - 6.3 Hillocks and Pyramids in Epitaxial Layers 316
  - 6.4 Dislocations and Slip 317
  - 6.5 Microprecipitates (S-pits) 320
- 7.0 Pattern Shift and Distortion 321
  - 7.1 Patterned Diffusions (Buried Layers) 321
  - 7.2 Pattern Distortion and Pattern Shift 322
  - 7.3 Pattern Shift Definitions 323
  - 7.4 Role of Crystallography of Surface 323
  - 7.5 Summary of Shift and Distortion Effects 324
- 8.0 Equipment for Epitaxy by CVD 325
  - 8.1 Classification of Commercial Reactors by Flow Geometry 326
    - 8.1.1 Horizontal Reactors 327
    - 8.1.2 Vertical Flow Reactors 329
    - 8.1.3 Cylinder Reactor Geometry 330
    - 8.1.4 Other Reactor Types 330
  - 8.2 Heating Techniques 331
    - 8.2.1 Resistance Heating 332
    - 8.2.2 R.F. Induction Heating 332
    - 8.2.3 Radiant Heating 334

xvi Contents

		8.2.4 Combined Mode Heating................ 334
	8.3	Operating Pressure of Reactors................. 334
9.0	Trends for the Future in Silicon Epitaxy............. 335
References.. 336

6. SILICON MATERIAL PROPERTIES 347
W. Murray Bullis

- 1.0 Introduction................................. 347
- 2.0 Crystallographic Properties 349
  - 2.1 Silicon Crystal Structure..................... 349
  - 2.2 Crystal Habit 352
  - 2.3 Crystal Orientation 355
  - 2.4 Crystal Defects........................... 362
    - 2.4.1 Point Defects 362
    - 2.4.2 Extended Defects 368
- 3.0 Electrical Properties........................... 371
  - 3.1 Bands and Bonds in Pure Silicon Crystals........ 371
  - 3.2 Dopant Impurities......................... 372
  - 3.3 Statistics 374
    - 3.3.1 Fermi Function...................... 375
    - 3.3.2 Density-of-States Function 376
    - 3.3.3 Intrinsic Carrier Density 378
    - 3.3.4 Qualitative Description of the Energy Structure of Silicon 379
    - 3.3.5 Actual Band Structure of Silicon 383
  - 3.4 Electronic Conduction 385
  - 3.5 Electrical Characterization 390
  - 3.6 Conduction at High Electric Fields 394
  - 3.7 Conduction In A Magnetic Field............... 397
  - 3.8 Deep-Level Impurities 399
  - 3.9 Rectification 405
  - 3.10 Thermoelectric Effects..................... 407
- 4.0 Optical Properties 409
  - 4.1 Index of Refraction and Reflectivity............ 410
  - 4.2 Antireflection Coatings..................... 410
  - 4.3 Relationships Between Wavelength, Wavenumber, and Photon Energy 412
  - 4.4 Absorption.............................. 412
    - 4.4.1 Photoconductivity.................... 413
    - 4.4.2 Lattice Absorption 416
    - 4.4.3 Impurity Absorption 417
    - 4.4.4 Free Carrier Absorption 419
  - 4.5 Optical Methods for Detecting Dopant Impurities ... 419
  - 4.6 Emissivity 420
- 5.0 Thermal and Mechanical Properties................ 422
  - 5.1 Elastic Constants.......................... 422
    - 5.1.1 Young's Modulus..................... 424
    - 5.1.2 Modulus of Compression............... 424

		5.1.3	Shear Modulus 425
		5.1.4	Poisson's Ratio 425
		5.1.5	Other Relationships..................425
	5.2	Piezoresistivity 425
	5.3	Mechanical Strength and Plastic Deformation 427
		5.3.1	Plastic Deformation...................428
		5.3.2	Warp................................430
		5.3.3	Fracture 430
	5.4	Thermal Expansion 431
	5.5	Thermal Conductivity 432
	5.6	Hardness 433
	5.7	Other Physical and Thermodynamic Properties 433
References..435

7. OXYGEN, CARBON AND NITROGEN IN SILICON............451
William C. O'Mara

- 1.0 Introduction....................................451
- 2.0 Properties of Dissolved Oxygen in Silicon...........452
    - 2.1 Solubility.................................453
    - 2.2 Diffusivity................................456
    - 2.3 Segregation Coefficient....................458
    - 2.4 Phase Diagram 460
- 3.0 Oxygen Cluster and Precipitate Formation...........463
    - 3.1 Precipitate Nucleation 463
    - 3.2 Precipitation from Solution.................467
    - 3.3 Influence of Thermal History on Precipitate Formation................................469
        - 3.3.1 Ingot Cooling History...................469
        - 3.3.2 Microscopic Growth Fluctuations.........471
        - 3.3.3 Retardation of Nucleation..............472
        - 3.3.4 Influence of Substrate Doping on Oxygen Precipitation.........................473
    - 3.4 Oxidation Induced Stacking Faults.............474
        - 3.4.1 Influence of Crystal Growth 476
        - 3.4.2 Influence of Dopant...................477
        - 3.4.3 Influence of Temperature and Time 478
        - 3.4.4 Elimination of Near Surface Stacking Faults..478
    - 3.5 Oxygen Out Diffusion and Wafer Surface Denuding..479
        - 3.5.1 Out Diffusion........................479
        - 3.5.2 Model for Out Diffusion...............480
        - 3.5.3 Measurement of Denuded Zone Depth......483
        - 3.5.4 Effect of Ambient....................484
    - 3.6 Intrinsic Gettering.........................484
        - 3.6.1 Internal Precipitation.................485
        - 3.6.2 Denuding and Precipitation for Device Processing 485
- 4.0 Quantitative and Qualitative Measurement of Oxygen.....487
    - 4.1 Quantitative Analysis by Chemical and Physical Methods 487

		4.1.1	Sample Fusion Analysis 488
		4.1.2	Activation Analysis 488
		4.1.3	Secondary Ion Mass Spectrometry 489
	4.2		Infrared Absorption....................... 489
		4.2.1	Absorption at 9 μm 489
		4.2.2	Interferences in Measurement of Oxygen Content. 492
	4.3		Interpretation of the Infrared Spectrum of Oxygen . . 495
		4.3.1	The Si_2O "Molecule". 495
		4.3.2	Low Temperature and Far Infrared Spectra . . 497
		4.3.3	Assignment of the 513 cm^{-1} Oxygen Absorption. 498
		4.3.4	Substitutional Oxygen in Silicon 499
		4.3.5	Properties of Substitutional Oxygen 503
		4.3.6	The A Center 506
5.0	Oxygen Thermal Donor 507		
	5.1		Occurrence and Properties. 507
		5.1.1	Kinetics of Formation 508
		5.1.2	Kinetic Models for Donor Formation 509
		5.1.3	Kinetics of Annihilation 514
	5.2		Influences on Donor Formation 516
		5.2.1	Effect of Dopants 516
		5.2.2	Effect of Carbon 516
		5.2.3	Oxygen Behavior at Donor Formation Temperature. 516
	5.3		Structural Models for Oxygen Thermal Donor. 517
		5.3.1	Thermal Donor in Germanium........... 517
		5.3.2	Infrared Absorption. 518
		5.3.3	Other Experimental Results. 519
		5.3.4	Model for the Oxygen Thermal Donor...... 519
6.0	Mechanical Strengthening and Wafer Warpage. 521		
	6.1		Dislocation Generation in Silicon 521
	6.2		Slip and Bow 522
	6.3		Role of Oxygen. 522
	6.4		Effects Due to Precipitation of Oxygen 522
7.0	Device Processing 524		
	7.1		Thermal Cycles and Process Simulation 524
	7.2		NMOS Circuits 526
	7.3		CMOS Circuits 526
	7.4		Bipolar Circuits. 526
	7.5		CCD Devices. 527
8.0	Carbon in Silicon. 527		
	8.1		Solubility. 527
	8.2		Segregation Coefficient. 527
	8.3		Diffusivity. 529
	8.4		State of Carbon in Silicon 529
	8.5		Complexes with Oxygen. 530
	8.6		Formation of Precipitates Due to Carbon. 531

9.0 Nitrogen in Silicon............................533
    9.1 Solubility and Phase Diagram................534
    9.2 Infrared Absorption.........................536
    9.3 Mechanical Strengthening....................536
 References...537

8. CARRIER LIFETIMES IN SILICON 550
   Dieter K. Schroder
   1.0 Introduction................................550
   2.0 Defects.....................................551
   3.0 Recombination Lifetime......................555
   4.0 Generation Lifetime.........................566
   5.0 The Role of Lifetime on Device Currents.....571
       5.1 Forward-Biased Diodes...................571
       5.2 Reverse-Biased Diodes...................574
       5.3 Non-Uniform Substrates..................578
   6.0 Lifetime Measurement Techniques.............581
       6.1 Recombination Lifetime..................582
           6.1.1 Photoconductive Decay............582
           6.1.2 Open Circuit Voltage Decay.......588
           6.1.3 Diode Reverse Recovery...........595
           6.1.4 Surface Photovoltage.............598
           6.1.5 Pulsed MOS Capacitor.............604
       6.2 Generation Lifetime.....................614
           6.2.1 Pulsed MOS Capacitor.............614
           6.2.2 Gate-Controlled Diode............617
   7.0 Summary.....................................624
   References......................................625
   List of Symbols.................................635

9. PREPARATION AND PROPERTIES OF POLYCRYSTALLINE-
   SILICON FILMS....................................640
   Theodore D. Kamins
   1.0 Introduction................................640
   2.0 Deposition..................................641
       2.1 Gas Dynamics............................643
       2.2 Wafer-to-Wafer Uniformity...............649
       2.3 Silicon Gas Sources.....................652
       2.4 Doping During Deposition................653
   3.0 Structure...................................655
       3.1 Nucleation and Surface Processes........655
       3.2 Evaluation Techniques...................660
       3.3 Dopant Diffusion........................665
       3.4 Grain Growth............................668
       3.5 Optical Reflection......................670
       3.6 Etch Rate...............................672
       3.7 Summary.................................672
   4.0 Oxidation 673

	4.1	Oxide Growth on Polysilicon..................673
	4.2	Oxide-Thickness Evaluation...................680
	4.3	Conduction Through Oxide Grown on Polysilicon...681
5.0	Conduction..684	
	5.1	Lightly and Moderately Doped Films...........685
	5.2	Heavily Doped Films..........................696
	5.3	Summary......................................698
6.0	Applications..698	
	6.1	Silicon-Gate MOS Technology..................699
	6.2	MOS Transistors in Polysilicon................706
	6.3	Bipolar-Circuit Applications..................711
	6.4	Polysilicon Diodes............................714
	6.5	Polysilicon Pressure Sensors...................715
	6.6	Device Isolation..............................715
	6.7	Summary......................................718

References..719
 Reviews...719
 Deposition..719
 Structure..720
 Oxidation...722
 Electrical Properties......................................723
 Applications..728

10. SILICON PHASE DIAGRAMS..................................731
Richard A. Seilheimer

1.0	Introduction...731		
2.0	Phase Diagrams...732		
	2.1	The Phase Rule..733	
	2.2	Free Energy of Alloy Systems.............................734	
		2.2.1	Complete Solubility....................736
		2.2.2	Eutectic and Peritectic Systems..........739
		2.2.3	Intermediate Phases....................746
	2.3	Solid Solutions..747	
		2.3.1	Limitations of Solubility................747
		2.3.2	Relative Size Factor....................747
		2.3.3	The Chemical Factor...................747
		2.3.4	Relative Valency......................751
		2.3.5	Lattice Type Factor....................751
3.0	Phase Changes..752		
	3.1	The Lever Rule..753	
	3.2	Intermediate Phases....................................753	
4.0	Techniques for Determination of Phase Diagrams.......754		
	4.1	Determination of Ternary Phase Diagrams..........755	
	4.2	Silicon Phase Diagrams..................................756	
5.0	Segregation Coefficient and Zone Refining...............756		
6.0	Retrograde Solubility...761		
7.0	Silicon Phase Diagrams...763		

References..787

INDEX..789

1

Silicon Precursors: Their Manufacture and Properties

Lee P. Hunt

1.0 INTRODUCTION

Silicon precursors are defined within this chapter to be silicon materials used to produce semiconductor silicon as either a bulk material or a thin film. Bulk silicon is polycrystalline silicon (polysilicon) that is converted into single crystalline ingots via Czochralski or float-zoning methods of crystal growth. Thin-film silicon — including epitaxial, amorphous, and polysilicon layers — is produced by various chemical-vapor-deposition processes during fabrication of devices.

The major precursor of industrial importance for the production of polysilicon is trichlorosilane ($SiHCl_3$). As of 1984, it was used for 98% of the polysilicon manufactured (1). The remaining 2% was produced from silicon tetrachloride ($SiCl_4$) and silane (SiH_4). However, these percentages are changing as more polysilicon is manufactured by Union Carbide and Ethyl corporations via new processes using SiH_4 for decomposition (2),(3).

Thin-film silicon is produced from dichlorosilane (SiH_2Cl_2), $SiHCl_3$, $SiCl_4$, and SiH_4.

This chapter describes manufacturing processes used to produce the silicon precursors. The chapter also details the physical and chemical properties of the precursors. The chemical properties are discussed more in terms of precursor handling and safety since the unique chemistry of each precursor is described in other chapters.

2.0 PRECURSOR MANUFACTURE

Silica, SiO_2, is the original source of silicon for all the precursors. It is a component of almost every rock formation accessible by mining. Quartz and quartzite are relatively pure forms of silica that are used in metallurgical processes (4),(5) as indicated in Figure 1. The impure products of these processes, metallurgical silicon, silicon carbide, and various grades of silicon carbide, are in turn converted to intermediate chemicals, as discussed in detail later. Figure 1 shows the relationship between the metallurgical, intermediate, and semiconductor device industries.

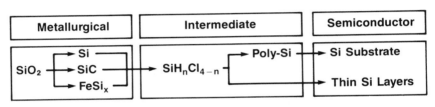

Figure 1. Flow of silicon across industries.

Several levels of the intermediate industry exist which consist of manufacturing of precursors, polysilicon, single-crystal ingots, and wafers for use as silicon substrates in device manufacturing. Companies have different levels of integration across these three industries.

2.1 Metallurgical-Grade Silicon (MG-Si)

The greatest demand (60%) for MG-Si is in the aluminum industry where it is an added in small (a few percent) amounts to improve machineability, castability, and corrosion resistance. The silicone industry has a smaller demand (25%) for MG-Si. The combined demand from these two industries is several hundred thousand tons per year. By contrast, the demand for MG-Si by the semiconductor industry is only a few thousand tons per years.

MG-Si is typically produced by the carbothermic reduction of quartzite in a 10,000 metric-ton-per-year, submerged-electrode arc furnace (6),(7) as indicated by the overall reaction:

$$SiO_2(s) + 2C(s) \rightarrow Si(l) + 2CO(g) \tag{1}$$

The major carbon sources are coal and coke while wood chips are used to add porosity to the reacting mass in order to allow escape of the product gases CO and SiO. Complex reactions take place between reactants and the SiC(s) and SiO(g) intermediates at temperatures that range up to 2300 K between adjacent electrodes (8),(9).

The manufactured MG-Si product is about 98 to 99-% pure after it is gas refined with chlorine or oxygen. Major impurities are aluminum and iron, as seen in Table 1.

Table 1

Impurities in Gas-Refined MG-Si Measured by Atomic Emission Spectroscopy

Impurity	Concentration (%)	
	Cl_2 Refined	O_2 Refined
Al	0.20	0.45
B	<0.002	<0.002
Ca	<0.05	0.07
Cr	0.034	0.030
Fe	0.46	0.65
Mn	0.025	0.024
Ni	0.017	0.015
Ti	0.035	0.045
V	0.032	0.024

<0.05 - 0.001 %: Ag, As, Ba, Bi, Cd, Cu, Mg, Mo, P, Pb, Sb, Sn, Zn, Zr

2.2 Trichlorosilane

Trichlorosilane is the major precursor for bulk polysilicon. Since its chemistry is intimately tied with that of $SiCl_4$, a process train is diagrammed in Figure 2 for the production of polysilicon. Other process-flow diagrams have been published (10)-(12).

MG-Si is ground in a ball mill until 75 % of the particles are <40 μm. The MG-Si particles are reacted with

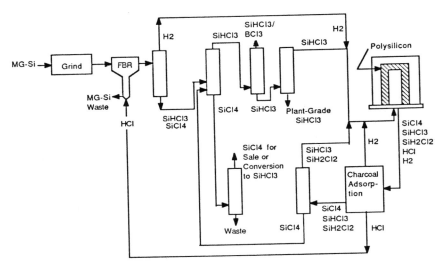

Figure 2. Process flow diagram for silicon involved in trichlorosilane decomposition process to produce semiconductor-grade polysilicon.

anhydrous hydrogen chloride in a fluidized bed reactor at 575 K under exothermic conditions to give approximately a 90-% yield of $SiHCl_3$, the rest being mainly $SiCl_4$:

$$Si(s) + 3HCl(g) \rightarrow SiHCl_3(g) + H_2(g) \qquad (2)$$

Small particles, such as $AlCl_3$, are filtered from the gas stream exiting the fluidized bed reactor. Such particles arise from impurities in MG-Si.

Trichlorosilane can be purified to a very high degree by fractional distillation because its boiling point differs from that of the impurity chlorides (see Table 2). Additional chemical purification can reduce electrically active impurities to a concentration ≤ 1 ppba (parts per billion atomic). Water treatment can be used to remove impurities such as BCl_3 by its adsorption on the hydrolysis products of $SiHCl_3$ (12). Ingle and Darnell (13) have shown the reduction of boron and phosphorus concentrations to <1 ppba by complexing them with $SiHCl_3$ that had been partially oxidized by oxygen.

Trichlorosilane can also be produced by the reaction of $SiCl_4$ with H_2 in a fluidized bed of MG-Si in the presence of a copper catalyst:

$$3SiCl_4(g) + 2H_2(g) + Si(s) \xrightarrow{Cu} 4SiHCl_3(g) \qquad (3)$$

Table 2

Comparison of Boiling Points of Silicon Precursors, HCl, and Impurity Chlorides

Compound	Boiling Point (K)
SiH_4	161.3
HCl	188.2
SiH_3Cl	242.8
SiH_2Cl_2	281.5
BCl_3	285.9
$SiHCl_3$	305.0
$SiCl_4$	330.5
PCl_3	347.4
$AsCl_3$	404.6
$SbCl_3$	492

The effects of temperature ($\approx$800 K) and pressure ($\approx$3 MPa) can produce up to 37-% conversions of $SiCl_4$ to $SiHCl_3$ (2), (14)-(16). The reaction is particularly significant for reducing the manufacturing cost of polysilicon since some polysilicon processes produce $SiCl_4$ as a by-product. This reaction allows a very efficient recycle of both $SiCl_4$ as well as H_2.

Weigert, Meyer-Simon, and Schwartz (17) have described a process that also has considerable potential for conversion of $SiCl_4$ to $SiHCl_3$. A mixture of $SiCl_4$ and H_2 is heated to about 1175 K. This temperature is sufficient to allow the materials to react and approach equilibrium, thereby producing a conversion to $SiHCl_3$ as high as 37 %. The mixture is quenched in a condenser in order to bring the high-temperature equilibrium distribution of products to room temperature.

2.3 Silicon Tetrachloride

Siemens Process: The Siemens process uses the hydrogen reduction of $SiHCl_3$ to produce bulk polysilicon (12),(18),(19):

$$SiHCl_3(g) + H_2(g) \xrightarrow{\approx 1375 \text{ K}} Si(s) + 3HCl(g) \qquad (4)$$

Since H_2 serves as a carrier gas for $SiHCl_3$, approximately a 10- to 20-fold excess of H_2 is used on a molar basis. Although the above reaction appears to be the reverse of the $SiHCl_3$ synthesis reaction (see Reaction 2), it is in fact an oversimplification. Since silicon rods are heated to a surface temperature of ≈ 1375 K, other reactions occur because of the large temperature gradients within the reactor system. Silicon tetrachloride is a major by-product of the silicon manufacturing process and is presumably produced by a reaction such as:

$$SiHCl_3(g) + HCl(g) \rightarrow SiCl_4(g) + H_2(g) \tag{5}$$

Through recycle of the $SiHCl_3$, approximately 2/3 of it is converted to $SiCl_4$ as indicated by the summary reaction:

$$3SiHCl_3(g) \rightarrow Si(s) + 2SiCl_4(g) + HCl(g) + H_2(g) \tag{6}$$

Since $SiCl_4$ is part of the overall recycle process that involves distillation, it can be easily purified and sold as a silicon source for depositing thin-film silicon. $SiCl_4$ is also sold to nonsemiconductor industries that convert it into SiO_2 for use as fumed silica or for fabrication of optical wave guides. $SiCl_4$ can also be converted into $SiHCl_3$, as discussed in the previous section.

The chemistry of the $SiHCl_3$ reduction process is more complex than is indicated above since the concentration of $SiCl_2(g)$ becomes significant at higher temperatures near the silicon rods (20). The importance of $SiCl_2$ in chlorosilane chemistry at high temperatures will become apparent in the section of free energies.

Processes Involving Metallurgical Products: Silicon tetrachloride can be produced by direct chlorination of silicon carbide (21):

$$SiC(s) + 2Cl_2(g) \rightarrow SiCl_4(g) + C(s) \tag{7}$$

Distillation is required to upgrade $SiCl_4$ to semiconductor quality.

Silicon tetrachloride is produced as a by-product during the manufacture of $ZrCl_4$ (22) and $TiCl_4$ (23). Since SiO_2 is an impurity in the titanium and zirconium ores, it is transformed into $SiCl_4$ by a reaction similar to the following reaction used to produce $TiCl_4$ in a fluidized bed reactor:

$$TiO_2(s) + 2Cl_2(g) + 2C(s) \xrightarrow{\approx 1175 \text{ K}} TiCl_4(g) + 2CO(g) \tag{8}$$

2.4 Dichlorosilane

Dichlorosilane can be readily manufactured by the catalytic rearrangement of $SiHCl_3$ at a temperature of about 350 K using polymeric resins as a catalyst (2),(24):

$$2SiHCl_3(g) \rightarrow SiH_2Cl_2(g) + SiCl_4(g) \qquad (9)$$

Approximately 10-mole-% SiH_2Cl_2 is produced per pass through the resin bed; the products are separated and purified by distillation. A small amount (≈ 0.3 mole %) of SiH_3Cl (monochlorosilane) is also produced due to the rearrangement of SiH_2Cl_2:

$$2SiH_2Cl_2(g) \rightarrow SiH_3Cl(g) + SiHCl_3(g) \qquad (10)$$

Details of this process are more completely described in the section on silane.

The Siemens process for producing polysilicon from $SiHCl_3$ results in up to 1 mole % of the $SiHCl_3$ being converted into SiH_2Cl_2 because of the high temperatures occurring in the reactor. The SiH_2Cl_2 can be either recycled to the decomposition reactor (see Figure 2) or separated during the distillation process as a by-product for sale.

2.5 Silane

Silane can be produced in relatively small quantities, sufficient for decomposition in thin-film processes, by reducing $SiCl_4$ with metallic hydrides. One molten salt process (25)-(26) involves dissolving LiH in a LiCl/KCl melt and reacting $SiCl_4$ according to the reaction:

$$4LiH(l) + SiCl_4(g) \xrightarrow{675 \text{ K}} SiH_4(g) + 4LiCl(l) \qquad (11)$$

There is potential of by-product recycling on a larger scale of production since the melt from the above reaction can be electrolyzed as shown below:

$$4LiCl(l) \rightarrow 4Li(l) + 2Cl_2(g). \qquad (12)$$

The chlorine could be reacted with silicon to produce more $SiCl_4$ and the lithium reacted with H_2 to regenerate LiH as follows:

$$Si(s) + 2Cl_2(g) \rightarrow SiCl_4(g) \tag{13}$$

$$4Li(l) + 2H_2(g) \rightarrow 4LiH(l) \tag{14}$$

The net reaction of this reaction sequence is:

$$Si(s) + 2H_2(g) \rightarrow SiH_4(g) \tag{15}$$

Magnesium silicide, produced from the elements, is used to produce larger quantities of SiH_4 via the reaction:

$$Mg_2Si + 4NH_4Cl \rightarrow SiH_4 + 2MgCl_2 + 4NH_3 \tag{16}$$

Since silane can be more easily purified to a higher degree than chlorosilanes, it can be decomposed in a Siemens-type reactor to produce very high-purity polysilicon. Compared to chlorosilanes, lower silicon deposition occurs due to the lower temperature and pressure that must be used to prevent gas phase nucleation of silane in the reactor.

A low-cost method has been developed to produce large quantities of SiH_4 for decomposition in either a Siemens-type reactor or in a fluidized bed reactor (2). The Union Carbide process uses several of the reactions described above and involves high recycling of the chlorosilanes. The process is represented by the following sequence of reactions:

$$3SiCl_4(g) + 2H_2(g) + Si(s) \xrightarrow{Cu} 4SiHCl_3(g) \tag{3}$$

$$2SiHCl_3(g) \rightarrow SiH_2Cl_2(g) + SiCl_4(g) \tag{9}$$

$$2SiH_2Cl_2(g) \rightarrow SiH_3Cl(g) + SiHCl_3(g) \tag{10}$$

$$2SiH_3Cl(g) \rightarrow SiH_4(g) + SiH_2Cl_2(g) \tag{17}$$

$$SiH_4(g) \rightarrow Si(s) + 2H_2(g) \tag{18}$$

The SiH_2Cl_2 and SiH_3Cl rearrangement steps are catalyzed reactions similar to that used to rearrange $SiHCl_3$.

Ethyl Corporation has developed a large-scale process to produce SiH_4 for a 1250 metric-ton/yr polysilicon plant based on fluidized bed technology (3). Their starting material is H_2SiF_6, a by-product from the superphosphate fertilizer industry. The fluorosilicic acid is reacted with concentrated sulfuric acid to liberate SiF_4 by the reaction:

$$H_2SiF_6 + H_2SO_4 \rightarrow SiF_4 + 2HF \tag{19}$$

The SiF_4 is reduced by LiH at about 525 K in diphenyl ether. The SiH_4 is then decomposed in a fluidized bed of silicon powder according to Reaction 18. Lithium hydride is produced at 90-% yield by reacting H_2 with molten lithium in mineral oil (27).

3.0 PHYSICAL PROPERTIES AND CRITICAL CONSTANTS

3.1 Non-Temperature Sensitive Properties and Constants

Most of the data for SiH_4, SiH_2Cl_2, $SiHCl_3$, $SiCl_4$, and Si in Table 3 are summarized from Yaws et al. (28)-(31). Data for HCl and H_2 were taken from Lange's Handbook of Chemistry (32).

3.2 Vapor Pressure

Yaws et al. (28)-(31) have reviewed vapor pressure data for the silanes of interest. For SiH_4, data were available over most of the range from melting point through boiling point to critical point. They correlated the data with the following 5-term temperature equation which resulted in an average deviation from the actual data of <3.5 %:

$$\log p = a + b/T + c\log T + dT + eT^2 \tag{20}$$

Vapor pressure data for SiH_2Cl_2 were available from only 190 to 300 K. The vapor pressure curve was extrapolated over the melting point to critical point range using the 5-term vapor pressure correlation equation. The average deviation of the correlated vapor pressures from the actual data was ≈ 1 %.

The correlation equation was also used to extend $SiHCl_3$ vapor pressures beyond the range of available experimental data (190 K to boiling point). The average deviation of the regression data from the experimental data was ≈ 0.8 %.

Vapor pressure data for $SiCl_4$ were available between the melting and boiling points and at the critical point. The average deviation of the correlated from the experimental data was ≈ 0.7 %.

Table 3

Physical Properties and Critical Constants

Property	SiH$_4$	SiH$_2$Cl$_2$	SiHCl$_3$	SiCl$_4$	HCl	H$_2$	Si
Molecular Weight (kg/mol)	32.12	101.01	135.45	169.90	36.47	2.016	28.09
Autoign. Temp (K)	-	331[a]	488[b]	none[b]	-	-	-
Flash Point (K)	-	-	245[b]	none[b]	-	-	-
Melting Point (K)	88.5	151.2	146.6	203.8	159.0	13.96	1685
Boiling Point (K)	161.3	281.5	305.0	330.5	188.2	20.39	3151[c]
Critical Temp. (K)	270[d]	452[c]	479	507.2	324.6	33.24	5159[c]
Critical Pres. (kPa)	4840[d]	4460[c]	4050[c]	3750	8270	1299	53700[c]
Critical Volume (m^3/kmol)	0.130[c]	0.228[c]	0.268	0.326	0.0868	0.0650	0.233[c]
Critical Density (kg/m^3)	247[c]	442[c]	505	521	420	31.0	121[c]
Crit. Compres. Factor. Z_c	0..281[c]	0.276[c]	0.273[c]	0.290	0.266	-	-
Accentric Factor	0.0774	0.1107	0.188[c]	0.2556	-	-	-

[a]Reference 24 and 47
[b]Reference 39
[c]Estimated (28)-(31)
[d]Questionable Value (28) and (29)

Vapor pressure data as a function of temperature is shown in Figure 3. Data cover the range from the melting to critical point of each compound.

3.3 Density (Liquid)

Yaws et al. (28)-(31) have reviewed the data in the literature for the densities of the liquid silanes. Data for the density of SiH$_4$ have been determined at temperatures between its melting and boiling points. Density data for SiH$_2$Cl$_2$ are available only at its melting point and at 280 K. Densities have been measured for SiHCl$_3$ between 263 K and its critical point. Measurements of the liquid density of $SiCl_4$ have been made between its melting and critical points.

Yaws et al. (28)-(31) extended liquid densities over the entire range from the melting point to the critical point for each of the compounds. They used the Yaws-Shaw equation in their calculations:

$$d = AB^{-(1-T/T_c)^{2/7}} \qquad (21)$$

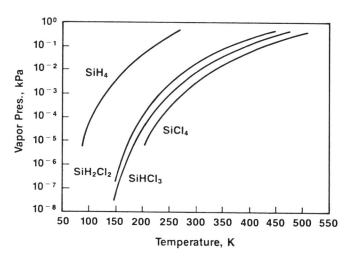

Figure 3. Estimated vapor pressures between melting and critical points.

A and B are constants determined by regression analysis and T_c is the critical temperature. A modified version of this equation was used for SiH_2Cl_2. Calculated densities varied from measured values by mean deviations of 1.5 % for SiH_4, 1 % for $SiHCl_3$, and 0.4 % for $SiCl_4$. Liquid densities appear as a function of temperature in Figure 4.

3.4 Heat Capacity (Liquid)

Measured liquid heat capacity data for the silanes have been reviewed by Yaws et al. (28)-(31). Heat capacity data are available for both SiH_4 and $SiCl_4$ in the range between their melting and boiling points. A constant value of 130 J/mol·K has been reported for $SiHCl_3$ between the temperatures of 298 and 333 K. No data appear in the literature for SiH_2Cl_2

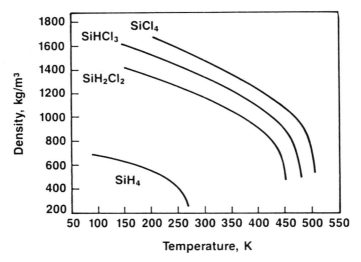

Figure 4. Estimated liquid densities between melting and critical points.

Yaws et al. extended liquid heat capacity data over the temperature range from the melting to the critical points of the silanes by assuming that the product of the heat capacity and the density is equal to a constant; or,

$$C_p(l) = A/d \tag{22}$$

where C_p is the heat capacity at constant pressure, A is a constant, and d is the density. More complex techniques were required to estimate the heat capacity of SiH_2Cl_2. The mean deviation of the calculated from the experimental data was 7 % for SiH_4, and 4 % for $SiHCl_3$ and $SiCl_4$. Plotted data appear in Figure 5. Yaws et al. reported that their data for SiH_2Cl_2 was an order-of-magnitude estimate. The error in the estimate is confirmed in Figure 5 where the SiH_2Cl_2 curve is seen to not fit within the envelop of the other curves in the homologous series of the silanes.

3.5 Heat Capacity (Gas)

Multiple regression analysis was performed on heat capacity data from the latest version of the JANAF Tables (33). Data are represented by the equation

$$C_p = a + bT + cT^2 + dT^{-2} \tag{23}$$

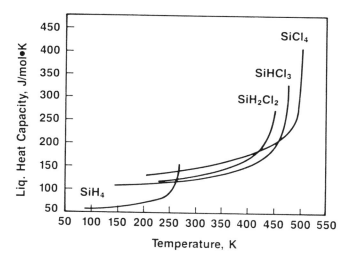

Figure 5. Estimated liquid heat capacities between melting and critical points.

over the temperature range from 298 to 2000 K. The equation contains an additional T^2 term, compared to that of a previous regression analysis reported by Hunt and Sirtl (34), which reduced the maximum error of the calculated from the tabulated data by a factor of between 2 and 4. The four constants in Equation 23 are listed in Table 4 for each of the silanes, as well as for $SiCl_2$, H_2, and HCl. The standard error of the regression (S_e) and the maximum percentage error (M_e) are also given. The maximum error in all cases occurs at the temperature of 400 K.

Table 4

Regression Data from the Gaseous Heat Capacity Equation

$$C_p = a + bT + cT^2 + dT^{-2} \text{ (J/mol·K)}$$

Compound	a	b x 10^3	c x 10^6	d x 10^{-6}	S_e	M_e
SiH_4	34.39	68.23	-17.79	-0.989	0.8	-1.2
SiH_2Cl_2	68.77	36.91	-9.73	-1.546	0.5	-0.8
$SiHCl_3$	86.18	20.77	-5.51	-1.488	0.3	-0.6
$SiCl_4$	103.91	4.80	-1.43	-1.345	0.2	-0.4
$SiCl_2$	56.80	1.68	-0.515	-0.531	0.06	-0.1
H_2	28.00	1.37	0.927	0.0438	0.1	0.2
HCl	24.43	8.48	-1.445	0.217	0.1	0.2

A plot of heat capacity for the silanes appears in Figure 6.

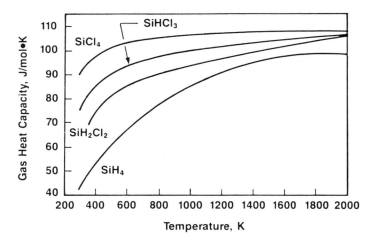

Figure 6. Gas heat capacities from 298 to 2000°K.

3.6 Free Energy, Standard Enthalpy, and Standard Entropy of Formation

The Gibb's free energy of formation, $\Delta Gf°_T$, for a compound can be calculated at a desired temperature, T, from the standard enthalpy of formation, $\Delta Hf°_{298}$, the standard entropy, $S°_{298}$, and the temperature function of the heat capacity, $C_p(T)$:

$$\Delta Gf°_T = \Delta Hf°_{298} + \int_{298}^{T} C_p dT - TS°_{298} - T\int_{298}^{T}(C_p/T)dT \qquad (24)$$

Heat capacities were expressed in the last section in equation form as function of temperature over the range from 298 to 2000 K. Consistent values of $Hf°_{298}$ and $S°_{298}$ were derived by Hunt and Sirtl (34) for the chlorosilanes in 1972. Since a more reliable value for $Hf°_{298}$ of $SiCl_4$ is now recognized (33), revisions were made in the standard enthalpies of formation of $SiHCl_3$ and $SiCl_2$ which have their values based upon that of $SiCl_4$. These revisions appear in the updated JANAF Tables (33). Walsh (35) reviewed $Hf°_{298}$ for the chlorosilanes and agreed that the JANAF data were reasonable except that for $SiHCl_3$. Three experimental values exist for $SiHCl_3$ in addition to the value reported by Hunt and Sirtl (34).

These are listed in Table 5. All values are reported on the same basis of Hf°$_{298}$ = -662.7 kJ/mol for SiCl$_4$(g).

Table 5

Standard Enthalpy of Formation Values for SiHCl$_3$(g) Based on

$$\Delta H f^{\circ}_{298} = -662.7 \, kJ/mol \, for \, SiCl_4(g)$$

$\Delta H f^{\circ}_{298}(kJ/mol)$	Reference
-494.1 ± 1.3	Schnegg, Rurlander, and Jacob (48)
-494.1 ± 4.2	Hunt and Sirtl (34); JANAF (33)
-498.7 ± 5.9	Wolf and Teichman (49)
-500.0 ± 6.3	Farber and Srivastava (50)
-497.3	Average

The JANAF value for SiHCl$_3$(g) appears to be a reasonable one based upon the other values and their associated errors. Therefore, it is recommended that the updated JANAF Tables (33) be used for all of the thermodynamic data for the chlorosilanes, with the exception of ΔHf°$_{298}$ for SiCl$_3$. The reason for not using the JANAF value for SiCl$_3$ is explained in an earlier paper (34) and is still believed to be valid. Standard enthalpy and standard entropy data for compounds at equilibrium in the Si-Cl-H system are summarized in Table 6.

Table 6

Standard Enthalpy and Entropy of Formation of Compounds at Equilibrium in the Si-Cl-H System

Compound	$\Delta H f^{\circ}_{298}(kJ/mol)$	$S^{\circ}_{298}(J/mol.K)$
SiH$_4$(g)	34	204.6
SiH$_2$Cl$_2$(g)	-320	286.6
SiHCl$_3$(g)	-496.2	313.6
SiCl$_4$(g)	-662.7	330.8
SiCl$_2$(g)	-169	281.2
H$_2$(g)	0	130.6
HCl(g)	-92.30	186.8
Si(s)	0	18.8

Free energies of formation from the JANAF Tables (33) are given in linear equation form as a function of temperature in Table 7. Standard errors (σ_e) and maximum residuals are listed for the linear regression equations.

Free energies of formation for silane and its selected derivatives are plotted in Figure 7. The compounds with the lowest free energies are the most stable. The $SiCl_2$ species is stable only at higher temperatures and can be responsible for significant and unexpected chemical changes in CVD systems.

Table 7

Linear Regression Data from Fitting JANAF (33) Free Energies between Temperatures of 400 and 1600 K to the Equation

$$\Delta G f_T^o (kJ/mol) = \frac{aT}{1000} + b$$

Species	a	b	σ_e	max.residual actual (%)
$SiCl_4(g)$	128.821	-660.549	0.3	0.1
$SiHCl_3(g)$	108.179	-497.612	0.1	0.1
$SiH_2Cl_2(g)$	96.455	-325.091	0.2	0.2
$SiH_4(g)$	97.666	24.732	0.5	1.8
$SiCl_2(g)$	-34.211	-171.537	0.5	0.4
$HCl(g)$	-6.653	-94.048	0.2	0.4
$H_2(g)$	0	0		
$Si(s)$	0	0		

Equilibrium Partial Pressure: The chemical equilibrium of a system can be calculated by minimizing its free energy. It is necessary to know ΔG_T for all significant chemical species at the temperature and pressures being considered. Equilibrium partial pressures have been calculated by Sirtl, Hunt, and Sawyer (20) over the temperature range of 300 to 1700 K while Herrick and Sanchez-Martinez (36) have extended the temperature range to 3000 K. An example is shown in Figure 8 of microcomputer calculations of equilibrium partial pressures as a function of temperature at Cl/H = 0.1 and at a total pressure of 1 bar. Calculations performed under other conditions for silicon deposition can be found elsewhere (37).

Silicon Precursors: Their Manufacture and Properties 17

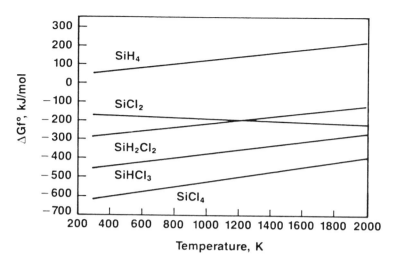

Figure 7. Gibb's free energies of formation from 298 to 2000°K.

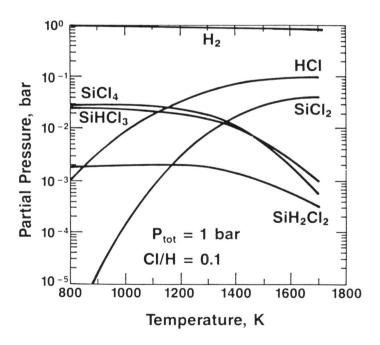

Figure 8. Equilibrium partial pressures in the Si-Cl-H system at Cl/H = 0.1 and at a total pressure of 1 bar.

It is necessary to be aware that equilibrium is not always reached under operating conditions. Equilibrium is reached only if favorable reaction rates exist. The degree to which a system approaches equilibrium must be determined through experimentation. High temperature is a favorable condition under which a system approaches equilibrium but it is not a sufficient condition.

Equilibrium Silicon Yield: The deposition and etching of silicon under given conditions can also be determined from equilibrium calculations. Both Hunt and Sirtl (34) and Herrick and Sanchez-Martinez (36) have presented Si/Cl versus Cl/H diagrams which allow, for example, calculation of the fractional amount of silicon that can be deposited from the Si-Cl-H system. The Cl/H atomic ratio uniquely defines the relative amounts of chlorosilane and hydrogen in a mixture of gases since neither Cl nor H leave the gas phase under normal conditions; see however, reference (38). The calculated Si/Cl atomic ratio in the gas phase is a measure of the amount of silicon in the gas phase at any time and the ratio changes as silicon leaves or enters the gas phase during either deposition or etching, respectively. Si/Cl versus Cl/H diagrams at given temperature and pressure conditions can be used to calculate the fractional silicon yield (η) of silicon between the initial (i) and final (f) conditions via the equation:

$$\eta = \frac{(Si/Cl)_i - (Si/Cl)_f}{(Si/Cl)_i} \qquad (25)$$

Figure 9 is an example of a microcomputer-calculated equilibrium diagram for Cl/H = 0.01 to 0.09 and for temperatures between 800 and 1500 K at 1 bar total pressure. If one were using $SiCl_4$ at 1100 K and Cl/H = 0.01, then $(Si/Cl)_i = 1/4 = 0.25$ and $(Si/Cl)_f = 0.18$ from the diagram. The calculated silicon yield is 0.28. Etching would occur for $SiCl_4$ if $(Si/Cl)_f > 0.25$.

Use of more negative values of $\Delta Hf°_{298}$ for $SiCl_4$, $SiHCl_3$, and SiH_2Cl_2, compared to those used previously by Hunt and Sirtl, results in these compounds having greater stability in the gas phase because their free energies are also more negative (see Equation (24)). Since more silicon exists in the gas phase, yields of deposited silicon are lower. An

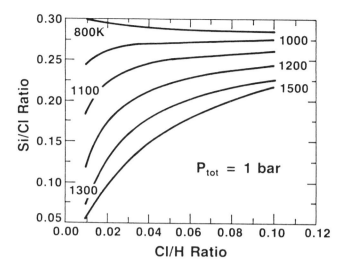

Figure 9. Si-Cl-H equilibrium diagram at various temperatures and at a total pressure of 1 bar.

example of sensitivity of equilibrium calculations of the $SiCl_4/SiHCl_3$ ratio to $\Delta Hf°_{298}$ of $SiHCl_3$ has been shown by Hunt and Sirtl. The Siemens process can be used as an example of the effect on silicon yield calculated with current thermodynamic data compared to that previously reported by Hunt and Sirtl. When the process is carried out at 1 bar total pressure, 1325 K, and Cl/H = 0.12, the yield using older thermodynamic data is 0.38, whereas, the yield with newer data is 0.31.

Detailed calculations of silicon yield are presented elsewhere (37).

3.7 Enthalpy of Vaporization

The enthalpy (heat) of vaporization has been measured only at the boiling point of each of the silanes. Yaws et al. (28)-(31) extended the data over the entire liquid range of each compound by using Watson's correlation:

$$\Delta H = \Delta H_b \left(\frac{T_c - T}{T_c - T_b} \right)^{0.38} \tag{26}$$

Enthalpy of vaporization for each of the silanes is graphically presented in Figure 10 for temperatures between their melting and critical points.

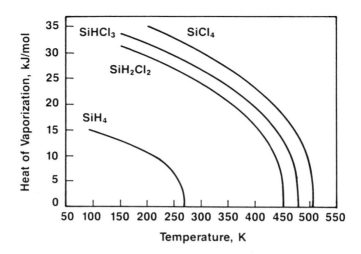

Figure 10. Estimated heats (enthalpies) of vaporization between melting and critical points.

3.8 Surface Tension

Surface tension data from the literature have been reviewed by Yaws et al. (28)-(31). Information exists for both SiH_4 and $SiCl_4$ between their melting and boiling points. The surface tension of $SiHCl_3$ has been measured over the temperature range from 273 to 313 K. No data exist for SiH_2Cl_2.

Yaws et al. extended surface tension data over the temperature range from the melting to the critical points of SiH_4, $SiHCl_3$, and $SiCl_4$ using the Othmer relationship:

$$\sigma = \sigma_1 \left[\frac{T_c - T}{T_c - T_1} \right]^n \tag{27}$$

where σ_1 is the surface tension (mN/m) at temperature T_1 (K). The correlation parameter, n, was 1.2 for both SiH_4 and $SiHCl_3$ and was 1.14 for $SiCl_4$. Correlated data agreed with experimental data within 1 % for both SiH_4 and $SiCl_4$ and to within 3 % for $SiHCl_3$.

A more complex technique was required to estimate the surface tension of SiH_2Cl_2. This technique showed good accuracy when used to compare estimated data for $SiHCl_3$ and $SiCl_4$ were compared to experimental data. Surface tension data for the compounds are plotted as a function of temperature in Figure 11.

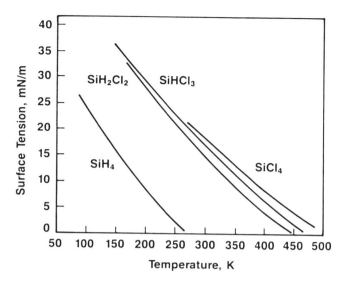

Figure 11. Estimated surface tensions between melting and critical points.

3.9 Viscosity (Gas)

Viscosity data were reviewed for the gaseous silanes by Yaws et al. (28)-(31). No viscosity data were available for SiH_2Cl_2 while only limited data appear in the literature for SiH_4, $SiHCl_3$, and $SiCl_4$. Various techniques were used to estimate viscosity data between temperatures of 273 and 1373 K and correlated data were generally within 2 % of the experimental data.

Gas viscosity data appear as a function of temperature in Figure 12. The average viscosities of the four compounds fall within ±1 μPa·s of one another, such that their values appear nearly as a band of curves in the figure. The data spread is ±10 % at 273 K and is ±3 % at 1373 K. The viscosities occur in a homologous sequence at 273 K. $SiCl_4$ exhibits the lowest viscosity over the entire temperature range. The viscosities of the other compounds are alternately higher or lower than one another at increasing temperatures due to the methods used in estimating the data.

3.10 Viscosity (Liquid)

Liquid viscosity data were reviewed by Yaws et al. (28)-(31). Data occur in the literature for SiH_4 between its melting and boiling points, for $SiHCl_3$ between 266 and 377

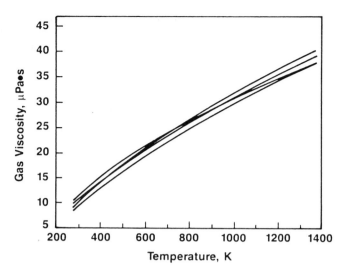

Figure 12. Estimated gas viscosities between 273 and 1373°K (see text).

K, and for $SiCl_4$ between its melting and critical points. No experimental viscosity data appear in the literature for liquid SiH_2Cl_2.

Yaws et al. used a 4-term equation to correlate the data for SiH_4 and $SiCl_4$ between their melting and critical points. The mean deviation of correlated from experimental data was within 1.4 % for SiH_4 and within less than 3 % for $SiCl_4$. Two equations were required to extend the liquid viscosity data for $SiHCl_3$ over the same range of temperature. Correlated data were within ±2 % of the experimental data. Since no experimental data were available for SiH_2Cl_2, the estimate was assumed to be only an order-of-magnitude approximation. Liquid viscosity data for the four compounds appear in Figure 13.

3.11 Thermal Conductivity (Gas)

Thermal conductivity data of the gaseous precursors were evaluated by Yaws et al. (28)-(31). While no data appear in the literature for SiH_4, data for the other compounds were available in the temperature range from about 300 to 600 K. These data were correlated and extended over the temperature range from 273 to 1373 K and gave mean deviations from experimental data within about 1.5%. Two equations were required to estimate the thermal conductivity of SiH_4 over the same temperature range; mean estimates agreed within 1%. Thermal conductivities of the gaseous compounds are given as a function of temperature in Figure 14.

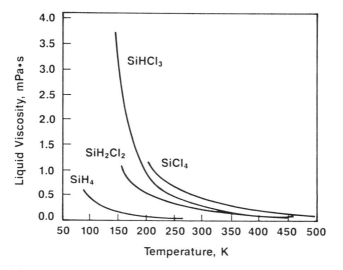

Figure 13. Estimated liquid viscosities between melting and critical points.

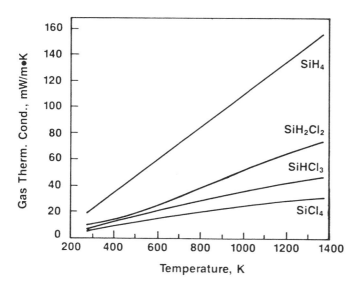

Figure 14. Estimated gas thermal conductivities between 273 and 1373°K.

3.12 Thermal Conductivity (Liquid)

No data appear in the literature for the thermal conductivities of the four compounds except for one measurement at 305 K for $SiCl_4$. Yaws et al. (28)-(31) made order-of-magnitude estimates of thermal conductivities, which appear in Figure 15, at temperatures between the melting and critical points of the liquids.

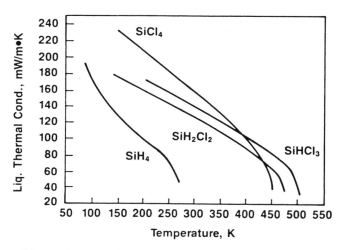

Figure 15. Order-of-magnitude estimates of liquid thermal conductivities between melting and critical points.

4.0 SAFETY

Much of the safety warnings about the chlorosilanes has been obtained from Dow Corning's handbook (39). This handbook also describes safe handling procedures which are not covered in this chapter.

4.1 Health Hazards

Chlorosilanes are stable in the absence of air, moisture, and catalytic agents such as Lewis acids (e.g., $AlCl_3$ and $FeCl_3$). They do not appreciably decompose when heated to 675 K for short periods. The chlorosilanes react violently with water and anhydrous bases such as ammonia and aliphatic amines. Primary alcohols react with the chlorosilanes almost as fast as water while secondary and tertiary alcohols react slower.

Liquid ingestion of the chlorosilanes may cause severe injury or death.

Liquid and gaseous chlorosilanes react with water whether present in its liquid or vapor state. The hydrolysis reaction produces hydrogen chloride vapor which has a maximum exposure limit of 5 parts per million. It is difficult to sustain an exposure at this limit since most people can detect HCl at a much lower concentration.

Contact with vapor clouds of the chlorosilanes produces corrosive injury to any body tissue. Eye contact may result in total loss of sight, whereas inhalation may cause inflammation of the upper and lower respiratory tracts. Systemic effects may result.

4.2 Fire and Explosion Hazards

$SiCl_4$ is not flammable in its liquid or vapor form and presents no known fire or explosion hazard.

$SiHCl_3$, SiH_2Cl_2, and SiH_4 are hazardous compounds. The chlorosilanes are nonconductors and can accumulate static electrical charge. Since vapor-air mixtures can be ignited by a static discharge, air should not be used to purge containers.

NFTA and the Department of Transportation classify $SiHCl_3$ as a flammable liquid. $SiHCl_3$ ignites with a very rapid flash-over at its surface and then burns with only a slightly visible flame and low evolution of heat while emitting copious amounts of white smoke. The flame is almost impossible to put out with water. $SiHCl_3$ can react with moisture or water vapor to generate hydrogen gas.

In the case of SiH_2Cl_2, measurements have been made of the autoignition temperature, explosive output and severity, and of ignition possibility under hydrolytic conditions by Sharp, Arvidson, and Elvey (40). The low boiling point, broad flammability range, low ignition temperature, and rapid rate of combustion combine to make SiH_2Cl_2 a much more hazardous material than $SiHCl_3$ or hydrogen. SiH_2Cl_2-air mixtures are both easily and sometimes unpredictably ignited and the explosive potential for such ignitions is unusually high. Partially confined or unconfined vapor clouds of SiH_2Cl_2 in air can detonate under simple conditions such as turning off a valve. Detonation of air-SiH_2Cl_2 mixtures can have a TNT equivalent of 1 kg of TNT per kg of SiH_2Cl_2 released as vapor (41). Detonations are accompanied by blast waves that can cause equipment and building damage. Nearly any thermal, mechanical, or chemical reaction energy could ignite SiH_2Cl_2-air mixtures. Spontaneous ignition is not observed when SiH_2Cl_2 is mixed with water or vice versa. However, when SiH_2Cl_2 is pumped into a pool of water, the resulting vapor cloud is ignitable (40). Normally-used fire extinguisher materials do not extinguish SiH_2Cl_2 fires nor do they neutralize spills (42); results from foam-type extinguishers have not been reported.

Haas (43) has studied the burning and explosivity hazards of SiH_4 in air by releasing the contents of different size cylinders of SiH_4 into static and dynamic air cabinets. It was found that a 0.006-cm restricting orifice kept the rate of SiH_4 release low, thereby preventing high temperatures that can cause an explosion. In the absence of an orifice, SiH_4 explosions were more severe in a dynamic air cabinet since SiH_4 vapors spread over a larger area. Haas (44) has detailed techniques for handling 100-% SiH_4. Cylinders of SiH_4 were stored in 25-m x 25-m fenced and sprinkled enclosures having an inclined roof to prevent SiH_4 accumulation. All valving was pneumatically controlled. Coaxial seamless tubing was used to transport 100-% SiH_4. Li and Gallagher (45) have reported that <2 % SiH_4 mixtures give better dielectric films than films prepared from 100-% SiH_4. Mixtures of <2% SiH_4 are nonflammable and safe for use (45).

4.3 Materials of Construction

Union Carbide tested various materials of construction under chlorosilane conditions (2). Samples of carbon steel and 316 stainless steel were exposed to liquid chlorosilanes for 6000 h. The carbon steel showed a uniform corrosion rate of 200 μm/y. The 316 stainless steel showed no detectable loss in weight but did exhibit pitting. Carbon steel was considered suitable for use with chlorosilanes under temperature conditions not limited by the mechanical properties of the steel.

Two studies have examined corrosion rates of materials of construction placed in a fluidized bed of silicon used to hydrogenate $SiCl_4$ to $SiHCl_3$ (see section on manufacture of $SiHCl_3$). In each case, a reactor was operated at a temperature of 773 K and with a $H_2/SiCl_4$ molar ratio of 2. Union Carbide (2) ran tests for 36 h at 1 MPa pressure, whereas Mui (46) performed tests for 87 h at 2.6 MPa. Union Carbide attributed scaling of test samples to reaction of HCl and nascent hydrogen; they observed no evidence of silicon penetration into the substrates. However, Mui found silicide layers on all the samples. He attributed this to chemical vapor deposition of silicon followed by its diffusion into, and reaction with, the samples. Mui reported that the silicide layers protected the materials from corrosion themselves. Corrosion rates are listed in Table 8. Corrosion rates for 304 stainless steel differ by a factor of 31 in the two studies.

Table 8

Corrosion rates of Materials of Construction at 773 K in a $SiCl_4$ Hydrogenation Reactor

Material of Construction	Penetration Rate (2) (mm/y)	Silicide Growth Rate(46) (mm/y)
Hastelloy B-2		0.7
Inconel 800H		0.8
304 S.S.[a]		1.2[a]
Alloy 400		2.3
Incoloy 800	4.1	
Carbon Steel		5.6
Incoloy 825	10	
Nickel		11
Copper		11
16 S.S.	21	
26-1 S.S. Alloy	23	
3 RE-60	25	
304 S.S.[a]	37[a]	
Chrome Plated Steel	40	
Carbon Steel	Heavy Scale	
Inconel 625	Heavy Scale	
Titanium	Reactive	

[a] Note Differences between references

Hemlock Semiconductor studied five materials of construction inside a Siemens reactor used to decompose SiH_2Cl_2/H_2 mixtures (47). All materials were suitably corrosion resistant but were ordered as follows from most to least corrosion resistant: Hastelloy B > 304 stainless steel > carbon steel > 316 stainless steel > Monel 400. None of the materials of construction contributed electrically active impurities to the silicon produced in the reactor but all contaminated the silicon with carbon at concentrations ranging from 0.1 to 10 ppma.

Appendix

Conversion Factors from SI units in this chapter to metric units are listed below

Property	From SI	To Metric	Multiply by
Critical Volume (liquid)	$m^3/kmol$	L/mol	1
Density (liquid)	kg/m^3	g/cm^3	10^{-3}
Enthalpy	kJ/mol	kcal/mol	1/4.184
Entropy	$J/mol \cdot K$	$cal/mol \cdot {}^oC$	1/4.184
Heat Capacity	$J/mol \cdot K$	$cal/mol \cdot {}^oC$	1/4.184
Molecular Weight	kg/kmol	g/mol	1
Pressure	kPa	mm Hg	7.501
	bar	atm	0.9869
Surface Tension	mN/m	dyne/cm	1
Thermal Conductivity	$mW/m \cdot K$	$cal/sec \cdot cm {}^oC$	$1/(4.184 \times 10^5)$
Viscosity	$\mu Pa \cdot s$	μP	10
	$mPa \cdot s$	cP	1
Temperature	${}^oC = K - 273.15$		

REFERENCES

1. Steele, R.V., "Strategies on Poly," *Photovoltaics Intern.* Vol 2, pp. 6-8 (1984)

2. Union Carbide Corp., "Experimental Process System Development Unit for Producing Semiconductor Silicon Using Silane-to-Silicon Process," *Final Report, DOE/JPL* Contract 954334, National Technical Information Center, Springfield, VA (1981)

3. Parkinson, G., Ushio, S., Short, H., Hunter, D. and Lewald, R., "New Ways to Make Crystals for Semiconductor Uses," *Chemical Engineering*, pp. 14-17, May 25 (1987)

4. Feldman, K.A. and Frank, K.-D., "Ferrosilicium," In: *Die Metallurgie der Ferrolegierungen*, 2nd ed., Durrer/Volkert eds, pp. 529-568, Springer-Verlag, New York (1972)

5. Müller, M.B., "Electrothermal Processes. Part 3: Silicon, Metals, and Alloys," Contents and Compendium on Subject 230.52: Extractive Metallurgy II (a course at the Tech. Univ. of Trondheim), Metallurgisk Institut, NTH Varsemesteret (1978)

6. Fairchild, W.T., "Technology of Silicon Metal Operation," *Proc. Elect. Furn. Conf.* Vol 21, pp. 277-288 (1964)

7. Dubrous, F.Y. and Septier, L.G., "Silicon Furnace Performance Related to the Carbon in the Load," *Trans. Met. Soc. AIME-Paper Selection*, Paper No. A 71-46, pp. 635-649 (1971)

8. Healy, G.W., "Silicon Metal from Volcano to Percolator, 1904-1969," *Earth and Min. Sci.* Vol 39, pp. 46-47 (1970)

9. Schei, A. and Larsen, K., "A Stoichiometric Model of the Ferro-Silicon Process," presented at the Iron and Steel Soc. of AIME, 39th Elect. Furn. Conf. (1981)

10. Gandel, M.G., Dillard, P.A., Sears, D.R., Ko, S.M. and Bourgeois, S.V., "Assessment of Large-Scale Photovoltaics Materials Production," *Report EPA*-600/7-77-087, U.S. Environmental Protection Agency, Cincinnati, OH (1977)

11. Runyan, W., "Silicon (Pure)," In: *Kirk-Othmer Encyclopedia of Chemical Technology* Vol. 20, pp. 826-845, John Wiley, New York (1982)

12. Herrman, H., Herzer, H. and Sirtl, E., "Modern Silicon Technology," In: *Festkörperprobleme XV Advances in Solid State Physics*, H.J. Queisser, ed. Vieweg & Sohn, GmbH., Braunschweig, West Germany (1975)

13. Ingle, W.M., and Darnell, R.D., "Oxidative Purification of Chlorosilane Source Materials," *J. Electrochem. Soc.* Vol 132, pp. 1240-1243 (1985)

14. Mui, J.Y.P., and Seyferth, D., "Investigation of the Hydrogenation of $SiCl_4$." *Final Report, DOE/JPL* Contract 955382, Nat. Tech. Infor. Center, Springfield, VA (1981)

15. Mui, J.Y.P., "Investigation of the Hydrochlorination of SiCl$_4$," *Final Report, DOE/JPL* Contract 956061, *ibid.* (1983)

16. Ingle, W.M., and Peffley, M.S., "Kinetics of the Hydrogenation of Silicon Tetrachloride," *J. Electrochem. Soc.* Vol 132, pp. 1236-1240 (1985)

17. Weigert, W., Meyer-Simon, E. and Schwartz, R., "Process for the Production of Chlorosilanes," *Ger. Pat. 2,209,267* (1973)

18. Spenke, E., "Silicon Technology," *Z. Werkstofftech.* Vol 10, pp. 262-275 (1979)

19. Crossman, L.D. and Baker, J.A., "Polysilicon Technology," In: *Semiconductor Silicon*, H.R. Huff and E. Sirtl, eds., The Electrochem. Soc., Pennington, New Jersey (1977)

20. Sirtl, E., Hunt, L.P. and Sawyer, D.H., "High-Temperature Reactions in the Silicon-Hydrogen-Chlorine System," *J. Electrochem. Soc.* Vol 123, pp. 919-925 (1974)

21. Hengge, E., "Inorganic Silicon Halides," In: *Halogen Chemistry*, Vol 2, V. Gutmann, ed. Academic Press, New York (1967)

22. Rosenbaum, J.B., "Vanadium Compounds," In: *Kirk-Othmer Encyclopedia of Chemical Technology*, Vol 23, pp. 688-704, John Wiley, New York (1983)

23. Whitehead, J., "Titanium Compounds (Inorganic)," *ibid.* pp. 131-176

24. Hemlock Semiconductor Corp., "Development of a Polysilicon Process Based on Chemical Vapor Deposition," *Final Report, DOE/JPL* Contract 955533, Nat. Tech. Inform. Center, Springfield, VA (1982)

25. Sundermeyer, W., "Preparation of Organosilicon Halides in Molten Salts as Reaction Media," *Pure Appl. Chem.* Vol 13, pp. 93-99 (1966)

26. Sundermeyer, W., "Chemische Reaktionen in geschmoltzen Salzen," *Chem. Zeit*, Vol 1, pp. 151-157 (1967)

27. Grayson, P.E. and Jaffe, J., "A Silane-Based Process," pp. 347-366, In "Proc. of the Flat-Plate Solar-Array Proj. on Low-Cost Polysilicon for Terrestrial Photovoltaic Solar-Cell Applications, *Report DOE/JPL*-1012-122, Nat. Tech. Inform. Center, Springfield, VA (1986)

28. Yaws, C.L., Li, K., Hopper, J.R., Fang, C.S. and Hansen, K.C., "Process Feasibility Study in Support of Silicon Material Task I," *Final Report, DOE/JPL* Contract 954343, *ibid.* (1981)

29. Borrenson, R.W., Yaws, C.L., Hsu, G. and Lutwack, R., "Physical and Thermodynamic Properties of Silane, *Sol. State Techn.* pp. 43-46 (1978)

30. Yaws, C.L., Shaw, P.N., Patel, P.M., Hsu, G. and Lutwack, R., "Physical and Thermodynamic Properties of Silicon Tetrachloride," *ibid.* pp. 65-70 (1979)

31. Cheng, J., Yaws, C.L., Dickens, L.L and Hooper, J.R., "Physical and Thermodynamic Properties of Dichlorosilane," *Ind. Eng. Chem. Process Des. Dev.* Vol 23, pp. 48-52 (1984)

32. *Lange's Handbook of Chemistry*, 12th ed., J.A. Dean, ed., McGraw-Hill, New York (1979)

33. *JANAF Thermochemical Tables*, 3rd ed., J. Phys. Chem. Ref. Data, Vol 14, Suppl. 1 (1986)

34. Hunt, L.P. and Sirtl, E., "A Thorough Thermodynamic Analysis of the Si-H-Cl System," *J. Electrochem. Soc.* Vol 119, pp. 1741-1745 (1972)

35. Walsh, R., "Thermochemistry of Silicon-Containing Compounds; Part 1-Silicon-Halogen Compounds, an Evaluation," *J. Chem. Soc.*, Faraday Trans. Vol 79, pp. 2233-2248 (1983)

36. Herrick, C.S. and Sanchez-Martinez, R.A., "Equilibrium Calculations for the Si-Cl-H System from 300-3000 K," *J. Electrochem. Soc.* Vol 131, pp. 455-458 (1984)

37. Hunt, L.P., "Thermodynamic Equilibria in the Si-H-Cl and Si-H-Br Systems", *ibid.* Vol 135, pp. 206-209 (1988)

38. Woodruff, D.W. and Sanchez-Martinez, R.A., *ibid.* Vol 132, pp. 706-708 (1985)

39. *Properties and Essential Information for Handling and Use of Chlorosilanes*, Dow Corning Corp., Midland, Michigan (1981)

40. Sharp, K.G., Arvidson, A. and Elvey, T.C., "Hazard Potential of Dichlorosilane," *J. Electrochem. Soc.* Vol 129, pp. 2346-2349 (1982)

41. Hemlock Semiconductor Corp., "Development of a Polysilicon Process Based on Chemical Vapor Deposition," *Quarterly Report DOE/JPL* Contract 955533-81-6, Nat. Tech. Inform. Center, Springfield, VA (1981)

42. Rainer, D., "Dichlorosilane-Leak Spill and Fire Control," In: *Extended Abstracts.* The Electrochem. Soc., Pennington, New Jersey, Vol 85-2, p. 454 (1985)

43. Haas, B.H., "Discharge of Silane in Open Air and in Toxic Gas Cabinets," *ibid.* p. 456

44. Haas, B.H., "Development of High Purity Gas Systems Programs," *ibid.* p. 453

45. Li, P.C. and Gallagher, J.P., "A Safe Dielectric Film Deposition Process," *ibid.* p. 451-452

46. Mui, J.Y.P., "Corrosion Mechanisms of Metals and alloys in the Silicon-Hydrogen Chlorine System at 500°C," *Corrosion NACE* Vol 41, pp. 63-69 (1985)

47. Hemlock Semiconductor Corp., "Development of a Polysilicon Process Based on Chemical Vapor Deposition of Dichlorosilane in an Advanced Siemens Reactor," *Final Report DOE/JPL* Contract 955533-81-6, Nat. Tech. Inform. Center, Springfield, VA (1983)

48. Schnegg, A., Rurländer, R. and Jacob, H., private communication (1986).

49. Wolf, E. and Teichmann, R., *Z. Anorg. Allg. Chem.* Vol. 460, p. 65 (1980)

50. Farber, M. and Srivastava, R.D., *Thermodynam.* Vol 11, p. 939-944 (1979)

2

Polysilicon Preparation

Leo C. Rogers

1.0 THE TECHNICAL HISTORY OF POLYCRYSTALLINE SILICON

1.1 Early And Present Polysilicon Manufacturers

Polycrystalline silicon (polysilicon) manufacturing has been shrouded in mystery from its beginning. Few scientific journal articles address the subject. And even today, nuclear power plants are more easily toured than semiconductor-grade polysilicon plants. Why all the secrecy?

Perhaps the answer is found in the fragmented beginnings of the semiconductor materials industry when most polysilicon manufacturers independently researched and developed their own polysilicon processes. The resultant concern of each manufacturer was, and remains, that its optimized technology could be combined with its competitors optimized technology to give the competitor a considerable advantage of product quality or costs. Thus, the sharing door was closed from the outset and remains shut today.

Du Pont, Bell Laboratories, Siemens, Union Carbide, Foote Mineral, Mallincrodt, International Telephone, Transitron, Tokai Denkyoko, Chisso, and Merck Chemical were early polysilicon manufacturers.

Many of the early participants were research oriented and soon retired from the mercantile market. Merck designed their plant using solid silver pipes; they retired. Union

Carbide's early efforts lacked reactor technology. Later entrants were Wacker and Westinghouse who became licensees of the Siemens' polysilicon process of silicon grown on silicon, yet they too developed their own technology. Westinghouse relicensed Siemens' technology to companies such as Dow Corning and Monsanto who again redesigned the systems to their own specifications.

In the 1960's, the semiconductor industry's demand for larger quantities, purer quality, and lower-cost polysilicon was not met by the mercantile producers. So Texas Instruments and Motorola, semiconductor device/circuit manufacturers, began to produce their own polysilicon. In 1980, Fairchild Semiconductor, another device manufacturer, began its own production of polysilicon through Great Western Silicon.

Semiconductor silicon usage has increased, on the average, about 15% per year beginning at less than 30 MT/y (metric tons per year) in 1965 and reaching 5,500 MT/y in 1988 (1),(2). Increased demand for polysilicon has been due to larger circuit chips and wafers and lower silicon utilization due to advanced fine-line circuit geometries, multiple circuit layering and tighter crystal specifications as well as the demand for more electronic circuits.

Yet the demand for semiconductor polysilicon pales when compared to the projections of 11,160 MT/y (1995) silicon for the photovoltaic (PV) industry (3).

The importance of polysilicon to the PV industry was emphasized when the U.S. Government, Department of Energy (DOE), spent over $800 million from 1978 to 1985 to discover lower-cost methods of producing polysilicon. Jet Propulsion Laboratory managed a large portion of this effort. Companies such as Westinghouse (reduction of silicon tetrachloride by sodium), Battelle (reduction of silicon tetrachloride by zinc), Schumacher (dissociation of tribromosilane), and Aerochem (reduction of silicon tetrachloride with sodium vapor), as well as others spent millions of dollars exploring old and new polysilicon technologies. But none of the DOE/JPL polysilicon projects merited commercialization. And the core polysilicon technology of the present-day manufacturers remained secret.

The late 1980's saw some reuse of established polysilicon technology. Union Carbide adopted and improved versions of the 1960's Komatsu metal bell-jar system for a 1,200-MT/y plant, and two Japanese companies, Tokoyama Soda and Nippon Kokan, separately purchased the quartz bell-jar reactor and chemical recovery technologies from Great Western

Silicon for their 1,000-MT/y plants. Only Ethyl Corporation announced a non-bell-jar reactor system.

A list of companies who have participated in the production of semiconductor-grade polysilicon are shown in Table 1, Polysilicon manufacturers Past And Present. In general, the list is chronological. It should be noted that the ownership or names of some polysilicon manufacturing plants change from time to time. Examples are Montecatini/ Smiel/Dynamit Nobel/Hüls AG; Great Western Silicon by Fairchild-Applied Materials/General Electric/Nippon Kokan; Dow Corning/Hemlock Semiconductor; Topsil/Motorola/Phoenix Materials/Topsil.

Table I

Polysilicon Manufacturers Past and Present

(Present Manufacturers Are Noted With An Asterisk)

Company	Company	Company
Du Pont (USA)	Siemens (W. Germany)	Int'l Tel. (USA)
Union Carbide[*] (USA)	Chisso (Japan)	Westinghouse (USA)
Hemlock Semiconductor[*] (USA)	Wacker[*] (W. Germany)	Shin Etsu Handotai (Japan)
Monsanto (USA)	Osaka Titanium[*] (Japan)	Komatsu[*] (Japan)
Great Western Silicon (USA)	Bell Laboratories (USA)	Mallinckrodt (USA)
Foote Mineral (USA)	Texas Instruments (USA)	Tokai Denkyoko (Japan)
Merck Chemical (USA)	Huels AG[*] (Italy)	Topsil (Denmark)
Mitsubishi[*] (Japan)	Motorola (USA)	Various[*] (East Block)
Various[*] (China)	Tokuyama Soda[*] (Japan)	

1.2 Semiconductor-Grade Polycrystalline Silicon Precursors

Three words are frequently confused for one another, silica, silicones, and silicon. To a degree, this confusion is understandable as their industrial activities are strongly intertwined.

1.2.1 Silica (SiO_2): Silica is the most abundant compound of the earth. Silica is the amethyst jewel, the quartz crystal figurine, the fire-starting flint, the tridymite near volcanoes, the sand at the beach, the glass window, and even the skeletons of diatomacae. Silica is low cost and readily available. Strip the oxygen from silica and you have low-purity products called ferrosilicon and metallurgical-grade silicon (MG-Si).

1.2.2 Silicones: Silicones are lubricating and sealing compounds. They are resistant to chemicals and offer flexibility over a broad temperature range. Because silicones are chemically compatible with organic tissue, implants to living organisms is a small application.

Silicones are a product of silica. Silica is converted to MG-Si which in turn is converted to silicon tetrachloride (STC). The STC is reacted with a Grignard reagent to give a long-chain chlorosilane which is further hydrolyzed to produce silicones.

1.2.3 MG-Si and Ferrosilicon: MG-Si and Ferrosilicon are manufactured by the direct reduction of SiO_2. By adding ferrosilicon to iron, the resulting steel becomes hard and corrosion resistant. The aluminum industry also requires low-grade silicon to form aluminum-silicon alloys.

Both ferrosilicon and MG-Si are produced in furnaces 30 feet in diameter by 30 feet tall at tens of thousands of metric tons per year per furnace. Through the open top of the furnace, tons of carbon (coke) and silica (quartzite) are loaded into a brick-lined, vertically cylindrical furnace. Extending downward from the top, large carbon electrodes reach near the bottom of the furnace. Between the electrodes, a high-temperature arc (>2,000°C) provides the energy to strip the oxygen from the silica resulting in carbon monoxide gas and molten silicon (4),(5),(6),(7).

Ferrosilicon quality approaches 93% silicon while MG-Si contains about 98% silicon. By careful selection of the furnace lining, the quality of the silica, and the purity of the carbon source, MG-Si purity above 99% can be obtained. The main impurities are iron, aluminum, carbon, and silicon carbide.

Further purification of the MG-Si has been achieved by second-stage, wet chemical treatment (8)(9)(10)(11)(12)(13). Elkem of Norway, the leader in this field, has manufactured 99.8+% pure polysilicon using wet chemical processing (14). The remaining impurities include tenths of percents of the silicates, carbon, and parts per million of the Group III and V elements of the Periodic Table (5).

When the electronics industry first began to produce semiconductor devices from silicon, they turned to the metallurgical industry to supply their polysilicon production needs. It was however, immediately evident that semiconductor-grade polysilicon with a twelve 9's purity was impossible to obtain by the metallurgical reduction of SiO_2. Thus the semiconductor industry had to explore new processes to obtain ppb-pure polysilicon. As will be seen, the parts-per billion technology was already in the wings of the silicone industry. And while silicone chemistry is not silicon chemistry, the raw feedstock is similar for both industries, a fortuitous situation that caused the semiconductor industry to jump forward at least a decade.

1.2.4 Silicon Purity Beyond MG-Si: Because the MG-Si industry was based on the conversion of a solid feedstock (silica) to a crude silicon, and because the improved quality MG-Si achieved by acid leaching of MG-Si was so drastically inadequate to meet the parts per-billion requirements of the semiconductor industry, it was logical for the semiconductor scientist to jump from the liquid chemical purification methods to gas-phase chemistry to achieve the ppb silicon purity. The presumption was that the further the molecules were separated, the greater the probability of ppb purification. Based on this presumption, scientists at Union Carbide, Wacker, Dynamit Nobel, and Dow Corning developed a reasonably pure polysilicon feedstock and with favorable production costs (15),(16),(17).

1.2.5 Feedstocks for Semiconductor-Grade Polysilicon: Chlorosilanes have dominated the industry as feedstock for the production of polysilicon. Some of the patents for the manufacture and purification of various chlorosilanes predate the mass production of polysilicon itself (15),(16),(17),(18) (19),(20),(21),(22).

Chlorine is inexpensive if purchased as hydrogen chloride (HCl). Silicon is inexpensive if purchased as MG-Si. These are the starting compounds for the chlorosilane production of polysilicon.

To produce a chlorosilane, HCl is passed through a heated fluidized bed of pulverized MG-Si. The resulting chemicals are a blend of chlorosilanes, H_2, HCl, and unreacted silicon. Two chlorosilane species predominate: trichlorosilane (TCS) (90%) and tetrachlorosilane, also referred to as silicon tetrachloride or STC (10%) (23). A distillation step is performed to isolate these species from the other products and from one another. Although dichlorosilane (DCS) can be formed from TCS by use of catalysts and H_2, DCS is not presently used to produce commercial polysilicon.

The available technology and favorable economics of STC and TCS production led to the selection of STC and then TCS as the feed stocks for the major polysilicon producers of the world. Since millions of pounds of STC were already produced to manufacture silicones and similar tonnages of TCS were produced to manufacture coupling agents, the high-volume, low-cost technology to manufacture STC and TCS was available for incremental expansion when the semiconductor industry was searching for a pure feedstock. And by adding one more purification step to the existing STC and TCS plants, the common feedstocks for the silicones and coupling-agent industries became the ultrapure feedstocks for the needs of the semiconductor industry.

Favorable economics needs to be emphasized. While fluorosilanes, iodosilanes, and bromosilanes could all be produced, profitable production was doubtful. On the other hand, experience had already proven the chlorosilane technology and its profitability. Union Carbide led the mercantile market in the USA by supplying Texas Instruments, Monsanto, Great Western Silicon, and Motorola. Dynamit Nobel and Wacker led the way in Europe. Other producers of semiconductor-grade TCS today include Dow Corning, Chisso, Hüls, Osaka Titanium, Shin Etsu Handotai, Tokuyama Soda, Chinese Companies, and Communist Companies.

The disadvantage of TCS and STC is that they are not usually manufactured at polysilicon sites. Transportation is thus required and involves the considerations of safety, cost, and maintaining TCS purity.

The advantages of chlorosilanes are: they are reasonably inexpensive to manufacture and, as liquids at at ambient temperature, they are relatively easy to handle. STC and TCS are commonly transported by train or truck trailer. TCS can be ignited to burn, but usually does not; STC is not usually flammable. Both TCS and STC may be stored in large quantities while maintaining semiconductor-grade purity.

1.3 Semiconductor-Grade Polysilicon

Ultimately, it was TCS that met the ppb-purity requirements and became the mainstay feedstock for polysilicon. While polysilicon specifications vary depending on its intended use, it is nonetheless appropriate to say that it is considerably more pure than MG-Si. Typical figures of merit for tolerable impurity concentrations in semiconductor-grade polysilicon are 0.3 ppba n-type impurities (phosphorous, arsenic) and 0.15 ppba p-type impurities (boron, aluminum, gallium, indium). Other impurities that affect the properties of the ultimate semiconductor device include: carbon with an acceptance range of 0.3-0.4 ppm; (carbon can cause defects in the crystal structure of the subsequent single crystal silicon); oxygen (can lower the yield of perfectly formed single crystal ingots, and it can distort the resistivity reading of the single crystal); heavy metals (can destroy single crystal growth and may provide deep-level band-gap irregularities); gold (can reduce the minority carrier life-time); and sodium (can cause charge migration at the surface of the silicon and degrade the properties of thermal oxides).

1.4 Potpourri Of Other Methods To Manufacture Polysilicon

Table II lists other generic methods that have been explored to produce polysilicon of usually lower quality than semiconductor grade and frequently at higher cost than the TCS method. These methods are listed without reference and indicate the variety of methods attempted to produce polysilicon.

Obtaining a purer grade of silicon has also been attempted by the reduction of silicon monoxide with silicon carbide (24), the reduction of silicon monoxide with hydrogen (25), and the reduction of silicates with molten aluminum (26). Another method reacted SiO_2 with Al_2O_3, removed most of the aluminum with HCl, then dissolved the silicon in tin. The silicon was finally precipitated leaving the remaining more soluble aluminum in the molten tin (27).

2.0 POLYSILICON PRODUCTION TECHNOLOGY (MOST PRACTICED)

Many technologies are practiced to produce polysilicon. Some use silane as a feedstock, others use STC; some use a fluidized-bed reactor (FBR), others use double-ended tube reactors; and depending on the feedstock technology, there are numerous by-product recovery schemes. However, the polysilicon technology most practiced today uses trichlorosilane as its feedstock, bell-jar reactors, and a somewhat elaborate system to recover the chemical by-products from

Table II

Potpourri of Historic Methods to Produce Polysilicon

Source Chemical	Reactant	Reaction
SiF_4	potassium	reduction
SiF_4	sodium	reduction
SiF_4	magnesium	reduction
$SiCl_4$	potassium	reduction
$SiCl_4$	sodium	reduction
$SiCl_4$	magnesium	reduction
$SiCl_4$	beryllium	reduction
$SiCl_4$	aluminum	reduction
$SiCl_4$	zinc	reduction
$SiCl_4$	sodium amide	reduction
SiO_2 (silica)	calcium	reduction
SiO_2 (silica)	magnesium	reduction
SiO_2 (silica)	aluminum	reduction
Silicon sulfide		hydrolysis
Potassium fluorosilicate	potassium fluoride	electrolysis
Silicofluoride	sodium	reduction
Silicofluoride	aluminum	reduction
SiO_2 (silica)	copper	reduction, followed by reaction with sulfur
Silicates	carbon	reduction
Silicates	aluminum	reduction
Silicon monoxide	silicon carbide	reduction

the reactors. It is this most-practiced technology which is described next. Later in the text, other technologies are discussed which are also being practiced and may become the technologies of the future.

The most-practiced technology usually includes two separate manufacturing plants: the first plant manufactures and purifies the TCS feedstock; the second plant manufactures the polysilicon and recovers the chemical by-products. The by-product plant usually includes: the storage of large quantities of TCS and by-product STC; a liquid-to-gaseous feedstock conversion system; a group of polysilicon reactors (10 to 200); an exhaust gas recovery system with purification augmentation; environmental control apparatus; a central logic control system; safety systems; and evaluations systems.

Polysilicon Preparation 41

In many instances today, the TCS plant is remote from the polysilicon plant and thus sophisticated transportation methods are required to maintain and deliver semiconductor-grade TCS to the polysilicon manufacturing site. Even large producers such as Hemlock Semiconductor in the USA and Hüls, AG in Italy have remote TCS plants.

However, as new 1000-MT/Y, and larger, polysilicon plants are constructed, it is expected that the TCS plants will likely be located at the polysilicon site to maintain TCS quality, to reduce transportation hazards, and to lower polysilicon manufacturing costs, as is the case for the on-site producers of TCS in Germany, Japan, and China. The manufacture and purification of TCS is noted by reference (20),(28),(29),(30),(6).

The polysilicon plant components of most interest are the feedstock system, the reactors, and the exhaust gas recovery system.

2.1 The Feedstock System

Chlorosilanes have been the predominate feedstocks for polysilicon plants. Most typically, ultrapure TCS is used (31),(32),(33),(34),(35),(36). Whether the TCS is delivered from an off-site location or produced on-site, the polysilicon production operation evaluates the trucked, railed, or piped TCS quality and then uses the TCS directly, or, if required, improves the quality by its own chemical purification methods. While complexing, liquid extraction, gas-absorption, and other techniques are known, distillation is the predominantly practiced art for impurity removal. TCS is conveniently stored as a liquid, is relatively safe, and maintains its ppb purity for months.

At the polysilicon plant site, pure TCS is blended with a carrier gas to form the gas-phase feedstock to the polysilicon reactors. Typical mole concentrations of TCS in the carrier gas are 5% to 20% with exceptions to 50% (37). The carrier gas is typically hydrogen.

2.2 Polysilicon Reactors

Polysilicon reactors are the point of elemental silicon formation. The gaseous feedstock enters the reactors to remain for a sufficient time (5-20 seconds) to allow the TCS to dissociate into elemental silicon and gaseous by-products.

The dissociation process occurs directly on resistance-heated filaments inside the reactor chamber. TCS dissociation occurs above 575°C (typically 1100°C is used). Between 10 and 200 polysilicon reactors may be located at one plant site. Polysilicon plant capacities vary from 10 to 2,000 MT/y. Figure 1 shows two rows of operating bell-jar reactors using TCS as feedstock.

Figure 1. Polysilicon reactors of the quartz bell jar type. (Note the quartz bell jar on pallet just visible after the second reactor on right.)

2.2.1 General Bell-Jar Reactor Design: Reactor designs vary, but have one common characteristic, most all are of the single-ended, bell-jar design. (Picture a water glass with its open end down.) The quartz bell-jar reactor is the offspring from the original Siemens "C" reactor (33),(34),(35),(36). The basic design is shown in Figure 2. The base plate is water cooled, has ports for inlet and exhaust gases, and has electrical posts to contact the initial seed rods, shown as the completed full-size rods. On the base plate rests an inverted quartz bell-jar. The bell jar is sealed to the base plate to create the gas-tight reaction chamber. In this chamber, the TCS (carried by hydrogen) comes into direct contact with the 1100°C rods, dissociates, and deposits silicon on the heated rods. The bell jar, if made of quartz, is surrounded by a thermal and personnel safety shield. Between the bell jar and the personnel shroud are the starting-rod preheaters which surround the bell jar. If the bell jar is made of metal, it is usually a double-walled bell jar using a water jacket for cooling. Each reactor has its own power source to heat the rods. Support systems for the reactor include a system to cool the bell jar and a system to remove the final silicon rods.

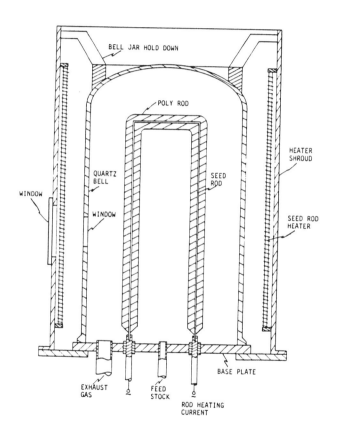

Figure 2.   Typical Siemens type quartz bell jar reactor. This depicts the Rogers-Heitz polysilicon reactor when wires replace the seed rods and seed rod heaters are removed. The latter configuration can be operated as either a metal or quartz bell jar system.

44 Handbook of Semiconductor Silicon Technology

Reactor bell jars vary in size from 18 inches to 10 feet in diameter and from 48 inches to 10 feet tall. Figure 3 shows a 50 inch diameter by 103 inch-tall quartz bell jar.

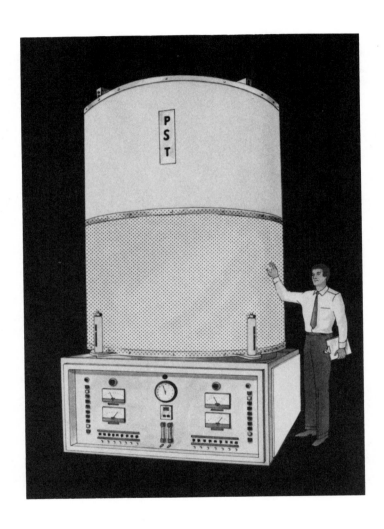

Figure 3. A high capacity (40 metric tons per year) polysilicon reactor using a quartz bell jar 50 inches in diameter by 103 inches in height.

2.2.2 General Bell-Jar Reactor Operation: Reactors are complex. They operate at 1100°C, can dissipate thousands of kilowatt hours of heat, may accept and exhaust thousands of standard cubic feet per hour of hydrogen and TCS, operate in the presence of corrosive HCl, and may produce a thousand kilograms of silicon in a single run.

Before a polysilicon production run, thin silicon starting rods are placed vertically on the electrical feed throughs. The bell jar is lowered over the silicon rods, and the reaction chamber is tested to assure gas tightness. The chamber is purged to displace the ambient air. The rods are heated, feedstock is introduced, and large-diameter polysilicon rods are grow.

Common to most bell-jar systems are one of three mechanisms to maintain rod temperature as the rods grow in diameter. The most straight-forward method uses an optical pyrometer to view the rods through a window in the base plate or in the bell jar itself. This method requires an operator to manually increase the power supply current to maintain the rod temperature. A second method programs the power supply current as a function of time. A third method detects the rod temperature with a sensing device which provides an electrical signal that raises the power supply current to maintain the desired rod temperature. All three systems are in common use. Each has its benefits. The manual system can grow rods faster than the other methods as an operator is present to adjust for abnormalities in the run, rod hot spots, arcing, bell-jar deposits, irregular silicon growth, and to maintain the rod temperature as high as possible, even near the rod center core melt-out point. However, this manual method requires an operator for every five reactors. As a compromise, conservative electrical power increases to the rods are frequently programed by time. (Conservative power increases avoid hot spots, arching, and bell-jar deposit, but decreases the polysilicon output of the reactor.) Finally, to obtain nearly the best temperature profile and to do so without constant operator attention, a multisensor current-adjustment method is used.

At the end of the run, the electrical power is removed from the rods, the feedstock is turned off, the reaction chamber is N_2 purged, and the bell is removed. After the silicon rods are removed, the base plate and the bell jar are cleaned and inspected and the process is repeated. This procedure is common to all runs using a bell jar.

Reactor Operation Using Siemens' Slim Rods: There are two operating modes for quartz bell-jar, polysilicon reactors. The most common mode uses silicon seed rods 5 to 10 mm in diameter (slim rods) on which to grow polysilicon. Two to seventy seed rods are used depending on the ultimate diameter of the rods and the diameter of the bell jar. The electrical power path is usually through a single-seed rod, across an electrically conductive bridge that connects the top of one seed rod with the top of another seed rod, and then through a second seed rod and to another power feed through. The bell jar is placed over the seed rods. The chamber is N_2 purged. With the purge gas remaining, or in the presence of the subsequent hydrogen, infrared heaters are energized outside the quartz bell jar to heat the intrinsically pure seed rods to near 400°C (See Figure 2). (Internal-to-the-chamber heaters are used for metal bell jars.) At 400°C, the intrinsic slim rods become conductive due to thermally-generated electrons within the silicon itself. At this point, high voltage, typically thousands of volts, is applied directly to the slim rods to increase the temperature. By 800°C, the silicon rods have reached the avalanche breakdown temperature and free electrons in the silicon permit the thousands of volts driving force to be replaced by a lower-voltage and higher-current power source. External heating is terminated. The electrical power continues to pass directly through the seed rods which are then elevated to 1100°C. This temperature is maintained by increasing the electrical current through the rods while the TCS and H_2 flow through the reaction chamber. Silicon grows on the intrinsic starting rods and their diameter increases up to 9 inches.

A seldom-used method to heat intrinsic starting rods is to pass a hot gas through the bell jar to raise the temperature of the rods. This heating method is slow and expensive.

An alternate starting method which is attractive when using metal bell-jar reactors is the use of doped silicon starting rods in place of the intrinsic rods. The doping lowers the cold resistivity of the slim rods such that external heating is not required to reach the initial 400°C. This initial temperature is reached by the direct application of high voltage to the silicon rods. However, the high-voltage power source and the high-to-low-voltage switching remain a part of the system. Also, the doped silicon rods are more expensive than the intrinsic rods, and there is the requirement that the doped rods in the same run have approximately the same resistivity so that all rods heat at near the same rate. Irregular heating causes starting-rod melt downs. (Molten silicon readily reacts with most base-plate materials.) Because the starting rods contain a dopant, the final polysilicon contains the same dopant diluted by the addition silicon. High-purity silicon is not grown using doped starting rods.

As mentioned, the doped-rod method avoids placing heaters inside the metal bell jar. However, not placing heaters inside the bell jar is desired as silicon deposition often occurs on all elements inside the chamber unless the elements are specifically cooled. (Water cooling inside the bell jar provides the occasion for a water leak into the 1100°C chamber, a condition to be avoided if possible.) In general, off-rod deposition (e.g., polysilicon deposition on internal heaters) is undesirable for several reasons: 1) it lowers the effective silicon deposition efficiency; 2) it can chemically react with quartz or metal, thus jeopardizing the integrity of the off-rod element; and 3) it can flake off the nonrod surface and attach to the rod surface, thus contributing impurities to the polysilicon rods.

Reactor Operation Using Rogers/Heitz Wires: The use of silicon slim rods is expensive, awkward, and slow. Thus in 1964, a nonsilicon starting-rod process was developed by Leo C. Rogers and Alfred J. Heitz. In place of either intrinsic or doped silicon rods, the Rogers/Heitz method uses wires on which to grow silicon. The wire process was commercialized in 1965 and has been used by Motorola for 27 years. Later, in 1980, Great Western Silicon (GWS) commercialized the process. GWS then licensed the process to two Japanese firms, Tokoyama Soda and Nippon Kokan.

Perhaps the most significant advantage of the Rogers/-Heitz process is the ability to obtain hyperpure polysilicon without the cost of a processing plant to manufacture the intrinsic or doped silicon starting rods. There are other benefits of the wire process. Using wires eliminates the need for considerable reactor equipment: the starting rod preheaters, the preheater electrical power supply, the high-voltage electrical power supply, and the accompanying high-voltage switch gear. Additional advantages are: the wires do not break during handling or during assembly into the reactor, as do the fragile silicon starting rods; the wires are all matched in resistance, but even if they were not, they are not susceptible to high-temperature, starting melt outs; the wires come to temperature faster so polysilicon growth begins sooner;

the wires may be used to preclean the bell enclosure prior to the silicon growth cycle; and the wires may be operated at higher temperatures to accelerate polysilicon growth (center-- core rod melt out occurs at a higher temperature for Rogers/- Heitz-grown silicon than for Siemens-grown silicon.) The resultant wire-grown polysilicon rods are cosmetically identical to the silicon-core polysilicon rods. Rods to nine inches in diameter have been grown by the wire method. There is no known limit to the diameter of silicon rods grown on wires. But if there were a limit, the size for wire rods would be larger than for polysilicon grown on silicon rods due to the center rod melt out occurring at a lower temperature for silicon grown on silicon slim rods than on wires.

The disadvantages of using wires as starting rods are: the requirement to remove the wire after the growth cycle and the possible dopant contribution of the wire to the grown silicon ingot. On the other hand, polysilicon grown on wires has served world-wide customers since 1964. The author is shown holding typical Rogers/Heitz polysilicon rods in Figure 4.

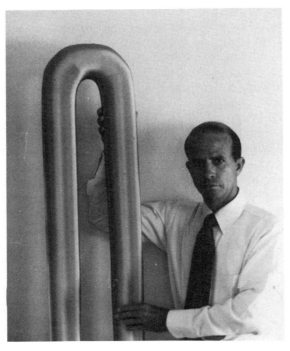

Figure 4. Typical inverted "U" shaped rods from a polysilicon bell jar reactor.

Polysilicon Preparation 49

Difficulties: Depending on reactor design and operating conditions, the following difficulties may arise when operating a bell-jar reactor:

1) Arcing may occur between the rods and the power feed-throughs and thus abort a run due to rod melting.

2) Due to uneven rod resistivity, one rod may increase in temperature faster than the rod being temperature monitored and thus a melt out may occur.

3) The rods may grow slowly due to low rod temperature or lack of feedstock and thus reduce plant production.

4) The rods may grow with undesirable "warts", known as "pop corn" (38), because the rods are above 1100°C and or the mole ratio and or mass flow of the feed stock are below design conditions. (The "pop corn" traps chemicals used in the final cleaning of the polysilicon and the trapped chemicals vaporize in the single-crystal puller and cause a loss of crystal structure).

5) The rods may tilt and strike each other or the bell jar. Striking each other may cause a melt down. Striking a quartz bell jar may or may not terminate a run. Striking a metal bell jar may or may not puncture the water jacket of the bell jar and introduce water into the 1100°C chamber and the chemical recovery loop.

6) The rods may grow in a nonuniform shape and thus cause the thinner portions to reach melt-down temperature while the larger-diameter portions of the rod grow at a slower rate due to their below-specification temperature. To some extent this is a self correcting situation since the hotter portion of the

50 Handbook of Semiconductor Silicon Technology

rod grows faster than the cooler, thicker portion resulting in an evening of diameters. This occurs if all rods are in series with the power source.

7) Excessive silicone oils may be formed. Since the oils are pyrophoric, they present a danger when the bell jar is opened to the atmosphere, and when the chemical recovery lines are opened for repair.

8) Silicon may be deposited on the interior of the bell jar. These deposits can break a quartz bell jar when it is cooled and can cause the IR absorption of a metal bell jar to exceed design limits. Deposits also fog the windows through which temperature measuring devices view the rods. The optical attenuation portrays a cold rod which in turn indicates the need for additional electrical power. Because the increased power is in excess of that required to keep the rods at 1100°C, the rods melt. All of the listed situations can be avoided by proper reactor design and operation.

2.3 Reactor Operation Criteria

There are general guide lines for TCS, quartz bell-jar reactor operation: keep the bell jar below 575°C, the rods near 1100°C, the TCS-to-hydrogen mole ratio between 5-15%, the bell-jar pressure less than 5 psig, and the mass flow in excess of the calculated values for the reactor size. The reason for the excess flow is to assure the highest grams-per-hour deposition possible. The high flow procedure assures high deposition rates by sweeping away the generated HCl and at the same time providing an abundance of TCS for dissociation. Low single-pass conversion efficiency accompanies this mode of reactor operation. While reaction 1 proceeds to the right at above 575°C for TCS (best near 1100°C), it proceeds faster if the HCl is removed so as to slow reaction 2 which occurs simultaneously.

Reaction 1: $2SiHCl_3 \rightarrow SiCl_4 + Si + 2HCl$

Reaction 2: $Si + 3HCl \rightarrow SiHCl_3 + H_2$

2.4 Deposition Objectives For Polysilicon

Achievement of optimum silicon deposition efficiency does not simultaneously provide for optimum economic production. In practice, deposition efficiency is sacrificed for deposition rate. The pragmatic goal is to deposit as many kilograms of polysilicon in the shortest time until the cost to achieve increased deposition rate lowers profitability. While high reactor efficiency lowers the number of times the TCS must be recycled, it is increased polysilicon deposition rate, not silicon conversion efficiency, that reduces the electrical energy requirements per ton of silicon and reduces the number of reactors required to produce a given tonnage. A possible design exception to low-efficiency polysilicon manufacturing would be a plant operating in a limited space (few reactors and small recovery system) and using near zero-cost electrical energy. A plant like this was once planned for operation on a sea ship.

The author's design of five generations of bell-jar reactors has been based in major part on the work of R.F. Lever (39). The thermodynamics of the deposition process also has been studied by Harper and Lewis (40) and by Sirtl, Hunt, and Sawyer (41) and--see also Chapter 1.

Increased deposition rate is achieved by: raising the rod temperature; increasing the TCS mole ratio; increasing the TCS mass flow over the rods; increasing the deposition surface area; removing the HCl from the rods faster than the TCS; and, according to Wacker Chemie (37), increasing the pressure in the reaction chamber. The combined implementation of these variables is complex. First, each change in one variable affects the other variables. Second, some of the changes alter the chemistry of short-lived intermediate compounds, such as the presence of dichlorosilane, which in turn affects deposition rate. Third, there are a plethora of negative aspects concerning the singular aggressive implementation of each of the variables. Nonetheless, the intelligent implementation of the deposition factors has increased deposition rates over the past 22 years from 100 grams per hour in 1960 to over 4,000 grams per hour in 1988. Large bell-jar reactors, the Wacker metal bell jar (37), the PST50103 quartz bell jar, and the larger Japanese bell-jar reactors near ten feet in diameter have claimed deposition rates in excess of 5,000 grams per hour.

2.5 Reactor Exhaust-Gas Recovery—General Practice

The major constituents of the exhaust gases are the carrier gas, hydrogen, and the unreacted TCS. The quantity of TCS which experiences dissociation in a single pass, as compared to the unreacted TCS, ranges from 8% to 25%. The by-product gases which are produced by the dissociation of the TCS include STC (about 45% of the dissociated TCS), HCl, and small amounts of other chlorosilanes including silicon-silicon-bonded polymers. The exhaust gases leaving the reactor chamber represent a changing blend of gases as the single-pass silicon deposition efficiency moves from less than 1% at the beginning of a run to perhaps 25% or more as the rods grow in diameter. The increase in efficiency results from the increase in rod surface area providing more dissociation sites for the TCS molecules. Poor reactor design causes the plateau of efficiency early in the run. The best reactor design permits increases in deposition efficiency until full rod diameter is reached.

Technologies vary considerably for chemical recovery of exhaust gas. Fundamentally, each system accomplishes the same goals: the TCS and STC are separated from the HCl and hydrogen gases; then the TCS and STC are separated from each other; and the TCS is repurified and returned as feedstock to the reactors. An operating polysilicon by-product plant designed and constructed by the author and Mr. Alfred Heitz is shown in Figure 5. This plant supports the production of 100 MT/y of semiconductor-grade polysilicon by the recovery of TCS, STC, HCl, and H_2.

By-product STC from polysilicon production is sold to the fumed silica, fiber optics, or semiconductor epitaxial industries. STC is also repurified and blended with TCS to be used as reactor feedstock. In addition, STC is converted to TCS which is repurified and used for feedstock to the reactors. The by-product HCl is sold to swimming pool or plating industries or used with MG-Si to manufacture additional TCS. The hydrogen is repurified and reused as the carrier gas.

The efficient chemical recovery of polysilicon by-products significantly reduces the cost to produce polysilicon, and it reduces the potential environmental problems to near zero.

Figure 5. Polysilicon level 1 chemical by-product recovery system. (Original design by author.)

2.6 Reactor Exhaust-Gas Recovery—General Design

Exhaust-gas recovery systems are designed to match the output from the reactors. The general design criteria must include many factors, but having chosen TCS as the feedstock and bell-jar systems as reactors, there are four general design considerations which are most important: the number of reactors to be in operation, the quantity of gas to flow through the reactors, the mole ratio of the feedstock, and the single-pass conversion efficiency of silicon within the reactors.

Small polysilicon plants often choose recovery designs which sell both the HCl and the STC by-products. This design choice reduces the initial capital cost of the polysilicon plant and, by the sale of the by-products, provides additional revenue to the plant's operation. However, when the smaller plants are expanded, and most all are expanded (Motorola grew from 10 to 100 MT/y; Hemlock Semiconductor from 100 to over 1000 MT/y), the general recovery design considerations must include an additional criteria, the safe transportation of the increased volume of incoming feedstock and outgoing by-products.

For a 10-MY/y polysilicon plant, the input trichlorosilane is a mere 250,000 pounds per year, or six truck loads. At 100 MT/y, incoming trucks arrive more than one per week. At 1,000 MT/y, incoming trucks arrive every 13 hours.

At the 10-MY/y level, 3 trucks per year of STC are sold to various industries. The anhydrous hydrogen chloride is sold to the semiconductor industry as an etching gas or converted to muriatic acid for the plating or swimming pool industries. At the 100-MT/y level, only the fumed silica industry can use the quantities of STC generated by the polysilicon plant. At the 1000-MT/y production level, there are few STC purchasers, and the cost and risk of moving 600 truck loads of TCS and 300 truck loads of STC per year is considerable. (HCl is always saleable.)

Thus, near 500 MT/y, the concept of by-product recovery begins to change from recover-to-sell to recover-to-reuse on site.

2.7 Specific Reactor Exhaust—Gas Recovery Designs

By-product chemical process methodology for polysilicon plants fall in three general levels of vertical integration. At the first level, polysilicon plants are designed to purchase TCS and sell polysilicon, STC, and HCl. The second level plant converts by-product STC to TCS on-site for reuse and sells the polysilicon and HCl. The final level of chemical methodology involves buying MG-Si, reacting it with by-product HCl to manufacture TCS, converting by-product STC to TCS on site, and selling semiconductor-grade polysilicon. Level three offers the safest and most economical methodology. It should be mentioned that there may be, some day, a fourth level of vertical integration. This stage may arise if the photovoltaic industry begins to use large 10,000- to 30,000-MT/y quantities of polysilicon per year. In the fourth level, the MG-Si is also manufactured on site.

Key to all levels of proper exhaust-gas, recovery-system design, sizing, and operation is the composition of the exhaust gas which includes H_2, HCl, TCS, STC, and some silicon oils. (The silicon oils are pyrophoric and their presence in the reactor and in the chemical recovery system is dangerous. Technology exists to suppress the formation of the oils and to handle the oils after formation.) It it important to know the quantities of the four by-product species. But to understand the details of the chemistry which generates the by-products, one must know the actual operating characteristics of the reactors and the philosophy of plant operation. The theoretical models of L.P. Hunt and others are considerably perturbed by the dynamics of reactor operation.

Two examples will illustrate the importance of knowing the quantities of the by-product gas species. Concerning the reactor's operation, some reactors operate with TCS concentrations near 5% while others operate near 50%. The effect of carrier gas concentration dramatically effects the sizing of pipes, the amounts of refrigeration, and even the methodology of by-product recovery. (This is mainly due to the high heat capacity of hydrogen and its ease of passing through a pipe—approximately 3 times—as compared to TCS and STC.) Again concerning reactors, some designs are clearly more efficient at converting TCS to silicon than others:

% Conversion Efficiency = 100 x Si-deposited/ Si-input

Reaction 3 typifies the dissociation reaction. The more efficient designs (23% conversion) will produce more by-products and less unreacted TCS, reaction (4), in the exhaust gas. Less efficient designs (6% conversion) will have an exhaust gas rich in TCS with less accompanying by-product gases.

(3) Reacted TCS: $2SiHCl_3 + H_2 \rightarrow Si + SiCl_4 + 2HCl + H_2$

(4) Unreacted TCS: $SiHCl_3 \rightarrow SiHCl_3$

Concerning the philosophy of operation, the plant design itself affects the quantities of the by-products. In plant designs where minimal capital costs are desired, high single-pass deposition efficiency is desired, and thus the by-product recovery system is small. In plant designs where the highest silicon production is desired, single-pass deposition efficiency is relatively low, and thus the recovery systems are relatively large.

There are different philosophies related to the operating utility of polysilicon plants, that is the percentage time a given reactor is to be on-line producing silicon. The conservative philosophy maintains 75% on-line time while the aggressive approach maintains over 95% on-line time. The recovery system must be full size for the aggressively-designed polysilicon plant.

2.7.1 Level-One Recovery System: A typical level-one chemical recovery system is shown in Figure 6. First, the exhaust gases are cooled (-40°C) to condense the STC and the TCS. The noncondensed STC, TCS, HCl, and H_2 are compressed to 80 psig and all gases are again cooled to -60°C to remove the remaining STC and TCS. Schemes for the removal of the HCl from the H_2 vary. One scheme uses activated charcoal to separate the H_2 and the HCl (the HCl is absorbed on to the charcoal while the H_2 passes through). Another scheme uses STC to wash the HCl from the H_2, and yet another uses H_2O to wash the HCl from the H_2. (The HCl is readily dissolved in both cold STC and room temperature H_2O.) The TCS is separated from the STC by distillation. The STC is usually sold without further processing and the HCl is sold as a gas or as water-formed muriatic acid. The recovered and frequently repurified TCS and H_2 are recycled to the reactors as hyperpure feedstock. (If the recovery system is clean, the by-products are of superior quality as many of the impurities in the virgin TCS preferentially deposit in the produced polysilicon.)

In practice, the chemical engineering to recover the reactor's chemical by-products is different for each polysilicon producer. These practices, as well as the technical methods and the know-how of maintaining the ppba purity of the chemicals are secrets of polysilicon production. (The science and art of maintaining ppb polysilicon manufacture is neither simple nor obvious. Several large polysilicon plants have been designed, constructed, and operated with only nominal high-grade polysilicon production because the art was not understood.)

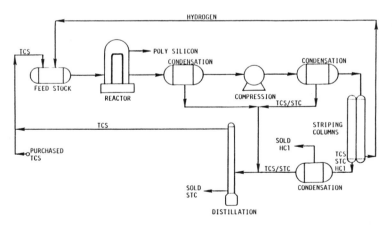

Figure 6. Level 1 chemical recovery system - polysilicon plant using purchased trichlorosilane as feed stock and selling silicon tetrachloride and hydrogen chloride.

2.7.2 Level-Two Recovery System: The second level of vertical integration design, a design used by medium-sized polysilicon plants, operates in a similar manner to the level-one design. However, the recovered by-product STC is not sold but becomes a feedstock to manufacture additional TCS at the polysilicon site. There are two reasons for this on-site secondary TCS manufacturing. Safety is the first reason. The transportation of millions of pounds of incoming virgin TCS and outgoing by-product STC is slow by rail and risky by truck trailer. Economics is the second reason. The STC is already ultrapure having just been purified by depositing many of its impurities in the deposited silicon, so the STC simply needs one of its chlorines replaced with a hydrogen to become useful TCS. (The reasons for not using the STC directly as a feedstock to produce polysilicon is discussed later in this chapter.) Unfortunately, the on-line practical single-pass conversion efficiency from STC to TCS is stated to be 18% and 32% at near 500°C, and at 73 psig and 300 psig, respectively. However, in both conversions the residence time required is nearly sixty seconds at a mole ratio of 2.8 $H_2/SiCl_4$ (42),(43) (44). If in practice, shorter residence times and richer blends of STC are used, the single-pass conversion efficiency of STC to TCS is usually reduced. Thus considerable energy is required to convert the STC to TCS, separate the two, and to then continue to reprocess the remaining STC to TCS. In addition, multiple passes often lower the purity of the STC and the resultant TCS. Thus, side streams of both STC and TCS are removed to withdraw the added impurities. Side-stream process losses, as well as the cost to waste-treat the impure side streams, adds to the cost of converting the by-product STC to TCS. All factors considered, the TCS converted from STC may be less expensive than purchased virgin TCS if only by the transportation charges for hauling both the otherwise purchased TCS and the resultant by-product STC to and from the polysilicon site. When the STC is converted to TCS at the polysilicon site, the plant becomes a nearly closed-loop chlorine system.

A polysilicon plant using level-two chemical recovery integration is shown in Figure 7.

2.7.3 Closed-Loop Recovery System: The final level of by-product vertical integration is that of closed-loop hydrogen and chlorine by-product recovery and reuse. This design reuses STC as in level-two integration to produce additional TCS and, in addition, uses the by-product HCl to react with

MG-Si in a FBR to produce both TCS and STC. This added step is definitely a cost saving step for a 1,000 to 2,000-MT/y plant since the HCl-to-TCS system is identical to the off-site TCS plant from which the virgin TCS (with profit added) was purchased. There are several design variations to the level-three vertical by-product integration. Figure 8 depicts one version, the closed-loop hydrogen and chlorine by-product recovery system to produce polysilicon (45).

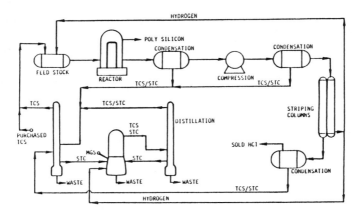

Figure 7. Level 2 chemical recovery system - polysilicon plant using purchased trichlorosilane as feed stock, converting silicon tetrachloride to trichlorosilane for reuse, and selling the hydrogen chloride.

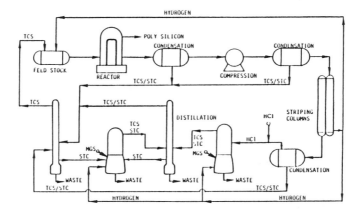

Figure 8. Level 3 chemical recovery system - polysilicon plant using purchased metallurgical grade silicon and hydrogen chloride and no by-product sales.

2.8 Capital and Operating Costs

As industry moves to level-three chemical by-product recovery and reuse, the initial capital investment to build a polysilicon plant increases, but the cost to produce polysilicon decreases. Table III indicates the capital costs for a 2000-MT/y polysilicon plant and Table IV indicates the expected operating costs for a 2000-MT/y polysilicon plant. These estimates are based on the use of 50-inch-diameter by 103-inch-tall, bell-jar reactor technology and the recovery and on-site reuse of nearly all reactor by-products.

Of note is that a 2,000-MT/y polysilicon plant which is ten years old, or one built in a country which does not use a depreciation schedule, can produce polysilicon for $14.23/kg in 1988 dollars.

Table III

Capital Cost (1988 $'s) of a TCS Polysilicon Plant

Using a Closed Loop Hydrogen and Chlorine Recovery System

Item	Cost ($ million)
Equipment	
Chemical Storage	
Feed System	
Reactors	
Hydrogen Recovery	
TCS Recovery	
STC Recovery, conversion to TCS	
HCl Recovery, conversion to TCS and STC	
Safety Systems	
Environmental Systems	$65.7
Installation	
Shop and Field Construction	$55.9
Engineering	
Design	
Construction	
Start-up	$6.5
Total	$128.0

Table IV

Operating Costs of a TCS Polysilicon Plant Using a Stage-Three, Closed Loop Hydrogen and Chlorine Recovery System

Item	Cost ($/kg Silicon)
Raw Materials	7.31
Utilities	2.23
Labor	2.45
Operating Supplies	1.74
Waste Treatment	0.20
Depreciation (10 y st. line)	6.40
Total	20.63

A 2,000 MT/y plant which is ten years old can produce semiconductor-grade polysilicon for $14.23/kg in 1988 dollars

3.0 ALTERNATIVE CHLORINE-BASED, SEMICONDUCTOR-GRADE, POLYSILICON FEEDSTOCKS

The four most-used polysilicon feedstocks are discussed: silane (SiH_4), dichlorosilane (DCS), trichlorosilane (TCS), and silicon tetrachloride (STC).

Feedstock selection is more than chemical theory brought to practice. Feedstock selection is the business of safety, quality, price, and dependability. At any period in history, the importance of these factors will depend on society's mood, economic condition, and available technology. In this text, 1988 is used as a snap shot in time to place a "desirability value" on each feedstock criterion considered. A score of ten—10—is the most desirable, while zero—0—is the least desirable.

Twelve criteria are considered: 1) Product Purity, 2) Manufacturing Cost, 3) Safety, 4) Alternative Sources, 5) Transportation, 6) Storage, 7) By-product Recovery, 8) By-Product Use, 9) Deposition Rate, 10) Construction Methods, 11) Reactor Choices, and 12) Electrical Energy Usage.

There are many specific conclusions that can be drawn from the grades of the four feedstocks. But the general conclusion is that more is known about STC than TCS than DCS than silane. As knowledge of the lesser chlorinated chlorosilanes increases, their ratings will likely increase unless safety risk makes the evaluation mute. Ratings for the feedstocks are summarized at the end of this section in Tables V and VI.

3.1 Silane As A Polysilicon Feedstock

3.1.1 Silane Purity—Grade 6 to 10 for Bell-Jar Reactor Polysilicon:
There are three basic manufacturing paths to obtain pure silane. The first (chlorosilane-to-silane) begins with STC and replaces chlorine atoms with hydrogen. Each step of hydrogenation provides a new species of molecule to be purified. The undesirable n-type chlorosilanes are frequently taken as undercuts and overcuts at the STC, TCS, and DCS distillation steps. Finally, molecular sieves may be used to purify the silane but distillation generally is used (46). The second method produces silane by the reaction of sodium hydride and silicon tetrafluoride (47), or by sodium and silicon tetrafluoride (48). In this manufacturing method, the n-type and p-type contaminates can be removed prior to the formation of the silicon tetrafluoride. The remaining trace impurities are treated in the same manner as the final step of the progressive hydrogenation of STC to form silane. The third method is to hydrolyze magnesium silicide with ammonium chloride to form silane. Purification follows.

In practice, the purity of chlorosilane-to-silane-produced polysilicon by the bell-jar process is reported to be near 1,000 ohm-cm in the USA (49) and is believed to be above 10,000 ohm-cm by the limited production process of Komatsu. Reported resistivity for FBR polysilicon is near 50 ohm-cm (50).

3.1.2 Silane Manufacturing Costs—Grade 4:
Manufacturing silane from metallurgical-grade silicon (MG-Si) is a long process. Purchased silicon tetrachloride (STC) is reacted with MG-Si. The products are STC and TCS (20%-22.5% conversion). The separated STC is recycled to the MG-Si reaction. The TCS is purified and converted to dichlorosilane (DCS) (9.5% conversion). The non-converted TCS is separated and returns to the DCS converter. The STC generated in the DCS converter is returned to the MG-Si reactor. The separated DCS is purified and converted to silane (14% conversion) (51). The by-products of the DCS-to-silane conversion, DCS, TCS and STC, are recycled to prior steps. The silane is purified. The entire process requires heating and cooling for the hydrogenation reactors, stills, and the final molecular-bed polishing of the silane. Lower yields from the converters

require more cycles through the converters, stills, and separation columns. Figure 9 indicates the process flow of a silane-fed polysilicon plant using purchased MG-Si and STC. This chlorosilane-to-silane process is one of the most involved polysilicon processes.

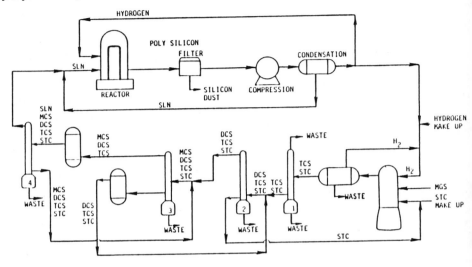

Figure 9. Polysilicon plant using purchased metallurgical grade silicon and silicon tetrachloride to manufacture silane for feed stock. No by-products are sold.

Union Carbide's direct manufacturing cost estimate for a 1,000-MY/y polysilicon plant is $6.41 (1980) per pound of silane (52). A lower silane manufacturing cost has been calculated by the author to be $3.73 per pound of silane. Side-stream loses, required to maintain silane purity, may account for differences in costs. The silicon tetrafluoride method uses elemental sodium and thus the silane costs are higher than for the chlorosilane-to-silane process. The magnesium silicide process is practiced on a limited and expensive basis with no published costs.

3.1.3 Silane Safety—Grade 4 for Chlorosilane-to-Silane, and 5 for Hexafluorosilicic-Silane: Silane ignites and can explode upon exposure to air. Silane explosions have occurred in polysilicon plants and have been forceful.

In the case of the silane produced by the chlorosilane-to-silane process, one of the steps is the production of DCS.

DCS is said by some to be more hazardous than silane. Thus the production of silane has the combined dangers of: powered metallurgical-grade silicon, DCS, and silane all of which can explode, and hydrogen which usually burns.

In the case of the silane produced using hexafluorosilicic acid to generate silicon tetrafluoride, there are the hazards of: liquid sodium and hydrogen which burn; the three acids hexafluorosilicic, sulfuric, and hydrofluoric which are hazardous; and sodium hydride and sodium which react violently with water.

3.1.4 Silane Alternative Sources—Grade 0:

A true alternative source is one which can be used when the primary source is unavailable. Since large quantities of silane can not be safely transported, the only second source of silane for a polysilicon plant is a second silane plant on the polysilicon site. This concept is not practiced for economic reasons. Another possible second source is the availability of large quantities of silane stored on-site. For safety reasons, this concept is not practiced. Thus a lack of on-site silane production causes a complete shut down of the polysilicon plant.

3.1.5 Silane Transportability—Grade 2:

Transportation is deemed very hazardous. Small quantities of silane are shipped throughout the world.

3.1.6 Silane Storage—Grade 4:

Silane can explode. Silane is stored as a refrigerated liquid or a pressurized gas. In either state, it takes energy to prepare the silane for storage. Compression processes often lower silane purity.

3.1.7 Silane By-Product Recovery—Grade 2:

Normally, one speaks of polysilicon by-product recovery as the recapture of by-products from the polysilicon reactor. In the case of the chlorosilane and hexafluorosilicic acid manufacturing processes for silane, it is also important to consider extensive by-product recovery during the manufacture of the feedstock before it enters the polysilicon reactor. There is also the loss of silicon dust.

3.1.8 Silane By-Product Use—Grade 5 for Chlorosilane-to-Silane; Grade 3 for Hexafluorosilicic-Silane:

Most of the by-products of the pre- and postreactor chlorosilane-to-silane system are reusable on-site. There is a loss of 10 to 20% of the silane due to dusting at polysilicon deposition rates equivalent to using TCS as the feedstock. The by-products of the hexafluorosilicic-silane process are sodium fluoride and hydrogen.

3.1.9 Silane Deposition Rate—Grade 3:

Silane's low decomposition rate is a problem. While the complex set of polysilicon reactor design criteria for TCS, bell-jar reactors offers a 999:1 desired deposition (on the rods) to non-desired deposition (anywhere but on the rods) ratio, it is believed that the silane, bell-jar, polysilicon reactors operate at ratios nearer 9:1 at a 3 to 8 μm/min deposition rate.

The crucial criteria for high deposition ratio is to keep those parts of the reactor hot where polysilicon production is desired, and keep those parts cool where no polysilicon deposition is desired (see Figure 10). To prevent off-rod silicon deposition when using silane, the metal bell jar is cooled to near 100°C, #3, as are the infrared shields between the rods, #5, and the base plate, #4 (53).

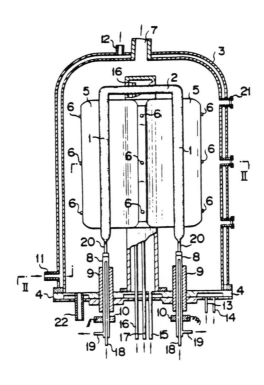

Figure 10. Polysilicon metal bell jar reactor design for silane feed stock-US patent 4,150,168 (Komatsu).

Yet another consideration for a favorable deposition ratio in a metal bell-jar reactor is to maintain the silane gas temperature sufficiently low to prevent free-space silicon formation as the silane moves from the inlet ports, #15 and #17, to the deposition rods, #1. Since 100% silane feedstock begins to dissociate at 300°C, and because the slim rods are at 800°C, it is difficult to prevent the open-space formation of silicon dust within the reaction chamber. The silicon dust clogs inlet and outlet ports, coats the bell-jar enclosure and reflector surfaces, sends silicon dust into the chemical recovery system, and lowers the yield of the saleable polysilicon. Having to lower the silane mole ratio from 100%, to reduce silicon dusting, lowers the polysilicon deposition rate and requires that the hydrogen by-product recovery system be increased in size.

3.1.10 Silane Construction—Grade 5 for Chlorosilane-to-Silane, and 3 for Hexafluorosilicic-Silane: The type of piping, compression, and cooling required for the chlorosilane--to-silane process entails all the special requirements of the STC, TCS, and DCS systems, and in addition, the special requirements for the silane process. Since the STC-to-MG-Si system operates at 500°C and 515 psia, these conditions require more expensive and sophisticated construction than for STC, TCS, and DCS construction. Also, silane, like DCS, ignites and may explode on contact with air. As a result, thicker-walled pipes, heavier-walled vessels, and special valves are required to accommodate the hazards of the DCS and silane converter.

The construction for the hexafluorosilicic-silane plant must include the handling of liquid sodium, hexafluorosilicic acid, sulfuric acid, hydrogen fluoride, hydrogen, and silane.

3.1.11 Silane Reactor Choices—Grade 5 going to 7: Free-space, polysilicon reactors produce unusable polysilicon dust. Silane is improving as a feedstock for FBRs. The FBR problems of varying silicon purity and silicon dusting, which lower production throughput, are engineering hurdles which experience may resolve. Silane is not as an effective feedstock for bell-jar reactors because its silicon deposition rate is low.

3.1.12 Silane Electrical Energy Usage—Grade 5/10: Due to the low polysilicon deposition rate in silane bell-jar reactors, the need for hydrogen-diluted silane feedstock, the unwanted production of silicon dust, and the need to cool the bell jar, the lower electrical energy advantage of operating the rods at a relatively cool 800°C abates considerably. One energy estimate for bell-jar-grown silicon from silane is 40 kWh/kg (54), but in practice the energy requirement could exceed 80 kWh/kg.

The electrical energy used to produce polysilicon from silane in a FBR may be the lowest of all reactors due to the possible use of 100% silane feedstock, no carrier gas to heat, and high deposition rates. For FBRs the electrical energy usage could be as low as 10 kWh/kg of polysilicon. (54) With feedstock mole ratios near 20% for the FBR, the electrical energy cost per kilogram of polysilicon is expected to rise considerably (55).

3.2 Dichlorosilane As A Polysilicon Feedstock

3.2.1 Dichlorosilane Purity—Grade 7: Polysilicon grown from DCS is reported to have a boron content of 0.07 ppba and a donor content of 1.22 ppba with a carbon level of 0.3 ppma (56). This 100 ohm-cm, n-type purity of polysilicon was achieved using a quartz bell-jar reactor. However, because polysilicon deposition is significant on the quartz bell-jar when using DCS as a feedstock, water-cooled, metal bell-jars may be a better operating choice. The polysilicon purity when using a metal bell-jar is not known.

3.2.2 Dichlorosilane Manufacturing Costs—Grade 6: The DCS manufacturing process is similar to the chlorosilane-to-silane process. Estimated yields are 30% for the conversion of STC to TCS (57). The estimated yield for the conversion of TCS to DCS is 11% (58). Hemlock Semiconductor estimates $0.59 per pound of DCS (59). The author's estimate is $0.80 per pound of DCS.

3.2.3 Dichlorosilane Safety—Grade 3: Hemlock semiconductor has conducted several DCS safety tests using a prototype polysilicon reactor. The explosion hazards of DCS were found to exist over a wide range of DCS in air mixtures. "... data obtained from the experimental program have well confirmed ... that DCS/air mixtures are both very easily (and sometimes unexpectedly) ignited, and that the explosive potential from such ignitions is usually high." (60),(61).

3.2.4 Dichlorosilane Alternative Sources—Grade 0: DCS is primarily obtained by the hydrogenation of TCS. Whereas silane has three practiced methods of production, DCS has only one. DCS, like silane, but unlike TCS and STC, is not used as a primary feedstock in another major industry. As in the case of silane, DCS is only manufactured at the polysilicon site due to safety reasons. Thus an alternative source would likely require a parallel, on-site, DCS production facility—an unlikely option due to capital costs.

3.2.5 Dichlorosilane Transportability—Grade 2:
DCS can be transported as a compressed gas or as a liquid under pressure. It is doubtful, based on the safety information presently available, that large quantities would be transported. Nonetheless, DCS is shipped to various industrial locations by Matheson in 10 to 90 pound quantities. DCS is piped in short runs to epitaxial reactors using normal methods.

3.2.6 Dichlorosilane Storage—Grade 3:
The practice by Hemlock Semiconductor, a leader in DCS polysilicon experiments, is not to store DCS.

3.2.7 Dichlorosilane By-Product Recovery—Grade 3:
The dissociation of DCS to form polysilicon gives the full range of chlorinated silane by-products: STC, TCS, DCS, and HCl and H_2. By-product recovery efforts are equivalent to that for the TCS and STC feedstock polysilicon reactors plus the addition of DCS recovery. There is also the loss of silicon dust.

Figure 11 indicates one method of DCS production, polysilicon production, and chemical by-product recovery.

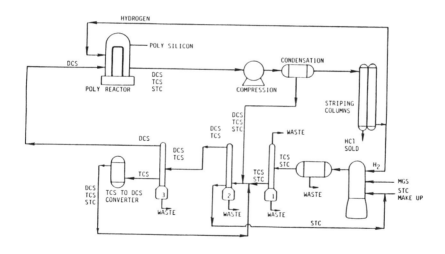

Figure 11. Polysilicon plant using purchased metallurgical grade silicon and silicon tetrachloride to manufacture dichlorosilane for feed stock. No by-products are sold.

68 Handbook of Semiconductor Silicon Technology

3.2.8 Dichlorosilane Use of By-Products—Grade 5: All of the by-products from the polysilicon reactors using DCS as a feedstock can be sold or reused. For an on-site DCS production plant, all the by-products can be recycled, except for the HCl when STC is used in the FBR reactor to form TCS from MG-Si. If HCl is used to form the TCS from MG-Si, then the HCl is also recycled. Silicon oils are generated in considerable quantity if operating conditions are not correct.

3.2.9 Dichlorosilane Deposition Rate—Grade 5: DCS, theoretically deposits silicon slower than silane but faster than STC and TCS. More silicon is available per pound of DCS feedstock (28%) delivered to the polysilicon reactor than for STC (16%) or TCS (21%). Lower deposition temperatures are required for the same polysilicon deposition rate, 1000°C for DCS versus 1200°C for STC and 1100°C for TCS. Also lower DCS mass flows are required for the same deposition rate. Thus, at reasonable deposition temperatures and feedstock flow rates, higher polysilicon deposition rates are expected than for STC and TCS. The Si(in) to Si(deposited) conversion efficiency is expected to be 48% for DCS (62), as compared to 20% for TCS. However, DCS has many of the same deposition-rate limiting factors as silane (63). A target-to-nontarget deposition ratio is about 50:1.

3.2.10 Dichlorosilane Construction Methods—Grade 5: TCS and STC construction methods are the bases for DCS construction. However, nickel/chrome/molybdenum steels are additionally used for DCS construction. Care must be taken to use construction methods that can contain or direct the potential explosions caused by DCS.

3.2.11 Dichlorosilane Reactor Choices—Grade 3: Free-space reactors using DCS would likely generate unusable silicon dust, and the by-product HCl would likely attack the hot injection nozzles.

DCS dissociation to polysilicon in FBRs may be an option. The drawbacks would be similar to those of using silane and TCS in FBRs. The silane FBR generates excessive silicon dust, and the TCS reactors generate HCl which reacts with the bed walls.

The quartz bell-jar walls of a DCS polysilicon reactor operate between 450° and 550°C. However, temperatures near 500°C foster excessive silicon deposition on the bell-jar walls (64). Silicon deposition often cracks the quartz. If a quartz bell jar fractures, 1000°C hydrogen and DCS can combine with air to cause an explosion. While the use of quartz bell jars should not be precluded categorically for use with DCS, especially for the use of high-purity polysilicon, their choice may be second to a metal bell jar.

Metal bell jars operate at temperatures near 40°C and thus provide a cold wall adjacent to hot rods. This practice permits isolated deposition of silicon on the rods. Since metal bell jars seldom crack, they reduce the probability of a hazardous reactor explosion. The present inclination is to use metal bell jars when manufacturing polysilicon using DCS as a feedstock.

3.2.12 Dichlorosilane Electrical Energy Usage—Grade 5: Because DCS deposits polysilicon at a lower temperature (1000°C) than does TCS (1100°C), it is logical to surmise that the DCS electrical energy costs to produce a kilogram of silicon are less than for TCS and STC. With the proper reactor design, this may be the case. About 80-100 kWh/kg is required to produce five-inch diameter rods from DCS. (65)

3.3 Trichlorosilane as a Polysilicon Feedstock

3.3.1 Trichlorosilane Purity—Grade 7 for Bell-Jar Polysilicon; Grade 5 for FBR Polysilicon: Typical specifications for TCS are 0.8×10^4 ppba carbon, 0.4 ppba boron, and 0.5 ppba donor. While the capture rate of impurities into polysilicon rods from the TCS gas phase is different for each reactor design and for each polysilicon plant's mode of operation, it is possible to produce 2,000 ohm-cm, n-type, uncompensated polysilicon in a quartz bell-jar reactor (Great Western Silicon). Typical polysilicon purity is about 300-500 ohm-cm.

Polysilicon produced from TCS in a FBR is of a lower purity than polysilicon produced in bell-jar-type reactors, perhaps near 50 ohm-cm. Since polysilicon pellets produced in FBRs are sometimes formed on fine seed particles of injected silicon, those particles must be of reasonably high quality if the resultant pellet is to be of high quality. Also, because the produced pellets of silicon contact the walls of the FBR, the pellets can abrade the wall material. If the wall is carbon, then carbon is likely to be found in the polysilicon. Wall reactions with the hot chlorosilanes and HCl also take place, and thus impurities in the wall material are transferred to the silicon by chemical transport.

FBR pellets may degrade after production due to the large surface area of the pellets relative to their weight and/or the microscopic dust that attaches itself to the pellets.

3.3.2 Trichlorosilane Manufacturing Cost—Grade 7:

The standard method of manufacturing TCS is to operate a FBR with feedstocks of HCl and MG-Si at 300°C and 65 psia. A design which optimizes TCS produces by-products of 5.2% STC, 1.4% DCS, and 1.9% heavies (66). The STC is used in the silicone and fumed silica industries; the TCS is used as a coupling agent, and when further refined, to manufacture semiconductor-grade-quality TCS. Figure 12 shows a TCS plant using purchased HCl and MG-Si as feedstocks. Union Carbide sells semiconductor-grade TCS for about $1.00/lb. of TCS. The manufacturing cost for TCS is estimated by the author to be $0.49/lb of TCS.

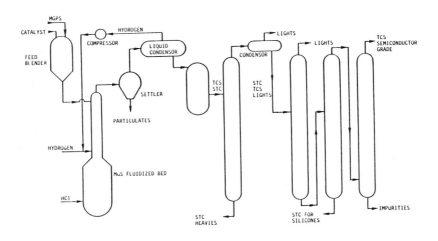

Figure 12. Semiconductor grade trichlorosilane manufacturing plant using purchased metallurgical grade silicon and hydrogen chloride. (Typical of remote manufacturing plant design.)

When by-product STC is used as feedstock with MG-Si to manufacture TCS, the TCS cost becomes very sensitive to the cost of the replacement STC. While the costs of HCl and MG-Si are stable, the historic bulk technical or lower-grade mercantile STC price has varied greatly in the U.S.A. from $0.09/lb to $0.35/lb STC. When STC ceases to be sold as a by-product from existing polysilicon plants and is sold primarily as a manufactured product, its cost will approach

$0.49/lb of STC. When the price of STC approaches TCS, then silane, DCS, and TCS manufactured from STC as a feedstock will likely cost more than TCS manufactured directly from HCl.

3.3.3 Trichlorosilane Safety—Grade 9:

TCS has been used to manufacture semiconductor—grade polysilicon for over twenty five years. The following incidents have been observed. Considerable quantities (hundreds of gallons) of TCS have been spilled with no subsequent ignition, smoke, or fire, only evaporation. Purges have been missed in TCS-fed reactors with no explosion: a reactor chamber, holding rods at 900°C has turned white inside when the TCS and H_2 combined with air, but no explosion occurred. Quartz bell jars containing H_2 and TCS surrounding 1100°C polysilicon rods have broken; the reactor gases have burned outside the cracked bell; there was no explosion. On other occasions, explosions of a bell-jar have been reported, but no injuries occurred as the shroud protected personnel. Over pressure in TCS reactors, if relieved at the reactor itself, does cause a loud report as the H_2 ignites, but damage seldom occurs.

3.3.4 Trichlorosilane Alternate Sources—Grade 10:

In the U.S.A., Union Carbide, Dow Corning, Texas Instruments, and General Electric produce TCS; in Europe: Wacker and Dynamit Nobel; in Japan: Shin Etsu Handotai, Osaka Titanium, Chisso, and Tokoyama Soda. TCS is also produced in China and eastern block countries.

TCS is used as a coupling agent for glass-to-metal bonding. Thus, polysilicon plants using TCS as a feedstock have other polysilicon suppliers as a second TCS source, and if required, the coupling-agent industry as a third TCS source.

3.3.5 Trichlorosilane Transportability—Grade 9:

TCS has been transported in 4,000-gallon trailers on a regular basis from West Virginia to Missouri, Texas, and Arizona and from Germany to Italy. The TCS is carried as a liquid in specially-prepared truck trailers. Ease of transportation is a quality that has fostered the use of TCS as the feedstock for five polysilicon plants in the U.S.A. and one in Europe.

3.3.6 Trichlorosilane Storage—Grade 9:

Storing 20,000 gallons of TCS is common. Properly prepared storage vessels can maintain near-virgin, semiconductor—grade TCS over a 60-day period. TCS boils at 33°C, thus storage is in the liquid state under most ambient conditions. TCS has a reasonable vapor pressure. Even in hot climates, venting of storage vessels is usually unnecessary. If a TCS storage

72 Handbook of Semiconductor Silicon Technology

vessel is opened, the TCS does not necessarily burn. Even in the event of a fire at the port of the storage vessel, the fire is relatively cool and the reclosure of the vessel is possible. Spilled TCS is drained from beneath storage vessels, contained in a diked area, and pumped into another vessel or allowed to evaporate.

3.3.7 Trichlorosilane By-Product Recovery—Grade 9:
The by-products of the TCS bell-jar reactor are: H_2, HCl, (DCS which dissolves in the TCS and STC) TCS, and STC. STC and TCS are removed from the reactor exhaust stream by low temperature condensation. HCl is removed from the H_2 by using activated carbon columns that are cooled to absorb HCl and then heated to remove HCl. In another process, HCl is removed from H_2 by the use of STC washing to capture HCl in a falling film tower. HCl can also be washed from H_2 by use of water. STC is separated from TCS by distillation. Frequently, HCl and STC are sold, and H_2 is recycled as the carrier gas for the TCS feedstock to the polysilicon reactors. Figure 5, shows a chemical plant that buys TCS, recovers HCl through gaseous absorption, and sells both the HCl and separated STC. Figure 6 shows a one-line drawing of the plant.

Figure 8 is representative of a fully closed-loop TCS process to manufacture polysilicon; this design recovers all of the by-products from the polysilicon reactor, uses the HCl to manufacture additional STC and TCS, and uses the STC to manufacture more TCS. The H_2 and TCS are recycled directly to the polysilicon reactors.

3.3.8 Trichlorosilane By-Product Use—Grade 8:
By-products from TCS polysilicon production are H_2, TCS, HCl and STC. The H_2 and TCS are reused to produce additional polysilicon, and STC is purchased by the silicone and fumed silica industries. In the U.S.A., the by-product STC from TCS polysilicon plants have generally matched the needs of the fumed silica industry. Value-worthy uses of STC are for the semiconductor industry as a feedstock for epitaxial reactors and for the fiber optics industry. Only minor amounts of silicon oils are formed in the advanced-design reactors.

3.3.9 Trichlorosilane Deposition Rate—Grade 7:
The deposition rate of polysilicon from TCS is average. It is, nonetheless, as rapid, or faster, than the mercantile production of polysilicon from silane or DCS-fed, bell-jar-type, polysilicon reactors. Reactor size governs polysilicon deposition rate.

In general, deposition rate increases with increased initial rod surface area. Thus, a reactor that holds 16 rods and receives a excess mass flow per rod surface area will produce more silicon per hour than the same flow over 8 rods at the same deposition temperature.

When using a quartz bell-jar reactor, TCS can cause polysilicon deposition on the wall of the bell jar. Rod temperature, closeness of the rods to the bell-jar walls, mass flow rates of TCS, and the mole ratio of TCS in H_2 all affect the deposition of silicon on the bell-jar wall.

Metal bell jars operate near 100°C, well below silicon deposition temperature, and are not subject to silicon deposition on the bell jar. Thus, the chemical deposition constraints for metal bell jars are relaxed as compared to quartz bell jars.

The experience of one mercantile polysilicon producer, using metal bell jars of similar size to quartz bell jars, indicated no appreciable gain in polysilicon deposition rate. The reason may be as follows. The rods in a metal bell jar face other rods at 1100°C as well as a 100°C bell-jar wall. Because the rods are resistively heated, the core of the rod is at an elevated temperature, say 1300°C, as compared to the surface temperature (it must dissipate its heat to the surface). If an attempt is made to raise the temperature of that portion of the rod facing the bell jar to 1100°C, then that portion of the same rod facing other hot rods may increase to 1300°C. This hotter surface temperature retards the dissipation of heat from the rod's core, and the core rises in temperature, say to 1423°C. This is the melt-out condition. Thus, to prevent melt-out, the rods in a metal bell jar are operated with a cold side and an at-temperature side. Because the cold side's deposition rate is below that of the hot side, the over all deposition rate of the run is reduced. If the rods are moved inward from the metal bell jar to lessen the cold-wall effect, then the number of rods in a given reactor must be smaller. Fewer rods lower the surface area on which to deposit polysilicon and, thus, lowers the effective deposition rate. Wacker's 49-inch-diameter, metal bell-jar design allows the rods to be considerably separated from the cold metal jar. As a result, with eight rods, high feedstock mole ratios, and a pressurized bell jar, the metal reactor is reported to reach deposition rates near 5 kg/h (37).

A large quartz bell jar, of similar size to the Wacker 49-inch by 103-inch metal bell jar, can operate at standard mole ratios and at a standard 0.5-psig pressure with more rods (16 or more as compared to 8 for the metal bell jar) in the same reactor because the rods can be placed closer to the bell jar. (The bell jar operates just below 600°C). A target-to-nontarget deposition ratio can vary from 999:1 to 9999:1.

The deposition rate of polysilicon from TCS in a FBR is a function of the reactor size and other chemical and thermodynamical factors. Because the FBR grows small pellets (1/8-in diameter), the effective growth rate must consider the initial size of the polysilicon seeds. Texas Instruments operated a commercial TCS-fed FBR in the 1980's.

3.3.10 Trichlorosilane Construction Methods—Grade 9:

While there are tricks of the trade in construction of polysilicon reactors and chemical systems, non-exotic materials of construction, except in certain limited instances, have proven appropriate to obtain 2,000 ohm-cm polysilicon (author's experience at GWS). The key to polysilicon production at the ppba level is the proper design of the reactor and the proper preparation of the reactor before deposition begins. Also the absence of leaks to and from the chemical system is critical.

In a polysilicon plant that purchases TCS, the operating pressures are typically below 85 psig with operating temperatures of the chemical plant's chemicals 100°C or lower. In a plant that manufactures its own TCS by the reaction of HCl with MG-Si, temperatures reach 300°C and construction requirements increase. Finally, in a plant that produces TCS from a reaction of STC with MG-Si, the conditions reach 500 psig and 500°C and construction requirements increase further. In no case, however, are the requirements as severe as for the production of DCS or silane.

3.3.11 Trichlorosilane Reactor Choices—Grade 9:

TCS has been used as a feedstock in the following reactors: hot wall with silicon deposition on the interior wall; hot wall with silicon deposition on the exterior wall; quartz bell jar with the deposition on internally heated rods; cold metal jar with deposition on internally heated rods; and fluidized bed deposition.

The most used TCS reactor is the single-ended, bell-jar system (double-ended systems also have been used). Quartz-wall, bell-jar reactors require a sophisticated design to keep the rods hot, the bell jar cool, to prevent bell jar breakage, and to form symmetrical and cosmetically acceptable rods. An excessively hot bell jar receives a silicon deposit which lowers silicon deposition efficiency and breaks the jar. Cold rods lower production throughput. Improper mechanical matching of the quartz bell with the metal base plate drastically lowers bell-jar life. Forced air cooling about the quartz bell assists bell-jar cooling while maintaining hot rods. Inventive mechanical and thermal reactor designs maintain bell-jar life. However, quartz bell-jar reactors are presently limited to 50-inch diameter and 103-inch height. The author has operated 36-inch quartz bell-jar reactors and consistently obtained 4,000 hours average bell-jar life. Cold-wall, metal, bell-jar reactors up to approximately ten feet in diameter are reported in use.

3.3.12 Trichlorosilane Electrical Energy Usage—Grade 4: Small, quartz, bell-jar reactors require 250 kWh/kg silicon. Small, cold-wall, metal, bell-jar reactors require 305 kWh/kg of silicon because the rods must be held at a reduced deposition temperature next to the cold bell-jar wall. In addition, more electrical energy is required to cool the metal bell jar than the small quartz bell jar. When larger, metal, bell-jar reactor systems are used, the rods can be moved inward, away from the cold bell, and the electrical energy requirements per kilogram of polysilicon decreases. Large polysilicon reactors, both quartz and metal wall, are estimated to use 100-130 kWh/kg including bell cooling for the cold-wall reactor. It is believed that a FBR with TCS uses 30 kWh/kg (67). This may not include requirements for cooling the FBR.

3.4 Silicon Tetrachloride As A Polysilicon Feedstock

3.4.1 Silicon Tetrachloride Purity—Grade 10: It is believed that STC has been refined by Topsil of Denmark to produce polysilicon of intrinsic resistivities greater than 10,000 ohm-cm. Typical values exceed 2000 ohm-cm.

3.4.2 Silicon Tetrachloride Manufacturing Costs—Grade 7: Semiconductor-grade STC can be manufactured for approximately the same price as TCS when HCl is used with MG-Si in a FBR: $0.49 per pound of STC.

3.4.3 Silicon Tetrachloride Safety—Grade 10: STC does not spontaneously ignite. It can, but usually does not sustain a flame. Its vapor pressure is low. When spilled, it frequently evaporates with no smoke. High humidity conditions may cause smoke generation.

3.4.4 Silicon Tetrachloride Transportation—Grade 9: STC is transported in 4,000-gallon trailers on a regular basis in Europe and the U.S.A. Its low vapor pressure and slow

76 Handbook of Semiconductor Silicon Technology

ignition makes it a convenient chemical to transport.

3.4.5 Silicon Tetrachloride Alternative Sources—Grade 9 going to 3: Topsil and Union Carbide use STC as a feedstock for their polysilicon plants. Hemlock may consider this option also. As more and more polysilicon plants move to the STC/MG-Si feed stock technology, and as the old polysilicon plants that had STC as a saleable by-product at $0.09 - $0.36 per pound disappear, the sole source of STC will be the directly manufactured from HCl and MG-Si.

3.4.6 Silicon Tetrachloride Storage—Grade 9: STC is stored in very large quantities, hundreds of thousands of gallons. In the 1970's, due to gross negligence, a large spill occurred in the U.S.A. While there was considerable HCl and SiO2 generated in the clean-up process, it was otherwise an uneventful mishap. Semiconductor-grade quality STC can be stored for months without quality degradation.

3.4.7 Silicon Tetrachloride By-Product Recovery—Grade 10: STC is easily condensed from the exhaust stream of a polysilicon reactor. Fifteen psig and -40°C are adequate for recovery. STC is easily separated from the TCS produced as a by-product in the polysilicon reactor. HCl is recovered as in the case for the by-products of TCS.

3.4.8 Silicon Tetrachloride By-Product Use—Grade 9: The by-products from a STC-fed polysilicon reactor are STC, TCS, HCl and H_2. Typically the TCS and H_2 are recycled with the STC to the reactor for additional deposition of silicon. The HCl is sold.

Silicon oils can be formed when using STC as a polysilicon feedstock, but proper reactor design minimizes the oils to an amount that can be easily extracted from the system.

3.4.9 Silicon Tetrachloride Deposition Rate—Grade 4: STC has the lowest silicon deposition rate of the chlorosilane feed stocks. Typically, if a reactor uses STC as feed stock, the growth rate can be doubled by using TCS. In addition to slow growth rates, the polysilicon deposition temperature for STC is 1250°C, or 100°C to 150°C above that of TCS. There is little that can be done to accelerate the 4- to 6-μm/min deposition rate of Si from STC in a bell-jar reactor.

Polysilicon from STC feed stock is easily deposited at a 9999:1 target-to-nontarget ratio as chlorine back etching of silicon occurs at temperatures only a few hundred degrees less than the silicon deposition temperature.

3.4.10 Silicon Tetrachloride Construction Methods—Grade 10: STC requires the least specialized equipment. While it is true that HCl is produced when the STC leaks to the atmosphere, the solution to that problem is to use construction techniques that prevent leaks. Such construction technology is available for all the chlorosilanes.

3.4.11 Silicon Tetrachloride Reactor Choices—Grade 9: In addition to all the TCS reactor choices (except the FBR), there are the metal reduction furnaces that convert STC to a metal chloride and silicon.

3.4.12 Silicon Tetrachloride Electrical Energy Usage—Grade 1: STC requires the highest temperature (1200° to 1250°C) to deposit silicon on a hot rod. This requirement along with the slow deposition rate of STC leads to the usage of 250 to 800 kWh/kg polysilicon depending on the size of the reactor.

Table V

Comparison of Feedstock Criteria for the Production of Semiconductor-Grade Polysilicon

	Feedstocks			
Feedstock Criteria	SiH_4	SiH_2Cl_2	$SiHCl_3$	$SiCl_4$
Polysilicon Purity (ohm-cm)	>1000	100	>300	>2000
Mfg. Cost ($/kg Si) (Reference)	16.12 (52)	1.29 (59)	-	-
Mfg. Cost ($/kg Si) (Author)	9.38	6.29	5.22	5.22
Safety	Great Care	Great Care	Care	Care
Alternate Sources	None	None	Other	Other
Transportation	Unlikely	Unlikely	Reasonable	Reasonable
Storage	Small Vol.	Not Suggest.	Months	Months
Prereactor Recovery	Yes	Yes	No	No
Postreactor Recovery	Yes	Yes	Yes	Yes
By-product Use	On Site	On Site	On Site/Sale	On Site/Sale
Depostion Rate (μm/min)	3-8	5-8	8-12	4-6
Surface Deposition (rod/nonrod)	9	99	999	9999
First Pass Conversion (%)	Unknown	17	5-20	2-10
Construction Methods	Exotic	Exotic	Average	Average
Reactor Choices	MBJ/FBR	MBJ	Many	Many
Electrical Energy Usage (kWh/kg Si)				
Bell Jar	40+	90+	120+	250+
Fluidized Bed	10+	-	30+	-

Note that the costs of manufacturing, as obtained from the references, are believed high for SiH_4 and low for SiH_2Cl_2.

Table VI

Comparison of Rankings of Feedstocks to Produce Semiconductor-Grade Polysilicon in Bell Jar Reactors

	Feedstocks			
Feedstock Criteria	SiH_4	SiH_2Cl_2	$SiHCl_3$	$SiCl_4$
1. Purity	8	8	7	10
2. Mfg. Cost	4	6	7	7
3. Safety	4	3	9	10
4. Alternate Sources	0	0	10	10
5. Transportation	2	2	8	9
6. Storage	4	2	9	9
7. By-Product Recovery	2	3	9	10
8. By-Product Use	5	5	8	9
9. Deposition Rate	3	5	7	4
10. Construction Methods	5	5	9	10
11. Reactor Choices	6	3	9	9
12. Elec. Energy Usage	6	5	4	1
Total	49	47	96	98

Note that the location of both the feedstock and polysilicon plant on the same site significantly reduces the importance of items 5 and 8.

4.0 ALTERNATE POLYSILICON REACTOR SELECTIONS

Semiconductor-grade polycrystalline silicon is manufactured in vapor-type reactor systems. The chemistry of the vapor-phase process is described by many renown scientists (68),(39),(69),(70),(71),(72),(73). Four common gaseous dissociation reactors are: free space, fluidized bed, hot-wall bell jar, and cold-wall bell jar. The hot-wall system is no longer used since removal of silicon from the wall could not be accomplished without contaminating the silicon with impurities from the wall. The cold-wall, bell-jar system has been previously described as quartz and metal bell-jar systems. This leaves two systems which yet offer promise, the free-space reactor and the FBR.

4.1 Free-Space Polysilicon Reactors

The free-space reactor is the simplest of all gaseous dissociation reactors (74). The gaseous feedstock enters a chamber and is raised to dissociation temperature in free space (that is in an open but controlled space). The gas receives sufficient energy from a thermal source to dissociate to silicon and by-products. Selection of the correct feedstock permits silicon to be formed in the solid phase and the by-products in the gas phase. The separation of the silicon from the by-products is accomplished by gravity in the chamber. Free-space reactors can operate using various feedstocks, but hydrogenated and halogenated silicon compounds have been the focus of most experiments. Tetraiodide feedstock has the disadvantage of solidifying under normal ambient operating conditions. Bromine, fluorine, and chlorine compounds have the disadvantage of producing by-products that are highly reactive with free-space injection nozzles (which are frequently near the dissociation temperature) and to some extent with the reactor containment system. Silane is the preferred free-space reactor feedstock because it is a gas at normal operating temperatures, it dissociates at a relatively low temperature, and it has a nonoxidizing by-product, hydrogen.

The free-space reaction chamber can be a room. Since chemical dissociation takes place in the open space of the room, the walls of the containment chamber can be made of ordinary construction materials. The chamber is gas tight to assure the pristine nature of the silicon and to permit collection of by-products without contamination. The tilted "V" shaped floor, onto which the silicon drops, is of a nonimpurity contributing material. The feedstock dissociation energy can be supplied by induction, radiation, plasma, conduction, or any noncontaminating source.

An auger removes silicon to an internal bagging compartment located within the dissociation chamber. The room operates slightly above ambient temperature. Hydrogen from the dissociated silane is recovered through a top exit port.

More compact, free-space reactor designs have the shape of ram jets (18). Compact reactor design problems include: need for reactor wall cooling; unwanted deposition of silicon on reactor walls; undesired reactor wall reaction with hot by-product gases; reactor wall liner failure; and inclusion of wall impurities in the polysilicon. Estimated polysilicon manufacturing cost for such a compact design was $7.56 per kilogram (1975 dollars) (18). However, polysilicon had a low density (0.2 g/cm^3) (75),(76) and thus did not provide a full crucible melt for a Czochralski puller. Also, submicron (0.3 to 0.4 μm) silicon dust floated inside the Czochralski puller and caused operating problems (18). The powder caused the resistivity of the crystal to vary from p-type to n-type.

Free-space reactors have the highest deposition rate of polysilicon reactors and thus use very little energy per kilogram of silicon. If a plasma is used for heating silane feedstock, dissociation occurs near 800°C (77). In general, only the silane needs to be heated since no carrier gas is required and since no filament rod or wall is kept at deposition temperature. Since no enclosure partition is near the dissociation reaction point, walls, base plate, and top plates do not need substantial cooling except in the ram-jet design.

Free-space reactors can operate near 100% conversion efficiency (18); that is, all the silicon in the feedstock can be converted to solid polysilicon in one pass through the reaction chamber. An advantage to the 100%-single-pass efficiency is the nominal energy and equipment needed to recover the H_2 by-product.

Free-space-produced polysilicon flows through pipes which can be used to continuously feed a Czochralski furnace (78). There are, however, a few problems using the powder. Exposure of the powder to the atmosphere causes the polysilicon to oxidize. When melted in a Czochralski furnace, the oxidized powder forms a slag which changes the crystallography of the single crystal ingot. Exposure of polysilicon powder to the atmosphere also raises dopant impurity levels in the silicon.

Because free-space polysilicon floats, powder disperses itself through out the furnace. This can cause the Czochralski heating system to arc even at low power levels and thus limits the electrical power to below that required to melt the polysilicon powder. The arcing problem is compounded by the requirement for more power to melt the free-space powder compared to that needed to melt rod-type, chunk polysilicon. Since the powder is many times less dense than standard polysilicon, the powder requires a one-time-charged crucible many times the size of the crucible used in a crystal puller using chunk polysilicon. To overcome the underloading of the crucible, an alternative is to feed the powder to the crucible during the melt-down process, that is to melt the powder so that more powder may be added to the crucible. Kayex Corporation has converted free-space polysilicon to single-crystal silicon with a resistivity of 55 ohm-cm, p-type (18). Another alternative is to convert the powder to pellets or ingots of dense polysilicon in a shot tower. Finally, silicon ingots can be formed by melting the powder and casting the molten silicon. The resultant shot or ingots can then be processed in existing crystal puller systems. However, the shot (0.4 to 3 mm) and ingot process, due to the shot melting sequence, further degrades the polysilicon (20 ohm-cm) and makes single crystalline structure difficult to maintain (79).

4.2 Polysilicon FBRs

A FBR can be described as a vertical pipe closed at both ends. Inlet feedstock is injected at the bottom and rises to a heated zone above the inlet. At the heated zone, the gaseous feedstock dissociates to solid silicon particles which grow larger by subsequent suspended epitaxial silicon deposition. The inlet gas velocity is sufficient to float the silicon particles until they grow to a sufficient weight to drop below the floating bed and into a receptacle. By-product gases exit the top of the pipe (80),(81),(82), (83),(84).

The FBR concept offers considerable theoretical advantages over bell-jar reactor design: continuous operation, large reactor size, safer reactor operation, lower electrical energy per kilogram of polysilicon, and a product that is pellet in nature rather than chunk. Polysilicon purity and product format remain a challenge.

As mentioned, the chemical and thermodynamic variables of a polysilicon reactor considerably affect the quality of the resultant polysilicon. These influences are more apparent in polysilicon produced in a FBR than a bell-jar reactor. The present fluidized-bed polysilicon from silane is said to be of lower and less **consistent resistivity** and conductivity type than polysilicon produced from the same feedstock in a bell-jar reactor.

FBRs are ideal for silane feedstock. Operating conditions are: bed temperatures 575°C-685°C, feed mole ratio 1 to 21 and 99.7% conversion efficiency (18). The resultant polysilicon is a pellet which has a reasonable density. Theoretically, the larger the diameter of a FBR, the higher the through-put. A FBR six inches in diameter has been operated by JPL (55),(85). The FBR operates by silicon deposition on silicon seed particles. Two modes of operation are possible: self seeding and injected seeding. Injected seeding using MG-Si is a practiced art. The pellets form on the seeds and drop out of the bed chamber. Their size varies from 150 to 1500 μm. The challenge is to obtain a continuously working and high through-put reactor that produces semiconductor-grade resistivity and carbon-free polysilicon. Companies that have worked to solve these problems include Union Carbide, General Atomic, Rhône-Poulenc, Ethyl, Texas Instruments, and others.

The thermodynamic problems, which lower deposition rate in the bell-jar reactor (nearby walls of the reactor) and cause dusting in the free-space reactor, combine in the FBR. Deposition rate can be enhanced by high temperature, high silane mole ratio, high silane mass flow, and a large diameter bed. Except for the larger-diameter option, the other options tend to increase silicon dusting. Figure 13 shows the fundamental design of a reactor. (The chamber used to collect the silicon pellets is not shown).

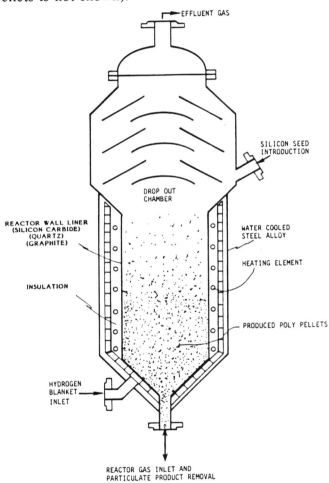

Figure 13. Polysilicon fluidized bed reactor. (The reactor size, shape, and internal mechanisms vary considerably depending on the type of feed stock and if self seeding or injected seed is used.)

Functionally, the expected continuous operation of a FBR, as compared to the batch operation of a bell-jar reactor, should increase the polysilicon plant's production through-put. There are reported difficulties, however, that suggest that high productivity has not been yet achieved.

First, FBRs need liners to protect polysilicon from becoming contaminated by the structural portion of the walls. These liners can be sleeves of a high-purity material (quartz), in situ liners formed by depositing silicon on the reactor's inner wall, or chemically-reacted liners which are formed by reacting a silicon compound with the wall itself (carbon reacting with silicon to form silicon carbide). Once heated to operational temperature, cooling of the liners can cause them to fail. Thus, the liner can cool and crack under thermal stress due to shut down of the electrical plant, heater malfunction, power controller malfunction, or other numerous difficulties. Replacement of the liner is expensive and time consuming. Whereas a bell-jar run termination due to power interruption is just a normal turn-around, such an interruption with a FBR is a major delay.

A second down-time difficulty with the FBR is its critical parameters of operation. Pressure and temperature must be maintained very accurately on the x- and y-axis of the bed. Temperature variations can dump the bed, a process where nearly all the particles fall to the bottom of the bed, or temperature variations can dust the bed, a process where the particles become so fine that they remain dust. Nozzle blockages and wall erosion are two other down-time contributors.

When achieved, continuous, fluidized-bed operation will reduce the labor to produce polysilicon as compared to the bell-jar reactor which requires clean ups, rod set ups, and opening and closing of the reactor. However, this labor savings is partially off-set by the additional labor required to handle (dedust, etch in some cases, purge pack, and double containerize) the fluidized-bed pellets.

Larger-size, FBRs may offer capital savings for a polysilicon plant. One polysilicon plant proponent suggested the use of six FBRs to supply 1,000 MT/y of polysilicon. The reactors were to be 30-in diameter. Remembering that Union Carbide remains unsuccessful in its operation of a 6-in-diameter FBR, a 30-in reactor is perhaps several generations away. It took four generations of quartz bell-jar designs and over 20 years

84 Handbook of Semiconductor Silicon Technology

of experience to achieve a successful 36-in diameter, quartz, bell-jar design.

Texas Instruments, Union Carbide, General Atomic, Ethyl, Schumacher, Rhône-Poulenc, JPL and others have developed FBRs. Texas Instruments has operated the largest polysilicon facility with 250 MT/y (estimate by author) of TCS-produced capacity. Thus, at this point, the FBR is yet an alternative to the large bell-jar reactors which, in 1988, had over 5,000 MT/y of polysilicon capacity.

Less than 40 kWh/kg Si energy consumption has been mentioned in connection with FBRs for TCS, and near 10 kWh/kg for silane. This significant reduction as compared to bell-jar designs is due, in major part, to the higher polysilicon deposition rates rather than to lower absolute use of electrical energy.

Fluidized-bed pellets offer an ideal feedstock for a Czochralski puller. The author has sampled pellets from General Atomic and Ethyl Corporation. Both pellets pack well in a Czochralski puller. However, the pellets require special treatment. Ethyl Corporation is attempting to resolve such difficulties by using special pellet cleaning and purged, double-container, pellet packaging. There are challenges of loading sufficient pellets to fill the Czochralski crucible with molten silicon and in melting the pellets uniformly (without silicon ice bridges and pellets clinging to crucible) in the Czochralski puller. Such problems require further R&D on the part of the fluidized-bed polysilicon supplier and co-operative R&D efforts with the potential fluidized-bed polysilicon customers. Good progress has already been reported.

5.0 EVALUATION OF SEMICONDUCTOR-GRADE POLYSILICON

Resistivity establishes the basic electrical conductivity of the substrate used for semiconductor circuits and devices. In general, p-type dopants should be less than 0.5 ppba (the resistivity is 550 ohm-cm in the case of boron) and n-type dopants should be less than 1 ppba (the resistivity is 100 ohm-cm in the case of phosphorus). Although resistivity measures only active electrical impurities, it can be used in conjunction with photoluminescence which measures both electrically and nonelectrically active impurities. Dopant impurities arise from the feedstock, the reactors, and the by-product recovery system.

Minority carrier life-time is important to the transport rate of electrons through a circuit's or device's crystallographic lattice. Gold is the major life-time killer. The concentration of gold in polysilicon must be less than about 1 ppba. There is no gold in the manufacturing process for polysilicon.

Oxygen is present at less than the ppba level in polysilicon manufactured in bell jars since conditions within the reactor result in oxygen being converted to SiO_2 separately from the polysilicon.

Carbon concentration in polysilicon is desired below 4 ppma. Carbon sources are carbon compounds in the feedstock, carbon parts in the reactor, and carbon compounds that are generated in the by-product recovery system.

Heavy metals in polysilicon are desired at the sub-ppba level. These can be determined by neutron activation analysis and inductively coupled plasma (ICP) spectroscopy. Also used is graphite furnace atomic adsorption spectroscopy.

The following methods are used to evaluate polysilicon:

1. Resistivity

    a. Convert polysilicon to a FZ crystal (ASTM F41)
    b. Measure electrical resistivity with a 4-point probe (ASTM F43)

2. Dopant concentration

    a. Convert polysilicon to a FZ crystal (ASTM F574)
    b. Measure electrical resistivity with a 4-point probe (ASTM F43)

    or

    c. Convert polysilicon to a FZ crystal (ASTM F41)
    d. Measure dopant concentrations by low-temperature photoluminescence or low-temperature Fourier-transform infrared (FTIR) spectroscopy (to be proposed by ASTM)

3. Carbon concentration

   a. Convert polysilicon to a FZ crystal (ASTM F41)
   b. Measure carbon by FTIR spectroscopy (ASTM F121)

4. Oxygen concentration

   a. Convert polysilicon to a FZ crystal (ASTM F41)
   b. Measure carbon by FTIR spectroscopy (ASTM F121)

5. Heavy metal concentrations

   a. Use neutron activation analysis
   b. Use ICP spectroscopy and/or graphite furnace atomic adsorption spectroscopy

6. Life-time

   a. Convert polysilicon to a FZ crystal (ASTM F41) or CZ crystal
   b. Measure life-time by photon injection (ASTM F28)

7. Conductivity type

   Convert polysilicon to a FZ crystal (ASTM F41) and measure conductivity with a "type" probe (ASTM F42)

6.0 FUTURE OF POLYSILICON

It is expected that future polysilicon plants will be of 1,000 to 3,000 MT/y capacity for mercantile sales-perhaps 5,000 MT/y for photovoltaic purposes-and 100 to 200 MT/y capacity for strategic manufacturing. Operational safety will increase because larger polysilicon plants can economically produce their own feedstock on-site and thus eliminate the transportation of hazardous feedstocks and by-products to and from the polysilicon plant. Even if conversion of STC to TCS becomes standard, the low 30%, single-pass, STC-to-TCS conversion efficiency may cause some operations to sell their STC to other industries. Silane may become the feedstock of the future if safety or noncompetitive polysilicon growth rates do not become an inhibiting factor. Finally, bell-jar production of polysilicon may be replaced by the fluidizedbed process which offers continuous polysilicon production.

Looking beyond polysilicon as the basic feedstock for the semiconductor and photovoltaic industries, there is on the horizon an innovative concept which may entirely replace the present two-step silicon manufacturing process of growing polysilicon and then CZ or FZ crystals. The process is a one-step concept that eliminates both polysilicon production and the melting steps now used. Present CZ manufacturing costs, including the cost of the polysilicon, are about $185/kg; float-zone costs are even higher. The one-step process will offer single crystal for $27/kg to $50/Kg depending on the cost of electrical power and the specification of the crystals (author's estimate). Figure 14 shows three prototype one-step single crystals (86). Crystals 1, 2, and 3 are two inches in diameter and are 80% single crystalline; crystal 4 is one inch in diameter and 100% single crystal with some imperfections; crystal 5 is one half inch in diameter and 100% single crystalline. All crystals were grown to seven-foot length.

The future of polysilicon remains dynamic. Higher silicon quality at lower cost remains the desire of the semiconductor integrated-circuit manufacturer and the demand of the photovoltaic cell manufacturer.

Figure 14. One-Step single crystal silicon ingots produced in a polysilicon reactor by L.C. Rogers and A.J. Heitz.

REFERENCES

1. Cosoque, E. and Pellin, R., "Polycrystalline Silicon Material Availability Study for 1980-1988," Report *DOE/JPL*-1012-79, Nat. Tech. Inform. Center, Springfield, VA (1983)

2. Pellin, R., "Semiconductor Market," Report *DOE/JPL*-1012-122, pp. 397-414, ibid. (1985)

3. Maycock, P., "Silicon Requirements of the Photovoltaic Solar Cell Market," Report *DOE/JPL*-1012-122, pp. 417-422 ibid.

4. Berezhnoi, A.S., *Silicon and Its Binary Systems*. Consultants Bureau, New York (1960)

5. Aulich, H.A., "Solar Grade Silicon Prepared from Carbothermic Production of Silicon," Report *DOE/JPL*-1012-122, pp. 267-275, Nat. Tech. Inform. Center, Springfield, VA (1985)

6. Crossman, L.D. and Baker, J.A., "Polysilicon Technology," in *Semiconductor Silicon 1977*, Vol 77-2, pp. 18-31 H.R. Huff and E. Sirtl, eds. The Electrochem. Soc., Pennington, New Jersey (1977)

7. Fairchild, W.T., paper number A70-36, *TMS-AIME Annual Meeting*, Denver (1970)

8. Torrey, H.C., *Crystal Rectifiers*, McGraw Hill Book Co., New York (1948)

9. Voss, W., *US Pat.* 2,972,521 (1961)

10. Tucker, N.P., *Iron and Steel Inst.*, Vol. 115, pp. 412-416 (1927)

11. Kinnzel, A.B., and T.R. Gunninghamm, *Metals Technology*, Tech. Paper 1138 (1939)

12. Murach, N.N., *Khim. Hauka i Prom*, No. 5, p. 492 (1956)

13. Stewart, J.. *Electroplat. and Metal Finish.*, vol. 9, pp. 212-219 (1956)

14. Ref. Aas, H., TMS Paper Selection A71-47, *The Metall. Soc. AIME*, NY, NY; pp 651-667 (1971)

15. Wolff, G.A., *US Pat.* 2,857,249 (1958)

16. Krchma, I.J, *US Pat.* 2,820,698 (1958)

17. Shalit, H., *US Pat.* 2,999,736 (1961)

18. Union Carbide Corp., *Final Report DOE/JPL*-954334-21:1, Nat. Tech. Inform. Center, Springfield, VA (1981)

19. Clasen, J., *US Pat.* 2,900,25 (1959)

20. Theuerer, H.C., *US Pat.* 3,071,444 (1963)

21. *US Pat.* 3,148,035 (1964)

22. Bracken, R.C., *US Pat.* 3,565,590 (1971)

23. Truitt, J.K. and Bawa, M.S., "Manufacture of High Purity Silicon," *AIChE Meeting*, Detroit (1981)

24. Potter, H.N., *US Pat.* 875,672 (1907)

25. Aries, R.S., *US Pat.* 3,010,797 (1961)

26. Smith, W.E., *US Pat.* 2,955,024 (1960)

27. Thurmond, C.D., *US Pat.* 2,904,405 (1959)

28. Bradley, H.B., *US Pat.* 3,188,168 (1965)

29. Bradley, H.B., *US Pat.* 3,540,861 (1968)

30. Bawa, M.S., *US Pat.* 3,704,104 (1972)

31. Alikberov, S.S., Shemlev, G.I., and Shklover, L.N., *Trans. 3rd Cong. Semiconducting Mater.*, in Russian (in press)

32. Lyon, D.W., Olson, C.M. and Lewis, E.D., "Preparation of Hyper Pure Silicon," *J. Electrochem. Soc.*, Vol. 96, pp. 359-363 (1949)

33. Spenke, E., *Seimens Rev.*, vol. XLVIII, pp. 4-9 (1981)

34. *US Pat.* 3,042,494 (1962)

35. *US Pat.* 3,146,123 (1964)

36. *US Pat.* 3,200,009 (1965)

37. Koppl, F., *US Pat.* 4,179,530 (1979)

38. Eagle Pitcher Research Laboratories, "Industrial Preparedness Study on High Purity Silicon," *4th Quarterly Report DA* 36-039-SC 66042 (1956)

39. Lever, R.F., *IBM J.*, Sept. (1964)

40. Harper M.J. and Lewis, T.J., "Thermodynamics of the Chlorine-Hydrogen-Silicon System," *UK Ministry of Aviation Report ERDE* 6/M/66, Clearinghouse Doc. AD 641 310 (1966)

41. Sirtl, E., Hunt, L.P. and Sawyer, D.H., "High-Temperature Reactions in the Silicon-Hydrogen-Chlorine System," *J. Electrochem. Soc.* Vol. 123, pp. 919-925 (1974)

42. Solarelectronics Inc., "Hydrochlorination Process," *DOE/JPL*-1012-71, p. 18 (1982)

43. Yaws, C.L., Li, K., Hopper, J.R., Fang, C.S. and Hansen, K.C., "Process Feasibility Study in Support of Silicon Material Task I," *Final Report, DOE/JPL* Contract 954343, Nat. Tech. Inform. Center, Springfield, VA (1981)

44. Hemlock Semiconductor Corp, "CVD of Polysilicon from DCS," *DOE/JPL*-1012-88, p. 228, ibid. (1983)

45. Padovani, F., *US Pat.* 4,092,446 (1978)

46. Union Carbide, "Experimental Process System Development," *Report DOE/JPL*-954334-21, Nat. Tech. Infor. Center, Springfield, VA (1981)

47. Jaffe, J., "A Silane-Based Polysilicon Process," *Report DOE/JPL*-1012-122, p. 347, ibid. (1985)

48. Sanjuro, A., Sancier, K.M., Emerson, R.M. and Leach, S.C., "Characterization of Solar-Grade Silicon Produced by the SiF_4 - Na Process," *ibid*. p. 325 (1985)

49. Iya, S., "Union Carbide; Development of the Silane Process for the Production of Low-Cost Polysilicon," *Report DOE/JPL*-1012-122, p. 139, ibid. (1985)

50. *Ibid.*, p. 141.

51. Union Carbide, "Base Case Conditions for UCC Silane Process," *Report DOE/JPL*-954343-21, p. 228, ibid. (1981)

52. *Ibid.*, p.436.

53. Yatsuriigi, Y., *US Pat.* 4,150,168 assigned to Komatsu.

54. Shimizu, Y., "Economics of Polysilicon Process — A View from Japan," *Report DOE/JPL*-1012-122, p. 62, Nat. Tech. Infor. Center, Springfield, VA (1985)

55. Hsu, G., "JPL Fluidized Bed Reactor Silicon Deposition Research," *Report DOE/JPL*-1012-88, p. 237, ibid. (1983)

56. "Silicon Material Task," Progress Report 19, *Report DOE/JPL*-1012-67, p. 59, ibid. (1981)

57. "DCS Process (Dichlorosilane), Feasibility Study in Support of Silicon Material Task I," *Report DOE/JPL*-954343-21, p. 287, ibid. (1981)

58. Hemlock Semiconductor, "Dichlorosilane Research Process," *Report DOE/JPL*-1012-94, p. 367, ibid. (1983)

59. *Ibid.*, p. 373.

60. Hazards Research Corp., "Dichlorosilane Experiments," Progress Report 16, *Report DOE/JPL* 1012-51, p. 178, ibid. (1980)

61. "Silicon Material Task," Progress Report 18, *Report DOE/JPL* 1012-58, p. 122, ibid. (1981)

62. "Silicon Material Task," Progress Report 16, *Report DOE/JPL* 1012-51, p. 176, ibid. (1980)

63. "Silicon Material Task," Progress Report 18, *Report DOE/JPL* 1012-58, p. 123, ibid. (1981)

64. Hemlock Semiconductor, "Dichlorosilane Research Process," *Report DOE/JPL*-1012-94, p. 358, ibid. (1983)

65. "Silicon Material Task," Progress Report 18, *Report DOE/JPL* 1012-58, p. 125, ibid. (1981)

66. "Conventional Polysilicon Process (Siemens Technology), Process Feasibility Study in Support of Silicon Material Task I, *Report DOE/JPL* 954343-21, p. 208, ibid. (1981)

67. "Economics of the Polysilicon Process; A View from Japan," *Report DOE/JPL*-1012-122, p. 59, ibid. (1985)

68. Erk, H.F., Monsanto Co., "Chlorosilane Thermodynamic Equilibria Calculations with applications to High Purity Silicon Preparation," *Report DOE/JPL*-1012-81, pp. 33-41, Nat. Tech. Inform. Center, Springfield, VA (1982)

69. Bawa, M.S., Goodman, R.C., Truitt, J.K, "Kinetics and Mechanism of Deposition of Silicon by Reduction of Chlorosilanes with Hydrogen," *SC Eng. J.*, 42 (1980)

70. Bloem, J., "Trends in the Chemical Vapor Deposition of Silicon," *Semiconductor Develop. Labs N.V.*, Philips Gloeilampenfabrieken, Nijmegen, The Netherlands.

71. Bailey, D., "Kinetics and Mechanisms of Chlorosilane Decomposition," *Report DOE/JPL*-1012-81, pp. 51, Nat. Tech. Inform. Center, Springfield, VA (1982)

72. Ring, M.A. and O'Neal, E., "Kinetics and Mechanism of Silane Decomposition," *ibid.* p. 63.

73. Dudukovic, M.P. and Milorad, P., "Reactor Models for CVD of Silicon," *ibid.* p. 199.

74. Flagan, R.C. and Alam, M.K., "Factors Governing Particle Size in the Free Space Reactor," *ibid.* p. 107.

75. Lorenz, J., "The Silicon Challenge," *Report DOE/JPL*-1012-81, p. 8, Nat. Tech. Inform. Center, Springfield, VA (1982)

76. "Silicon Material Task, Polycrystalline Silicon," *Report DOE/JPL*-1012-51, ibid., p. 152 (1980)

77. Lorenz, J., op. cit.

78. Lutwack, R., "Silicon Material Task," Progress Report 19, *Report DOE/JPL*-1012-67, p. 50, Nat. Tech. Inform. Center, Springfield, VA (1981)

79. Ibid. p. 49.

80. Grimmett, E.S., "An Update on a Mathematical Model Which Predicts the Particle Size Distribution in a Fluidized-Bed Process," *Report DOE/JPL*-1012-81, op cit. p. 171.

81. Kayihan, F., "Steady-State and Transient Particle Size Distribution Calculations for Fluidized Beds," *ibid.* p. 159.

82. Fitzgerald, T.A., "A model for the Growth of Dense Silicon Particles from Silane Pyrolysis in a Fluidized Bed," *ibid.* p. 141.

83. Kaae, J.L., "The Mechanism of the Chemical Vapor Deposition of Carbon in a Fluidized Bed of Particles," *ibid.* p. 127.

84. Lutwack, R., "Flat-Plate Collector Research Area, Silicon Material Task," Progress Report 20, *Report DOE/JPL*-1012-71, ibid. p. 180.

85. Hsu, G., "JPL In-House Fluidized-Bed Reactor Research," *Report DOE/JPL*-1012-94, ibid., pp. 371-379 (1983)

86. Rogers, L., "An Alternative Method to Produce Silicon Crystals," *ASRE 86 Egyptian Section*, March (1986)

3

Crystal Growth of Silicon

H. Ming Liaw

1.0 INTRODUCTION

The majority of single-crystal silicon ingots (>75%) used today by the semiconductor industry are grown by the Czochralski (Cz) method. The application of this method to silicon ingot growth was first reported by Teal and Buehler (1) in 1952. Cylindrical crystals 1" in diameter and up to 5" in length were obtained. Growth techniques since then have been steadily improving for achieving a greater degree of crystal perfection and larger ingot sizes. Dash (2) was the first to introduce a technique for the growth of dislocation-free silicon ingots. Silicon ingots perhaps exhibit the best crystal perfection among all man-made crystalline materials. The size of the crystal diameter has been steadily increased from 1" in 1950's to 10" (in the laboratory) today. The melt has also been correspondingly increased in weight from several hundred grams to 60,000 grams in production. The new pullers being designed are aimed for the melt sizes of 100,000 grams and larger. The rate of increase in crystal diameter perhaps will be slowing down in the future. This is not because of limitations in the technique of large diameter growth, but of wafer processing capabilities such as slicing. The mechanical strength of wafers also needs to be improved in order to prevent the bending which may occur in large diameter wafers. Improvement of mechanical strength may rely on the modification of the crystal growth process. A method such as nitrogen doping into the silicon melt has been suggested (3). Work in this area remains to be explored.

Automation of crystal growth requires further development. Historically, the growth of silicon ingots involves more art than science. A typical crystal grower requires a lot of hands-on training before he or she can achieve single crystal growth and then achieve dislocation-free growth. Development of the automatic diameter control (only in the constant diameter section of ingots) in the early 1970's greatly improved crystal production yields. Automation for an entire crystal growth cycle which includes the neck-in, shoulder, main body growth, and taper growth was implemented in the mid 1980's. However, techniques for automatic control of dislocation-free growth remain to be developed.

"Continuous" growth techniques have been investigated but have not been used for production. They include (a) growth by multiple replenishments of the silicon melt, and (b) continuous melt replenishment. The machine harvesting of grown ingots becomes necessary since the crystal weight and thermal mass have passed beyond the limit of manual operation. It is foreseen that further mechanization will lead to robots loading the silicon charge, harvesting the ingots, and cleaning the pullers at the end of each crystal growth operation.

Crystal quality requires further improvement to meet the needs of advanced integrated circuit (IC) device requirements. Examples of the requirements are (a) precise control of oxygen and carbon in silicon, (b) minimization of metallic impurities, (c) control of the doping uniformity both in radial and axial directions, and (d) reduction of oxidation-induced surface defects such as hillocks and stacking faults.

Another application of silicon crystal growth development is in solar cells. The silicon requirements for this application are high purity and low cost. Three approaches have been pursued. The first approach is to grow silicon crystals in sheet form so that the cutting and polishing operations are no longer needed. The edge-defined, film-fed growth (EFG) and dentric web techniques are the most successful sheet growth processes. The second approach is to develop low cost Czochralski processes by using a continuous or semicontinuous method of crystal pulling. Metallurgical grade (instead of semiconductor grade) silicon has been suggested (4) for use as feedstock to reduce the starting material cost. However, a disadvantage of this is that the pulled ingots need to be repulled and remelted more than once until their purity meets the specifications. The pull of square ingots has also been suggested. This will enable the achievement of a high packing density of silicon wafers in the solar panels. The third approach is to cast the silicon melt into square ingots. This method can produce silicon ingots at a lower cost than the Cz method because of the simplicity in equipment and processing. However, the cast method produces polycrystalline ingots rather than single crystals.

Single crystals can be grown from the melt without using a crucible. The pedestal and floating zone techniques are two such methods. The floating zone [FZ] technique was first applied to silicon by Petritz (5) in 1962 and since then has been used in production. The FZ crystals are purer than Cz crystals since they are free from contamination by crucible material. In the late 1960's, the industry forecasted that integrated circuit devices would require the use of FZ silicon wafers because of their superior purity. On the contrary, the devices built on FZ wafers are often lower in yield than those fabricated on Cz wafers (6). Two unique properties of FZ wafers may possibly contribute to this adverse result. One is that FZ wafers are more susceptible to the formation of thermal-induced slip since they lack oxygen to harden the crystal lattice. The other is that the FZ wafers do not have enough oxygen to form SiO_2 precipitates in the bulk, which are essential for intrinsic gettering. However, FZ crystals are still the only material suitable for the fabrication of high voltage power devices.

Ingot growth from the melt by the Cz or FZ technique requires semiconductor grade polycrystalline silicon as feedstock. The polycrystalline silicon is deposited from silicon compounds existing in the vapor phase. Attempts have been made to grow single crystal silicon ingots directly from the vapor phase so that the process of crystal growth from the melt can be eliminated. This has been achieved by using a single crystal as the seed and carrying out deposition at a slower rate. Sangster et al (7) reported the deposition of single crystal silicon rods by the hydrogen reduction of silicon tetrabromide. Sandmann (8) later demonstrated that single crystal ingots up to 4" in diameter could be deposited. However, the crystallographic defects in the deposited ingots are much higher in density than those in melt grown ingots. The deposited ingots are polygonal rods that require extensive surface grinding to convert to circular ingots. The elimination of the melt growth step is a trade off for (a) a greater material loss due to grinding, (b) reduced throughput of ingot production due to the nature of vapor deposition, and (c) poor crystal quality. Therefore, the vapor phase growth of single crystal silicon ingots has never been implemented into production.

One purpose of this chapter is to provide some insight for understanding the melt growth of single crystal silicon ingots. The discussions will focus on the Czochralski technique although other techniques will also be briefly introduced. The basic theories involved in Czochralski growth will be discussed first, followed by the practical aspects. Another purpose of this chapter is to review the key areas of crystal growth technology which could affect the properties of silicon wafers that are used for the fabrication of semiconductor devices.

2.0 MELT GROWTH THEORY

2.1 Thermodynamic Consideration

Crystal growth from the melt is a phase transition from liquid to single crystal solid due to the decrease in melt temperature. The most significant difference between the melt and single crystal solid is the rearrangement of atoms from a random structure to a high degree of order with a characteristic symmetry. This corresponds to a decrease in entropy (S) and Gibbs free energy (G). The phase transition is accompanied by a change in the physical properties of the material. Examples of these are changes in density, thermal and electrical conductivities, optical reflectivity, and heat capacity. At the melting temperature (T_M) liquid and solid can co-exist in equilibrium and the Gibbs free energy of the two phases are equal.

$$\Delta G = G_L - G_S = \Delta H - T_M \Delta S = 0, \quad (1)$$

or

$$\Delta S = \Delta H/T_M \quad (2)$$

Equation (2) states that the entropy change (ΔS) due to freezing is a function of the heat of fusion ΔH (or enthalpy) and melting temperature T_M, which are constants for a given material. As the melt cools below the melting point which is required for crystal growth, the change in Gibbs free energy associated with the phase transition is:

$$\Delta G = \Delta H \, ((T_M - T)/T_M) = (\Delta H \, \Delta T/T_M) = \Delta S \, \Delta T \quad (3)$$

where T is the melt temperature at the melt-solid interface, and ΔT is the degree of undercooling (or supercooling) of the melt. ΔS is a constant with a negative value for the freezing. ΔG is also negative and its value is directly proportional to the degree of undercooling (ΔT). ΔT is therefore the only driving force for phase transition and is a predominate factor in determining the crystal growth rate.

The equation of Gibbs free energy can also be used for the discussion of defect formation in the grown crystals. The total change of free energy of the crystal when generating n number of a given defect type is given by

$$\Delta G = nH - nTS_f - TS_m \qquad (4)$$

where S_f and S_m are the entropies of defect formation and mixing in the crystal lattice, respectively. Formation of point defects such as vacancies and interstitials can increase the disorder in the lattice so that TS_m becomes a significantly large term in the equation. The enthalpy (H) for the formation of point defects on the other hand is relatively small.

The formation of point defects is favorable since it results in a decrease in the free energy of crystals. The equilibrium number (n) of a type of point defect is given by Seeger (9):

$$n = N_o \exp (S_f/k) \exp (-H/kT) \qquad (5)$$

where N_o is the number of sites unoccupied by the defect, and k is Boltzmann's constant. Point defects cannot be eliminated during crystal growth and the equilibrium number is exponentially proportional to the crystal growth temperature. The diffusivities of the point defects are very high at an elevated temperature so these defects very actively interact and can transform to other types of defects, such as the formation of oxygen precipitates and oxidation-induced stacking faults. The ability to control the point defect interaction has become a critical factor in silicon wafer processing.

The formation of line defects such as dislocations increases very little in ΔS but very greatly in ΔH. The Gibbs free energy for their formation is likely to be a positive value. Therefore the dislocations are not equilibrium defects, and they can be reduced or eliminated if proper crystal growth parameters are used. Techniques for growing dislocation-free crystals will be discussed in a later section.

2.2 Heat Balance In Crystal Growth

The melt growth starts with heating the polycrystalline silicon charge to above its melting point to form a complete liquid in the crucible. In a typical Cz puller, thermal insulation is applied to the bottom of the crucible and outside surfaces of the heater. This creates vertical and radial temperature gradients in the melt. The lowest temperature of the melt is preferably located at the center of its free surface. To initiate crystal growth, the melt temperature is reduced until the center of the melt surface has reached the melting point. A single crystal seed is dipped into it until an equilibrium condition is established. The crystal pulling can be initiated by lowering the melt temperature further. Once the freezing occurs on the seed, it can be gradually withdrawn from the melt and grows as an ingot. The freezing of molten silicon is accompanied by generating the latent heat of solidification. This latent heat needs to be removed or else it would prevent further freezing of silicon. It is dissipated into the crystal via conduction. The crystal in turn dissipates heat to the ambient through radiation and convection. Figure 1 depicts a typical heat flow pattern during the crystal pulling. Under a steady-state condition, the heat input to the system should equal the heat output. The heat input consists of heat supplied by

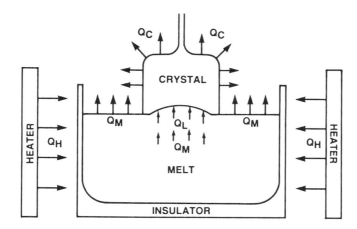

Figure 1: A heat flow pattern in the furnace during crystal pulling.

the heater (Q_H) and latent heat of solidification (Q_L). The heat output includes heat loss from the melt (Q_M) and heat loss from the crystal (Q_C).

$$Q_H + Q_L = Q_M + Q_C \tag{6}$$

Heat loss of the melt (Q_M) occurs primarily from its free surface to the ambient gas and from the melt-crystal interface to the crystal. The latent heat (Q_L) generated at the interface mainly dissipates to the crystal by conduction. The crystal in turn dissipates the heat to cooler portions of the puller interior by radiation and to the ambient gas by convection. The heat-transfer behavior through the crystal-melt interface is a critical parameter which affects the crystal growth rate and crystal perfection. At a steady state, the heat balance at the interface can be expressed as (10):

$$K_s G_s A - K_l G_l A = Q_L = LV \tag{7}$$

or

$$K_s(dT/dx)_s - K_l(dT/dx)_l = L\,V/A \tag{8}$$

where Gs (or $(dT/dx)_s$) and G_l (or $(dTdx)_l$) are the vertical temperature gradients in the crystal and melt, respectively, at the vicinity of the interface, x is distance in the vertical direction, K_s and K_L the thermal conductivity of solid and liquid silicon respectively, L the latent heat of solidification, V the crystal pull rate in the unit of mass per unit time, A the area of the solid-liquid interface, and T the temperature.

Equation (8) states that the crystal pull rate is determined by the two variables $(dT/dx)_s$ and $(dT/dx)_l$. The vertical temperature gradient of the melt $(dT/dx)_l$ cannot be a negative value, otherwise the freezing would occur below the solid-melt interface and prevent the growth of a single crystal. When $(dT/dx)_l$ is equal to zero, the growth rate becomes maximum and is directly proportional to $(dT/dx)_s$. The magnitude of $(dT/dx)_s$ has been a subject of many studies. Theoretically it has been evaluated from the following differential Equation (10, 11):

$$d^2T/dx^2 = BT^4 \tag{9}$$

where $B = 2k_r e/k_s R$. Here e is the emissivity, k_s the thermal conductivity of the crystal, k_r the radiation constant, R the radius of the crystal and x the distance in the ingot from the cold end. The ratio of radiation to conduction heat loss is referred to as the emission power of the ingot and is a constant (A) for a given material. Its value is in the range of 10^{-9} to 10^{-12} for most materials. The differential equation (9) has been solved under the boundary conditions in which one end of the crystal is at the melting point and other end is maintained at zero. The solution is:

$$(dT/dx)_s = C\, B^{1/2} T_m^{5/2} \tag{10}$$

where T_m is the melt temperature of the crystal, and C is a numerical constant. Equation (10) suggests that the magnitude of $(dT/dx)_s$ depends only upon the emission power of the ingot. This simple model uses the following assumptions which may oversimplify the real crystal growth process: (a) one dimensional heat flow used in Equation (9), and (b) ignorances of (i) convective heat loss of the crystal, (ii) radiation heat exchanges between the crystal and surrounding bodies, and (iii) a temperature gradient in the melt. Nevertheless, Equation (10) can be used for a rough estimate of temperature gradient from which the maximum growth rate can be calculated by substituting it into Equation (8) and letting $(dT/dx)_l=0$. The result gives

$$V_{max} = \frac{CA}{L}\left(\frac{2\, K_s\, K_r e\, T_M^5}{R}\right)^{1/2} \tag{11}$$

Another approach for the calculation of temperature gradients in the crystal has been by using a three dimensional model. The Laplace equation which describes the heat flow in a cylindrical solid (i.e. ingot) was used (12-16):

$$\frac{d^2T}{dr^2} + \frac{1}{r}\frac{dT}{dr} + \frac{d^2T}{dz^2} = 0 \tag{12}$$

Arizumi and Kobayashi (12) used this model and made the following assumptions: (a) The crystal pulling process is under the steady state condition so that there is no change in temperature with respect to time $(dT/dt=0)$. (b) The crystal pull rate is constant. (c) The heat transfer in the melt is by conduction only. (d) The thermal conductivity is constant and isotropic. (e) The crucible temperature is constant and the heat carried by the crystal is independent of the growth rate. In addition, three boundary conditions were used for solving the Laplace equation.

(i) Temperature is symmetrical around the crystal pull axis.

$$dT/dr = 0 \qquad (13)$$

(ii) Heat conducted through the crystal is transferred to the surrounding area by radiation and convection.

Heat loss through the side surface:

$$K_s (dT/dr) = q_{rad}^{net} + q_{conv} \qquad (14)$$

Heat loss through the top surface:

$$K_s(dT/dx) = q_{rad}^{net} + q_{conv} \qquad (15)$$

(iii) Temperature at the crystal-melt interface is equal to the melting point and Equation (8) is applicable at the interface.

The numerical solution is made and the results are presented in terms of temperature isotherm curves. The temperature isotherm curve for T at the melting point is equivalent to the contour of the crystal-melt interface (i.e. interface shape). The results from this model have shown that the interface shapes are generally concave toward the melt. Recent studies of Srivastava et al (17) have considered additional terms such as radiation interaction and convection heat loss for silicon ingot growth. Figure 2 shows an example of isotherm profiles obtained by Srivastava et al for a pull rate of 7.5 cm/hr and crucible temperature of 1790 K. The predicted interface shape is fairly flat in this case.

The interface shape can be qualitatively predicted by considering whether there is any deviation from the heat balance at the interface. When the heat dissipation rate (i.e. the $K_s(dT/dx)_s$ term in Equation(8)) is greater than the receiving rate (i.e. $K_l(dT/dx)_l$ + LV/A terms in Equation (8)), the interface shape will increasingly become more convex toward the melt. The interface shape will become concave if the rate of heat dissipation is slower than the receiving rate. Let us assume that the crystal pull rate and axial temperature gradient in the melt are kept constant. The rate of heat receiving will be constant. However, the rate of heat dissipation will be progressively reduced as the crystal length

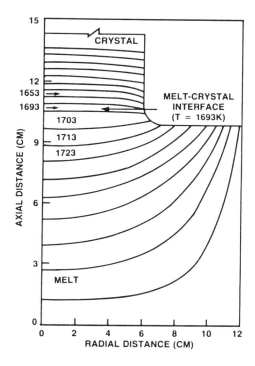

Figure 2: A profile of temperature isotherms in the melt and crystal obtained from numerical simulation [From Srivastava (17).]

increases. Therefore, the interface shape will change during the course of the crystal growth process. This has been verified experimentally (18).

The interface shapes can vary widely during the growth of the crystal shoulder that is also referred to as the crown. The tapered angle of the crystal shoulder is a predominate contributing factor for variation in interface shape (18). Figures 3(a) through 3(d) show the interface shapes at the crystal shoulder respectively for 30°, 66°, 110°, and 170° tapered angles. Figures 3(a) and 3(b) are viewed from the side while Figures 3(c) and 3(d) are viewed from the bottom of the crystals. Figures 3(e) and (f) show the interface shapes at the very upper section of the crystal main body when the tapered angle at the shoulder are 105° and 157°. Note that there is a drop of silicon with a pointed tip which attached to the interface during the withdrawal of the crystal from the melt. This drop of silicon should not be considered

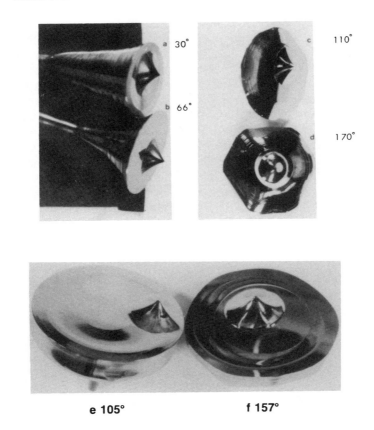

Figure 3: Shapes of the solid-liquid interface for crystals pulled with different tapered angles.

as a part of the interface. Since these crystals are grown in a <111> direction and are free of dislocations, a facet which is perpendicular to the pull axis exists in each crystal. The interface shape for discussion includes the faceted and non-faceted regions. These figures show that the interface shapes change from concave to planar, and then to convex with increase of the taper angle during shoulder growth. The greater the taper angle is, the greater the portion of crystal surface facing away from heater for heat dissipation. Therefore, the interface tends to be more convex. However, when the crystal growth has rolled over the shoulder, the interface becomes concave. The center facets shown in (e) and (f) will eventually disappear when the crystal grows longer and the interface will look more like (a).

2.3 Crystal Growth Mechanisms

The terrace-step-kink growth model has been widely accepted for the interpretation of crystal growth. In this model, the atoms or molecules from the melt in contact with the solid surface (i.e.interface) are adsorbed on the surface terrace. The adsorbed atoms then diffuse to the surface steps,if any, and along the steps to the kinks for incorporation into the crystal lattice. The type of steps existing on the solid surface is an important factor in determining the mechanism of crystal growth. Two mechanisms of the terrace-step-kink growth model have been proposed to account for the different types of surface steps. One is referred to as two-dimensional nucleation growth and the other is screw-dislocation growth. When a surface contains no crystallographic defects and is atomically flat such as one shown in Figure 4(a), the crystal will grow by the two-dimensional nucleation growth mechanism. The formation of a new layer requires a high degree of supersaturation (or undercooling) since the adsorbed atoms could not find energetically favorable sites (such as steps and kinks) for incorporation into the lattice. The high undercooling will force the adsorbed atoms to form nuclei on the surface. Once a nucleus is formed, surface steps are created as shown in figure 4(b). The crystal growth will proceed rapidly with the steps extending laterally until completion of the entire surface layer. The growth cycle repeats by the formation of two-dimensional nuclei on the new surface. However, when a surface contains a screw dislocation as one sketched in Figure 4(c), the crystal will grow by the screw dislocation growth mechanism. The growth occurs by diffusion of the adsorbed atoms to the step and incorporation into the crystal lattice. The step will move as a spiral in a counter-clockwise

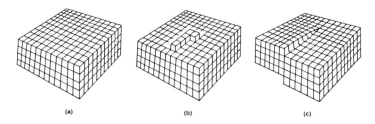

Figure 4: A model showing dependence of growth mechanisms on surface morphology of the seed. (a) Two-dimensional nucleation on atomically flat and defect-free substrate. (b) Lateral growth on 2-dimensional nucleus, (c) Screw dislocation growth mechanism on the dislocated substrate.

direction for the surface shown in Figure 4(c). When the step has completed one revolution, the solid surface advances upward by one atomic layer. Since the surface step remains after completion of a layer growth, no new nuclei will be needed for the initiation of new layers. Growth by this mechanism requires only slight undercooling and is expected to proceed in steady rates. Experimentally, supercooling in the range of 3.7—9.0 °K is observed for dislocation-free growth and the range of 0.32-0.8 °K for dislocated growth of silicon crystals (19, 20). Kinetically, the number of the surface steps (f) available at the interface can affect the growth rate. The f value in turn is determined by the degree of undercooling (ΔT). For the growth sites provided by the screw dislocations, f is proportional to ΔT, and for growth by the two-dimensional nucleation mechanism, f is proportional to $\exp(-T^*/\Delta T)$, where T^* is a constant which is a dependent on the material (21).

The crystal surfaces at the interfaces can be classified into singular and diffused, or faceted and non-faceted interfaces. A singular or faceted interface in silicon crystals is crystallographically a perfect surface. The diffused or non-faceted region does not need two-dimensional nucleation since the kinks and steps are conserved during the crystal growth. When dislocations are present, however, the crystal may grow by the screw-dislocation mechanism. The advance of singular or faceted interface is by two-dimensional nucleation growth. It is very common to see that an interface may consist of a singular surface in one region and a diffused surface in the remaining area (eg. see Figure 3). The closely-packed crystallographic planes such as {111} favor the formation of singular or faceted surfaces. The diffused or non-faceted surfaces at the interface are made of less closely-packed planes such as {100}. The {111} planes may become diffused surfaces when they contain dislocations.

2.4 Mass Transport of Impurities

Silicon ingots are typically doped with a group-III or group-V element for p-type or n-type conduction respectively. A dopant is added into the silicon melt prior to the onset of crystal pulling. Silicon melts also contain unintentionally "doped" impurities such as oxygen and carbon. Thus, the melt growth of silicon ingots involves the solidification of a multicomponent solution. In the molten solution growth, the impurity concentration in the frozen solid differs from that in the solution. The phase diagrams depicted in Figures 5(a) and 5(b) illustrate this relationship, in which the impurity concentrations are C_s and C_o in the solid and liquid respectively at temperature T_l. Note that this relation is applicable only for solutions which dissolve with low concentrations of an impurity. Figure 5(a) shows the melting point of the solution is decreased by the increase in impurity concentration, and in Figure 5(b) the melting point is increased by the impurity concentration. The ratio of the impurity concentration in the solid C_s to that in the liquid C_o at a given temperature obtained from the phase diagram is referred to as the equilibrium distribution coefficient K_o. The distribution coefficient is also referred to as the segregation or partition coefficient. The impurity distribution shown in Figure 5(a) is for K_o less than one. The dopants from groups III and V, and most metallic impurities belong in this category. Figure 5(b) shows the case in which the Ko is greater than one.

The equilibrium distribution coefficient can be estimated theoretically from thermodynamic consideration (22-24). The following equation is derived from the fact that in an equilibrium condition, the partial molar Gibbs free energies of the solute in solid solution is equal to that in liquid solution:

$$\ln k_o = \frac{\Delta H^f - \Delta H^s}{RT} + \frac{\sigma - \Delta S^f}{R} + \ln \gamma \qquad (16)$$

where ΔH^f is the heat of fusion between the melting point of the impurity and that of the host crystal, ΔH^s is the differential heat of solution of the impurity in the solid host crystal, ΔS^f is the entropy of fusion of impurity at its melting point, σ is the change in vibrational entropy as a result of transferring the impurity from its own lattice to that of the host crystal, and γ is an activity coefficient. Among them the heat of solid solution is a major factor which determines the distribution coefficient. The heat of solution can be equated to the addition of (a) the difference in bond energies in the host crystal and in the impurity crystal, and (b) the strain energy related to the insertion of the impurity into the lattice. The distribution coefficients calculated by this method agree fairly well with experimental values.

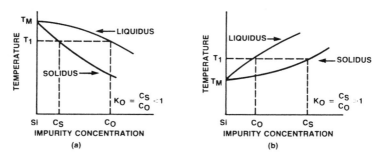

Figure 5: Demonstration of the distribution coefficient using a phase diagram. (a) for $k_0 < 1$, and (b) for $k_0 > 1$.

The distribution coefficient for a doping or metallic impurity in Ge and Si is found to relate to the maximum molar solid solubility (x_M). The empirical relationship referred to as the Fischler rule can be expressed as (25):

$$x_M = 0.1 \, k_o \tag{17}$$

When the unit of atoms/cm3 is used for the maximum solubility (C_M), Equation (17) can be re-written as:

$$C_M = (5.2 \times 10^{21}) \, k_o \tag{18}$$

The plot of Equation (18) is shown in Figure 6 (upper line).

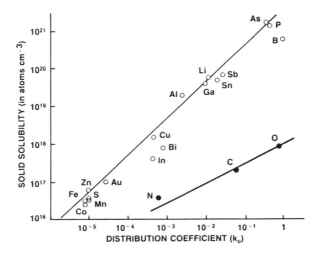

Figure 6: Plot of distribution coefficient vs. maximum solubility [From Trumbore (24), and Jaccodine and Pearce (26).]

Table 1

Distribution Coefficient (k_O) at the Melting Point of Silicon

Element	k_O	Element	k_O
Lithium	0.01	Nitrogen	$< 10^{-7}$ (?)
Copper	4×10^{-4}	Phosphorus	0.35
Silver	-	Arsenic	0.3
Gold	2.5×10^{-5}	Antimony	0.023
Zinc	$\sim 1 \times 10^{-5}$	Bismuth	7×10^{-4}
Cadmium	-	Oxygen	0.5 (?)
Boron	0.80	Sulfur	10^{-5}
Aluminum	0.0020	Tellurium	-
Gallium	0.0080	Vanadium	-
Indium	4×10^{-4}	Manganese	$\sim 10^{-5}$
Thallium	-	Iron	8×10^{-6}
Silicon	1	Cobalt	8×10^{-6}
Germanium	0.33	Nickel	-
Tin	0.016	Tantalum	10^{-7}
Lead	-	Platinum	-

The dotted points are measured values adopted from Trumbore (26). It can be seen that the agreement between the empirical and measured values is fairly good. However, Jaccodine and Pearce (24) have observed the anomaly from the Fischler rule for nitrogen, carbon and oxygen. This is shown in the lower line of Figure 6. Table 1 lists the values of equilibrium distribution coefficient of various impurities in silicon.

The crystal pulling process is a progressive freezing of the melt at a localized region which is referred to as the interface. The impurity concentration at the growth interface or in the melt is varied with the fraction of melt solidified and the K_o value. The impurity will progressively accumulate at the interface when K_o is less than one, and deplete when K_o is greater than one. Figure 7(a) is a sketch of a crystal being pulled from the melt. The vertical distance of this figure is used as the y-axis in Figures 7(b) and 7(c). They show schematically the impurity concentration vs. distance measured from the crystal to the melt for $K_o<1$ and $K_o>1$,

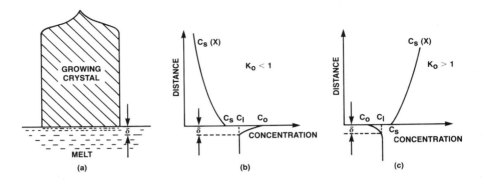

Figure 7: (a) A sketch of the solid-liquid interface during crystal pulling. (b) Plot of distance along the crystal axis vs. impurity concentration in the crystal and in the melt for $k_o<1$. (c) The same type of plot as (b) for $k_o>1$.

respectively. These plots have assumed that the melt is thoroughly stirred either by crystal rotation or a combination of crystal and crucible rotations so that the impurity concentration in the bulk solution C_b is uniform. However, a thin stagnant layer δ of the melt at the boundary of the solid-liquid interface will not be thoroughly mixed. A positive or negative impurity concentration gradient exists across this boundary layer depending on whether K_o is greater or less than one. The thickness of this boundary layer is an important parameter that can affect the impurity distribution in the crystal. Since crystal growth occurs in a non-equilibrium condition, the actual distribution of an impurity differs from the equilibrium value and is referred to as the effective distribution coefficient (K). The equilibrium coefficient can be expressed according to Figure 5 as,

$$K_o = C_s/C_o \tag{19}$$

while the effective distribution coefficient is expressed as

$$K = C_s/c_l \tag{20}$$

where C_l is the impurity concentration of the liquid at the interface.

The transport of impurities from melt to crystal involves the following steps: (a) diffusion through the boundary layer at the solid-liquid interface, and (b) adsorption on the crystal surface and migration to the lowest energy sites for incorporation into the crystal lattice. The first step is referred to as diffusion while the second is called surface reaction. It has been postulated that the surface reactions occur almost instantaneously at a high temperature particularly at the melting point of silicon. Thus, diffusion through the boundary layer is the rate limiting step. Based on this assumption, Burton et al (27) have derived an equation to relate the effective distribution coefficient as a function of the crystal growth rate, boundary layer thickness, and the diffusion coefficient of impurity. This equation is derived from the continuity equation which describes the change of impurity concentration with time under a mass conservation system:

$$\partial C/\partial t = -\nabla \cdot (CV_l - D\nabla C) \tag{21}$$

where V_l is the vector fluid velocity, and D and C the diffusion coefficients and concentration of the impurity, respectively. It is assumed that the liquid is incompressible and that the impurity concentration is uniform in the radial direction. Under this condition, the mass conservation can be expressed by the one-dimensional continuity equation:

$$D\frac{d^2C}{dx^2} - V_{lx}\frac{dC}{dx} = 0 \tag{22}$$

where V_{lx} is the fluid velocity in the x-direction. It is further assumed that the V_{lx} is the sum of the normal fluid velocity and the crystal growth velocity (v). When the fluid velocity is negligible, Equation (22) can be re-written as:

$$D\frac{d^2C}{dx^2} - v\frac{dC}{dx} = 0 \tag{23}$$

Equation (23) can be solved by using the following boundary conditions:

(a) $C=C_l$ at $x = \delta$, and

(b) $(C_o-C_s) v + D(dC/dx) = 0$ at $x = 0$

The solution is:

$$\frac{C_o - C_s}{C_l - C_s} = \exp(\Delta) \qquad (24)$$

where $\Delta = v\delta/D$

Substitute Equations (19) and (20) into Equation (24) and solve for K as

$$K = \frac{K_o}{K_o + (1 - K_o)\exp(-v\delta/D)} \qquad (25)$$

Equation (25) can be rearranged into

$$\ln\left(\frac{1}{K} - 1\right) = \ln\left(\frac{1}{K_o} - 1\right) - \frac{\delta v}{D} \qquad (26)$$

Equation (25) or (26) shows that the effective distribution coefficient is a function of v and (δ/D). It also shows that when v becomes a very large number, K approaches unit, and that when v approaches zero, K is equal to K_o. The value of δ is affected by the flow dynamics of the melt. When the convection of the melt is suppressed by the influence of a magnetic field, the delta value increases and leads to a decrease of the value of exp($-v\delta/D$) in Equation (25). Therefore, K approaches unit when the crystal is grown under a strong magnetic field (28). δ and D cannot be measured directly from experiments. Analytically, Button et al have related them to the flow parameters of the melt as follows:

$$\delta = 1.6\, D^{1/3} k_v^{1/6} w^{-1/2} \qquad (27)$$

where k_v is the kinematic viscosity and w is the crystal rotation rate.

An attempt has been made to calculate (δ/D) from Equation (25) or Equation (26) if the growth rate, v_o, and K are known. Experimentally, K can be calculated from the measured doping concentration in the crystal (converted from resistivity measurements) if (a) the crystal is pulled with a constant growth rate from a melt of a known doping concentration, and (b) K_o of the doping impurity is known. After (δ/D) has been evaluated, it can be substituted into quation (27) and solved for δ and D. Kodera (29) has used this method to obtain δ/D and D for melts doped with various impurities under the application of different crystal rotation rates. The results show that δ/D is inversely proportional to the square root of the rotation rate as shown in Table 2.

Table 2

Values of $\frac{\delta}{D}$ for Silicon

Impurity Element	Rotation Rate rpm	$\frac{\delta}{Dl}$ s/cm	Diffusion Coefficient, Dl, cm²/s
B	10	170 ± 19	$(2.4 \pm 0.7) \times 10^{-4}$
	60	84 ± 37	$(2.4 \pm 0.7) \times 10^{-4}$
	200	43 ± 18	$(2.4 \pm 0.7) \times 10^{-4}$
Al	10	86 ± 34	$(7.0 \pm 3.1) \times 10^{-4}$
	60	40 ± 17	$(7.0 \pm 3.1) \times 10^{-4}$
Ga	5	144 ± 54	$(4.8 \pm 1.5) \times 10^{-4}$
	55	71 ± 26	$(4.8 \pm 1.5) \times 10^{-4}$
	200	24 ± 8	$(4.8 \pm 1.5) \times 10^{-4}$
In	10	84 ± 15	$(6.9 \pm 1.2) \times 10^{-4}$
	60	43 ± 5	$(6.9 \pm 1.2) \times 10^{-4}$
P	5	127 ± 36	$(5.1 \pm 1.7) \times 10^{-4}$
	55	60 ± 19	$(5.1 \pm 1.7) \times 10^{-4}$
As	5	190 ± 53	$(3.3 \pm 0.9) \times 10^{-4}$
	55	79 ± 16	$(3.3 \pm 0.9) \times 10^{-4}$
Sb	5	283 ± 55	$(1.5 \pm 0.5) \times 10^{-4}$
	55	157 ± 57	$(1.5 \pm 0.5) \times 10^{-4}$
	200	62 ± 37	$(1.5 \pm 0.5) \times 10^{-4}$

Microscopic Variations of Doping Concentration: In melt growth, the growth rate is always fluctuating on a microscopic scale even though the crystal pull rate is kept constant. The growth rate fluctuation causes the formation of doping concentration striations in the crystal since the effective distribution coefficient is a function of the growth rate. The growth rate fluctuation can be the results of (a) the crystal rotation being off-centered from the concentric axis of melt isotherms, and/or (b) instability of the melt temperature. The doping (or impurity) inhomogeneities in crystals due to the former cause is referred to as rotational striation, and if due to the latter cause is called as thermal striation. The thermal striation in essence is the result of the variation in melt temperature with time. The instability of the furnace temperature and the turbulent flow of the melt are two examples that contribute to thermal striation. The rotational striation results from the fact that the crystal rotation axis does not coincide with the axis of the thermal symmetry in the melt. Figure 8(a) depicts that the axis of thermal symmetry is at the center of the crucible while the crystal rotation axis is off-center from the thermal axis. The concentric circles indicate temperature isotherms of the melt. Let us consider point A at the crystal-melt interface as an example. When the crystal has completed a revolution in the clock-wise direction, the temperature at point A will vary through the sequence $T_4, T_3, T_2, T_1, T_2, T_3$ and back to T_4. Figure 8(b) depicts the temperature versus the angle of crystal rotation. This shows that the temperature at a given point in the interface varies sinusoidally as the crystal rotates. The growth rates in turn will also vary sinusoidally since they are inversely proportional to the temperature. The instantaneous growth rate (V_{inst}) fluctuating with this pattern can be expressed mathematically as (30):

$$V_{inst} = V(1 - a \cos(2\pi wt)) \tag{28}$$

where V is the crystal pull rate, $a = 2\pi \Delta Tw/GV$, ΔT the degree of undercooling (i.e. $T_m - T$), w the rotation rate, G the temperature gradient of the melt at the interface, and t the time.

Witt et. al. (31) have used the demarcation technique to determine the instantaneous growth rate. This technique applies current pulses at a constant rate through the crystal--melt interface. The electrical current flowing through the interface induces the Peltier cooling effect and alters the

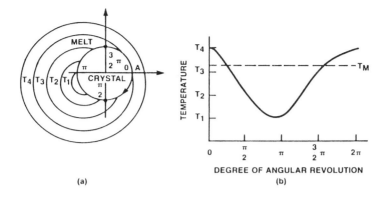

Figure 8: (a) A sketch of relative locations between temperature isotherms in the melt and crystal rotation axis. (b) Plot of the crystal temperature at the solid-liquid interface vs. its angular rotation position.

impurity segregation. Thus each electric pulse results with one impurity striation that can be revealed by preferential etching of the polished surface of the crystal cross-section. The distances of the consecutive striations are measured with the aid of an optical microscope. The instantaneous growth rate between the two consecutive pulses is calculated from the distance divided by the time of pulse interval. They have found that in each rotation cycle the instantaneous growth rate varied continuously from the minimum of 15μm/sec to the maximum of 39μm/sec for the pull rate of 21μm/sec, although the curves are not exactly like a sinusoidal pattern as shown in Equation (28). The spreading resistance measurements also showed that doping concentrations fluctuated cyclically. Each cycle of the doping variations is equivalent to a striation band.

Convection Instability: The instability of melt temperature can be caused by convection flow of the melt. The driving force for the convection flow comes from two sources: (a) the thermal-driven convection and (b) the force convection due to rotation of the crystal. The existence of temperature gradients in the melt is the main cause of the thermal-driven convection. In a typical temperature distribution in the melt, the crucible sidewall is hotter than the center, and the bottom is warmer than the free surface. Figure 9(a) shows the radial temperature gradient of the melt expressed in terms of the voltage output from the sensing pyrometer.

Figure 9(b) shows temperature of the crucible bottom located at various positions with respect to the heater height. Since the density of the melt is inversely proportional to the temperature, the density of melt at the sidewall is less than that at the center and that in the bottom is less than at the top. This density difference causes the buoyant effect that forces the melt to flow. A typical flow pattern of the thermal-driven convection current is shown in Figure 10(a) when no force convection either from the crystal or crucible rotation is applied. The magnitude of this thermal-driven convection is related to the dimensionless Grashof number N_{Gr} (32) or Rayleigh number N_{Ra} (33)

$$N_{Gr} = g\,(\alpha)\,(\Delta T)\,R^3 / K_v^2 \qquad (29)$$

$$N_{Ra} = N_{Gr}\,N_{pr} = N_{Gr}\,(K_v/k) = g\,(\alpha)\,(\Delta T)\,R^3/k\,K_v \qquad (30)$$

where g is acceleration due to gravity, α is the melt thermal expansion coefficient, ΔT the temperature difference between the highest and lowest points, R the crucible radius, k the thermal diffusivity, and K_v the kinematic viscosity. The ratio of K_v to k in Equation (30) is referred to as the Prandtl number, N_{pr}. In the case of silicon, $a = 1.43 \times 10^{-4}$ (°C)$^{-1}$, $K_v = 3 \times 10^{-3}$ cm^2/sec and $N_{Gr} = 1.56 \times 10^4\,\Delta T\,R^3$. The typical values for N_{Ra} and N_{Gr} are in the range of 10^7 to 10^8.

The convection current in the melt flows steadily when the N_{Gr} or N_{Ra} is below a critical value. The critical values for N_{Gr} and N_{Ra} are 10^5 and 10^3 respectively. The melt becomes a time-dependent turbulent flow when N_{Gr} or N_{Ra} is greater than this critical value. The turbulent flow in turn causes the temperature at the interface to be time-dependent and leads to the fluctuations in growth rate and impurity inhomogeneities in the crystal. Equations (29) and (30) show that crystal growth parameters affecting the N_{Gr} and N_{Ra} values are ΔT and R. R is a predominate factor since its effect is to the third power. This suggests that the problem of the thermal striation in large diameter crystals will increase drastically as the melt volume continues to increase.

We have pointed out that crystal rotation can cause impurity striations in a crystal when the rotation axis does not coincide with the thermal axis. However, the crystal rotation can be beneficial to the doping uniformity because: (a) it causes stirring and reduces the temperature asymmetry in the melt, and (b) it propels the melt and produces a counter flow to the thermal-driven convection current. A

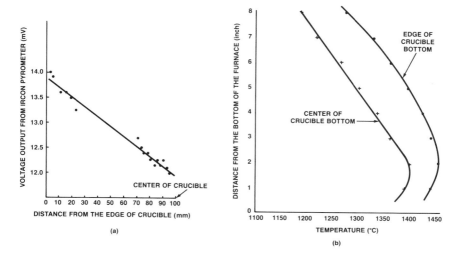

Figure 9: (a) Plot of temperature of the melt vs. its location along the crucible radius. (b) Plot of crucible temperature vs. its vertical position inside the heater.

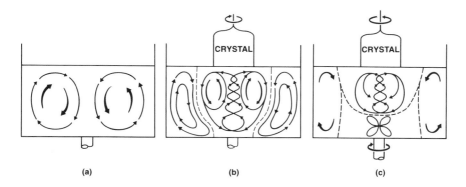

Figure 10: (a) A symmetrical convection flow pattern of the melt without the presence of crystal and crucible rotations. (b) A flow pattern in the melt with crystal rotation. (c) A flow pattern with counter rotation of crystal and crucible.

typical flow pattern of the melt resulting from the crystal rotation and thermal-driven convection is shown in Figure 10(b). The dotted line shows the boundary of the two flows. The area inside the dotted line is the cell in which the melt flow is driven by the crystal rotation. Outside the dotted line is another cell in which the melt flow is driven by the thermal-driven convection. The extent of this boundary line depends on the crystal rotation rate, temperature gradient, and melt height. The magnitude of the force-convection flow due to the crystal rotation is characterized by the dimensionless Reynolds number, N_{Re}.

$$N_{Re} = w\, r^2/K_v \tag{31}$$

where r is the crystal radius and w the crystal rotation rate. It has been shown by Carruther et at. (32) that when $(N_{Re})^2$ is greater than N_{Gr}, the crystal rotation can effectively isolate the thermal-driven convection from contribution to impurity striation. Equation (31) shows that N_{Re} is a function of the crystal radius squared. As melt sizes increase for the growth of large diameter crystals, the increase in N_{Gr} will outpace the increase in N_{Re}. If we want to keep $(N_{Re})^2 > N_{Gr}$, the only option left is to increase the crystal rotation rate.

Figure 10(a) shows the convection flow pattern of a melt that has an ideal thermal symmetry. Such a perfect thermal symmetry is seldomly achieved in practice. Thermal symmetry in the melt can be improved by rotation of the crucible. The crucible rotation can provide additional effects: (a) it forces the thermal convective flow of the melt into spiral paths and increases the radial temperature gradient (33), and (b) it tends to form an additional cell underneath that is created by the crystal rotation. Figure 10(c) shows an example of melt flow patterns when the crucible rotation is applied in a counter direction to the crystal rotation. The detailed change in flow patterns resulting from the variations of crystal and crucible rotation rates have been given by Carruthers and Nassau (34). The first effect is harmful to the doping uniformity since the increase in the radial temperature gradient can enhance the thermal-driven convective flow. The second effect, however, is beneficial since it isolates the growing interface from instabilities in the bulk melt. The degree of the convective flow driven by the crucible rotation is given by the Taylor number (N_{Ta}):

$$N_{Ta} = (2\, w'\, h^2/k_v)^2 \tag{32}$$

where w' is the crucible rotation rate and h is the melt depth.

The thermal convection current can be suppressed by

applying a magnetic field to the melt. A dimensionless magnetic parameter which is referred to as the Hartmann number (N_H) is used to describe its magnitude (35).

$$(N_H)^2 = 4(\mu HR)^2 \sigma/\rho k_v \qquad (33)$$

where μ is the magnetic permeability, σ is the electrical conductivity, ρ is the density of the silicon melt, H is the magnetic field strength, and R the crucible diameter. The value of the Hartmann number for a typical sized silicon melt (R=32cm) is in the order of 10^{-1}H (33). Suzuki (35) have suggested that when $(\pi N_H)^2$ reaches 10^7, the thermal convective current will be suppressed. This is equivalent to applying a magnetic field of 1500 gauss or higher to the silicon melt.

Other thermohydrodynamic parameters which can affect the flow of the melt are the convective flows that are caused by surface tension and the electric field. The Marangoni number (N_M) is used to describe the magnitude of the flow caused by surface tension:

$$N_M = \frac{\Delta TR}{\rho k_v k}\left(\frac{\partial k_v}{\partial T}\right) \qquad (34)$$

where ($\partial k_v/\partial T$) is equivalent to the temperature coefficient of surface tension. The N_M is typically in the order of 10^4 for the Cz melts. The convective current caused by the electric field becomes significant when a 3-phase electric heater is used. The heater current exerts a rotational force on the melt. It has been reported that a melt rotation as high as 20 rpm has been measured in certain systems. This rotational current can reduce the thermally induced convective current. Therefore, the 3-phase electric heaters have been suggested to use for the purpose of improving the doping uniformity in crystals.

Striations of doping non-uniformities in crystals can be revealed by preferential etching when the crystal is doped with a medium-high concentration (1×10^{17} to 1×10^{18} atoms/cc). The rotational striations are generally easier to observe. Their striation periods should equal v/w, where v is the crystal growth rate and w is the crystal rotation rate. The periods of the thermal striations are varied and more difficult to define.

Macroscopic Distributions of Impurities: We have shown

in Figure 7(b) that impurities will progressively accumulate at the interface during the course of crystal pulling if the K_o is less than one. Thus, the impurity concentrations in the crystal progressively increase from the seed to tang ends. It has been seen that the reverse trend occurs when the K_o is greater than one. The K value should be used in a non-equilibrium condition although it is only a slight deviation from K_o in a typical Czochralski crystal growth condition. Mathematically, the concentration of an impurity in the crystal C_s can be expressed as a function of K and the fraction of melt pulled (g):

$$C_s = k\, C_o\, (1-g)^{k-1} \tag{35}$$

where C_o is the initial impurity concentration in the melt. Curves of C_s/C_o as a function of g for various K are plotted in Figure 11.

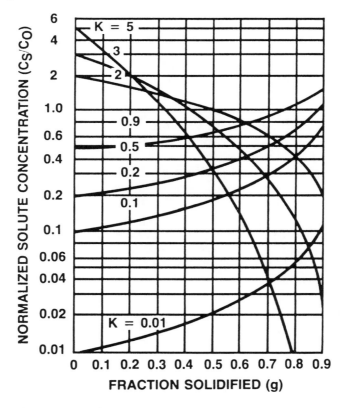

Figure 11: Plot of C_s/C_o vs. g for various values of k using Equation (35).

Equation (35) can be used to determine the effective distribution coefficient (K) of an impurity if (a) C_s and its corresponding g values are measured experimentally, and (b) the mass of the impurity is conserved during crystal growth. Mass conservation means that the total mass of the impurity in the system which includes the melt and crystal is constant. For K>1, the [1-(K-1)g] is approximately equal to $(1-g)^{k-1}$. Therefore, Equation (35) can be expressed as $C_s = k\ C_o [1-(K-1)g]$. The plot of C_s versus g will yield a straight line and its slope is equal to kC_o (K-1). The K can be evaluated from the value of the slope. For K<1, 1/(1-g) is approximately equal to $(1-g)^{k-1}$ in Equation (31). In this case one should plot C_s versus 1/(1-g). The slope of the plot is equal to kC_o from which the K value can be determined. Alternatively, K can be evaluated directly from Equation (35) by plotting $\log(C_s/C_o)$ versus $\log(1-g)$.

Equation (35) needs to be modified for the crystal grown under a non-conservative system condition. An example of non-conservative conditions is the loss of dopant by evaporation from the melt. Bradshaw and Mlavsky (36) have added two more terms into Eq. (35).

$$C_s = k\ C_o\ (1-g)^{k+AEV-1} \qquad (36)$$

where A is the area of evaporation surface, E is the evaporation rate constant, and V is the crystal pull rate. E is a function of temperature, partial molar heat of solution, and molar free-energy of evaporation. The characteristic values of E for certain impurities are listed in Table 1.

2.5 Constitutional Supercooling

A melt for the Cz silicon growth is unintentionally contaminated with impurities such as oxygen and carbon. We have shown in Figure 7 that an impurity concentration gradient exists at the interface because their K values are not equal to one. Thus the freezing point of the melt at the vicinity of the interface varies with distance from the interface. Figure 12(a) is a re-sketch of Figure 7(b) showing the impurity concentration distribution vs. distance from the interface for an impurity with K<1. The corresponding plot of the melting point vs. distance is shown in Figure 12(b), which is deduced from the phase diagram shown in Figure 5. A constitutional supercooling exists in the system if the temperature at a given point of the melt in the interface is

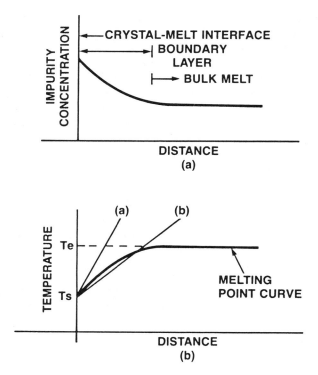

Figure 12: (a) Plot of impurity concentration in the liquid vs. distance from the solid-liquid interface. (b) Plot of melting point of the melt in crucible vs. distance from the solid-liquid interface.

below its freezing point. Whether a constitutional supercooling occurs or not is determined by the following: (a) the melting point variation across the boundary layer at the interface, and (b) the temperature gradient of the melt at that layer. The former item is in terms determined by the impurity distribution coefficient and crystal pull rate. The latter is determined by the furnace set-up and crucible position with respect to the heater height. Lines (a) and (b) in Figure 12(b) show an example of two temperature gradients of the melt at the interface and resemble two different crystal growth conditions. No portion of Line (a) is below the melting point curve and no constitutional supercooling exists. However, a portion of Line (b) is below the melting point curve and therefore, the crystal growth is under the existence of constitutional supercoolings. It is important to differentiate constitutional supercooling from super cooling.

Crystal growth from a multicomponent melt does not require constitutional super-cooling as a driving force. Supercooling of a binary-element melt can exist without the onset of constitutional super-cooling. It can be seen in Figure 12(b) that the melt temperature at the crystal surface Ts is below the melting point of the bulk liquid. In fact, constitutional supercooling is not desirable for single crystal growth. Tiller (37) has given the crystal growth conditions under which constitutional supercooling can be avoided:

$$\frac{G}{V} < \frac{mC_s(0)(1-K_o)}{K_oD + (1-K_o)e^{-\Delta}} \qquad (37)$$

where G is the temperature gradient, V, the pull rate, m the liquidus slope, $C_s(0)$ the solute concentration in the solid at the interface, D the diffusivity of the impurity in the melt, and $\Delta = v\sigma/D$ (where σ is the thickness of boundary layer).

The consequence of constitutional supercooling is the occurrence of freezing ahead of the growth interface. This can distort the interface shape from a planar to rough boundary. The severe roughness of the interface causes the formation of the cellular morphology in the crystal. The variation of cell structures is a function of the degree of constitutional supercooling (37). Constitutional supercooling has rarely occurred in lightly-doped melts. However, it can occur for doping concentration greater than 1×10^{19} atoms/cc. The critical concentration for the onset of constitutional supercooling is (38)

$$C^*_L = \frac{D}{-m}\left(\frac{K_s\,G_s}{K_L\,V} - \frac{L}{K_L}\right)e^{-\Delta} \qquad (38)$$

Equation (38) can be simplified by substituting into it with $G_s=(dT/dx)_s=C(2K_reT^5_m/K_sR)^{\frac{1}{2}}$, $C=(1/2)^{0.5}$, and with numerical values for all physical constants. Hopkins et al (38) have obtained the following equation:

$$C^*_L = \frac{D}{-m}\left(\frac{A}{R^{1/2}V} - B\right) \qquad (39)$$

where A=92.44 and B=6.88×10³ with R and V in the unit of cm and cm/sec respectively. Figure 13 shows the plot of critical impurity concentration (C^*_L) for interfacial breakdown versus pull rate (V) for various crystal radii (R) assuming $D=10^{-4}$ cm^2/sec.

A crystal grown under the condition of constitutional supercooling can be revealed by a preferential etching. Figure 14(a) shows the cell structure that exists in the center portion of a (100) silicon ingot pulled under the existence of constitutional supercooling. Figure 14(b) is a higher magnification of (a) and shows a better view of the rectangular cells. An extremely high constitutional supercooling eventually leads to polycrystalline growth.

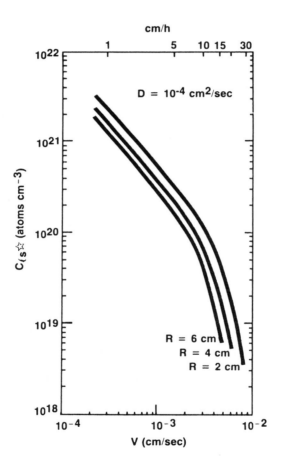

Figure 13: Critical impurity concentration for interfacial breakdown vs. pull rate for various crystal radii. [From Hopkins et al. (38).]

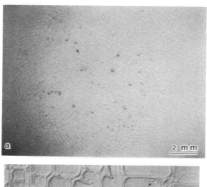
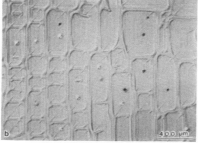

Figure 14: (a) A cell structure in the center portion of an ingot pulled under a constitutional supercooling. (b) A higher magnification of (a).

3.0. PRACTICAL ASPECT OF CZ CRYSTAL GROWTH

3.1 Crystal Pullers

A typical Czochralski silicon crystal puller is shown in Figure 15. It consists of an air-tight, water-cooled growth chamber in which a heater and a graphite cup are placed. The puller is capable of operating in a reduced pressure (50-100 torr) by the attachment of a mechanical vacuum pump. A quartz crucible is placed inside the graphite cup as a liner to contain silicon charge. The graphite cup is placed on a graphite pedestal that is attached to the lower rotation shaft. A single crystal seed is attached to the seed holder which is held by an upper rotation shaft (or cable). The pulling and rotation mechanisms are placed on top of the crystal harvesting chamber. Both rotation shafts can move transversely in the vertical direction during rotation. An optional isolation gate valve can be placed between the growth and harvesting chambers. The puller is purged with

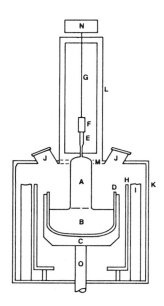

Figure 15: Sketch of a Czochralski puller for the growth of silicon crystals.

an inert gas. The flow of the purging gas is from the top to the bottom of the puller. This reduces the chance of evaporated SiO particles from falling back onto the melt. Some purging gas is also injected at the view ports to keep that area free from deposition of SiO. Sensing devices can be used to measure the following items: (a) temperature of the heater, graphite cup or the melt, (b) diameter of the crystal or the position of meniscus, (c) pressure of the chamber, and (d) intensity of infrared emission from the melt or meniscus.

Melting: Two heating methods have been used. In the 1960's when crucibles were less than 6" in diameter (and the melt size was less than 3 kg/charge), rf heating was commonly used. The rf coils are wrapped around a quartz pipe which serves as a growth chamber. The rf energy is transmitted to the graphite cup which generates heat for the melt. When the crucible size was increased to greater than 6" in diameter, difficulty arose in fabrication and maintenance of quartz chambers. Current silicon crystal pullers use stainless steel growth chambers exclusively and resistance heated graphite heaters. The graphite heater is placed inside the growth chamber as shown in Figure 15. The two-terminal DC heaters are still more popular although the 3-phase AC heaters are suggested for the suppression of the thermal driven convective current in the melt.

The graphite heaters can be made of either extruded or molded graphite cylinders. The extruded graphite is relatively porous with apparent grain textures while the molded graphite gives a higher density and smoother surfaces. Slots are cut out of cylinder to form a zigzag conductive path as shown in Figure 16. The cross-section area of the conductive paths should be kept constant in order to obtain a uniform radial temperature distribution. Arcing in the furnace can be prevented by the use of a high electrical conductivity graphite. The typical voltage between two terminals of a heater is less than 100 volts (DC). The height to the diameter ratio of a heater is an important parameter which affects the temperature profile in the furnace. The smaller the ratio the greater the vertical temperature gradient. Typically, the ratio is in the range from 1.0 to 1.4. Ratios between 1.1 to 1.2 are most commonly used.

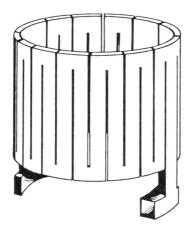

Figure 16: Sketch of a graphite heater used in a silicon crystal puller.

Graphite cups have been made with thick bottoms for the purpose of thermal insulation. This is to insure that the melt will have a positive temperature gradient so that the melt in the bottom of the crucible is hotter than on the top (see Figure 9(b)). Since silicon expands approximately 9% in volume by solidification, the graphite cup is likely to crack when a certain amount of molten silicon remains in the quartz crucible after the crystal pulling. To prevent this, the graphite cups are made to be "expandable". This is achieved by cutting a slit in the sidewall of a graphite cup.

Alternatively, it can be assembled from two parts; a split graphite cylinder as the top and a shallow graphite dish as the bottom. The top section is easily separated from the bottom section when the melt expands after the freezing. This prevents the total loss of graphite cup if only one section of the cup is cracked.

Crucible: Quartz crucibles are made of high purity fused quartz which softens at 1670°C and fuses at 1800°C. Two types of quartz crucibles are commercially available. One is completely fused quartz which is transparent and bubble-free, and the other is partially fused and is opaque. However, the inner surfaces of the opaque crucibles are fused by fire-polishing and are also smooth. Both types of crucibles are made of semiconductor-grade quartz and they yield no significant difference in the impurity content in crystals. The crucible bottom profile and its aspect ratio can affect the convection flow pattern of the melt. The author prefers crucibles with semiflat rather than round bottoms since they produce a better radial resistivity in the crystals (see later section for details).

Recently, a-Si_3N_4 crucibles have been developed for the growth of oxygen-free silicon ingots (39). Pure and dense a-Si_3N_4 was deposited by the chemical vapor reaction of $SiCl_4$ and NH_3 at 1400°C and 30 torr. Carbon or pressed Si_3N_4 crucibles are used as substrates. Thickness of up to 1mm of a-Si_3N_4 films have been deposited without cracking. When the carbon substrate is used, it can be removed later by burning off in the air. Single crystal silicon ingots have been grown from a-Si_3N_4 crucibles and their oxygen contents are below 2×10^{16} atoms/cc. However, a nitrogen concentration of 4×10^{15} atoms/cc has been found.

3.2 Dislocation-Free Growth

As pointed out in the section of thermodynamic consideration, dislocation generation is not favorable because of the increase in Gibbs free energy in the crystals. Therefore, no equilibrium numbers of dislocations can be given in a crystal. The actual numbers found in silicon ingots can vary widely from zero to 10^6/cc depending on the thermal conditions of crystal growth. One of the reasons for their formation during the crystal growth is to relieve the lattice strains resulting from thermal and mechanical stresses. Thermal stresses also occur during the seed dipping that exerts a great thermal shock due to a large temperature difference between the melt and seed. Original dislocations existing in the seed either from bulk imperfections or surface damages are also sources that may propagate to the crystal. Furthermore, multiplication of dislocations may occur during the course of crystal growth.

It is preferable for dislocations in silicon crystals and other diamond-cubic or face-center-cubic materials to propagate in a (111) plane. When a silicon crystal is pulled in a <100> or <111> direction, all {111} planes are oblique to the pull axis. Therefore, all the dislocations with Burgers vectors of (a/2) [101] type will grow out and terminate at the crystal surface provided that (a) the crystal diameter is reduced to a very small size as compared to crystal length, and (b) no new dislocations are generated at the interface. This principle was used by Dash (40) to eliminate all the dislocations during the seeding process and he was the first person to report the growth of dislocation-free silicon ingots. Today the dislocation--free crystals can be routinely grown using the following steps: (a) melt back a portion of the seed before pulling, (b) neck-in the seed to reduce its diameter by at least one-half, (c) use high pull rates up to 25 cm/hr for the neck-in process, and (d) grow the crystal with appropriate temperature gradients to maintain facets in the interface and to maintain as near a flat interface as possible. The purpose of step (a) is to remove surface damages and to create a clean substrate surface. This is to ensure that no new dislocations will be generated during the initial growth on the seed. Step (b) is for the dislocations to emerge and terminate at the ingot surface. Step (c) is to let the dislocations propagate behind the growth interface so that they will be eliminated by pairing or bending to form complete loops. Purposes of step (d) are to provide thermal requirements for the two-dimensional growth mechanism and for avoidance of thermal stress.

Figure 17 shows the X-ray topographs taken at the neck (a), shoulder (b) and main-body (c) of silicon ingots. Figure 17 (a) shows that the seed originally contained very few dislocations. A high number of dislocations were introduced at the initial growth because there was no melt-back. However, the dislocations were totally eliminated toward completion of the neck-in process. Figure 17(b) shows the same crystal through the upper portion of the crystal shoulder. There many dislocations were suddenly re-introduced probably because of high thermal stresses in the crystal. Figure 17(c) shows another ingot that remained free of dislocations throughout the entire growth process.

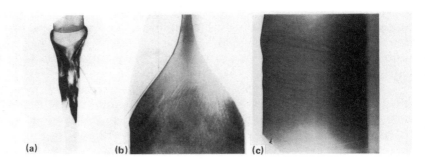

Figure 17: X-ray topographs showing the propagation and termination of dislocations at the crystal neck (a), crystal shoulder (b), and at the crystal main body (c). [a [220] diffraction.]

Experienced silicon crystal growers are able to recognize whether or not the growing crystal is free of dislocations. Dislocation-free growth requires the existence of singular or faceted surface(s) at the interface for two-dimensional nucleation. In a (100) crystal, four side-facets that are 90° apart are present in the interface. Since these side facets are very small, they were not seen by other investigators (19) who used the x-ray topographic technique for detection. Figure 18(a) shows the top view of a dislocation-free (100) ingot. The facets form four ridges that are uninterrupted from the top to the side surface of the ingot. Figure 18(b) shows the sketch of facets and meniscus at the interface. The side facets in a (100) crystal are likely to consist of (111) planes which are 54.5° from the melt surface. The facets protrude beyond the circular circumference of the ingot as they can be seen in Figure 18 (a). Thus, facets can be clearly identified in the meniscus during the crystal growth. These facets reflect light that irradiates from the heater through the crucible with a greater intensity than the non-faceted regions. This is confirmed by the measurement of the light intensity emitting from the halo of the meniscus. Figure 19 shows a plot of the output voltage of an infrared pyrometer vs. time. The pyrometer was focusing on the meniscus while the crystal was rotating during crystal growth. The output voltage is directly proportional to the brightness of the meniscus. The strong peaks in Figure 19 are corresponding to the reflection from the facets. If not all of the facets are present at the interface, it is an indication of the introduction of dislocations into the crystal. The presence or lack of dislocations is apparent after the crystal has been harvested from the puller. Figure 18(c) is a photograph which shows the side surface of a (100) crystal. The side ridge resulting from the facets in this crystal is not continuous. This indicates that facets were not present all the time during crystal growth. Therefore, the crystal is not free of dislocations. In fact, the crystal contains twin planes as the twin lines can be seen in this photograph.

The ingot that grows in a [111] direction has (111) facet(s) parallel to the melt surface if the crystal is free of dislocations. In addition, the crystal also has (111) side facets in <211> directions 120° apart. Figure 20(a) sketches

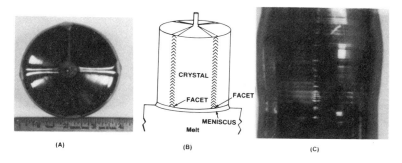

Figure 18: (a) Facets forming four continuous ridges from the seed on the top of a dislocation-free crystal. (b) A sketch showing the formation of continuous ridge side facets on a dislocation-free (100) crystal. (c) A dislocated crystal with discontinuous ridges.

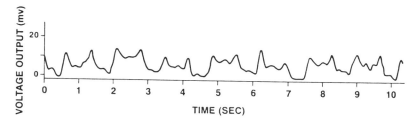

Figure 19: A plot of light intensity (in mV) emitting from the meniscus vs. time. Peaks exhibiting a constant period are corresponding to the reflection from four side facets in the solid-liquid interface during the growth of a dislocation-free crystal.

132 Handbook of Semiconductor Silicon Technology

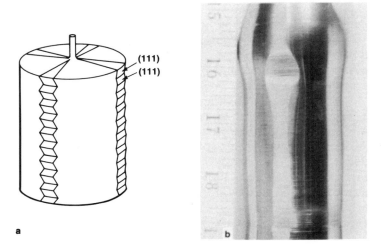

Figure 20: (a) A sketch of side facets in a dislocation-free (111) crystal. (b) A photograph showing uninterrupted side facets in a dislocation-free (111) crystal.

two strips of side facets in a (111) crystal. Figure 20(b) is a photograph showing a dislocation-free (111) crystal with un-interrupted strips of side facets. These facets can also be seen at the meniscus during the crystal growth.

We have discussed the techniques to obtain dislocation-free growth (DFG) during the seeding and neck-in processes. The DFG may not be necessarily maintained throughout the entire crystal length as illustrated in Figure 17(b). The chance of losing DFG becomes greater when the interface is excessively concave toward the melt. In this case, the periphery of the ingot has already become solid when the center just starts to freeze. Thus the melt freezing in the center does not have free space to expand (Si expands 9% in volume upon freezing). This creates severe mechanical stress to the crystal. The mechanical stress in turn can generate dislocations. Dislocations are also very easy to introduce at the conclusion of crystal pulling. This occurs when the grown ingot is retracted from the remaining melt. Similar to seed dipping, the retraction (or separation) also exerts a great deal of thermal shock to the crystal. The dislocations generated by the rapid cooling of a retracted crystal always pile up in [110] directions of (111) planes and are referred to as slip dislocations (or simply called slip). The thermal shock can be reduced by tapering down the crystal diameter prior to retraction. Figures 21(a) to 21(c) shows the extent of the slip generation for three different retraction procedures. A sharp taper with a pointed tip as shown in Figure 21(c) is clearly a preferred procedure. The rule of thumb is that the distance for slip to propagate upward is equal the diameter at which the ingot is separated from the melt.

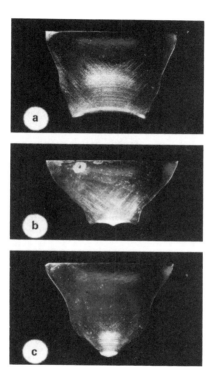

Figure 21: The tang end of (100) silicon ingots being cross-sectioned and chemically etched to reveal the slip-dislocations induced by thermal shock. Extent of the slip lines were reduced from (a) to (c) by decrease in crystal diameter for withdrawal from the melt.

3.3 Growth Forms and Habits

The growth forms and habits of silicon ingots can be affected by the growth direction, growth rate, and crystal perfection. It has been shown in the previous sections that a dislocation-free ingot exhibits non-interrupted lines of ridges or strips of facets on the crystal surface, while a dislocated ingot shows discontinued facets. The form or shape of a crystal crown varies somewhat with growth condition. For a (111) ingot the crown can change from a circular to a hexagonal shape when the cooling rate of the melt is increased. Figure 22(a) is a photograph of the crown and Figure 22(b) is a sketch of Figure 22(a) superimposed by stereographic projection of the crystal. It shows that six corners are in <110> directions and six edges are in <211>. The three {111} planes shown in Figure 22(b) constitute the formation of three side facets. Figure 22(c) is a sketch of the ingot and shows that six edges at the crown reduce to three side strips in the main body. The width of the side facets can be increased by the combination of a high axial temperature gradient and a high pull rate.

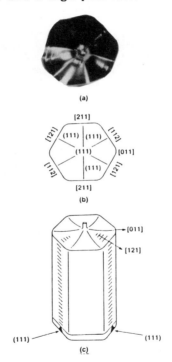

Figure 22: A "hexagonal" growth habit of a (111) ingot grown under a high degree of supercooling. (a) A photograph of top view, (b) Sketch of a top view with stereographic projection of key atomic planes, (c) Sketch of a prospect view.

The growth form of the (100) ingots has four-fold symmetry. The crown is typically circular in shape as shown in Figure 18(a). It can be converted to a square shape (Figure 23 (a)) when a high supercooling is applied. The directions of the square edges are <110> while corners are in <001> directions. Figure 23(b) shows the stereographic projection of a square crown. All four ridges on the crown continue on to the side surface of the main body (Figure 23 (c)). However, the square cross-section in the crown always converts to a circular in the main body growth. The growth of square shaped ingots requires specific techniques. We will discuss this in a later section.

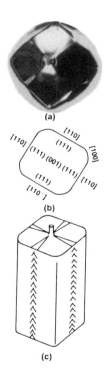

Figure 23: A "square" habit of a (100) ingot grown under a high degree of cooling. (a) A photograph of top view. (b) Sketch of a top view together with stereographic projection of key atomic planes. (c) Sketch of a prospect view.

3.4 Automatic Diameter Control

Constant Melt-Level Control: During the Czochralski crystal pulling process the melt height will progressively decrease. Consequently, the crystal-melt interface will also progressively move to higher temperature regions in the furnace since the temperature is higher toward the bottom of the heater (Figure 9(b)). Therefore, the crystal diameter will decrease with the increase in crystal length if the growth parameters are not adjusted accordingly. The reduction of the furnace temperature can be applied but can only solve parts of this problem. This is because the axial temperature gradient continues to decrease as the melt level drops. The fact is that the heat loss of the melt and crystal at the vicinity of the interface will continue to reduce with drop of the melt level since the heat radiation to the cooler objects in the puller is increasingly restricted. Furthermore, the optical pyrometer, if it is used for sensing the meniscus, will be out of the focus by the change in melt level. Therefore, it is desirable to maintain a constant melt level. The melt level can be kept constant by lifting the crucible. We assume that the crucible is cylindrical and the diameters of the crystal and crucible are constants. The cup-lifting rate (l) needed to keep the melt level constant for a given crystal pull rate (v) can be calculated from the mass balance between the melt and crystal:

$$l = \frac{\rho_s}{\rho_l} \frac{r^2}{R^2} v = 0.9 \ l \frac{r^2}{R^2} v \qquad (40)$$

Most of the crystal pullers used in production can be set with a crucible-lift to crystal-pull ratio (l/v). This means that the crucible lift rate is automatically changed with the crystal pull rate for maintaining a constant melt level. Selection of this value can be calculated from eq. (40). When a crucible with semispherical (instead of flat) bottom is used, the cup lift ratio should be progressively increased from that of Equation (40) when the melt level has reached semispherical region of the crucible.

Diameter Control: Common techniques used for the control of crystal diameter include: (a) sensing the optical reflection from the meniscus (41),(42); (b) weighing the crystal or crucible (43),(44); and (c) imaging the crystal diameter using an IR TV scan (45)-(48). Review of these techniques has been given by Hurle (49). He has concluded that the best control system is the one that can detect changes in meniscus shape rather than that seeks to measure the diameter. We will restrict our discussion to the optical technique for sensing the change in meniscus shape.

During the crystal pulling, the meniscus is formed because of the surface tension of the melt. The height of the meniscus is determined by the balance between the gravitational force of the liquid and its surface tension. It has been found that the height (h) of the meniscus increases with the diameter (d) and reaches a plateau rapidly resembling the relationship of $h = \sqrt{d}$ (49). The diameter of the meniscus is slightly larger than that of the crystal (see Figure 18(b)). The meniscus is also brighter than the free surface of the melt, particularly at the facets. The brightness was ascribed to be the result of releasing the latent heat from the crystal-melt interface (41). Digges et al (50) have measured the melt temperature and found that the halo is not warmer than the melt that is only a fraction of inch away from it. They concluded that the formation of the halo results from the reflections of the crucible which is brighter than the melt.

Brightness of the halo is not uniformly distributed. The brightness decreases as the distance increases from the crystal toward the melt free surface. The characteristic brightness distribution in the meniscus has been used for the control of crystal diameter. Figure 24(a) sketches that an optical pyrometer is in focus at the meniscus for its diameter control. Figure 24(b) plots the output voltage of a pyrometer (or brightness) vs. distance from the crystal toward the melt main surface. Let us assume that the diameter of crystal B in Figure 24(a) is what we want to keep (middle line). In this case the pyrometer is sensing at B' in the meniscus. The output voltage of the pyrometer will be at Y in Figure 24(b). The Y value is then used as the set point for the diameter control. If the diameter of the crystal increases to the size of crystal A in Figure 24(a), the position of the meniscus is also changed (to the top line). The pyrometer becomes in focus at A' that is mush closer to the crystal A. Its corresponding output voltage would be at X in Figure 24(b). The X value is greater than set point so that the adjustment is made accordingly to reduce it. On the contrary, if the diameter decreases to the size of crystal C, the pyrometer becomes in focus at C' in meniscus (bottom line) and its output voltage will be Z. The Z value is smaller than the set point and the adjustment is made to increase its value

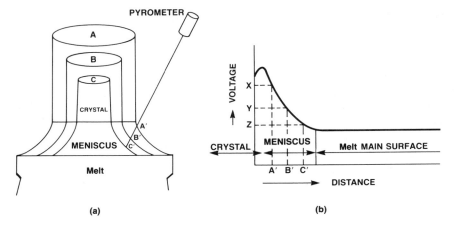

Figure 24: (a) Illustration of the change in crystal diameter affecting the sensing position in meniscus by a pyrometer in a fixed location. (b) A plot of sensing output from the pyrometer as a function of position in meniscus.

(i.e., diameter). The positive deviation from the set point value is commonly adjusted by the increase in pull rate or power input, and vice versa.

The pyrometer used for diameter control is placed at a fixed position and focuses at the meniscus with a fixed angle. When the control is for varying the diameter, the pyrometer location is varied with time while the focusing angle is kept constant. For example, when the diameter needs to be reduced, the pyrometer is programed to move toward the crystal. The rate of the pyrometer translation determines the tapered angle of the crystal. Alternatively, the pyrometer position is fixed while its sensing angle is varied.

3.5 Doping Techniques

Silicon crystals for semiconductor device applications require doping with a dopant to form an extrinsic semiconductor. An element from group III of the periodic table is used to form p-type (positive charge) conduction and from a group V to form n-type. The dopant concentrations ranging from 10^{14} to 10^{19} atoms/cm are used to meet the needs of various device applications. A pure dopant element can be directly added into the silicon melt for producing a heavily-doped crystal. However, an impurity-silicon alloy is preferred when a lightly-doped crystal is needed. An alloy for this application is in fine grains which are ground from the solidified silicon melt that contain a known concentration of the dopant.

Boron is the most commonly used p-type dopant. Elemental boron is dissolved slowly into the silicon melt. It is necessary to stir the melt and allow enough time for complete dissolution. Antimony, arsenic, and phosphorus are three commonly used n-type dopants. Antimony can be added with a solid silicon charge in the crucible for melting. However, arsenic has to be added after the silicon charge has melted. This prevents the loss of arsenic by evaporation during the melting. Since phosphorus is quite reactive and volatile, it is difficult to add into the silicon charge in the elemental form. A phosphate can be used as a phosphorus source for making silicon alloy.

The weight x (in grams) of pure elemental dopant needed to be added to the silicon charge can be calculated by the following equation:

$$X = M \left(\frac{w}{\rho}\right)\left(\frac{N}{A}\right) = 7.12 \, M \, W \, N \times 10^{-25} \tag{41}$$

where M is the atomic weight of the impurity, w the weight of the silicon melt, the ρ density of silicon (2.331 gm/cm^3), A Avogadro's number (6.023 × 10^{23}), and N the desired doping concentration (in atoms/cm^3) in the melt.

The amount of an impurity that can be added into the silicon melt is limited by its solubility. Figure 25 shows the plot of solubility (in atoms/cc) of various impurities as a function of temperature. Several interesting points in this plot are worth mentioning. (a) Antimony is the most desirable n-type dopant for heavily doped crystals because of its low volatility. However, its solubilities are more than one order of magnitude lower than those of phosphorus and arsenic. (b) Boron is only one p-type dopant which can dissolve in silicon in excess of 1×10^{20} atoms/cc. (c) Both n and p-type dopants are less temperature dependent than transition metals which are lifetime killers of minority carriers in silicon. (d) Fortunately, the transition metals and oxygen are less soluble in silicon than groups III and V dopants.

The doping concentration in the crystal can be calculated from the concentration in the melt and its effective distribution coefficient using Equation (35). We can consider that the doping impurity is completely ionized in the crystal lattice at room temperature so that the concentration of the majority carrier is equal to that of the doping impurity. In the doped crystal, the concentration of the minority carrier

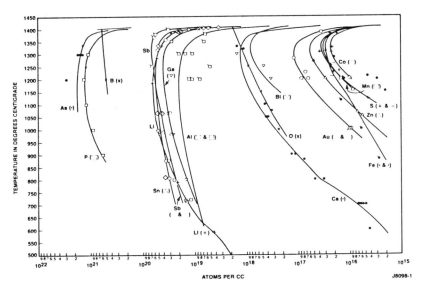

Figure 25: Plot of solubility of various impurities in silicon as a function of temperature. [From Trumbore (26).]

is very low as compared to the majority carrier and we can neglect its contribution to the electrical conductivity. Therefore the electrical resistivity of the silicon can be simply calculated from the equation:

$$1/(\rho) = e N \mu \qquad (42)$$

where ρ is the resistivity, e is the electric charges of the ionized dopant, N is the majority carrier (or doping) concentration, and μ is the drift mobility. Electron and hole mobilities in doped silicon have been evaluated by Wolfstirn (51). By knowing the mobility, the resistivity as a function of the carrier concentration can be calculated from eq. (42). Figure 26 shows this plot compiled by Irvin (52).

3.6 Variations in Radial Resistivity

Resistivity of silicon wafers is an important parameter that can affect the device characteristics such as depletion width and breakdown voltage of a p-n junction. A uniform radial resistivity distribution in wafers can improve certain device manufacturing yields. In the previous sections we have discussed the sources that can cause longitudinal doping

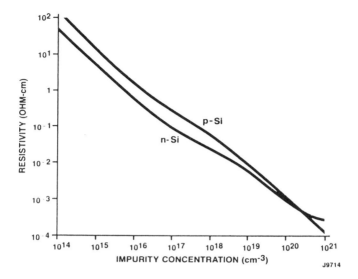

Figure 26: Plot of electrical resistivity vs. doping impurity concentration in silicon. [From Irvin (52).]

striations (or inhomogeneities) in crystals. These same sources can also cause radial doping inhomogeneities when the growth interface is not planar. Figure 27 is a sketch of wafers to be cut from lower portion of a crystal with concave interfaces. The traces of interfaces from a series of consecutive growth steps transfer to concentric striation rings in a wafer. The outermost ring was solidified first and was followed by the next rings toward the center of the wafer. Therefore,

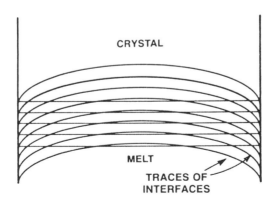

Figure 27: Illustration of wafers cut from the crystal with a concave interface.

the impurity concentration could increase macroscopically from edge toward center of the wafer. In addition, this macroscopic variations are superimposed with microscopic variations which are the result of time-dependent turbulent flow of the melt. Thus any longitudinal doping inhomogeneities in the crystal are directly translated into the radial resistivity striations. Experimentally, we have confirmed the increase in radial resistivity variation by the increase of interface curvature (or decrease of radius of curvature). This is given in Table 3 (18). The data in the first four rows of the table were taken from the crystal shown in Figure 3. Note that when an interface consists of double facets (e.g., Figs. 3(e) and (f)), the curvature values are taken from the annular groove.

Table 3

Crystal Radius of Curvature, Dopant Variation, and Interface Shape

Crystal shown in	Radius of Curvature	Spreading Resistance Variation	Resistivity Variation	Interface Shape
Fig. 3(b)	48.3 cm	±12%	-	planar
Fig. 3(a)	5.3 cm	±14%	-	concave
Fig. 3(e)	1.9 cm	±80%	-	double facets
Fig. 3(f)	0.7 cm	±90%	-	double facets
Not shown	28.0 cm	-	13%	concave
Not shown	6.9 cm	-	20%	convex
Not shown	6.9 cm	-	21%	convex

The effect of the radius of curvature on radial resistivity variations in dislocation-free (111) wafers can also be explained by the degree of supercooling. The interface of a dislocation-free (111) ingot consists of curved and faceted surfaces. The curved region represents the melting point isotherm while the flat region or the facet is a supercool from the isotherm (53). Figures 28(a) and (b) are the sketches of the interfaces with different curvatures. The dotted line in each figure is the isotherm of the freezing point. The longer the distance between the dotted line and the interface, the greater the supercooling at that point in the interface. These two figures illustrate that the smaller the radius of curvature the greater the ΔT. The greater ΔT, in turn, the greater the impurity incorporation at the facet. This can be

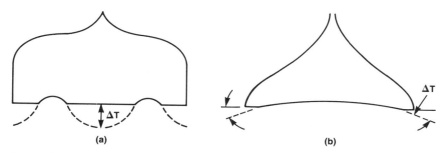

Figure 28: Schematic illustration of interface shape and degree of supercooling. (a) A small radius of curvature at the interface results with a high degree of supercooling. (b) Large radius of curvature with small degree of supercooling.

visualized from both kinetic and thermodynamic points of view. Kinetically, a high ΔT can cause a high lateral growth at the facet and lead to the increase in impurity incorporation. Dikhoff (54) has suggested that the effective distribution of impurities on the faceted region is greater than on non-faceted regions. Thermodynamically, the phase diagram shown in Figure 5(a) indicates that the solute concentration in the solid is increased by a lower solidification temperature (i.e., higher ΔT).

The radial inhomogeneity can be caused by other factors which are fixed in a growth system. They are (a) crystal orientation, and (b) the ratio of crystal to crucible diameter. Crystals with a (111) orientation have greater radial resistivity variations than a (100) orientation. This is because of the strong (111) facet existing at the growth interface when the crystal is pulled in a (111) direction. Table 4 lists the percent resistivity variations in crystals with (100),(110),(211), and (111) growth directions. The first three columns are for the Sb-doped crystals taken from Benson (55). The remaining are for 2" diameter phosphorus-doped crystals grown from 5" (4th & 5th columns) and 6" diameter cups by the author. It can be seen that the crystals growing in a <111> or <110> direction are greater in radial resistivity variations than those growing in <100> or <211>. The radial resistivity variation can be decreased by the increase in the ratio of the crystal to crucible diameter (55) for a constant crystal rotation rate. This is due to the fact that as the diameter increases, the flow rate of the melt propelled by the crystal's centrifugal action will also increase. This in turn will decrease the boundary layer thickness and improve its uniformity.

Table 4

Effect of Orientation on the Radial Resistivity of Sb-doped Silicon Crystals Grown at 11.2 cm/hr with Seed Rotation of 50 rpm Counter to Crucible Rotation of 2 rpm

(Maximum % Change in Radial Resistivity)

Orientation	Benson (55)			5" cup [b]		6" cup [b]	
	g=0.1	g=0.5	g=0.9	g=.10	g=0.7	g=.10	g=0.7
<111>[a]	7.9	56*	10.0	18	9.1	10.2	6.0
<100>	6.3	7.0	6.7	1.0	5.1	3.0	1.4
<110>	10.9	14.3	12.4	-	-	-	-
<211>	-	-	-	5.2	9.1	-	-

[a] This crystal was selected since it contained a region that showed the central facet and other regions did not. It is not typical of the crystals used in Figure 5.
* Exhibited the "facet effect".
[b] This work

However, Table 4 shows that the crystals grown from a 6" cup are better in radial resistivity uniformity than from a 5" cup for the same crystal diameter (2") and the same charge size (2 kg) of polysilicon. The melt height in the 6" diameter crucible is lower than that in 5" crucible. The low melt height contributes to the reduction of the thermal convection current of the melt and improves the radial doping uniformity.

The adjustable variables that can affect the radial resistivity variation are crystal and crucible rotation, and growth rate. Detailed studies of these variables have been made by Benson (55). For the individual effect of crystal or crucible rotation, he has found that the radial resistivity variations increase with the rotation rate and reaches a peak value at 25 and 10 rpm, respectively for crucible and crystal rotations. From there the radial resistivity variations decrease with the increase in rotation rates. Similar results have been found for the combined effect of crystal and crucible rotations. Figure 29 shows the plot of the maximum percent change in radial resistivity vs. crystal rotation rate for various growth rates. Benson (55) has explained this as the combined effect of crystal and crucible rotations on the flow pattern of the melt. The melt flow to the interface by crystal rotation is uniform over the radius, whereas that flow due to thermal convection will follow gradual curved streamlines and possess a definite stagnation point near the center of the interface. At low rotation rates the boundary layer thickness near the stagnation point remains unchanged, while at the outer regions of the interface the thickness is reduced by the additive effect of the tangential velocity and the thermal convection velocity. This causes an increase in the radial resistivity gradient as the crystal rotation is in the low range. Figure 29 also shows that the decrease in growth rate (v) reduces the radial resistivity variation, particularly under a low crystal rotation rate. Figure 30 is also the plot of the radial resistivity variation vs. crystal rotation rate obtained by the author. We do not see the resistivity variation peak in this plot. This could be either due to the fact that our increment in the variation of the rotation rate is not as fine a division as Benson's or due to the difference in temperature profiles in the melt.

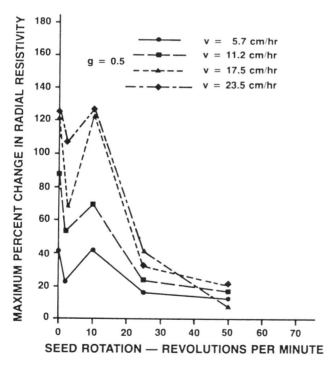

Figure 29: Plot of the maximum percent change in radial resistivity in crystal vs. crystal rotation rate. [From Benson (55).]

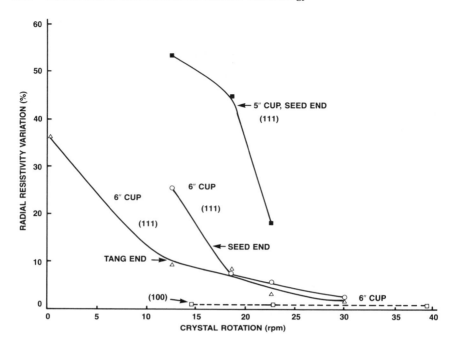

Figure 30: Plot of radial resistivity variation in crystal vs. crystal rotation rate.

Other techniques for the improvement of doping uniformity include the use of a double crucible (56),(57), and use of a three-phase AC heater. The three phase furnace technique can also improve the axial oxygen distribution. Discussion of this technique will be given later under the section of oxygen control. The double crucible technique can improve both axial and radial doping uniformity. Figure 31(a) shows that the inner crucible is inserted into the melt which is contained by the outer crucible. An opening toward the bottom of the inner crucible provides the passage for replenishment of the melt from the outer to inner crucible during crystal pulling. When we use C_o/K and C_o as doping concentrations of the melts for the inner and outer crucible respectively, the pulled crystal will have constant solute concentration C_o as shown in the left segment of Figure 31(c). This results from the fact that build-up of the doping concentration in the inner crucible melt due to the segregation (k<1) is diluted by the inflow of the melt from the outer crucible. Therefore, the dopant concentrations of the melt in the inner crucible will remain constants until the bottoms of both crucibles come in contact as shown in Figure 31(b). The solute distribution in the remaining section of crystal becomes identical to that is pulled from the single crucible technique which is shown in the right segment of Figure 31(c). This technique is particularly useful for the growth of heavy antimony-doped crystals which give low crystal yields by the single crucible method due to low segregation of antimony.

Improvement of radial resistivity uniformity in the crystal by the double crucible method is in parts due to the baffle action of the inner crucible. The thermal-driven convection current will be confined to the melt in the outer crucible. Thus temperature at the melt/crystal interface will become more stable.

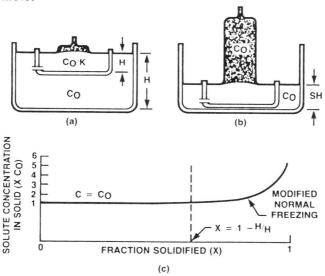

Figure 31: A crystal pulling using double crucibles. (a) At the stage of the pulled crystal with constant axial doping concentration. (b) Doping concentration in the crystal begins to change with time. (c) Plot of solute concentration vs. fraction of melt pulled. [From Lin and Hill (57).]

3.7 Oxygen and Carbon in Silicon

Oxygen and carbon are two major unintentional-doped impurities in the silicon crystals grown by the Czochralski method. The quartz (SiO_2) crucible is the oxygen source of contamination. Oxygen is transported to the silicon melt by SiO which is the reaction product of quartz crucible and molten silicon.

148   Handbook of Semiconductor Silicon Technology

SiO₂ (43)

The SiO   the melting
point of   eric pressure
(58). Tl   es from the
free surf   ion dissolves
into the   solidification.
The solut   the melting
point is   asionally we
do see th   gots exceeds
this solub   is that the
evaporatin   n incorporates into tl
    Oxyg   distributed
as single   tes.   Each
atom form   )).   Oxygen
with this   However,
four oxygen atoms can also jointly make bondings with one
silicon atom (i.e. SiO₄) or other possible four oxygen atom
complexes. Oxygen complexes with this type of bonding is
electrically active and contributes to n-type conduction.
They are referred to as oxygen or thermal donors. The
thermal donors can be generated from the interstitial oxygen
by heat treatment at 450°C. They can be annihilated by a
higher temperature heat treatment, typically at 650°C for 30
minutes. A prolonged heat treatment at temperatures between
650 to 750°C can form SiO₂ nuclei. Heat treatment at a
higher temperature between 750-1200°C enhances the diffusion
of oxygen. Oxygen atoms in the sub-surface regions can
diffuse to the surfaces and evaporation away from silicon
wafers. Those in the bulk region diffuse to the nucleation
sites and form SiO₂ precipitates in the bulk. Thus, the heat
treatment results in the silicon wafers to exhibiting a unique
structure in which the sub-surface regions are free of defects
(or called the denuded zones) and the bulk region is loaded
with a high density of crystallographic defects. The strain
fields surrounding the SiO₂ precipitates act as sinks for
impurity gettering. The gettering by the SiO₂ precipitates is
referred to as intrinsic gettering. This denuded structure is
beneficial for the performance of MOS devices that utilize
only the upper silicon layer for electrical transport. However,
when the oxygen concentration is too low, the SiO₂ precipitates can not form and the silicon wafers will not be capable
of intrinsic gettering. On the other hand, when the silicon
wafers contain an excessive oxygen concentration (e.g. > 1.24

× 10^{18} atoms/cm^3), the SiO$_2$ precipitates also occur at the sub-surface regions and the defect-free, denuded layers can not be obtained. Thus the control of oxygen concentration in silicon wafers becomes critical for IC devices in general and MOS devices in particular. There are other reasons which require precise control of oxygen concentration in silicon. One is that the initial oxygen concentration can affect the kinetics of the SiO$_2$ precipitation. In other words the degree of the SiO$_2$ precipitation is directly proportional to the initial oxygen concentration. The other is that excessive precipitation can lead to the warpage of wafers (62). The total time and temperature of the heat cycles during wafer processing also determine oxygen behavior in silicon. Thus the optimum concentration of oxygen in silicon is dictated by the specific wafer processing procedures. Typically, the desired initial oxygen concentration in the wafers are in the range of 0.91×10^{18} to 1.11×10^{18} atoms/cm^3 (63).

The concentration of interstitial oxygen in lightly-doped silicon is most commonly measured by infrared absorption at the 9μm peak. The oxygen concentration is calculated by multiplying the absorption peak at the 9μm with a conversion factor. The value of the conversion factor has been revised several times by ASTM as well as other technical communities (64). Confusion may occur in regard to the oxygen concentration published in literature by various authors since different values of the conversion factor have been used. This paper uses the recent JEIDA value (64) of 3.14×10^{17} or 6.28 for obtaining the oxygen concentration with the unit of atoms/cm^3 or ppma respectively. The quoted values from literature used in this paper have been converted to the same factor if different factors were used originally.

Oxygen in heavily-doped silicon can not be measured directly by the IR absorbtion technique since a high fraction of infrared is also absorbed by free carriers resulting from heavy doping. Recently, two techniques have been developed to measure the oxygen in heavily-doped silicon. One technique uses the gas fusion method (65). The other technique is to bombard the silicon with high energy electrons. This makes heavily-doped silicon transparent to infrared and restores the oxygen absorption peak (66).

The graphite heater and accessory graphite parts such as graphite cup and heat shields are the main sources of carbon supply into the silicon melt. It is likely that trace amounts of graphite are converted into gaseous compounds by the reactions either with O_2, H_2O and SiO or with the quartz due to the direct contact between the graphite cup and quartz crucible. The carbon compounds subsequently dissolve into the melt via gas phase transport. Another possible source of carbon supply to the silicon crystal is from polycrystalline silicon charges. However, the contribution from this source is insignificant when a typical polysilicon charge is used. This is because that the equilibrium distribution coefficient of carbon is only 0.07 (67), while the carbon concentration in the single crystal (the range of 1×10^{16} to 4×10^{17} atoms/cm^3 (68)) is greater than that in the polysilicon charge (in the range from 2×10^{15} atoms/cm^3 to 8×10^{16} atoms/cm^3) (67),(69).

Carbon atoms dissolved in solid silicon occupy the substitutional sites. The diffusivity of carbon is therefore much lower than that of oxygen. Unlike oxygen, it is very difficult to form carbon precipitates in silicon. Since the atomic radius of carbon is considerably smaller than that of silicon, the presence of carbon in the silicon lattice decreases the lattice parameter and induces local stresses. These stresses can be revealed by the x-ray double crystal topographic technique and are exhibited as striation patterns. Carbon in silicon is electrically inactive. However, its presence can affect the oxygen behavior in silicon (70). Carbon inhibits thermal donor generation at 450°C, but forms complexes with oxygen to generate stable donors at 750°C. Carbon can enhance SiO_2 nucleation and precipitation rates in silicon. Overall, it is thought that the presence of carbon in silicon is not desirable. Therefore, the effort has been made to reduce the carbon content to a minimum.

Distribution of Oxygen and Carbon: The transport of oxygen and carbon into the silicon melt occurs throughout the entire course of the crystal growth process. In the same time, some of the oxygen continuously evaporates from the melt. Therefore, the total mass of oxygen and carbon is not conserved and their contents in the crystal are not determined solely by the distribution coefficients as shown in eq. (35). Instead, the crystal growth parameters are the dominant factors in determining the concentrations and distribution of oxygen and carbon in the grown crystals. Under a typical crystal growth condition in which the rotation rates of the crystal and crucible are constant, the concentration of oxygen is higher at the seed than at the tang end of ingots, and higher in the center than in edge of wafers. Figure 32 shows the plot oxygen versus the ingot length and fraction of melt pulled (g). It can be seen that the oxygen concentration decreases from the seed end and reaches to its minimum value at a g value of 0.6-0.7. From there the concentration increases towards the tang end. Figure 33 shows the radial distribution, where curves 1 and 2 are for wafers cut from the crystal shoulder right below the crown, curves 3 and 4 are wafers cut one to two inches below the crown, and curve 5 is for a wafer cut from one inch above the crystal tail. It shows that the seed end gives a higher radial gradient than the tang end. The distribution of carbon is an inverse of oxygen both in the radial and axial directions. Figure 34 shows the plot of carbon concentration versus $1/(1-g)$.

Attempts have been made (68),(69) to evaluate the effective segregation coefficient of oxygen and carbon from the normal freeze curve shown in eq. (35). Series and Barraclough (69) have used the least square method and yield a negative value for K and C_0. Liaw (68) has found non-linearity in the C versus $1/(1-g)$ plot as shown in Figure 34. Figure 32 also shows that the non-linearity between the plot of concentration versus g for oxygen in entire length of the ingot. All these results confirm the non-conservative nature of the mass transfer of oxygen and carbon in the crystal growth system.

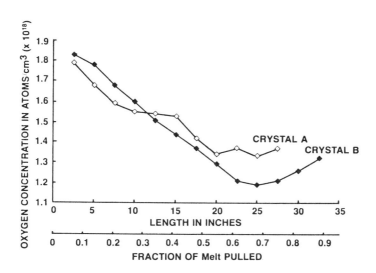

Figure 32: Plot of oxygen concentration in the crystal vs. ingot length and fraction of melt pulled.

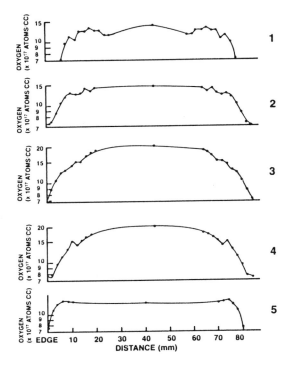

Figure 33: Plot of radial resistivity distribution along the wafer diameter. Curves 1 and 2 are for wafers cut right below the crown, Curve 3 and 4 for wafers from 1 to 2 inches below the crown, and Curve 5 for wafer one inch above the crystal tail.

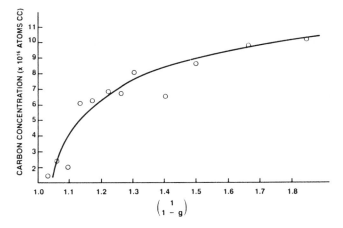

Figure 34: Plot of carbon concentration in silicon vs. $1/(1-g)$.

Factors Affecting Oxygen Concentration: The oxygen concentration in the crystal is determined by its concentration in the melt and the effective distribution coefficient. The concentration in the silicon melt is the difference between the dissolution rate of the quartz crucible and evaporation rate of silicon monoxide. Assuming that the diffusion of oxygen through the boundary layers at the interfaces are the rate limiting steps for the dissolution and evaporation, the oxygen concentration in the crystal $[O]_{Si}$ can be expressed as (34),(71),(72):

$$[O]_{Si} = A_i v C_m = A_c D (C_c - C_m)/(\delta)_c - A_m D C_m/(\delta)_s \qquad (44)$$

where A_i is the cross-section area of the ingot, v is the pull rate of the ingot, A_c is the area of the crucible/melt interface, D is the diffusion coefficient of oxygen, C_c, C_m, are oxygen concentrations at the crucible wall, and melt respectively, $(\delta)_c$ and $(\delta)_s$ are the boundary layer thickness at the crucible/melt and melt surface respectively, and A_m is the area of the melt free surface. The first term in the right hand side of the equation is the mass transfer rate from the crucible surface to the melt. The second term is the mass transfer rate from the melt to the gas ambient.

The primary factors determining the dissolution rate are the temperature at the crucible wall and the surface area of the melt/crucible interface. The dissolution, as most chemical reactions, is exponentially proportional to the temperature. The increase in crucible wall temperature will increase the C_c value in Eq. (44). The increase in the crucible wall temperature is equivalent to the increase in the temperature gradient in the melt since the center of the melt surface is required to remain at the melting point. There are several occasions in which the melt is under a high temperature gradient: (a) a severe heat loss from the melt free surface, and (b) poor thermal conduction in the melt. The former could result from the excessive stirring of the melt either due to high crystal or crucible rotation. The latter could be due to lack of melt convection resulting from the influence of a magnetic field.

The ratio of surface area between the crucible/melt interface and melt free surface (A_c/A_m) is another important parameter which determines the C_m value in Equation (44). This area ratio is primarily determined by the weight of the polysilicon charge for a given crucible size. The larger the polysilicon charge, the greater the A_c/A_m. This is why the oxygen concentration in the seed end of a crystal is increased by the charge size while the concentration in the tang end remains the same. For example, we have observed 1.24×10^{18} and 1.11×10^{18} atoms/cm^3 for the 6 kg and 4 kg charge respectively when 8" diameter quartz crucibles are used. The (A_c/A_m) ratio decreases with time during the crystal pulling. This is the primary reason that the oxygen concentration decreases with the length of the ingot until g reaches a value between 0.6 and 0.7 as shown in Figure 32. The boundary layer thickness of the melt/crucible interface is primary affected by the convection flow of the melt. The boundary layer thickness is decreased by the increase in the convection flow. As pointed out earlier, the contribution to the melt flow includes thermal, force, magnetic field-driven and surface tension-driven convections. Some of these convection currents counter-balance each other. This causes the difficulty in predicting the effect of each component on oxygen concentration in silicon.

3.8 Techniques for Control of Oxygen

Low or Medium Oxygen: We have pointed out that the quartz crucible is the oxygen source which contributes to silicon. Therefore, silicon ingots with a low oxygen content can be obtained if the quartz crucible is replaced by a non-oxide material. Recently, Watanabe et al (39),(73) have developed alpha silicon nitride crucibles and applied them to the Cz growth of silicon ingots. Single crystal ingots have been successfully grown with an oxygen concentration below 1.3×10^{16} atoms/cm^3. They also developed a controlled oxygen doping technique for the growth of medium oxygen ingots. This technique involving adding fused quartz into the melt. Since quartz will float on silicon melt and can interfere with the single crystal growth, it needs to be welded at the bottom of the crucible. The concentration of the oxygen in silicon ingots is controlled by adjusting the ratio of surface area between the fused quartz and the melt/crucible interface. Oxygen concentrations ranging from 1.3×10^{17} to 4.56×10^{17} atoms/cm^3 can be obtained by varying the surface area ratio from 3.6% to 17.8%. This technique also yields very uniform axial distributions of oxygen concentration in silicon ingots. However, the ingots are contaminated with nitrogen which is dissolved from silicon nitride crucible. The nitrogen concentration increases with the fraction of melt pulled. The tang end of ingots reached a maximum nitrogen concentration of

4×10^{15} atoms/cm^3 which is approximately two orders of magnitude less than oxygen in silicon ingots when they are grown from a quartz crucible.

Several other techniques have also been developed for the growth of medium oxygen ingots. They include (a) magnetic Czochralski method, (b) three-phase furnace method, and (c) shallow melt method. The magnetic Cz method will be discussed in the later section. The three phase furnace method uses a ladder type cylindrical heater with a delta connection. This can generate a rotational magnetic field at the center of the heater. The magnetic field can force the melt to flow circularly with rotation rates of 20 to 30 rpm (74). When the seed and crucible rotations are in the same direction as the magnetic-field-driven rotation, the thermal-driven convection current in the melt can be suppressed. The low net flow of the melt increases the boundary layer thickness at the melt/crystal interface and the diffusion of oxygen through this layer is limited. The ingot grown by this technique while linearly decreasing the seed rotation from 20 to 10 rpm has produced crystals with medium oxygen concentration with uniform axial distributions (3.26×10^{17} to 4.56×10^{17} atoms/cm^3). The three phase furnace method can also produce crystals with high oxygen concentrations and graded axial concentration profiles when the seed rotation is constant and its direction is opposite to that of the magnetic-field-driven rotation.

The control of oxygen by the shallow melt method is essentially to reduce A_c value in eq. (44). The shallower the melt height, the lower the A_c value and lower the oxygen content in the crystal. The shallow melt is typically obtained by reducing the load size of polysilicon charge in a given size of crucibles. From the practical point of view, the melt height can not be too shallow otherwise the productivity of the crystal puller will be dramatically reduced. Therefore, the oxygen concentrations in the ingots grown by this technique are only slightly reduced, for example, to 0.91×10^{18} atoms/cm^3 with a 4 kg melt from 1.17×10^{18} atoms/cm^3 with a 6 kg melt in 8" diameter crucible. A flat profile of the axial oxygen distribution is also difficult to achieve, although improvement can be made by a programmed increase in the rate of crystal and/or crucible rotation.

High Oxygen: One effective technique to produce high oxygen concentrations with uniform axial distribution in an ingot is to weld a quartz ring or quartz rods at the bottom of the quartz crucible (75). This is to increase the ratio of

A_c/A_m in Equation (44) and hence $[O]_{Si}$ is also increased. The diameter of the quartz ring should be smaller than that of the crystal so that the additional oxygen dissolved from this quartz ring will trapped at the crystal/melt interface. The trapping efficiency is improved by decrease in the distance between the quartz ring and the crystal/melt interface. Since this distance is automatically decreased as the increase in the fraction of melt pulled (g). This can compensate the progressive reduction in oxygen concentration with increase in g which occurs without insertion of a quartz ring. Therefore, the axial distribution of oxygen concentration in the silicon ingot becomes independent of g. Figure 35 shows the plot of oxygen concentration versus the length of ingots grown by this technique. This figure also shows the reproducibility of this technique for obtaining flat axial oxygen profiles.

An alternative technique for controlling the axial oxygen concentration profile is to program the increase in crystal or/and crucible rotation rates. This technique can produce axial profiles somewhat flatter than the "standard" technique which keeps both rotation rates constant. However, it has never been as flat as those grown by welding a quartz ring to the crucible. This is due to the fact that the rotation variations is limited to a narrow span. Attempt to increase the span by starting with a low rotation rate can result with

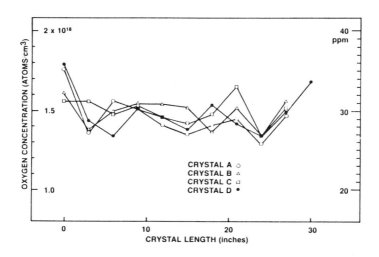

Figure 35: Plot of oxygen concentration vs. crystal length for ingots grown by a controlled oxygen technique.

a high radial resistivity gradient in the top end of the crystal.

3.9 Control of Carbon Content

In the previous section, we have discussed that the contribution of carbon to silicon ingots from a typical polysilicon source is insignificant, while the main source is from the graphite parts used in the crystal puller. We have carried out experiments to verify the above arguments and to find a means to reduce the carbon concentration in silicon ingots. The experiments involved the growth of crystals from a crystal puller in which all the graphite parts were coated with a layer of silicon carbide film by the chemical vapor deposition technique. Table 5 shows the experimental results which compare the carbon concentration in the tang end of ingots grown from different polysilicon lots using SiC coated and non-coated graphite parts. It shows that the polysilicon lot B can yield the carbon concentrations either non-detectable (less than 1×10^{16} atoms/cm^3) or above 4.5×10^{16} atoms/cm^3 solely dependent on whether the graphite parts were coated or not. The carbon concentration in crystals grown from polysilicon lots A and B using the non-coated graphite parts are representative of typical values. This result reconfirms that polysilicon is not a major factor contributing to the carbon concentration in silicon ingots. Table 5 also shows that the partial coating of the graphite part yields a higher carbon concentration average than that of completely coated graphite. This concludes that the coating of the graphite parts is an effective way to reduce the carbon concentration in silicon ingots. However, there is a problem associated with this technique. The problem is the cracking of graphite parts, predominately occurring in the graphite ring which was placed above the graphite heater as a heat shield. Occasionally, the graphite cups also cracked after several crystal growth runs. Therefore, work remains to be done to overcome the cracking problem.

4.0 NOVEL CZOCHRALSKI CRYSTAL GROWTH

4.1 Semicontinuous and Continuous Cz

The conventional Czochralski crystal growth is a batch process and includes the following pitfalls: (a) It consumes one quartz crucible per run since the crucible cracks during

Table 5

Carbon Concentration Measured at the Tang End of Ingots Grown From Different Poly Lots Using SiC Coated or Non-coated Graphite Parts in the Furnace

Crystal Number	Poly lot	SiC Coating of Graphite Parts	Carbon Concentration at Tang End of Crystal (atoms/cm^3)
1	A		5.7×10^{16}
2	A		1.4×10^{17}
3	A	Non Coated	6.8×10^{16}
4	B		4.5×10^{16}
5	B		7.0×10^{16}
6	B		4.7×10^{16}
7	B		$<1.0 \times 10^{16}$
8	B		$<1.0 \times 10^{16}$
9	B		$<1.0 \times 10^{16}$
10	unidentified	All graphite	$<1.0 \times 10^{16}$
11	unidentified	parts	1.8×10^{16}
12	unidentified	coated	$<1.0 \times 10^{16}$
13	unidentified		$<1.0 \times 10^{16}$
14	unidentified		$<1.0 \times 10^{16}$
15	unidentified		3.2×10^{16}
16	unidentified		$<1.0 \times 10^{16}$
17	unidentified		$<1.0 \times 10^{16}$
18			5.5×10^{16}
19			$<1.0 \times 10^{16}$
20		All parts	$<1.0 \times 10^{16}$
21		coated except	$<1.0 \times 10^{16}$
22	unidentified	cup	$<1.0 \times 10^{16}$
23			4.6×10^{16}
24			2.0×10^{16}
25			2.2×10^{16}
26			1.2×10^{16}

Courtesy of Frank Secco d'Aragona, Motorola Inc., Semiconductor Products Sector, Phoenix, AZ.

cooling off the furnace for crystal harvesting, (b) It produces crystals with a large difference in doping concentration between the seed and tang ends of ingots, (c) It requires a long machine idletime to dismantle and set-up of furnace for each crystal growth run. These pitfalls can be reduced or eliminated if the crystal growth is converted to a semicontinuous or continuous process. The work in these areas had been investigated by the author (76), Lane and Kachare (77), and Fiegl (78). These new processes are particularly appealing to the photovoltaic industry. It requires low cost silicon wafers for the fabrication of terrestrial solar cells.

The semicontinuous process can use a conventional Czochralski puller. However, a gate valve is required between the growth and harvesting chambers as shown in Figure 15. The growth of the first ingot by this process is identical to that by the batch process. After the grown ingot is separated from the melt and raised to the harvest chamber, the melt should be kept molten and the gate valve is closed. The crystal is then removed from the puller and is replaced with a polysilicon charge. After a few minutes of purging, the gate valve can be open and the polysilicon is loaded into the crucible. After the recharging of the crucible with the polysilicon, the growth of the second ingot can be initiated. The process can continue to alternate growth and recharging for several times from a single crucible without cooling the furnace. Author has used an atmospheric crystal puller and reproduciblely demonstrated the growth of three dislocation-free ingots with total weight of 31-32 kg from a 8" diameter crucible by two rechargings. Lane and Kachare (77) have used a reduced pressure puller that contained a vacuum-tight gate valve. They have demonstrated the growth of 5 single crystal ingots from a 12" diameter crucible by four rechargings. The total weight of five crystals were approximately 100 kg.

Several methods have been developed for the recharging of the polysilicon. Lane and Kachare used a long polysilicon rod which was about a one-half section of the U-shaped polysilicon rod harvested from a Siemens reactor. The polysilicon rod is attached to the recharging mechanism which they added to the puller. The recharging mechanism was incorporated with a weighting device so that the amount of each recharging from this rod could be controlled. Helda and Liaw (76) used a charge container to hold the preweighed polysilicon cylinders. The dopant can be placed between two poly cylinders for addition into the melt. The bottom of the container consists of heat deformable support members. At low temperatures the support members rigidly hold the polysilicon cylinders in the container. When the container is lowered to about 1-2 inches above the melt, the heat of the furnace and the weight of the polysilicon force the support member to deform, as shown in Figure 36. This opens the bottom of the container and allows the polysilicon to descend into the crucible. During the melting of the polysilicon, the container is gradually withdrawn from the furnace and removed from the puller.

One concern about the recharging technique is increase of impurity concentration in the melts with the number of rechargings. The impurity concentrations can be calculated by the repeated application of Equation (35). Let us assume that the concentration of an impurity in the polysilicon or the initial melt is C_o, and that the g fraction of the melt is pulled and an equal weight of polysilicon is recharged in each cycle. The impurity concentration in the recharged melt at the beginning of the n-th pull, $(C^i_L(n))$ has been deduced (38):

$$C^i_L(n) = C_o\, p^{n-1} + C_o g(p^{n-2} + p^{n-3} + \ldots + 1)$$
$$= C_o\,[p^{n-1} + g\,(p^{n-1} - 1)/(p-1)] \qquad (45)$$

where $p=(1-g)^k$. At the end of the n-th growth run, the impurity concentration of the melt left in the crucible before the recharge is

$$C^f_L(n) = (1 - g)^{k-1} C^i_L(n) = (p/(1 - g))C^i_L(n) \qquad (46)$$

If $K<<1$, then $P=1$ and Eqs. (45) and (46) can be approximated by

$$C^i_L(n) = C_o[1 + g(n-1)] \qquad (47)$$

$$C^f_L(n) = C_o[1 + ng/(1-g)] \qquad (48)$$

The impurity concentration in the seed and tang ends of the n-th pull ingot can be calculated from Equations (47) and (48) respectively by multiplying them by k.

To visualize the build-up of impurities in the multiple recharging process, let us assume that g equals 0.9 and n equals 5 as an example. The build-up of an impurity in the melt at the beginning and the end of the 5th pull are respectively 4.6 and 46 times of the initial value. This seems to be a considerable increase in concentration. However, its impact

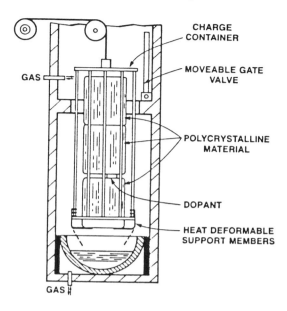

Figure 36: A device used for recharging polysilicon into the melt for the multiple ingot growth from a single crucible.

on the formation of constitutional supercooling and acceptable impurity concentration in the crystals is still negligible. This is because the increase of the concentration in the melt is still below the critical concentration for onset of the constitutional supercooling (77). Therefore, stable growth of the crystals may still be achieved. This has been verified by experiments that have obtained dislocation-free single crystals from the first to fifth pulls. In addition, the very low distribution coefficient of metallic impurities in silicon also makes the build-up of the impurity in the melt of little concern. For example, experimental results have shown that no detectable increase in the concentration of most metallic impurities could be found from the first to fourth pulled ingots. One exception is aluminum, which has a higher distribution coefficient than other metallic impurities. The increase in aluminum concentration with the number of pulls is shown in Table 6. The concentration of aluminum at 1.7×10^{15} atoms/cm^3 does not affect the electrical property of silicon as measured by the minority carrier lifetime or solar cell efficiency (38). The concentration of carbon and oxygen in the multiple-pulled ingots are also listed in Table 6. It shows that the carbon concentration in the seed end of the first three pull ingots which are less than 2×10^{16} atoms/cm^3 are not detectable. However, the carbon concentration in the tang end of the ingots are slightly increased with the number of pulls. This can be understood from the fact that the chance of the graphite parts being exposed to trace amount of air increases with the number of recharging. Table 6 also shows that the variation of carbon concentration in the crystals is affected to a greater extent by air-tightness of the gate valve than by the number of pulls. The oxygen concentration shown in Table 6 does not follow a clear trend although one of the experiments indicates a decrease with the number of pulls.

Table 6

Concentration (atoms/cm^3) of Al[a], C[b], and O[b]

in Multiple-pulled Silicon Ingots

	first pull		second pull	
	seed	tang	seed	tang
Al	-	-	3.03×10^{14}	10.3×10^{14}
C (a)	$<2.0 \times 10^{16}$	2.2×10^{16}	$<2.0 \times 10^{16}$	4.84×10^{16}
(b)	-	2.7×10^{17}	-	4.60×10^{17}
O (a)	1.99×10^{18}	1.43×10^{18}	1.88×10^{18}	1.37×10^{18}
(b)	-	1.30×10^{18}	-	1.10×10^{18}
	third pull		fourth pull	
	seed	tang	seed	tang
Al	1.54×10^{14}	-	16.3×10^{14}	17.7×10^{14}
C (a)	$<2 \times 10^{16}$	8.6×10^{16}	-	-
(b)	-	4.7×10^{17}	-	-
O (a)	1.42×10^{18}	1.33×10^{18}	-	-
(b)	-	1.40×10^{18}	-	-

[a] From Lane and Kachare (1980), measured by neutron activation and spark source spectrometry.

[b] From Liaw (1978), measured by infrared absorption.

Continuous Czochralski growth of silicon crystals has been developed by Fiegl (78),(79). The puller that they used is shown in Figure 37. This continuous crystal puller consists of two separated furnaces connected by a continuous liquid feed quartz tube. One furnace is for crystal pulling and the

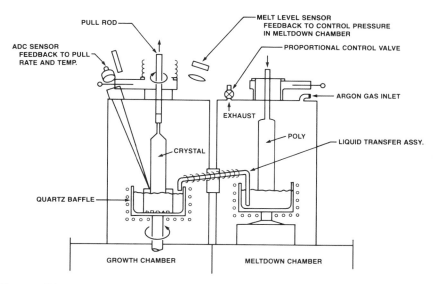

Figure 37: An arrangement of two crystal pullers for continuous growth of silicon crystals. [From Fiegle (78)].

other for the melting of polysilicon. The silicon melt is transferred by siphon action. The crucible in the growth chamber contains a quartz baffle which dampens the melt vibration caused by the melt feeding. The melt is fed at such a rate that a constant melt level in the growth chamber is maintained.

One advantage of this method is the uniform impurity distribution in the axial direction of the grown crystals. The distribution of the impurity can be derived from the following differential equation which describes the overall conservation of solute in a system (38).

$$V_o dC_L = (C_o - kC_L) dV_c \tag{49}$$

where V_o is the melt volume and V_c is the volume of crystal which has been grown. With the boundary condition of $C_L = C_o$ when $V_c = 0$, the solution of Equation (49) is

$$\frac{C_L}{C_o} = \frac{1}{k}\left[1 - (1-k)\exp\left(-\frac{kV_c}{V_o}\right)\right] \tag{50}$$

The plot of C_L/C_o versus V_c/V_o is shown in Figure 38. The

C_L/C_0 from semicontinuous processes for various number of pulls (n) is also shown in this figure for comparison. This graph clearly shows that the axial distribution of impurities in the continuous process is much more uniform than by that from the semicontinuous process. The axial impurity distribution in the ingots grown by the semicontinuous process can be improved if only a small fraction of the melt is pulled in each recharged cycle (except the last cycle). This will not increase the silicon loss since the leftover melt is re-used by adding new polysilicon charge.

Drawbacks of the continuous growth process are the complexities in equipment and processing. The major process problems are the transfer of the melt from one chamber to another and the control of equality between the feed and pull rates. Nevertheless, crystal sizes up to 65 kg have been grown by this process. It has also demonstrated the capability for growing dislocation-free ingots.

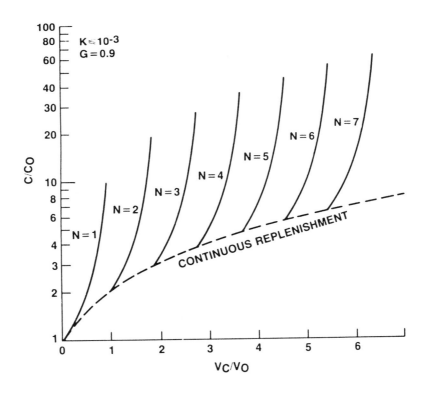

Figure 38: Plot of CL/Co vs. Vc/Vo.

4.2 Magnetic Czochralski (MCz) Crystal Growth

The electrical conductivity of silicon increases with the temperature. The conductivity further increases to 12300 ohm^{-1}cm^{-1} when it transforms into a molten state at its melting temperature (80). This number is within the same range of conductivity values for many metals. Electrically, the molten silicon can be considered as a metal. We have pointed out that the molten silicon flows in the crucible due to the thermal-driven convection. In other words, a "moving metal" is confined to circulate inside the crucible during the crystal growth. Therefore, application of a magnetic field into the silicon melt can result with a force that retards its flow (i.e. Lenz law). The distribution of impurities into the crystal can also be altered by the magnetic field.

Two types of magnetic fields have been applied to the Cz growth of silicon crystals. They are the transverse (horizontal) and axial (vertical) fields. Figures 39(a) and 39(b) show distributions of the magnetic flux generated by the axial and transverse superconductive magnets respectively (81). Few detailed studies have been made on the effect of a magnetic field on heat and mass transfer of solutes in a conductive solution during freezing. Therefore, understanding of impurity distribution in MCz crystal growth is still lacking.

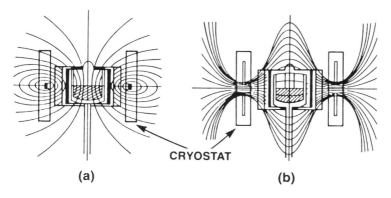

Figure 39: Distribution of magnetic flux in a crystal puller. (a) An axial magnetic field. (b) A transverse magnetic field. [From Ohwa (81).]

Experimentally, it has been found that the silicon crystals grown under the influence of a transverse magnetic field provide better physical characteristics. For example, the presence of a transverse field (>100 gauss) considerably decreases incorporation of oxygen in the crystals and improves the uniformity of axial and radial distributions of dopant and oxygen. On the contrary, the presence of an axial field (also >100 gauss) increases incorporation of oxygen, carbon and phosphorus in the crystals, and increases the non-uniformity of the impurity distributions (82).

Interpretation of these results can partially be made by the observation of the change in melt temperature after the application of a magnetic field. Temperature of the melt at the center surface of the crucible is decreased by the axial field but not by the transverse field although both magnetic fields can reduce the temperature fluctuation. In order to maintain the temperature of the melt at the melting point, the heater temperature has to be increased when an axial field is used. This also increases the crucible temperature and increases the dissolution rate of oxygen from quartz crucible. This explains the high oxygen content in crystals grown under the influence of an axial magnetic field. It has also been suggested (82) that the axial magnetic field strongly suppresses the radial outflow of melt at the interface. This creates non-mixing cells in the melt and causes a higher radial non-uniformity in the crystal.

Suzuki et al. (33),(83) have shown that crystals grown under a transverse field result in very uniformly impurity distribution both in the longitudinal and radial directions. Micro-inhomogeneities such as striations have been eliminated. Bulk stacking faults in the crystal have also been eliminated by applying high pull rates.

4.3 Square Ingot Growth

Two approaches have been used to grow square silicon ingots from a Czochralski puller. One approach is to enhance the formation of natural crystal habits. This requires a melt with an extremely good radial symmetry and stable temperatures. Under these conditions, the fast growth portions (or directions) of the ingot will not be melted back during the crystal rotation and the ingot will maintain a natural crystal habit. Kuroda et al. (84) have demonstrated the growth of (100) square ingots from charge sizes of 1 to 1.5 kg melt. Continuous seed and crucible rotations both at 10 rpm were applied in opposite directions. The temperature fluctuation at a given point was kept below ±2°C. They have found that a square ingot was obtained when the temperature variations along a circular contour in the melt were less than ±2.5°C, and that a circular ingot was grown when the variations were ±10 to 15°C. The square ingots exhibited such a crystal habit that the diagonals of the square were along <100> directions and four edges of the square were perpendicular to <110>. They have been able to maintain square cross-sections throughout the length of the ingots. The ingots were single crystals with etch pit density variations from zero at the ingot center to $1-2 \times 10^3/cm^3$ along the crystal periphery. Resistivity measurements on the wafers showed that the iso-resistivity contours were parallel to the edges of the wafers.

Another approach is to shape the temperature profile of the melt into a square configuration. Liaw (85) used a thermal insulation plate suspended above the melt surface. The gap between the melt and the plate was approximately 2 cm or less. The center of the plate was cut in a square opening. A single crystal seed was dipped into the center of melt surface through this opening. During the crown portion of the crystal growth, a high supercooling was applied to the melt. The high supercooling forces the crystal to grow faster in <110> direction than in <100>. Thus the crystal crown will grow into a square with diagonals in <110> and edges in <100>. Once the size of the crystal crown was approaching that of the opening, the crystal rotation was paused in such a way that the four edges of the crystal crown were parallel to the four sides of the opening. Then the crystal was pulled with a discontinuous rotation during the growth of main crystal body. Each rotation applies a 90 degree turn of the crystal. The purpose of pause rotation is to keep the ingot growing straight since the temperature of the melt was not perfectly symmetrical with respect to the pull axis. The time interval between each rotation is determined by the symmetry of the melt. The poorer symmetry in the melt, the shorter the pause time is needed between each rotation and the less square will be in the ingot. Figure 40(a) shows a square ingot grown by this technique and Figure 40(b) shows wafers cut from a square ingot.

4.4 Web and EFG Techniques

Several techniques have been investigated for the growth of silicon crystals in a flat sheet form. Two of the most well developed techniques are the dendritic web growth and edge-defined, film-fed growth (EFG). The web technique was first developed by Bennett and Longini (86) for Ge growth in

Figure 40: (a) A "square" silicon ingot. (b) Wafers cut from the "square" ingot.

1959 and was later applied to silicon in 1962 by their coworkers. The EFG technique was first applied to grow sapphire filaments, tubes, and ribbons in 1971 (87),(88), then applied to grow silicon ribbons in 1972 (89). Since the mid 1970's these two techniques, particularly EFG, have produced a considerable amount of silicon sheets for solar cell applications.

Web Technique: A silicon web can be pulled from a melt by surface tension between two coplanar dendrites of the same seed as shown in Figure 41 (90). The two parallel dendrites act as a solid frame which holds the web until it solidifies. The frozen web exhibits the same crystallographic orientation as the dendrites except that it is thinner and smoother on he surfaces. The main surfaces of the web consist of (111) planes. The preferred propagation directions of the dendrites are [211]. During pulling, the dendritic tips extend below the melt surface while the web/melt interface is on or above the melt surface.

The growth mechanisms of semiconductor webs have been studied extensively (91),(92). Growth is initiated by the dendritic growth at both edges of the web. The dendritic growth is controlled by the twin plan reentrant growth mechanism. Figure 42 shows sketches of twin planes in a platelet habit that is often observed in diamond cubic materials (93). The main surfaces of the platelet are bounded by {111} planes on which nucleation is difficult. However, the twin planes provide 141° reentrant grooves for easy nucleation. The nucleation is followed by rapid lateral growth that terminates when it has reached the main (111) surfaces. The

Crystal Growth of Silicon 169

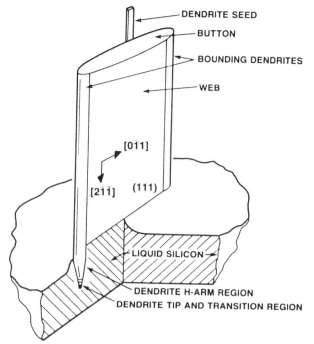

Figure 41: A silicon web grown by freezing a thin sheet of melt supported by two coplanar silicon dendrites. [From Seidensticker (90).]

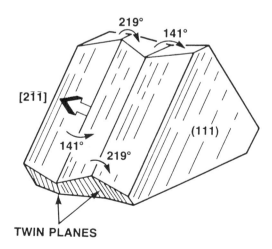

Figure 42: Dendritic growth by the twin plane reentrant growth mechanism. [From Faust and John (91).]

repeated nucleation at the reentrant grooves and lateral growth cause the platelet to become a dendrite which advances in a [211] direction. If two or more twin planes are present, the reentrant grooves will serve as perpetual nucleation sites. Then the growth of dendrites can continue and remain as a flat strip rather than a rod (93). For web growth, it is also required that the two bounding dendrites contain no branches. Branching of dendrites can be avoided by using a three-twin seed that keeps the branching directions from emerging into the melt (90).

A web crystal grown in the early stages of development always contained twin planes which were extended from the bounding dendrites (Figure 43a). The twin planes in the dendrites are located at the center and parallel to the main surfaces. The web can be free of twin planes as shown in Figure 43b if the web is not coinciding with the center plane of two bounding dendrites. This can be achieved by pulling the web from a melt of asymmetrical temperature distribution with respect to the pulling plane. Web crystals can also be grown free of dislocations. This is achieved by using a flat interface (94).

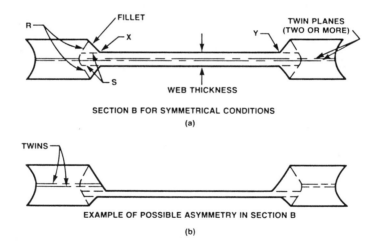

Figure 43: (a) Web contains twin planes when the thermal symmetric plane is coincided with a twin plane in dendrites. (b) Free of twin planes in the web when the plane of thermal symmetry is not overlaid with any twin plane between two dendrites. [From O'Hara and Bennett (92).]

EFG Technique: The EFG technique (95) utilizes a die that is inserted into the melt. The construction of the die is the predominate entity that governs the growth of ribbon crystals. The die parameters include the shape, temperature distribution, and the material of construction. Figures 44(a) and (b) show examples of two types of die designs. Figure 44(a) is a die which contains a narrow slot, while Figure 44(b) is another die with multiple holes in it. When a die is made of a material such as graphite that can be wetted by silicon melt, the melt will flow into the holes or slot and rise above the melt surface by means of capillary action. The height, h, of the capillary rise can be calculated from :

$$h = (2\gamma \cos\theta)/\rho g\, d) \tag{51}$$

where γ is the interfacial surface tension, θ is the wetting angle of the liquid silicon on the die, ρ is the density of silicon, g is the gravitational constant, and d is the diameter of the capillary, or the distance of the capillary spacing of the slot in the die. Figure 45 shows the plot of theoretical capillary rise versus capillary spacing for dies made of fused silica and graphite. Although low capillary rise in a silica die can be improved by using extremely thin capillaries, it is extremely difficult to control the dimensions in quartzware. The high plasticity of fused silica at the melting temperature of silicon can induce both stress on the ribbon and momentary freezing of the ribbon to the die during the ribbon pulling. Therefore, the use of a fused silica die has never been popular.

A die should be so designed that when it is placed in the melt, the vertical distance of the die that floats above the melt surface should be less than the theoretical capillary rise. When this condition is met, the melt will feed to the top of the die. The top of a die is shaped with a shallow groove for melt confinement. If the temperature of the die is kept above the melting point of silicon, the silicon in the groove will remain as liquid film. A flat thin silicon seed can be made to contact the liquid film in the groove and a silicon ribbon can be pulled from it. The width of the groove and other factors such as pull rate and melt temperature can affect the thickness of the ribbon. During the ribbon pulling process, the liquid silicon film in the groove is continuously fed from the melt outside the die through capillary holes. Multiple silicon ribbons have been pulled from the melt in a single crucible that contained multiple

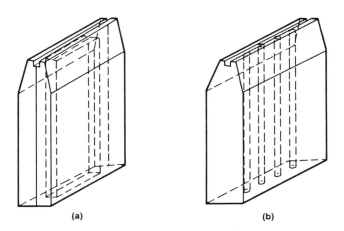

Figure 44: Examples of die design. (a) A die contains a narrow slot. (b) A die contains multiple holes. [From Ravi (95).]

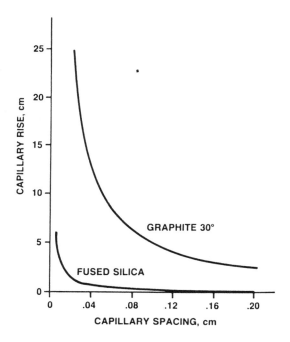

Figure 45: Plot of Capillary rise vs Capillary space in a die. [From Ravi (95).]

dies. Continuous ribbon growth process has been demonstrated. This is achieved by (a) the use of a spool for winding of the pulled ribbon, and (b) continuous melt replenishment of the crucible.

EFG ribbons prefer to grow in a <211> direction. The main surfaces are bounded by the planes approaching {110}. This growth orientation is naturally formed and maintained regardless of the seed orientation. Dislocation-free ribbons are difficult to obtain. The magnitude of dislocation densities is typically at $10^5/cm^2$. The ribbons also contain multiple close-spaced twin planes that are parallel to the growth directions. Approximately 10 ppma of carbon is also contained in the ribbons because of the use of a graphite die. Typical minority carrier diffusion lengths in the EGF ribbon range from 30 to 80 μm as compared to 100 or higher in the Czochralski wafers. However, solar cells with efficiency in excess of 10% have been fabricated from EFG ribbons.

4.5 The Float-Zone Technique

Application of the float-zone (FZ) technique to grow silicon single crystals was first reported by Keck and Golay (96). The FZ crystals are purer than Cz ingots since the silicon melt used for crystal growth is not contained by a crucible. Large diameter silicon ingots are more difficult to grow by the FZ than by Cz technique. However, the progress of the FZ technique has been able to keep pace with the Cz, and crystals with diameters of up to 125mm have been grown in production. Figures 46(a) to 46(c) sketch the main growth procedures used in the FZ technique. First, the bottom end of a polysilicon rod is preheated as shown in Figure 46(a). This end of the rod is ground into a v-shape and is placed in the center of a water cooled, single-turned copper coil. A graphite susceptor is then placed underneath the polysilicon rod with a minimal gap. When an rf current is applied to the copper coil, an electrical eddy current is induced in the graphite susceptor and the temperature of the susceptor increases. The heat is then transferred to the polysilicon rod by radiation. Once the portion of polysilicon in close proximity to the susceptor starts to glow, the eddy current can be induced in this segment of silicon by the rf energy. The graphite susceptor is no longer needed and is removed from the rf coil. The heat is continuously applied until the cone segment of the polysilicon rod becomes molten. Subsequently, a seed is dipped into the molten silicon from below as shown in Figure 46(b). Once the seed is wetted by the molten silicon, the growth of a crystal can be initiated by lowering the seed. The polysilicon rod also need to be lowered, but with a much slower rate. Typically, the first zone pass of a polysilicon rod always yields a polycrystalline ingot. The purpose of the first zone pass is for zone refining and is preferably carried out in a vacuum. A single crystal can be obtained by a second zone pass. As in the Cz technique, dislocation-free growth should be initiated during the seeding process by using very fast pull rates. Once the dislocation-free structure is observed (due to the appearance of strong side facets), the ratio of pull rates between the seed and polyrod is gradually decreased so that the crystal diameter will gradually increase. When the diameter of the single crystal needs to be larger than that of the polysilicon rod, this can be achieved by gradually off-setting the axis of the bottom pull rod from the upper pull rod. Figure 46(c) sketches the result of this off-setting and the ratio of the pull rates between the seed and polyrod being less than 1. This off-set alignment can help mixing of the melt and improves the doping uniformity. A dopant gas is also introduced during the second zone pass.

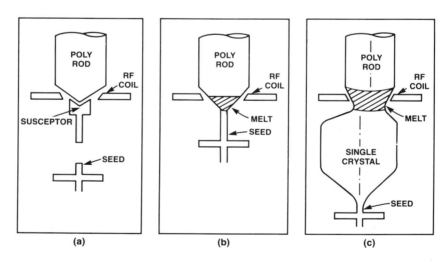

Figure 46: A procedure used in float zone growth of a silicon crystal. (a) Preheat of polysilicon rod by a graphite susceptor. (b) Initiate growth by dipping single crystal seed. (c) Growth of an ingot with diameter greater than that of molten silicon by off-setting the pull axis.

A critical factor for successful float-zone growth is maintaining the stability of the molten zone. The zone is stable when the inward pressure of the zone is greater than the outward pressure. The inward pressure includes surface tension, cohesion between the solid and liquid, and levitation due to the radio frequency field. The latter two terms are relatively small as compared to the surface tension. The outward pressure mainly includes the hydrostatic pressure resulting from the gravitational force of the molten zone. The hydrostatic pressure is directly proportional to the zone height. Therefore, the molten zone should be kept as narrow as possible. Frequency of the rf energy is also an important parameter. It is inversely proportional to the penetration depth of the rf energy and levitation force of the melt. Gupta and Gregory (97) have suggested that the optimum frequency is between 2 and 3 MHz. When the frequency is below 500 kHz, an undesirable surface melting can occur. On the other hand, a frequency higher than 3 MHz tends to increase the susceptibility for arcing.

The radial doping distributions in float-zone ingots are generally poor as compared to those in the Cz-ingots. This is due to the fact that the temperature gradients in the FZ-melts are very steep and the melt exhibits a severe turbulent flow. In addition to the parameters which affect thermal-driven convection currents in the Cz melt, the surface tension-driven convection current can play an important role in doping non-uniformity in the FZ-crystals. Convection arising from surface tension gradients due to the temperature gradient is known as the thermocapillary flow or Marangoni flow. The magnitude of the Marangoni flow is characterized by the Marangoni number which is shown in Equation (34). Chang and Wilcox (98) have shown that FZ melts of silicon exhibit significant thermocapillary flow and a uniform doping in crystals is difficult to obtain. The following will discuss an alternative doping technique which transmutes parts of silicon into phosphorus after the silicon crystals has been grown.

Transmutation Doping: Transmutation doping is transforming one element to another by means of nuclear reactions. The transmutation of silicon into phosphorus was first experimentally demonstrated by Tanenbuam and Mills in 1961 (99) using thermal neutrons to irradiate silicon in a nuclear reactor. Silicon crystals consist approximately 92.2% of Si^{28}, 4.7% of Si^{29} and 3.1% of Si^{30} isotopes. The isotope Si^{30} can capture one thermal neutron and become radioactive Si. Si then transmutes to P accompanied by the emission of a β particle. The half-life of the radioactive Si^{31} isotope is only 2.6 hours. The neutron irradiated silicon will become non-radioactive in several days if the original crystals have not been contaminated with other impurities. The silicon isotopes distribute uniformly in the melt-grown crystals. Therefore, transmutation doping can also result with a very uniform phosphorus distribution if the crystals are irradiated by a uniform thermal neutron flux.

The concentration of phosphorus donors (ΔN_D) generated by thermal neutron irradiation can be calculated from the equation (100)

$$\Delta N_D = \rho Si^{30} \phi t \sigma \tag{52}$$

where ρ is the atomic density of silicon, σ is the neutron capture cross section of Si^{30} (0.107 barns), ϕ is the thermal neutron flux, and t the irradiation time.

The crystal lattice of the as-irradiated silicon is highly damaged, and electrical resistivity of the silicon is therefore very high. Electrical conductivity and crystal perfection can be restored by thermal annealing. The required temperature and time for the annealing depend on the degree of damage, which in turn is determined by the ratio of thermal neutrons to the fast neutrons flux (the so-called cadmium ratio). Herzer (101) has shown that when the cadmium ratio was 1000, annealing needed only to be at 900°C for 4 hours, while when the ratio was less than 10, a 1200°C temperature for 4 hours was required.

The transmutation doping technique has been used commercially since the mid 1970's. The transmutation-doped silicon wafers are primarily used for the fabrication of thyristors, power transistors, and rectifiers. Recently, this technique has also been used for compensation overdoping of p-type silicon, and has produced very high resistivity (up to 10 k ohm-cm) silicon. High resistivity materials are required for fabrication of devices for detecting nuclear particles, electrons, and gamma-rays (102).

Crystallographic Defects in FZ Crystals: The dominant crystallographic imperfections in FZ crystals are swirl-type defects. The swirls are point defect clusters or agglomerates which distribute inhomogeneously with a swirl pattern in wafers. The swirls typically consist of two types of clusters which are referred to as A-defects and B-defects (103). Both types of swirls can be revealed by preferential chemical etching or by x-ray topographs taken from Cu decorated wafers. Defects revealed by these techniques can differentiate A-defects from B-defects by their size. A-defects are larger than B-defects, and their densities are typically in the ranges of 10^6 to $10^7 cm^{-3}$ and 10^7 to $10^8 cm^{-3}$ respectively.

A swirl formation model was first proposed by Plaskett (104) as the result of vacancy condensation. Later, work by deKock (103) concluded that the condensation of silicon interstitials is the cause of A-defect formations. TEM studies have found that A-defects can be composed of perfect dislocation loops, single loops, and loop clusters. B-defects are too small to be identified by TEM. However, it has been observed that they are associated with carbon since their densities increase with the carbon concentration in silicon.

It is undesirable for FZ crystals to contain swirl-type defects. Defect sites can easily collect metallic impurities which in turn can degrade the device performance. Several techniques have been developed to eliminate the swirls. An early technique is to introduce a few percent hydrogen into the growth ambient during crystal pulling (105),(106). The absorption of hydrogen into the silicon lattice can suppress the formation of A- and B-defects. Presence of hydrogen can embrittle silicon when the densities of hydrogen precipitates have exceeded 10^{10} cm^{-3}. The hot spots in high-power devices have been suggested to be the result of hydrogen precipitation. A recent approach to eliminate the swirls is to carry out the crystal growth with high pull rates, typically greater than 5 mm/min. The use of very low pull rates (<0.2 mm/min) can also avoid the formation of swirls. But this low pull rate is not feasible for crystal production.

5.0 TRENDS IN SILICON CRYSTAL GROWTH

Trends in silicon crystal growth technology are dictated by the evolution of semiconductor devices. The semiconductor devices are heading toward (a) higher operation speeds, (b) smaller sizes of each individual device, (c) a larger scale of integration (ie toward ULSI), and (d) lower cost of manufacturing. High speed operations of devices require silicon with low capacitance (ie high resistivity), although the speed can also be improved significantly by shrinkage of device dimensions. Resistivity in the range of 15 to 30 ohm-cm has been widely used for both MOS and Bipolar integrated circuits. These IC devices may not require wafer resistivity greater than 100 ohm-cm. This is because the advantage of capacitance reduction is traded off for adverse factors such as the increase in substrate currents (noise) resulting from higher minority carrier lifetimes in the extrinsic regions of the devices. However, high resistivity wafers will be continuously needed by power devices. Fabrications of these devices are primarily provided by the FZ crystals. Cz crystals have not been able to meet a high resistivity specification because of the presence of oxygen donors. Recent development of the transverse magnetic Cz technique may potentially produce wafers with stable resistivities of up to 200 ohm-cm.

Shrinkage of device dimensions requires lower leakage currents in the devices. Electrical leakages, particularly the junction leakages, are due to the existence of crystallographic defects in the active regions of devices. A trend in silicon wafer technology is to continuously reduce the defect densities in these areas. One of the current methods is to form a denuded zone based on controlled oxygen precipitation. Better control of oxygen precipitation will be needed for the fabrication of smaller devices. To meet this need, crystal growth techniques such as transverse MCz growth and oxygen doping in FZ growth have to be further developed.

Greater integration of devices requires smaller power dissipation per device. CMOS circuits consume the least amount of power and are the best candidate for ULSI. One of the disadvantages of the CMOS circuits is the ease in latchup of the adjacent devices through the formation of parasitic transistors. Latchup can be reduced or eliminated by using a lightly doped epitaxial layer on heavily-doped substrates. Demand for heavily-doped crystals will increase. However, research and development on heavily doped crystals has not been very active since these materials were not used for IC manufacturing. Quality improvement on heavily doped crystals is needed particularly when Sb is used as a dopant. Recently, it has been found that defect densities in the epitaxial layers are always high when a heavily doped Sb wafer is used as the substrate. Several causes have been suggested for the result of this finding. (a) Heavily-doped Sb crystals contain less oxygen, and, therefore, lack intrinsic gettering capability. (b) Sb dopant is less pure than other dopants so that the greater amount of metallic impurities can diffuse to the epitaxial layer to form epitaxial haze (i.e. shallow etch pits). Further understanding of heavily doped Sb crystals is needed in order to solve this problem.

Lowering of the manufacturing cost of devices requires

use of larger diameter silicon wafers. Although the cost saved due to the increase in diameter from 6" to 8" is not as dramatic as from 2" to 3", the trend still heads for larger diameter wafers. Problems arising from large diameter wafers are (a) increase of wafer warpage, and (b) increase of wafer batch weight, which can exceed the yield point of quartz tubes used for high temperature processing of silicon wafers. The first problem can be solved if the hardness of wafer is increased. The solution to the latter problem is to reduce the thickness of the wafers. This in turn also requires improvement of wafer hardness to prevent bending and warpage. The improvement of silicon hardness has been unintentionally carried out by incorporation of oxygen into silicon lattice during crystal growth. However, oxygen can easily form SiO_2 precipitates which, contrary to oxygen in a solid solution, can enhance the wafer warpage. Therefore, we need a reliable wafer hardening technique. Nitrogen dopings in FZ crystals have been considered as an alternative to oxygen in Cz crystals. Further development is needed in this area.

The size of crystals will continue to increase in the next few years. The growth of dislocation-free Cz ingots of up to 10" in diameter can be obtained by the extension of current growth technology. However, the growth processes require further automation which may include control of ingot diameter, dislocation-free structures, and oxygen and dopant concentration targets. The batch growth process should also be converted to a continuous process in order to reduce the cost as well as improve axial resistivity uniformity.

6.0 SUMMARY AND CONCLUSION

This chapter has covered both the theoretical as well as practical aspects of growing silicon single-crystal from the melt. The melt growth is essentially a process of liquid-to-solid phase transition with stringent control of mass and heat transfer. The heat transfer behavior can influence the shape of grown crystals and affect distribution of host and impurity atoms in crystal lattice. Achievable crystal perfections are limited by the thermodynamic rule. The formation of line defects such as grain boundaries, twinings, and dislocations would increase the free energy of the crystals. Therefore, they are not equilibrium defects and can be eliminated. Techniques used for eliminating line defects have been reviewed.

Contrary to line defects, formation of point defects would not necessarily increase the free energy of the crystals. Point defects such as vacancies, self-interstitials, and impurities are thermodynamically favorable and cannot be easily eliminated. These point defects are very dynamic and will interact with each other at high temperatures. The results lead to the formation of larger defects such as dislocation loops, impurity clusters, stacking faults, or second-phase precipitates. The fast pull technique which provides greater deviations from equilibrium conditions is an example in FZ growth which suppresses the formation of these point defects.

Silicon melts used for the growth of extrinsic semiconductor crystals are doped intentionally with a dopant and unintentionally (in case of Cz growth) with other impurities. Melt growth of a silicon crystal involves the mass transport of multicomponent elements from liquid to solid. Theories for the solute distributions in the crystals have been reviewed. The factors contributing to microscopic inhomogeneities are also given. Techniques to improve the axial doping homogeneity include the double crucible method and continuous crystal growth. The most effective method is applying a transverse magnetic field into the Cz melt and transmutation doping in FZ crystals.

Single crystal silicon will remain the most important material for the majority of semiconductor device applications. The ingot growth techniques which include the Cz and FZ methods have been well established. We do not foresee any difficulty for this technique to grow the even larger diameter crystals which will be needed in future. However, the quality of crystals grown by current techniques is by no means certain for meeting future device requirements. Continued research and development are needed to focus on the understanding of point defect interactions, and then tailor the growth conditions to produce crystals in which the unfavorable interactions of point defects can be avoided.

ACKNOWLEDGMENTS

Author would like to thank Howard Norman and Clarence Tracy for their reading of the manuscript.

REFERENCES

1. Teal, G.K. and Buehler, E., Growth of Silicon Single Crystals and Single Crystal Silicon p-n Junctions. *Phys. Rev.* 87:190 (1952)

2. Dash, W.C., The Growth of Silicon Crystals Free from Dislocations, in *Growth and Perfection of Crystals* (R.H. Doremus, B.W. Roberts, and D. Turnbull, eds.) pp. 361-382, Wiley, New York (1958)

3. Abe, T., Masui, T. and Harada, H., The Characteristics of Nitrogen in Silicon Crystals. *The 16th Conf. on Solid State Devices and Materials*, Late News Abstrates, pp. 60-61 (1984)

4. Liaw, H.M. and Secco d'Aragona, F., Purification of Metallurgical-grade Silicon by Slagging and Impurity Redistribution. *Solar Cells* 10:109-118 (1983)

5. Petritz, R.L., Contribution of Materials Technology to Semiconductor Devices. *Proc. IRE* 50:1025-1038 (1962)

6. Steinbech, H.H., Impact of Device Leakage by Different Wafer Materials. *The Electrochem Soc. Fall Meeting*, Extended Abstracts, pp. 1325-1326 Hollywood, Florida (1980)

7. Sangster, R.C., Maverick, E.F. and Croutch, M.L., Growth of Silicon Crystals by a Vapor Phase Pyrolytic Deposition Method. *J. Electrochem. Soc.* 104:317-319 (1957)

8. Sandmann, H., Comparison of Three Single Crystal Techniques, in: *Semiconductor Silicon*, (R.R. Haberecht and E. Kern, eds.), pp. 124-131, The Electrochem. Society, Princeton, N.Y. (1969)

9. Seeger, A., Point Defects in Metals. in: *Theory of Crystal Defects*, (B. Gruber, ed.) pp. 37-55, Academic Press, New York (1966)

10. Runyan, W.R., *Silicon Semiconductor Technology*, McGraw-Hill, New York (1965)

11. Reed, T.B., Heat Flow in High-Temperature Crystal Growth, in: *Crystal Growth*, (H.S. Peiser, ed.), pp. 39-43, Pergamon, Oxford (1967)

12. Arizumi, T. and Kobayashi, N., The Solid-Liquid Interface Shape during Crystal Growth by the Czochralski Method, *Japan J. Appl. Phys.* 8:1091-1097 (1969)

13. Kobayashi, N. and Arizumi, T., The Numerical Analyses of the Solid-Liquid Interface Shape during Crystal Growth by the Czochralski Method. *Japan J. Appl. Phys.* 9:361-367 (1970)

14. Kobayashi, N. and Arizumi, T., The Numerical Analyses of the Solid-Liquid Interface Shape during Crystal Growth by the Czochralski Method. part II. Effect of the Crucible Rotation. *Japan J. Appl. Phys.* 9:1255-1259 (1970)

15. Arizumi, T. and Kobayashi, N., Theoretical studies of the Temperature Distribution in a Crystal being Grown by the Czochralski Method. *J. Crystal Growth* 13/14:615-618 (1972)

16. Kuo, V.H.S. and Wilcox, W.R., Influence of Crystal Dimensions on the Interfacial Temperature Gradient. *J. Crystal Growth* 12:191-194 (1972)

17. Srivastava, R.K., Ramachandran, P.A. and Dudukovic, M.P., Interface Shape in Czochralski Grown Crystals: Effect of Conduction and Radiation. *J. Crystal Growth* 41:487-504 (1985)

18. Liaw, H.M., Interface Shape and Radial Distribution of Impurities in <111> Silicon. *Crystal Growth* 67:261-270 (1984)

19. Abe, T., The Growth of Si Single Crystals from the Melt and Impurity Incorporation Mechanisms. *J. Crystal Growth* 24/25:463-467 (1974)

20. Chikawa, J. and Sato, F., In Situ X-ray Study of Dislocations in Silicon Crystal Growing from the Melt. *Inst. Phys. Conf. Ser.* NO 59, ICDRES-11:95-109 (1980)

21. Jackson, K.A., in: *Crystal Growth and Characterization*, (R. Ueda and J. B. Mullin, eds.), pp. 21-32, North-Holland, Amsterdam (1975)

22. Thurmond, C.D. and Struthers, J.D., Equilibrium Thermochemistry of Solid and Liquid Alloys of Germanium and Silicon, II. The Retrograde Solid Solubilities of Sb in Ge, Cu in Ge and Cu in Si, *J. Phys. Chem.* 57:831-835 (1953)

23. Weiser, K., Theoretical Calculation of Distribution Coefficients of Impurities in Germanium and Silicon, Heat of Solution. *J. Phys. Chem. Solids* 7: 118-126 (1958)

24. Jaccodine, R.J. and Pearce, C.W., The Segregation Coefficient of Oxygen in Silicon, in: *Defects in Silicon* (W.M. Bullis, L.C. Kimerling, eds.), pp. 115-119, The Electrochem. Society, Pennington, N.J. (1983)

25. Fischler, S., Correlation between Maximum Solid Solubility and Distribution Coefficient for Impurities in Ge and Si, *J. Appl. Phys.* 33:1615-1615 (1962)

26. Trumbore, F.A., Solid Solubilities of Impurities in Germanium and Silicon. *Bell System Tech. Journal* 39:205-233 (1960)

27. Burton, J.A., Kolb, E.D., Slichter, W.P. and Struthers, J.D., The Distribution of Solute in Crystals Grown from the Melt. Part II. Experimental, *J. Chem. Phys.* 21:1991-1996 (1953)

28. Hirata, H. and Inoue, N., Macroscopic Axial Doping Distribution in Czochralski Silicon Crystal Grown in a Vertical Magnetic Field, *Japan J. Appl. Phys.* 24:1399-1403 (1985)

29. Kodera, H., Diffusion Coefficients of Impurities in Silicon Melt, *Japan J. Appl. Phys.* 2:212-219 (1963)

30. Morizane, K., Witt, A.F. and Gatos, H.C., Impurity Distribution in Single Crystals, III. Impurity Heterogeneities in Single Crystals Rotated during Pulling from the Melt, *J. Electrochem. Soc.* 114:738-742 (1967)

31. Witt, A.F., Lichtensteiger, M. and Gatos, H.C., Experimental Approach to the Quantitative Determination of Dopant Segregation during Crystal Growth on a Microscale: Ga Doped Ge, *J. Electrochem. Soc.* 120:1119-1123, (1973)

32. Carruthers, J.R., Witt, A.F. and Reusser, R.E., Czochralski Growth of Large Diameter Silicon Crystals - Convection and Segregation, in: *Semiconductor Silicon*, H.R. Huff and E. Sirtl, eds.), pp. 61-71, The Electrochem. Society, Princeton (1977)

33. Moody, J.W., Oxygen in Czochralski Crystals and Melts- A Review, in: *Semiconductor Silicon*, (H.R. Huff, T. Abe, and B. Kolbesen, eds.) pp. 100-116, The Electrochem. Society. Pennington, N.J. (1986)

34. Carruthers, J.R. and Nassau, K., Non Mixing Cells Due to Crucible Rotation during Czochralski Crystal Growth, *J. Appl. Phys.* 39:5205-5214 (1968)

35. Suzuki, T., Isawa, N., Okubo, Y. and Hoshi, K., Cz Silicon Crystals Grown in a Tranverse Magnetic Field, in: *Semiconductor Silicon*, (H.R. Huff, R.J. Kriegler, and Y. Takeishi, eds.), pp. 90-100, The Electrochem. Society, Pennington, N.J. (1981)

36. Bradshaw, S.E. and Mlavsky, A.I., The Evaporation of Impurities from Silicon, *J. Electronics and Control* 2:134-144 (1956)

37. Tiller, W.A., Principles of Solidification, in: *The Art and Science of Growing Crystals* (J.J. Gilman, ed.), pp. 276-312, John Wiley & Sons, New York (1963)

38. Hopkins, R.H., Seidensticker, R.G., Davis, J.R., Rai-Choudhury, P., Blais, P.D. and McCormick, J.R., Crystal Growth Considerations in the Use of Solar Grade Silicon, *J. Crystal Growth* 42:493-498 (1977)

39. Watanabe, M., Usami, T., Muraoka, H., Matsuo, S., Imanishi, Y. and Nagashima, H., Oxygen-free Silicon Single Crystal Grown from Silicon Nitride Crucible. in: *Semiconductor Silicon*, (H.R. Huff, R.J. Kriegler, and Y. Takeishi, eds.), pp. 126-137, *The Electrochem. Society*, Pennington, N.J. (1981)

40. Dash, W.C., Silicon Crystals Free of Dislocations, *J. Appl. Phys.* 29:736-737 (1958)

41. Patzner, E.J., Dessauer, R.G. and Poponiak, M.R., Automatic Diameter Control of Czochralski Crystals. *Solid State Tech.* 10(10):25-30 (1967)

42. Kim, K.M., Kran, A., Riedling, K., and Smetana, P., Digital Control of Czochralski Silicon Crystal Growth, *Solid State Tech.* 28(1):165-168 (1985)

43. Zinnes, A.E., Nevis, B.E. and Brandle, C.D., Automatic Diameter Control of Czochralski Grown Crystals, *J. Crystal Growth* 19:187-192 (1973)

44. Bardsley, W., Green, G.W., Holiday, C.H. and Hurle, D.T.J., Automatic Control of Czochralski Crystal Growth, *J. Crystal Growth* 16:277-279 (1972)

45. Gartner, K.J., Rittinghaus, K.F., Seeger, A. and Uelhoff, W., An Electronic Device Including a TV-System for Controlling the Crystal Diameter during Czochralski Growth, *J. Crystal Growth* 13/14:619-623 (1972)

46. O'Kane, D.F., Kwap, T.W., Gulitz, L. and Bednowitz, A.L., Infrared TV System of Computer Controlled Czochralski Crystal Growth, *J. Crystal Growth* 13/14:624-628 (1972)

47. Van Dijk, H.J.A., Jochem, C.M.G., Scholl, G.J. and van der Werf, P., Diameter Control of LEC Grown GaP Crystals, *J. Crystal Growth* 21:310-312 (1974)

48. Geil, W., Malitzki, H. and Tanzer, D., The IR Television Scanning Technique as Temperature Field Control for Growing Silicon Crystals, *Crystal Res. & Technol.* 17:723--728 (1982)

49. Hurle, D.T.J., Control of Diameter in Czochralski and Related Crystal Growth Techniques, *J. Crystal Growth* 42:473-482 (1977)

50. Digges, Jr., T.G., Hopkins, R.H. and Seidensticker, R.G., The Basis of Automatic Diameter Control Utilizing "Bright Ring" Meniscus Reflections, *J. Crystal Growth* 29:326-328 (1975)

51. Wolfstirn, K.B., Hole and Electron Mobilities in Doped Silicon from Radiochemical and Conductivity Measurements, *J. Phys. Chem. Solids* 16:279-284 (1960)

52. Irvin, J.C., Resistivity of Bulk Silicon and of Diffused Layer in Silicon, *The Bell System Tech. Journal* XLI:387-410 (1962)

53. Morizane, K., Witt, A.F., and Gatos, H.C., Impurity Distribution in Single Crystals, IV. Growth Characteristics and Impurity Incorporation during Facet Growth, *J. Electrochem. Soc.* 115:747-749 (1968)

54. Dikhoff, J.A.M., Inhomogeneities on Doped Germanium and Silicon Crystals, *Philips Technical Review* 25:195-206 (1963/64)

55. Benson, K.E., Growth of Silicon Crystals. in: *Semiconductor Silicon*, (R. R. Haberecht and E. Kern, eds.), pp. 97-123, The Electrochem. Society, Princeton, NJ (1969)

56. Benson, K.E., Kin, W., Martin, E.P., Fundamental Aspects of Czochralski Silicon Crystal Growth for VLSI. in: *Semiconductor Silicon*, (H.R. Huff, R.J. Kriegler, and Y. Takeishi, eds.), pp. 33-48, The Electrochem. Society, Pennington, NJ (1981)

57. Lin, W., Hill, D.W., Large-Diameter Czochralski Silicon Crystal Growth, in: *Silicon Processing*, ASTM STP 804, (D.C. Gupta, ed.) pp. 24-39, American Soceity for Testing and Materials (1983)

58. Kubaschewski, O. and Chart, T.G., Silicon Monooxide Pressure due to the Reaction between Solid Silicon and Silica, *J. Chem. Thermodynamics* 6:467-476 (1974)

59. Kaiser, W., Keck, P.H. and Lange, C.F., Infrared Absorption and Oxygen Content in Silicon and Germanium, *Phys. Rev.* 101:1264-1268 (1956)

60. Kaiser, W. and Keck, P.H., Oxygen Content of Silicon Single Crystals, *J. Appl. Phys.* 28:882-887 (1957)

61. Yatsurugi, Y., Akiyama, N., Endo, Y. and Nozaki, T., Concentration, Solubility, and Equilibrium Distribution

Coefficient of Nitrogen and Oxygen in Semiconductor Silicon, *J. Electrochem. Soc.* 120:975-979 (1973)

62. Leroy, B. and Plougonven, C., Warpage of Silicon Wafers, *J. Electrochem. Soc.* 127:961-970 (1980)

63. Secco d'Aragona, F., Tsui, R.K., Liaw, H.M. and Fejes, P.L., Thermal Annealing of Silicon Wafers for Intrinsic Gettering, in: *Defects in Silicon* (W.M. Bullis, and L.C. Kimerling, eds.), pp. 166-179, The Electrochem. Society, Pennington, NJ (1983)

64. Bullis, W.M., Watanabe, M., Badhdadi, A., Li, Y.Z., Scace, R.I., Series, R.W., and Stallhofer, P., Calibration of Infrared Absorption Measurements of Interstitial Oxygen Concentration in Silicon, *The Electrochem. Soc. Meeting*, Extended Abstract, 86-1:pp. 196-197 (1986)

65. Walitzki, H., Rath, H.-J., Reffle, J., Pahlke, S. and Blatte, M., Control of Oxygen and Precipitation Behavior of Heavily Doped Silicon Substrate Materials, in: *Semiconductor Silicon*, (H.R. Huff, T. Abe, and B. Kobesen, eds.), pp. 86-99, The Electrochem. Society, Pennington, NJ (1986)

66. Tsuya, H., Kanamori, M., Takeda, M. and Yasuda, K., Infrared Optical Measurement of Interstitial Oxygen Content in Heavily Doped Silicon Crystals, in: *VLSI Science and Technology* (W.M. Bullis, and S. Broydo eds), pp. 517-525 (1985)

67. Nozaki, T., Yatsurugi, Y., Akiyama, N., Concentration and Behavior of Carbon in Semiconductor Silicon, *J. Electrochem. Soc.* 117:1566-1568 (1970)

68. Liaw, H.M., Oxygen and Carbon in Czochralski-Grown Silicon, *Microelectronics Journal* 12:33-36 (1981)

69. Series, R.W. and Barraclough, K.G., Carbon Contamination during Growth of Czochralski Silicon, *J. Crystal Growth* 60:212-218 (1982)

70. Leroueille, J., Influence of Carbon on Oxygen Behavior in Silicon, *Phys. Stat. Sol.* (a) 67:177-181 (1981)

71. Hoshikawa, K., Hirata, H., Nakanishi, H. and Ikuta, K., Control of Oxygen Concentration in Cz Silicon Growth, in: *Semiconductor Silicon* (H.R. Huff, R.J. Kriegler, and Y. Takeishi, eds.), pp. 101-112, The Electrochem. Society, Pennington, NJ (1981)

72. Carlberg, T., King, B. and Witt, A.F., Dynamic Oxygen Equilibrium in Silicon Melts during Crystal Growth by the Czochralski Technique, *J. Electrochem. Soc.* 129:189--193 (1982)

73. Watanabe, M., Usami, T.,Takasu S., Matsuo, S. and Toji, E., Controlled Oxygen Doping in Silicon, *Japan J. Appl. Phys.* 22:Supplement 22-1, pp. 185-189 (1982)

74. Hoshikawa, K., Kohda, H., Hirata, H. and Nakanishi, H., Low Oxygen Content Czochralski Silicon Crystal Growth, *Japan J. Appl. Phys.* 19:L33-L36 (1980)

75. Secco d'Aragona, F., *U.S. Patent* 4,545,849; assigned to Motorola Inc. (1985)

76. Liaw, H.M., *U.S. Patent* 4,394,532; assigned to Motorola Inc. (1983)

77. Lane, R.L. and Kachare, A.H., Multiple Czochralski Growth of Silicon Crystals from a Single Crucible, *J. Crystal Growth* 50:437-444 (1980)

78. Fiegl, G., Recent Advances and Future Directions in Cz-Silicon Crystal Growth Technology, *Solid State Technology* 26(8);121-131 (1983)

79. Bonora, A.C., Silicon Crystal Growth and Processing Technology: A Review, in: *Silicon Processing*, ASTM STP 804 (D.C. Gupta, ed.) pp. 5-15, Am. Soc. for Testings and Materials (1983)

80. Allgaier, R.S., Interpretation of Transport Measurements in Electrically Conducting Liquids, *Phys. Rev.* 185:227-244 (1969)

81. Ohwa, M., Higuchi, T., Toji, E., Watanabe, M., Homma, K. and Takasu, S., Growth of Large Diameter Silicon Single Crystal under Horizontal or Vertical Magnetic

Field, in: *Semiconductor Silicon* (H.R. Huff, T. Abe, and B. Kolbesen, eds.) pp. 117-128, The Electrochem. Society, Pennington, NJ (1986)

82. Barroclough, K.C., Series, R.W., Bae, G.J., and Kemp, D.S., Axial Magnetic Czochralski Silicon Growth, in: *Semiconductor Silicon* (H.R. Huff, T. Abe, and B. Kolbesen, eds.) pp. 129-141, The Electrochem. Society, Pennington, NJ (1986)

83. Suzuki, T., Isawa, N., Hoshi, K., Kato, Y. and Okubo, Y., MCz Silicon Crystals Grown at High Pulling Rates, in: *Semiconductor Silicon* (H.R. Huff, T. Abe, and B. Kolbesen, eds.) pp. 142-152, The Electrochem. Society. Pennington, NJ (1986)

84. Kuroda, E., Matsubara, S., Saitoh, T., Czochralski Growth of Square Silicon Single Crystals, *Japan J. Appl. Phys.* 19:L361-L364 (1980)

85. Liaw, H.M., Growth of Single Crystal Silicon Square Ingots, *The Eelectrochem. Soc. Meeting*, Extended Abstract, 80-1:806-807 (1980)

86. Bennett, A.I. and Logini, R.L., Dendritic Growth of Germanium Crystals, *Phy. Rev.* 116:53-61 (1959)

87. LaBelle, H.E., Jr. and Mlavsky, A.I., Growth of Controlled Profile Crystals from the Melts: part I- Sapphire Filaments, *Mater. Res. Bull.* 6:571-580 (1971)

88. LaBelle, H.E., Jr., Growth of Controlled Profile Crystals from the Melts: part II-Edge-Defined, Film-Fed Growth (EFG), *Mater. Res. Bull.* 6:581-590 (1971)

89. Ciszek, T.F., Edge-Defined, Film-feed Growth (EFG) of Silicon Ribbon, *Mat. Res. Bull.* 7:731-737 (1972)

90. Seidensticker, R.G., Dendritic Web Silicon for Solar Cell Application, *J. Crystal Growth* 39:17-22 (1977)

91. Faust, J.W., Jr. and John, H.F., Germanium Dendrite Studies, I. Studies of Twin Structures and the Seeding Mechanism, *J. Electrochem. Soc.* 108:855-860 (1961)

92. O'Hara, S. and Bennett, A.I., Web Growth of Semiconductors, *J. Appl. Phys.* 35:686-693 (1964)

93. Hamilton, D.R. and Seidensticker, R.G., Propagation Mechanism of Germanium Dendrites, *J. Appl. Phys.* 31:1165-1168 (1960)

94. Tucker, T.N., and Schwuttke, G.H., Growth of Dislocation-Free Silicon Web Crystals, *Appl. Phys. Lett.* 9:219-221 (1966)

95. Ravi, K.V., The Growth of EFG Silicon Ribbons, *J. Crystal Growth* 39:1-16 (1977)

96. Keck, P.H., Golay, M.J.E., Crystallization of Silicon from a Floating Liquid Zone, *Phys. Rev.* 89:1297-1297 (1953)

97. Gupta, K.P., and Gregory, R.O., Dependence of Silicon Float-Zone Refining Parameters on Frequency, in: *Silicon Processing*, ASTM STP 804 (D.C. Gupta ed.), American Society for Testing and Materials, pp. 50-61 (1983)

98. Chang, C.E. and Wilcox, W.R., Inhomogeneities due to Thermacapilary Flow in Floating Zone Melting, *J. Crystal Growth* 28:8-12 (1975)

99. Tanenbaum, M. and Mills, A.D., Preparation of Uniform Resistivity in n-type Silicon by Nuclear Tranmutation, *J. Electrochem. Soc.* 108:171-176 (1961)

100. Liaw, H.M. and Varker, C.J., Phosphorus Donor Homogenization of Czochralski Crystals by Thermal Neutron Irradiation Overdoping, in: *Semiconductor Silicon* (H.R. Huff and E. Sirtl, eds.), pp. 116-125, The Electrochem. Society, Princeton, NJ (1977)

101. Herzer, H., Neutron Transmutation Doping, in: *Semiconductor Silicon* (H.R. Huff and E. Sirtl, eds.), pp. 106-115, The Electrochem. Society, Princeton, NJ (1977)

102. Ammon, W.V., and Kemmer, J., Production of Detection-Grade Silicon by Neutron Transmutation, in: *Semiconductor Processing*, ASTM STP 850 (D.C. Gupta ed.), American Society for Testing and Materials, pp. 546-557 (1984)

103. de Kock, A.J.R., Vacancy Clusters in Dislocation-Free Silicon, *Appl. Phys. Lett.* 16:100-102 (1970)

104. Plaskett, T.S., Evidence of Vacancy Clusters in Dislocation-Free Float-Zone Silicon, *Trans. AIME* 233:809-812 (1965)

105. de Kock, A.J.R., The Elimination of Vacancy-cluster Formation in Dislocation-Free Silicon Crystals, *J. Electrochem. Soc.* 118:1851-1856 (1971)

106. de Kock, A.J.R., Vacancy Clusters in Dislocation-Free Silicon and Germanium, *Philips Tech. Rev.* 34:244-254 (1974)

4

Silicon Wafer Preparation

Richard L. Lane

1.0 INTRODUCTION

Integrated circuits and discrete solid state devices are manufactured on wafers made of silicon single crystal material, using a very complex series of processing steps. In order to obtain high yields and good device performance, it is important that the starting wafers be of reproducibly high quality because high resolution patterns are optically formed on the wafer, and the front surface must be smooth and flat on both a macro- and a microscale. The electrical and chemical properties of the wafer surface must also be well controlled and therefore the preparation of starting silicon wafers is a crucial portion of IC and device manufacture. This discussion will focus primarily on technical principles and theories that are applicable to silicon wafer preparation, and the practical application of those theories. Details, such as formulas, recipes, and specific process parameters are not given, because they vary considerably between different producers; likewise, specific equipment is not recommended, because the proper choice depends upon factors which are outside the scope of this discussion, such as plant capacity, labor costs, product specifications, and budget restrictions. For current equipment information, the reader is referred to the several directories (1-4).

Wafer preparation is the conversion of grown crystal

ingots into wafers which are ready for the wafer fab line. Each wafer must have at least one surface which is clean, flat and damage-free. In principle, a suitable wafer could be produced by cutting a crystal into thin slices and polishing one side of the slice until all of the saw marks are removed and the surface appears smooth and glossy. For some non-critical applications, such a wafer might be of sufficient quality. However, today's device and circuit designs are often more complex, and as circuit pattern geometries become smaller and wafer dimensions larger, the demands on wafer preparation become more stringent.

Additional preparation steps along with tighter process controls are required in order to achieve the required wafer quality.

Thus, silicon wafer preparation usually encompasses an extensive sequence of steps, starting with the as-grown crystal and ending with the polished wafer, ready for the fab line.

1.1 Wafer Preparation Processes

Figure 1 is a typical process flow diagram for wafer preparation. The three major categories, i.e. crystal shaping, wafer shaping, and wafer finishing are each represented by a series of process steps, some of which may be optional as indicated, depending upon the final wafer application. The right hand column includes some typical in-process control measurements that may be performed as the wafer preparation processes proceed.

These measurement techniques are discussed in the later section on in-process measurements. A brief statements of the purpose of each of the separate process steps follows below:

* Crystal cropping — Removal of the seed and tang ends of the crystal and any out-of-spec portions, using a circular diamond saw.

* Crystal grinding — The diameter of the crystal is reduced to the specified value and tolerance, usually with a diamond grinding wheel.

* Flat grinding — Flat areas are ground lengthwise along the crystal which serve later as identification of the wafer type and orientation.

* Crystal etching — The crystal is immersed in an etch bath which removes most of the surface getting from the grinding operation.

* Wafering — The crystal is cut into thin wafers, using a machine with an annular diamond blade (ID blade).

* Heat treatment — Crystals or sawn wafers are heat treated to eliminate oxygen donors, thereby normalizing resistivity.

* Edge Rounding — The as-sawn wafers are ground on the periphery with a diamond form wheel to remove the square corners and to produce the desired edge contour.

* Wafer Lapping — The thickness uniformity and flatness of the wafers are improved by the use of an abrasive slurry on a large flat lapping plate. Lapping may be either single or double sided.

* Wafer grinding — An alternative to lapping, using a diamond wheel instead of a lapping slurry. Commercial wafer grinders grind only one side of the wafer at a time.

* Wafer etching — Chemical etchants are used to remove damaged surface layers of the wafer from the sawing, lapping and edge profiling operations.

* Polishing — The matte surface of the lapped or sawn wafer is converted to a damage-free, specular surface. Polishing may be either single or double sided.

* Back side damage — A controlled amount of surface damage is applied to the back side of the wafer for gettering purposes.

* Cleaning — Contaminants in the form of particulates, organics, and inorganics are removed from the wafer surface by a series of chemical and mechanical cleaning operations, preparing the wafer for the device fab line.

* Marking — Wafers are marked individually for the purpose of identification and traceablility.

Silicon Wafer Preparation   195

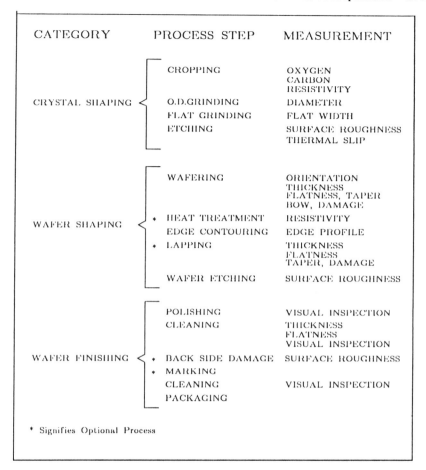

Figure 1:   Process Flow Diagram for Silicon Wafer Preparation

It should be noted that there is no "standard" sequence of steps for wafer preparation. The steps used in a typical operation will depend upon the use to which the wafers will be put and upon the specific procedures and quality control used in each successive step. For example, in some wafer preparation facilities polishing is performed directly on the as-sawn and etched wafers. In other operations, lapping is performed directly after wafering. Whether to use lapping depends upon (i) the final specification for the wafer (such as flatness) and (ii) the quality of the slicing operation.

Thus when lapping is performed after sawing, surface quality and thickness uniformity from the sawing operation may be relaxed somewhat, and the demands on the subsequent polishing process may also be less critical.

1.2 Silicon Removal Principles

The key processes listed above require the removal of silicon material to produce wafers of precise dimensions, and therefore wafer preparation is sometimes called "wafer shaping" in the industry. Material is removed from the crystal or the wafer by various removal processes, including mechanical, chemical, or a combination of mechanical and chemical means. Mechanical removal processes are sawing, lapping, and grinding using abrasives such as diamond, silicon carbide or aluminum oxide. These are frequently referred to as abrasive machining operations. Etching is a chemical removal process, whereas polishing combines both chemical and mechanical action to remove silicon.

According to Gielisse and Stanislao (5), material can be removed from a surface by one of two ways; (a) on a macroscale where the removed particles are much larger than atomic or molecular dimensions and (b) on a microscale in which the material is removed atom by atom or molecule by molecule. The distinction between macro- and microscale removal leads to the two fundamental types of removal processes: grinding and polishing. Macroscale removal requires penetration of the surface with an abrasive grit which produces localized forces in excess of the yield strength of the material. With grinding, the abrasive must be harder than the material being removed and the localized pressures are high in order to obtain penetration. Microscale removal leads to a very smooth surface and is called polishing. Polishing does not require such penetration and is by definition a low pressure process. The polishing agent is nearly always of lower hardness than the material being polished. Note: In this chapter the term "abrasive" is meant to be the particulate material which is used in a purely mechanical macroscale removal operation such as slicing, lapping or grinding. In the polishing operation, the term "polishing agent" is used rather than "abrasive" to distinguish between the two and to signify that polishing is more complex than a simple mechanical removal process. The differences between abrasive removal and removal by polishing are vividly evidenced by the roughness of the resulting surface, and the remaining lattice damage to be discussed later.

1.2.1 Mechanical Removal: At this point we limit the discussion to the abrasive machining of silicon. Unlike metals, plastics, and some amorphous materials, there is no plastic flow associated with abrasive machining of silicon. Silicon is a hard, brittle substance, and the penetration of abrasive particles establishes a field of damage in the form of cracks extending into the material from the point of penetration Figure 2. Intersection of the cracks leads to material removal by release of particles (6). Abrasive machining necessarily produces a rough surface and leaves sub-surface damage consisting of microcracks, dislocations and stresses. The magnitude of the roughness and the damage is directly related to the abrasive grit size. The speed of material removal is also related to grit size, (large grits remove material faster) and therefore as the wafer approaches final dimension, finer grit size may be used to more precisely control removal and to minimize remaining subsurface damage.

There are two modes of mechanical removal used in machining silicon: (i) bonded abrasive machining and (ii) free abrasive machining. Bonded abrasives are used on silicon for cropping, wafering, crystal and wafer grinding, and edge contouring. Virtually all bonded abrasive processes for silicon utilize diamond particles in a metallic or resin matrix. The diamond tool (wheel or blade) is fed into the workpiece (the silicon crystal) and the diamond abrasive grains which are held tightly in the matrix of the tool are forced to penetrate into the workpiece. The abrasive moves at high speed and intermittently impacts the workpiece as it moves through it. The appearance of a surface produced by bonded abrasives is usually a directional scratch pattern, depending upon the relative motions of the tool and the workpiece.

In free abrasive removal, a suspended abrasive in the form of a slurry is fed between the workpiece and a tool which is usually made of a hard material such as cast iron. Pressure applied to the tool forces the abrasive into the workpiece. Relative motion between the workpiece and the tool creates a rolling motion to the abrasive and a crushing action on the workpiece. The resulting surface has a uniformly matte character.

In silicon wafer preparation, it is conventional to use the term "grinding" when referring to the shaping of wafer surfaces with fixed abrasives, and "lapping" when indicating the free abrasive method, whereas in other industries, namely glass lens shaping, free abrasive removal is often called grinding.

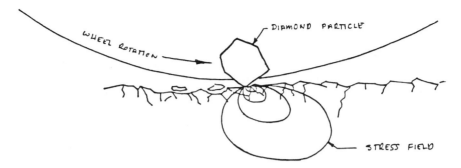

Figure 2: Abrasive Removal of Silicon

1.2.2 Chemical Removal: Etching is a chemical removal process whose principal application is to remove damaged surface material and the resulting stress caused by abrasive achining operations. The residual sub-surface damage in the form of dislocations and microcracks can propogate further into the bulk material if the wafer is placed under mechanical or thermal stress. Extensive propogation of cracks of course leads to fracture of the wafer, whereas extensive propogation of dislocations is called slip. Removal of the damaged layer by chemical means results in a stronger wafer which is resistant to chipping, breakage and slip from thermal stress.

Etching solutions either dissolve the silicon or react with it to form a soluble silicon compound. Etching is beneficial after any abrasive operation, however it is most common after crystal grinding, slicing, and lapping. Etching after lapping speeds the polishing process by removing material that would otherwise have to be removed by polishing. It also removes sawing or lapping damage from the back side of the wafer. If a wafer is polished without etching, the stress associated with the back side damage remains. When the stress is only removed on one side, the wafer may warp or bow.

Etching rate is a sensitive function of temperature, concentration and agitation of the etch solution. As an etchant is used, its effectiveness diminishes. These variables combine to make etching a very difficult process to control. Most etchants tend to attack the silicon preferentially at the damaged locations where there is lattice strain, dislocations, or other discontinuities where the lattice is distorted and therefore more reactive, leaving surfaces that are not flat.

Some etchants, however are more selective than others. The use of an etchant of low selectivity produces a "chemically polished" surface with a shiny "orange peel" appearance. The SEM photomicrographs of Figure 3 illustrate such a surface as it is progressively etched. The as-sawn surface of the wafer is shown in Figure 3a in which the rough surface and oriented damage pattern from the diamond abrasive is evident. The wafer was step-etched for various times to remove damaged surface material to the depths indicated in Figure 3b. The removal of damage is indicated by the decrease in the number of etch pits as material is removed. The final picture after 48 micrometers of silicon have been removed is a typical chemically polished surface.

Etchants for silicon can be either acids or bases. Acid etchants are usually mixtures of various ratios of concentrated nitric, hydrofluoric, and acetic acid. The nitric acid initiates the reaction by forming a layer of silicon dioxide on the surface which is subsequently removed by the hydrofluoric acid. Acetic acid acts as a buffer to control the dissociation of the nitric acid. A common etchant for chemically polishing silicon is a mixture of 5 parts nitric, 1 part hydrofluoric, and 1 part acetic, by volume. When the wafers are immersed in this solution, usually in teflon carriers of 25 wafers, the reaction starts immediately, evolving a large amount of brown noxious, corrosive gases. As the reaction proceeds, the solution heats up, accelerating the etch rate. As more wafers are etched, the solution becomes depleted and the etching rate decreases. Eventually the acid is discarded and a new batch is prepared.

Disposal of both the evolved gases and the spent etchant is a serious problem. In a production facility, the evolved noxious gases must be removed from the work area but cannot be released into the atmosphere. Typically the fume hood used for acid etching must be connected to scrubbing and neutralizing equipment. Spent etchant is still highly acidic and requires complete neutralization before discharge into the sewer. Codes may vary in different localities, for example in the Silicon Valley area, the discharge of fluorides is restricted to 2-10 ppm., and therefore, simple neutralization will not be sufficient. In this case, fluoride must be precipitated to calcium fluoride, and the precipitated wastes hauled to an approved dump site (7). The proper handling of waste materials has become a major concern and proper adherance to federal, state and local codes is mandatory.

200 Handbook of Semiconductor Silicon Technology

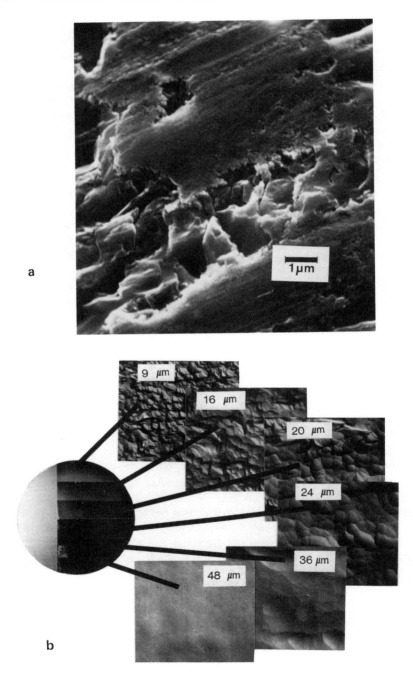

Figure 3: Photomicrographs of Etched and Unetched Silicon

The storage, handling and use of hydrofluoric acid also presents an important safety hazard because of the severe nature of the resulting burns. HF penetrates quickly and deeply into tissues, attacking the calcium and magnesium in bones, causing severe, slow-healing damage. If an HF burn is not quickly and properly treated, it can cause death.

Basic (or caustic) etchants have recently become more common in wafer preparation because of advantages in terms of cost, safety, and environmental problems. Potassium and sodium hydroxide are the most common caustic etchants, with potassium the preferred one. A 10 to 50% solution (by weight) of KOH in deionized water at 65-80 deg. C. removes silicon from the wafer surface at a reasonable rate. Because caustic etchants are more selective than acids, the resulting surface is not as smooth, however the damage and stress are removed as effectively as with acidic etching. Disposal of the spent caustic by-products is less of a problem than with acids.

1.2.3 Chemical-Mechanical Removal (Polishing): The polishing of silicon is in principle, similar to polishing other hard materials such as glass and ceramics. A polishing agent in a suspending fluid is applied to a soft pad such as felt or polyurethane and the workpiece is pressed against the rotating pad. A complete understanding of the polishing process has not been achieved, however it is generally agreed that it is not simply an extension of grinding or lapping with finer abrasives. There is ample evidence that polishing consists of interacting chemical and mechanical processes, and material removal rate is very slow compared to abrasive methods, because it is on an atomic or molecular scale. It is generally agreed however, that a substantial amount of material must be removed in order to obtain a satisfactory wafer surface. A discussion on polishing theory is presented below in the polishing section of this chapter.

2.0 CRYSTAL SHAPING

As-grown crystals have conical shaped ends. The seed end is quite short, perhaps with a height of one-fourth of the diameter, whereas the tang end of the crystal has a length approximately equal to the diameter. Automatic diameter control systems on crystal growth equipment are not capable of meeting the tight wafer diameter specifications that are required, so the crystals are usually grown one to two mm. oversize and reduced to the proper diameter by grinding. Finally, the crystal may be cut into shorter lengths before slicing. The operations required to prepare the crystal for the ID slicing operation are collectively referred to as crystal shaping Figure 4.

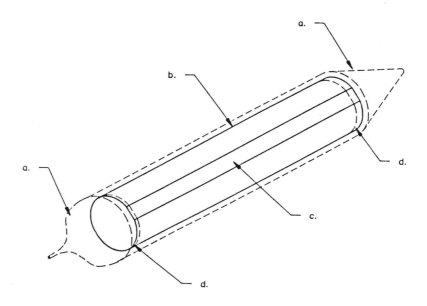

Figure 4: Crystal shaping operations.

(a) Remove crown and taper,
(b) Grind diameter,
(c) Grind flats,
(d) Slice samples for measurements.

2.1 Cropping

The first step in preparing the crystal for wafering is to remove the ends using a circular diamond saw. Another use for crystal cropping is to obtain samples for oxygen and carbon measurements. A final reason for cropping is to cut crystals to a suitable length to fit the saw capacity, to select portions of the crystal which meet desired resistivity specifications, or to remove portions of the crystal which do not meet specifications.

Commercial cropping equipment ranges from simple hand-operated cut-off machines to those which hydraulically clamp the crystal and move the blade through the crystal. The blade diameter must be sufficiently large to cut completely through the crystal in one pass in order to obtain a smooth, straight cut. Therefore the blade diameter must be at least equal to two times the crystal diameter plus the blade hub diameter. For today's large crystals, the equipment is large and is driven by powerful motors and hydraulic systems. Copious amounts of water coolant must be used to prevent damage to the crystal or to the blade from heat. A typical diamond blade is a 100 grit metal bonded (sintered) OD blade, 0.080 in (2 mm) thick. The thickness and coarseness of the required diamond blade is a problem because it uses up a large amount of crystal from kerf loss and lattice damage. There is a need in the industry for a cropping saw based upon ID technology which would minimize the kerf and damage losses in cropping and improve upon the present method of preparing samples for oxygen and carbon analysis.

2.2 Grinding

The primary purpose of crystal grinding is to obtain wafers of precise diameter because the crystal growth process does not produce sufficient diameter accuracy. Crystal grinding is a straightforward process using a fixed abrasive grinding wheel, however, it must be well controlled in order to avoid problems in subsequent operations. Exit chipping in wafering and lattice slip in thermal processing are problems often resulting from improper crystal grinding (6),(8). A water-based coolant is used in grinding operations.

There are two methods for grinding crystals, (a) "centerless" grinding and (b) grinding "on centers" similar to turning on a lathe. Centerless grinding, Figure 5a uses equipment developed in the metalworking industry for grinding cylinders that were too long to be easily supported on centers, for example, long steel rods. The workpiece (crystal) is held between the grinding wheel and a drive wheel by a bevelled supporting post. The relative rotations of the grinding wheel, (fast) and the drive wheel (slow) cause the workpiece to rotate slowly. The support post holds the workpiece centered between the wheels. The result is that the highest points on the workpiece are removed initially and it is slowly shaped into a perfect cylinder. Since the cylinder is longer than the width of the grinding wheel, the drive wheel is

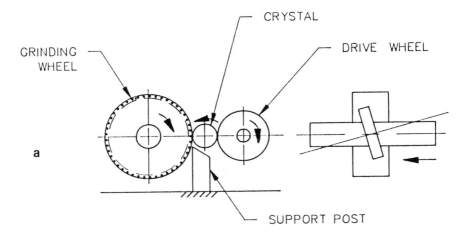

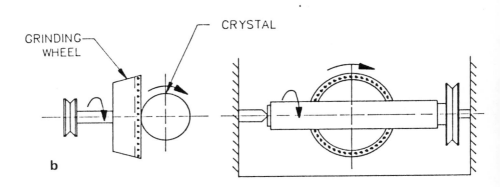

Figure 5: Crystal grinding.

(a) Centerless grinding,
(b) Grinding on centers.

slightly angled in order to feed the workpiece completely through the grinder. Reversing the angle of the drive wheel feeds it back through the machine. Repeated passes through the centerless grinder bring the workpiece to the specified diameter which is determined by the distance between the wheels. In order for productivity to be high and large amounts of silicon removed quickly from crystals, the grinding wheel is usually a fairly coarse one and therefore the resulting lattice damage to the crystal is high. These and other reasons have resulted in the obsolescence of centerless grinding of silicon crystals in favor of grinding on centers.

When grinding on centers Figure 5b, the crystal is held between two rotating supports and is turned slowly while a high rpm diamond grinding wheel is moved back and forth along the length of the crystal on a sliding work table. The wheel performs a lathe-like cut, removing as much as 0.2 mm (0.08 in) per pass, creating a spiral pitch along the crystal surface. With successive passes, the crystal is brought to the correct diameter. The grinding wheel is a diamond cup wheel. Grinding damage may be reduced in the final passes by removing less material per pass and by lowering the traverse rate. Damage may also be reduced by changing the wheel to one of finer grit size for the final few passes. Grinding the crystal on centers requires that the operator locate the crystal rotating supports on the proper crystal axis in order to obtain the best yield from the grinding operation. The center grinder is much smaller than a centerless grinder of the same diameter capacity, and it can be used for grinding the identification flats as well.

2.3 Orientation/Identification Flats

The identification flats (there may be one or two on a crystal) are ground lengthwise along the crystal according the to the crystal orientation and the dopant type. The largest flat is called the primary flat, and is used for positioning the wafer for front end processing such as patterning or dicing. A secondary flat may also be ground on the crystal. The specific arrangements of flats make it easy to identify the orientation (111 or 100) and the material (n- or p-type). The flats Figure 6 located according to a SEMI standard (9), and are ground to specific widths, depending upon crystal diameter. After grinding the crystal on centers, the crystal is rotated to the proper orientation, and the wheel is positioned with its axis of rotation perpendicular to the crystal axis

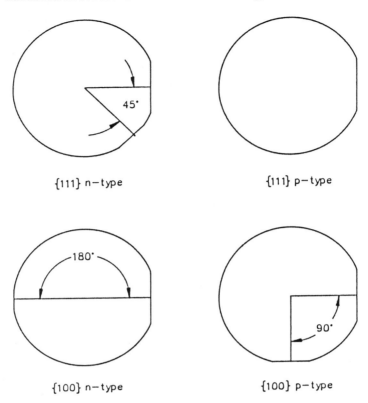

Figure 6: SEMI locations for orientation/identification flats (Ref. 9).

and moved along the crystal from end to end until the appropriate flat size is obtained. An optical or x-ray orientation fixture may be used in conjunction with the crystal mounting means to facilitate the proper orientation of the crystal on the grinder.

2.4 Etching

The cropping and grinding processes are performed with relatively coarse abrasives and therefore a lot of subsurface damage results, which will affect all of the wafers from the crystal. Pits, chips, cracks and stresses are all present. The usual method of removing the damage is to etch the crystal in a hot solution of sodium or potassium hydroxide. As has been described above, this etchant is a selective etch and attacks the silicon primarily at the damage. Regions of stress are removed and cracks will no longer propogate, resulting in higher yields in wafering and lapping due to reduced exit chipping and wafer breakage. After etching, the crystal is transferred to the slicing preparation area.

3.0 WAFERING

The purpose of wafering is to saw the crystal into thin slices with precise geometric dimensions. By far, the most common method of wafering semiconductor silicon is the use of the annular, or inner diameter (ID) diamond saw blade. The terms "wafering" "slicing", "sawing", and "cutting" are used interchangeably in the industry.

3.1 Historical

Although annular blades with sharpened cutting teeth on the ID were reported as early as 1908 in the patent literature (10), the use of ID blades for hard materials based upon abrasive cutting was first reported in 1950 by Jansen (11), who used the technique for the serial slicing of dental tissues. Jansen's ID blade, which utilized an abrasive slurry, produced significantly more slices per mm. than could be obtained with OD slicing.

A decade later, ID slicing technology was applied to germanium and silicon and soon came to play a prominent role in the advances in semiconductor wafer preparation. The technique has been developed into a proven production method, and has the main advantages of low kerf and straight cuts, combined with ease of automation. Although the cost of ID wafering is high in comparison to conventional OD methods (12), this is more than offset by the material saved with the low kerf.

3.2 The ID Blade

The heart of an ID slicing machine is the blade. The ID machine concept and blade configuration are illustrated in Figure 7a and 7b. The core of the blade is high tensile strength stainless steel. On the inside diameter of the hole, an electroplated coating of nickel is deposited (13), and diamond particles are firmly imbedded in this coating. These fixed abrasive diamond particles are the cutting "teeth" of the blade. The technology of simultaneously depositing abrasives with electroplated metal layers was known before the advent of the ID blade (14), however when the diamond plating technology was applied to ID blade manufacture, several new patents issued relating to process improvements for making blades, as well as new configurations for blades, for example those of Weiss (15) and Lane (16). The first blade used commercially in large quantities had an outside diameter of 8 in. (200 mm.) with a 3.25 in. (83 mm.) hole. It was capable of slicing up to 1.5 in. (38 mm.) crystals. Slicing technology has kept up with the demand for larger wafers, and today the largest blades in commercial use have an inner diameter of 12 in. (305 mm.) and an outside diameter of 34 in. (860 mm.) and are capable of slicing up to 8 in. (200 mm.) diameter crystals (17).

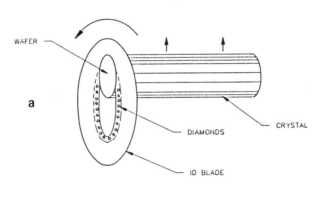

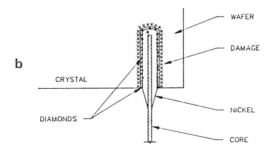

Figure 7: Schematic diagrams of ID slicing technology.

(a) Slicing machine,
(b) ID blade.

3.3 Blade Tensioning

The blade core material is very thin, in the range of 0.004 to 0.006 in. (0.1 to 0.15 mm.) depending upon blade diameter and it is made rigid by placing it in tension. The outer periphery of the blade is clamped between two steel rings and it is tensioned much like a drum head is tightened. The two most important considerations in tensioning are obtaining a concentric, true-running cutting edge and maintaining adequate tension throughout the life of the blade. A measure of the amount of tension may be determined by applying a known force perpendicular to the blade and measuring the deflection with a dial indicator (18). A more repeatable method of measuring tension is to measure the inner diameter, which typically increases 0.6 to 0.8% during initial tensioning. This amount of tension brings the blade very near to its yield point. Additional stretching only serves to permanently distort the blade core and adds the risk that damage to the diamond plated cutting edge or tearing of the core will occur. Concentricity is best measured by placing the tensioned blade directly on the machine, slowly rotating the blade and measuring the runout of the cutting edge with a microscope or a dial indicator. A runout of 0.001 in. (0.025 mm.) or less is desirable.

A number of tension head designs have been used in the industry, however most are similar in principal to those shown in Figure 8. There are two basic tensioning methods; hydraulic and mechanical. In the hydraulic method shown in Figure 8a a fluid (grease) is forced into a cavity in the tension mount at high pressure. This is usually applied with a conventional grease gun. The pressure acts on a piston which forces the blade into a recess thereby creating tension on the blade. Alternatively, the fluid can act directly on the blade provided there is adequate sealing to prevent leaks. Mechanical methods of tensioning utilize screws to pull the blade over a lip or press it into a recess Figure 8b.

The hydraulic technique is fast and requires less skill to perform, however the quality of blade centering depends collectively upon the accuracy of the tension mount, the blade locating pins and the blade dimensional accuracy, as well as the uniformity of the blade tensile strength. Operator technique has little or no effect upon centering accuracy. The mechanical tensioning process is a manual operation, and the skill of the operator is quite important in obtaining good results. With mechanical tension mounts, the operator is able

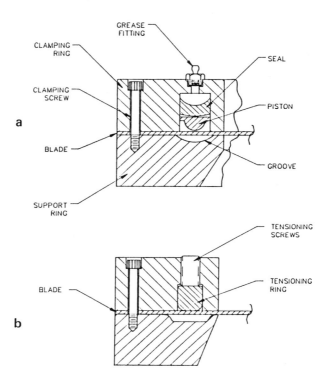

Figure 8: Typical tension mount designs for ID blades.

    (a) Hydraulic,
    (b) Mechanical.

to independently control centering and tension by selectively tightening the tensioning screws and observing the concentricity and inside diameter of the blade.

Early ID slicing machines used mechanical tensioning (19), then for a period of time in the 1970's when there was a strong emphasis on cost reduction and producivity, hydraulic tensioning was frequently used, particularily in the USA (20), but most recently the increased demands upon wafer quality, (flatness, chips, thickness, and damage) have dictated the use of mechanical tensioning, primarily because of its superior blade centering, and the resulting improved yields which more than offset the additional cost.

3.4 Process

Slicing with an ID blade is a fixed abrasive machining operation. The diamond particles which are firmly imbedded in the nickel matrix are forced into the silicon at pressures which exceed the compressive strength of the crystal. Microcracks are formed ahead of the blade due to the high stress fields. The microcracks propogate into the crystal generally along lattice planes. As these microcracks intersect, particles of silicon are released. The high speed blade and coolant motion carries the swarf out of the cut. It can be seen Figure 7b that microcracks also propogate laterally as the blade is forced into the material, and this lateral damage which remains in the wafer is called sawing damage. It can also be demonstrated that blade runout, both horizontal and vertical, as well as machine vibrations will cause increased damage due to the pounding effect of the blade on the material. Sometimes this excessive damage is easily visible to the unaided eye, appearing as a "sparkle" on the wafer surface when observed under bright illumination. Saw marks will have the characteristic curved shape of the cutting edge. Other yield problems such as exit chips, cracks, and breakage can often be related back to excessive sub-surface damage in the form of microcracks. Reference (8) describes the type of damage which can lead to rejects. Achieving acceptable wafer dimensional quality requires a constant attention to the blade, the machine and the process parameters. The productivity of an ID slicing machine is more often a function of the blade condition than it is the machine. Furthermore, it is a function of the care in using the blade/machine combination, — the care exercised in blade mounting, tensioning, dressing, coolant flow, feed rate, etc. Substantial losses can be experienced when a saw produces reject wafers because the accumulated value added in crystal growth, crystal shaping, and wafering is lost.

3.4.1 Crystal Mounting:
The silicon crystal, when it arrives at the sawing area, has been ground to diameter, flatted, and etched. In order to slice it, the crystal must be firmly mounted in such a way that it can be completely converted to wafers without waste. This is accomplished by attaching the crystal with wax or epoxy to a mounting block which is usually a cylindrical shape of the same diameter. Also a mounting beam or strip Figure 9 is attached along the length of the crystal at the break-out point of the saw blade. This reduces exit chipping and also provides support

212 Handbook of Semiconductor Silicon Technology

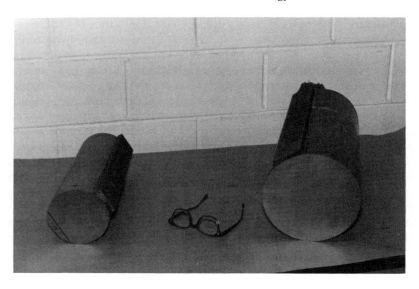

Figure 9: Photograph of 150 and 200 mm. crystals with slicing beams mounted, ready for the saw. (Courtesy Kayex-Capco, a Unit of General Signal).

for the sawn wafer until it is retrieved. Graphite or phenolic are common materials for the mounting block and beams. The saw manufacturer provides fixtures for holding the crystal and mounting materials during the mounting process. Epoxy formulations are available which can be readily removed from the wafers after slicing.

3.4.2 Orientation: Silicon crystals are grown with either the crystallographic <100> or <111> direction parallel to the cylindrical axis of the crystal. Wafers, however, may be cut either exactly perpendicular to the crystallographic axis, or deliberately off-axis by several degrees. In order to obtain the proper wafer orientation, the crystal must be properly oriented on the saw. All production slicing machines have adjustments for orientation of the crystal, however, it is usually necessary to check the orientation of the first slice in order to assure that all subsequent slices will be properly oriented.

3.4.3 Blade Condition: As pointed out above, the blade must be properly tensioned and centered. An eccentric blade edge results in pounding and vibrations which cause damage to the wafer such as chipping, and poor surfaces.

The blade must also cut freely. This means that there should be a low cutting force as the blade moves through the material. Excessive force will cause the blade to wander and result in tapered and bowed wafers. Edge chipping is also likely to occur due to the pressure and possible heat build-up. A free cutting blade is one in which there are many exposed and sharp diamonds available at the edge of the blade. As the blade is used, the diamonds may become dull or may be torn out, in which case the nickel matrix may begin to contact the silicon. In this situation, heat begins to develop, the nickel softens, and the remaining diamonds are driven into the metal or pulled out. If corrective action is not taken, catastrophic blade failure will result. The usual corrective action is to dress the blade. Dressing is performed by actually slicing into an abrasive stick of aluminum oxide or silicon carbide which is loosely sintered to a hardness of G, H, or I. As the blade cuts into the abrasive stick, the abrasive swarf from the stick removes the nickel matrix and dull diamonds, leaving new, sharp diamonds exposed. After a few cuts into the dressing stick, a wafer is sliced. The quality of the wafer will determine whether additional dressing is needed. Dressing also corrects for blade "loading", an accumulation of the mounting wax or epoxy at the blade edge that prevents the diamonds from penetrating into the crystal. Finally, an experienced operator can preferentially dress one side of the blade to correct for blade deflection caused by an assymetric cutting edge. Excessive dressing however reduces the life of the blade by removing diamonds. The thickness of the diamond layer on the leading edge of the blade is usually less than 0.2 mm., which is considerably less than the typical OD blade.

Bow is one of the more common problems in sawing. The bowed wafer has the shape of a saucer and is usually caused by lateral deflection of the blade as it cuts. It can easily be seen that any lateral deflection of the blade during slicing will cause it to deviate from a straight path through the material. Cutting force is greatest when the greatest portion of blade edge is in contact with the crystal, which is near the center. It is at this point therefore, that any side forces on the blade will have their greatest effect. At the entrance and exit regions of the cut where cutting forces are small, the blade will be deflected less. It is often observed that the condition of the blade changes during the cut, causing its path through the material to change. Thus taper, thickness variations, and warp are are often associated with bow, that is, they are caused by blade wander. Of course, improper alignment or worn ways in the equipment can result in a non-straight cutting path through the material, however this is generally easy to detect and correct. The more likely causes of blade wander are inadequate tension, assymetric blade edge shape, damage to one side of the blade core from rubbing, or any blade condition which will result in high cutting forces, such as a dull blade, excessive feed rate, or physical damage to the cutting edge or the core material. Fortunately, lateral blade deflection can be measured during the slicing process. An electronic sensor near the blade detects blade deflection, which can be plotted on a graph in real time. This information can assist the operator in determining when blade dressing or replacement is necessary, and can avoid yield losses. Figure 10 illustrates the effect of blade lateral deflection in creating bow, taper and warp.

3.4.4 Wafering Variables: The lack of scientific data on the ID wafering method makes it difficult to optimize the process or to diagnose the causes of wafering problems. To a large extent, wafering quality depends upon the level of attention to the machine and process, as well as upon operator skill and experience. The variables are interdependent and one should consider the process from a systems point of view. The obvious variables which can be easily controlled and evaluated are the cutting feed rate, wheel speed, and coolant flow rate. Unfortunately, these are not usually the cause of problems. The condition of the blade is the most important variable in practice, followed by the condition of the machine, such as alignment and vibration. Reference (18) is a comprehensive practical analysis of the slicing practice which includes recommended procedures for solving slicing problems.

3.5 Equipment

The ID slicing machine is simply a device designed to accurately position and move the material in relation to the blade to perform cutting. This is usually accomplished with a precision slideway for the cutting motion and an accurate crystal indexing system to obtain uniform wafer thickness. The typical saw holds the crystal horizontally and performs a vertical cut, either by moving the crystal up against the blade or the blade down into the crystal. Alternatively, the crystal can be held vertically, with the blade spinning on a

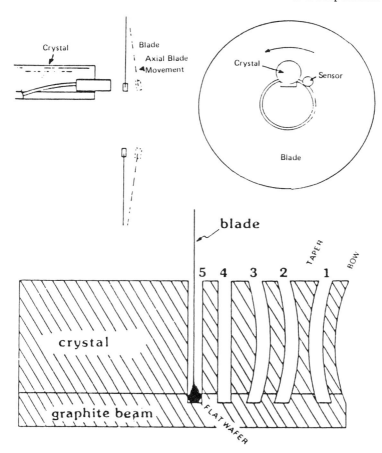

Figure 10: The Consequences of Blade Deflection during Wafering.

vertical axis and performing a horizontal cut. There does not appear to be any significant difference in the achievable wafer quality between the two methods. After one wafer is cut and the blade retracts, an index system advances the crystal a precise distance before the next cut is started.

Silicon wafering equipment such as the one illustrated in Figure 11 have evolved over the years into sophisticated production machines that usually operate continuously, and the industry has come to expect minimal downtime. The most advanced machines automatically slice and retrieve the

Figure 11: Modern ID wafering machine. (Courtesy Kayex-Capco, a Unit of General Signal).

wafers and load them into cassettes. With such automatic wafer retrieval, the crystal can be moved away from the blade during retract, avoiding any possibility of saw marks or damage as the blade retracts. Additionally, most slicing machines monitor the condition of the blade by measuring blade lateral deflection during slicing, helping to avoid reject wafers. Saws can be programmed to slice a predetermined number of wafers, or to stop slicing in the case of excessive blade deflection, the end of the crystal, etc. Some equipment manufacturers have offered devices to measure thickness and bow of each wafer immediately after slicing, with alarms if the wafer is out of specification, however this accessory has not been widely used in the industry. Some equipment has the capability of automatically varying the feed rate, a feature that can improve wafer quality, for example reduced breakout edge chipping. The purchaser of equipment will most likely make a decision to purchase based not only upon demonstrated quality of cutting, but also upon features such as convenience, safety, ease of maintenance and repair, and the like.

3.6 Unconventional Wafering Methods

3.6.1 Slurry Sawing: Two techniques, multiblade slurry sawing (21) and wire sawing (22),(23) are worthy of mention for slicing silicon because of their characteristically low kerf and low damage. Using the same principle as lapping, slicing is performed by applying an abrasive slurry to an oscillating wire or strip of metal under tension. The pressure of the metal tool on the workpiece forces the abrasive against the workpiece, thereby removing material. Because localized stress forces in the material are lower and damage is less than fixed abrasive methods, slurry sawing is capable of producing thinner wafers than ID saws. Furthermore kerf loss with wire sawing may be as small as about 0.0036 in. (0.09 mm.) using the smallest available wire and grit size, however one sacrifices cutting speed and risk of breaking wires or blades in these ranges, so that a more practical arrangement results in a kerf loss roughly equivalent to ID sawing. Although the slurry cutting process is very slow, blade ganging makes it possible for many wafers to be sliced simultaneously, thereby resulting in an acceptably fast production process. Present commercial applications are quartz and ceramic slicing, with some potential for slicing semiconductor and solar grade silicon.

3.6.2 Other Methods: Many other slicing methods for silicon have been proposed which would offer specific advantages, however all have major problems which must be solved before they could be utilized commercially. Examples are electrochemical (24), electro-photochemical (25), molten salt (26), fixed abrasive wire (27), acid string, laser, and high pressure water (28).

4.0 EDGE CONTOURING

The rounding of the edge of the silicon wafer to a specific contour is a fairly recent development in the technology of wafer preparation.

4.1 Background

It was known by the early seventies that a significant number of device yield problems could be traced to the physical condition of the wafer edge. The edge affected the wafer strength and its resistance to thermal stress, it was the source of particles, and it influenced the thickness of photoresist and epitaxial coatings. Chemical etching of wafers resulted in a degree of edge rounding, but control of the edge contour was difficult. Mechanical edge contouring was developed during that time and the result was a dramatic improvement in yields in downstream wafer processing. Losses due to wafer breakage were also reduced. By 1978, virtually all wafers produced by the major merchant supply houses had a contoured edge (29).

4.2 Reasons For Edge Contouring

In order to properly discuss the important aspects of edge rounding, one must review the condition of a wafer which has not been edge rounded Figure 12. In this figure, the wafer is illustrated just after slicing. The edge is generally square, but there is some edge chipping caused by the saw. Severe edge chipping caused by poor slicing techniques will of course be rejected, however such chipping is usually that which is easily visible to the naked eye (8). Microscopic edge chipping is always present. Sub-surface damage from the saw operation exists along the surface of the wafer on both sides, extending into the wafer for a few microns. Finally, there may be some residual damage from the crystal grinding operation which preceeds the slicing process. The grinding damage will have been largely removed by the crystal etch step before slicing, however, depending upon the grinding and etching process controls, it may not be possible to fully remove such damage (6).

The cross section of a wafer after double-side lapping Figure 12b is schematically the same except that the amount of damage due to lapping is less than that from slicing. The corners are somewhat sharper. Double side lapping, if performed on as-sawn wafers may cause excessive edge chipping that cannot be subsequently removed (30). (This is because the typical lapping plate is serrated with vertical cuts for proper slurry fow, and these cuts develop very sharp corners as the plate wears which impact on the sharp wafer edge.)

Silicon Wafer Preparation   219

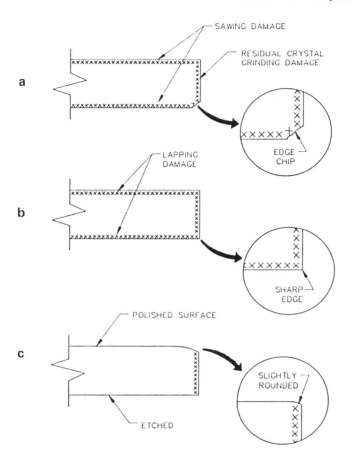

Figure 12: Cross sections of wafers that have not been edge rounded.

    (a) As-sliced,
    (b) After lapping,
    (c) After etching and polishing.

Thus, without edge rounding we have a wafer that contains lapping damage and accumulated edge damage from grinding, sawing and lapping. Etching and polishing will remove the damage from the surfaces, but will not remove all of the edge damage nor will it substantially remove the sharp corners Figure 12c. Because of these and other reasons to be discussed below, mechanical edge contouring is performed immediately after slicing, and the resulting contour remains essentially intact on the finished wafer.

4.2.1 Silicon Chips and Wafer Breakage:

Wafers go through many handling steps and processes, both manual and automatic. As they move through these steps, the wafer edge comes in contact with various materials, such as plastic carriers or quartz boats. The dump transfer step is an example of a handling process in which groups of wafers are moved from one carrier to another by simply inverting the two mating carriers. The contact with the carrier is usually in the form of an impact at the edge of the wafer. This impact can cause extremely high localized stress resulting in edge fracture and the formation of silicon chips. Chips can be a major source of circuit defects and device failures. Likewise, a wafer will be less resistant to breakage when edge chips form, as the discontinuity at the chip becomes a stress point for the propogation of fractures. Edge rounded wafers are significantly more resistant to edge chipping and breakage (31).

4.2.2 Lattice Damage:

Wafers are subjected to various thermal cycles in device processing, for example, oxidation, epitaxy, and diffusion. The heating and cooling cycles are often relatively rapid and cause temporary stresses to be built up because of the thermal expansion of the wafer and non-uniform heating across the wafer. If these stresses exceed the strength of the lattice, damage in the form of dislocations and slip occurs. Sharp corners as well as cracks at the edges become local stress points for the nucleation of lattice damage. Wafers with rounded edges are significantly more resistant to thermally generated stresses.

4.2.3 Epitaxial Edge Crown:

In the epitaxial process, the deposition tends to proceed faster at a sharp corner than on a flat surface, resulting in a ridge at the wafer edge. This ridge interferes with subsequent photomasking steps. Properly edge coutoured wafers minimize epitaxial edge crown.

4.2.4 Photoresist Edge Bead:

A photoresist layer is applied by spinning the wafer after applying the liquid photoresist. Surface tension causes the photoresist to build up a thicker coating at the edge. The photoresist edge bead interferes with photomask operations similar to the epitaxial crown. In contact masking the bead prevents the mask from making uniform contact with the wafer, thereby reducing optical resolution of the image. Also, the soft-baked photoresist bead can form a seal around the photomask and adhere to it, making separation difficult. Portions of the bead may stick to the mask, adding to the problem with subsequent wafers. After exposure, development, and hard baking, the photoresist is brittle, and the bead may chip off during handling, resulting in contamination of other wafers or photomasks. In proximity or projection aligners, the equipment positions the wafer by sensing the front surface near the wafer edge, so that the edge bead causes a focal plane error in imaging. Photoresist edge bead problems are effectively avoided when wafer edge rounding is used.

4.3 Commercial Equipment

Most commercial edge rounding equipment is capable of contouring wafers up to 150 mm. diameter, and can operate in a fully automatic mode from cassette to cassette. Figure 13 illustrates schematically the usual method of edge contouring. A small diamond form wheel which has been manufactured to the proper contour is held against the wafer edge. The wheel spins at high speed and the wafer is held on a vacuum chuck and is rotated at low speed. Since the contact area between wheel and wafer edge is very small, the equipment must be capable of reliably controlling the grinding force. A cam follower system follows the wafer edge so that the edges along the flats are also contoured.

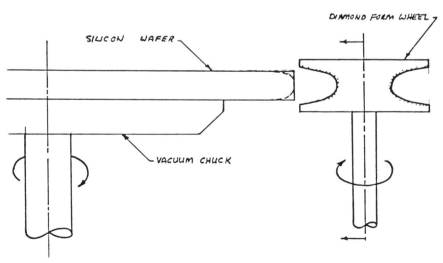

Figure 13: Schematic Illustration of Edge Contouring.

5.0 LAPPING

The slicing operation does not consistently produce the required flatness and parallelism for many wafer specifications. Since conventional polishing is not a flatness or thickness correcting process (see section 6, Polishing), lapping prior to polishing must often be used, because it is capable of achieving very precise thickness uniformity, flatness and parallelism. Lapping also prepares the surface for polishing by removing the sub-surface sawing damage, replacing it with a more uniform and smaller lapping damage.

5.1 Background

Lapping has long been used as a process to shape hard materials such as glass, stone, gems and ceramics, and to prepare them for the final polishing step. Even today the lapping technique is as much an art as it is a technology (32). The processes used for lapping and polishing silicon wafers evolved from the lens manufacturing industry which uses principles which were developed over several hundred years. A lapping wheel with the proper curvature is used as the tool to generate the desired lens shape. The machine holds the lens blank against the rotating lapping wheel and a silicon carbide slurry is fed onto the wheel while the lens is rotated and oscillated. If a fast removal rate is required, a coarse abrasive slurry is used, however the resulting surface will be rough and difficult to polish. Conversely, a fine abrasive removes material slowly, but the resulting surface is smoother and therefore the required polishing time will be shorter. Final lens polishing is performed on a similar machine, however the contoured wheel surface is covered with a polishing pad and the abrasive slurry is replaced with a polishing agent.

Silicon wafers are flat, and therefore the equipment is mechanically somewhat simpler than lens processing machines. During the last two decades, silicon wafer lapping techniques and materials have developed according to the needs of the semiconductor industry, and significant differences now exist between silicon and glass processing (33). For example, machines for lapping silicon wafers are large with many wafers lapped at a time. Wafers are thin, flexible and fragile compared to glass lenses, and therefore pressures, speeds, and abrasive materials are tailored to the material. Lens shaping usually requires single side processing because curvatures are often different on the two sides of the lens. Silicon wafers may be lapped by either a single or double side process. Double side processing is gaining in usage because improved wafer flatness can be achieved. Flatness of both the front and the back side is important since wafer fab lines use vacuum chucks to hold wafers for processing. With the trend toward larger diameter wafers and tighter dimensional tolerances, not only is double side lapping usually required but careful process control in the lapping step is necessary to obtain the required final wafer quality.

5.2 Current Technology

In its simplest form, a double side lapping machine consists of two very flat counter-rotating plates, carriers to hold and move the wafers between the plates, and a device to steadily feed abrasive slurry between the plates. The carriers are usually plastic templates of slightly smaller thickness than the wafers, with openings of a correct diameter to fit the wafers closely. The carriers are driven externally by gear rings which cause the wafers to move around the surface of the lapping plates in a planetary motion, thus the double side lapping machine is called a planetary lapper. A detailed description of the lapping process has recently been given by Dudley (34).

The lapping plates of the typical lapper are usually made of cast iron. Although other materials have been used, cast iron appears to have the right hardness for lapping silicon bacause the particles are not fully embedded in the lap, but move with a rolling, crushing action on the wafer to effect silicon removal. A softer lap would allow the abrasive to fully imbed in it which would cause the wafers to be scratched, whereas a hard lap would force the abrasive into the wafers, thereby causing the lapping plate to be worn rapidly, and the wafers to be damaged excessively.

Since the upper plate of the machine is floating, its axis of rotation adjusts to the upper surfaces of the wafers. Because of this, it is usually necessary to pre-sort the wafers before loading the machine, otherwise the parts may be tapered. If the sawing operation is closely controlled, sorting may not be required, except to reject the occasional out-of-spec wafer.

The wafers are manually loaded into the carriers, and they are flooded with slurry and the carriers are rotated in order to fully coat the wafers before lowering the upper plate and starting the machine. This prevents breakage of wafers by providing an initial lubricity. The abrasive is typically a nine micrometer aluminum oxide grit. Aluminum oxide has a more blocky particle shape combined with a greater toughness than silicon carbide, resulting in a more controllable removal rate with less damage to the wafer. Commercial abrasives are suspended in water with proprietary additives to assist in suspension and dispersion of the particles, to improve the flow properties of the slurry, and to prevent corrosion of the lapping machine.

Lapping pressure of about 2 to 3 psi. (14 to 21 kpa.) is applied to the plates usually by hydraulic or air cylinders through most of the process, however starting pressure is kept low for 2 to 5 min. in order to allow the wafers to settle into the carriers, to allow the slurry to distribute uniformly, and to remove the high points of the wafers where the initial lapping forces are concentrated. Most lapping machines automatically ramp from the starting pressure up to the final pressure at a predetermined time in the lapping cycle.

The completion of lapping may be determined by elapsed time, or by a device on the machine which senses the actual wafer thickness during the process and stops the machine when the wafers have achieved that desired thickness. A sensing device has been reported (35) which is based upon piezoelectric material that is lapped simultaneously with the wafers. The material emits an electrical signal whose frequency is a function of the its thickness. Ultrasonic (36) and electronic (37) methods have also been reported. Mechanical methods to sense wafer thickness are usually not accurate enough to achieve the thickness tolerances required. Thickness tolerance achievable within a batch is +/- 0.0002 in. (+/- 0.005 mm.) for incoming wafers that are within +/- 0.001 in. (+/- 0.025 mm.).

Alhough lapping would appear to be simple in concept, the successful implementation of a production lapping operation requires the development of a technique and experience to achieve acceptable quality with good yields. Since flatness of the wafers is of primary importance, maintaining the flatness of the lapping plates is crucial. Small adjustments to the rotation rates of the plates and carriers will cause the plates to wear concave, convex or flat. Similarily, wafers can be made to be lapped convex or concave by adjusting rotation parameters (38). From time to time the plates will have to be brought back to a flat condition by the use of special conditioning fixtures. As with the other processes used in wafer preparation, lapping will benefit by attention to optimizing process parameters, and the employment of trained and experienced personnel.

Commercially available equipment ranges over a wide size range, although all sizes are the same in principle. Machine manufacturers offer various features to improve productivity and yields and to reduce labor costs. The cost versus benefit tradeoff must be analyzed by the purchaser. Machine manufacturers also provide a base line process to the customer from which specific requirements can be developed. Figure 14 illustrates a commercial double side lapping machine.

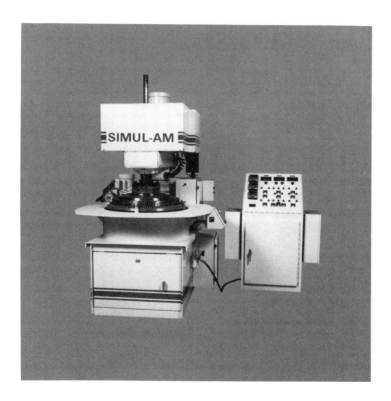

Figure 14: Commercial double side lapping machine. (Courtesy Kayex-Spitfire, a Unit of General Signal)

226 Handbook of Semiconductor Silicon Technology

5.3 Wafer Grinding

Lapping is a messy process at best, and there has always been a desire to avoid it or to substitute an alternative operation for it. The most likely approach is wafer grinding. In grinding, the wafer is held on a vacuum chuck and a series of progressively finer diamond wheels is moved over the wafer while it is rotated on a turntable. The first wheel, a coarse one, removes the most material, and the subsequent wheels remove less material, produce less damage and create a smoother surface. Grinding leaves the wafers cleaner than lapping. It produces an oriented scratch pattern on the wafers which is noticeable to the eye when compared to lapping, however, there is reason to believe that the damage is no greater than that of a comparable sized lapping abrasive. Certainly there is little risk of imbedding particles of abrasive in the wafer. The most important drawback of wafer grinding is that only one side can be ground at a time. Because of this it may not be possible to obtain ground wafers that are as flat as lapped ones. Machines designed specifically for grinding silicon wafers are available in the market.

6.0 POLISHING

The purpose of polishing a silicon wafer is to produce a smooth surface that has properties which are as close as possible to those of the bulk material. The process must therefore take place uniformly across the surface, and it must not leave residual contamination or surface damage. Silicon wafer polishing derives from the glass lens polishing industry which, of course is very old. In principle, the same techniques continue to be used, however some important modifications of the process have been developed to accommodate the special requirements of silicon wafer polishing.

6.1 Description of Polishing

If a lapped or ground silicon surface is examined with an electron microscope, it is found to contain a system of cracks, ridges and valleys closely resembling a rugged mountain terrain. Figure 15 schematically illustrates such an abraded surface. The peaks and valleys form a relief layer. Underlying the relief layer is a damaged layer characterized by microcr-

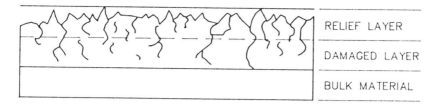

Figure 15: Cross sectional view of an abraded silicon surface.

acks, dislocations, slip, and stress. Both of these layers must be removed completely to expose the properties of the bulk material. When a much finer abrasive is used for lapping, the surface appearance is very much the same, only on a smaller scale. No matter how fine an abrasive is used, the lapped or ground surface will not achieve the smooth appearance of a polished one. This is the characteristic of brittle fracture of silicon which occurs in lapping and grinding. It indicates that the abrasive grains are moved across the surface under extremely high pressure, exceeding the elastic limit of the silicon.

Examinations of polished silicon or glass surfaces with the electron microscope have not shown any evidence of a relief surface such as that produced by lapping. Even at the highest resolution of the electron microscope which is about 15 angstroms, no scratches or ridges can be observed in a completely polished surface. It is therefore concluded that the polishing process is inherently different from the grinding process and that brittle fracture does not take place in polishing.

6.2 Historical Background

The bulk of the literature on polishing relates to the polishing of metals and glass. The polishing of metals is used principally to obtain the desired surface appearance as fast as possible and little concern is held for damage to the underlying bulk material. Metals are usually polished to produce a surface that is aesthetically pleasing, or to generate a certain roughness, (or smoothness), for example for the sealing surfaces in vacuum systems. Polishing of metals is considered to be simply an extension of fine grinding (39).

Glass and silicon polishing are quite different. Although

optical polishing experts might dispute it, it is believed that the surface quality required for silicon wafers exceeds that required for high quality optical glass, at least on a microscopic scale. In glass polishing, surface geometry is the principal objective, with damage and cleanliness of lower priority. Silicon wafers must be completely damage and haze-free, whereas geometry is not as critical. Typical wafer flatness is in the 1 to 2 micrometer range.

Although the objectives are different, the polishing of silicon resembles that of glass in many aspects, and therefore it is appropriate to review three theories that have been proposed for explaining the polishing mechanisms for glass:

Glass Polishing Theories

1. Glass polishing is purely a physical phenomenon, with the particles of polishing agent mechanically wearing away the material, much like grinding but on a smaller scale.

2. Polishing takes place because the surface exhibits a degree of plasticity. It can flow or be displaced when subjected to pressure, and in this way, a smoothing of the surface can take place. The plasticity is a result of thermal softening due to pressure and friction, or swelling due to the attack of water solutions which may produce a silica jel layer on the surface.

3. Polishing is possible because a great number of chemical and mechanical interactions occur. The chemical/mechanical characteristics of the polishing agent, the pad and the glass are affected by the suspending water solution. Low pressure contact results in atomic bonding between the polishing agent and the surface which removes material on a nearly atomic scale.

Herschel, (40) proposed the first theory by writing in 1830 that polishing was simply an extension of grinding. Beilby in 1903 (41) put forth the idea that polishing took place by flow of material from the peaks into the valleys due to thermoplastic deformation. He observed grooves in a seemingly structureless polished surface after etching with hydrofluoric acid and concluded that polishing does not require removal of material, but occurs as a leveling process. Early observations by Lord Rayleigh in 1901 (42) were the first important ones which led to the conclusion that the polishing agent particles did not simply act as tiny cutters as Herschel had suggested. Klemm and Smekal, (43) demonstrated conclusively that a steel needle moved across a glass surface caused surface flow, and that this flow could fill in previously made grooves, however, there was not proof that such high pressures occur in the normal glass polishing process. A large portion of the literature on glass polishing deals with the question of whether glass flow is a significant effect in polishing. The work of French (44) is an example. French described a "beta" layer which is a distorted surface layer which can reach a thickness of about four micrometers. The beta layer corresponds to Beilby's flowed layer. The investigations of Bruche and Poppa, (45) and Kaller, (46) have indicated that the glass polishing phenomenon is probably explained by a combination of all three of these theories, each operating to a greater or lesser degree.

It is clear, in any case, that the removal of material in significant amounts is required to achieve high quality polishing. Just how this removal takes place is still not well understood. Conditions of extreme pressure and speed of travel have been required to demonstrate observable flow. It does not seem likely that plastic flow can be an important part of normal glass polishing, especially thermoplastic flow. In the case of polishing silicon, surface flow is even more improbable because of the crystalline nature of the material and its inability to soften like glass. Nevertheless, the current literature continues to refer to the "Bielby Layer" on a polished surface as a disordered or damaged layer resulting directly from the polishing process. The ideal polishing process for silicon should remove material without creating additional damage and without leaving a residual contaminated or disturbed layer. The current belief is that the optimum process is one in which a monomolecular layer of reaction product is produced on the surface and the polishing materials (pad, slurry, fluid) are able to remove the reaction material but must not abrade or otherwise damage the underlying material. Current literature on silicon polishing lends support to this theory (47).

Like glass surfaces, etching with a selective etchant such as Sirtl etch (48) will reveal subsurface damage (often appearing as scratches) which are otherwise invisible. However, if the stock removal is sufficient to remove the original lapping damage, such a test only determines the quality of the polishing process (49). A high quality polished silicon wafer will be completely free of scratches or other polishing--related marks after etching.

6.3 Current Polishing Practice

Silicon wafers can be polished readily by the same pads and polishing agents that are used to polish glass, such as wool felt with iron oxide, (rouge) cerium oxide, or zirconium oxide (50). These agents, having a particle size in the 1 to 10 micrometer range, polish fast and produce a fairly good surface. Control of the pH of the slurry to a high value (around 9 to 10) gives the fastest removal rate, however the materials often produce a light haze on the surface that is difficult to remove.

Virtually all silicon wafers today are polished using a colloidal dispersion of silica in water (silica sol). The colloidal silica is an amorphous, synthetic sol produced by a gaseous reaction or by precipitation from a liquid, resulting in a sub-micron particulate material which is mixed with a high pH aqueous suspending medium. Typical particle size is in the range of 4 to 100 millimicrons (51). Additions of proprietary suspending agents are usually added to prevent settling of the particles. Slurries are shipped in concentrated form, and are diluted prior to use. The concentration of solids for polishing, (3 to 4%) is typically lower than with the fast glass polishing agents (52), and the removal rate achieved is lower, but the quality of the surface exceeds that of any other particulate polishing agent. In order to achieve a reasonable removal rate and still obtain the highest quality surface, polishing with colloidal silica is performed in two steps; stock removal and haze removal. As the names imply, the former is carried out with a higher concentration slurry and may proceed for about 30 minutes at a removal rate of about one micrometer per minute. The latter is done with a very dilute slurry, a softer pad, and for a short time, about 5 to 10 minutes. The amount of material removed is very small, usually less than one micrometer. Because of the active chemical reaction with colloidal silica agents, the wafers must be rinsed in deionized water immediately after polishing is complete to prevent haze or stains from forming again.

6.3.1 Polishing Variables:

There are many variables which influence the rate and quality of polishing. Many of these are obvious. Increased pressure increases the removal rate, but excessive pressure leads to non-uniform removal, excessive heat generation, and fast pad wear. Removal rate is increased with high polishing temperatures, but also leads to haze formation due to the high chemical activity. High wheel speeds (expressed in feet per minute of travel of the wafers across the pad) increase removal rate, but also can lead to excessive temperatures, as well as problems in maintaining a uniform flow of slurry across the pad. High slurry concentration increases polishing rate, but is more costly and the resulting higher chemical/mechanical activity may result in haze formation. As the pH of the slurry increases, the removal rate increases gradually, until at a pH of about 12, the polishing rate falls off dramatically (49),(53). Further, at this critical pH the surface of the wafer becomes hydrophilic rather than the characteristic hydrophobic surface obtained at lower pH values.

Flatness: The commercial polishing process for silicon wafers is not a flatness improving process. At best it will not degrade the wafer flatness achieved in the lapping operation. A typical polishing pad may be 1000 micrometers thick, and is quite resilient. Pressure from the polishing machine compresses the pad as much as 250 micrometers. The flatness of the average wafer is in the one to two micrometer range. The pad therefore contacts the complete wafer surface at all times with nearly the same pressure, so that the removal rate is essentially uniform across the wafer. Since the total amount removed is 25 to 30 micrometers, an optimum polishing process is therefore one that maintains both chemical and mechanical uniformity of polishing conditions across the wafer. This is achieved by careful attention to all phases of polishing, including mounting of the pad, the slurry flow and concentration, alignment and flatness of the polishing plates, temperature control, wafer mounting procedure and the like.

Polishing Pads: Pad materials used for silicon are usually poromeric synthetic materials such as polyurethane. Many types are available with different thickness, hardness and surface roughness. The choice of the best pad material is a function of the chosen process variables. A very soft pad may have a tendency to produce an "orange peel" surface where a hard pad may have a tendency to glaze, remove material slower, and scratch the wafer. One usually attempts to use the hardest pad possible in order to obtain long pad life, but soft enough to have a reasonably consistent removal rate.

Wafer Mounting: Wafers must be held in the polishing process in such a way as to firmly hold them against the polishing pad without bending, breaking, edge chipping or other damage. The wafer support must be firm because they are thin and flexible. There are two basic methods of mounting wafers for polishing; wax and waxless.

In wax mounting the wafers are attached to a flat carrier plate using a thermoplastic mounting wax. The plate is heated and the wax is coated on the plate. Great care must be exercised in assuring that the wax layer is of uniform thickness to fully support the wafers. Problems in achieving a uniform adhesive wax layer include entrapped air bubbles, foreign particles, and uneven coating of the wafer carrier plate. Bubbles and particles provide a source of localized pressure which tends to bow the wafer during the polishing process. When the wafer is released from the carrier plate there will be small regions where the wafer surface is not flat and the wafer thickness will vary in those regions. The patent by Walsh (54) describes a typical method and apparatus for wax mounting silicon wafers on a carrier plate.

After polishing, the wafers are removed from the carrier plate by heating until the wax is melted and carefully lifting them with tweezers. After removal, the wax and any polishing debris must be cleaned off. Needless to say, the wax mount/-demount process is time consuming and costly. Wafer breakage is often a problem because of the required handling. Thus the industry has sought for and attempted a number of "waxless" mounting techniques, some of which are in production.

The wafer mounting method must not only support the wafer uniformly against the polishing pressure, it must also resist the strong lateral force from the moving polishing pad. Wax mounting, of course accomplishes both of these objectives because the wax is an adhesive. Polishing with a waxless mounting method, sometimes called "free polishing" has to utilize some other means of resisting the lateral forces. Templates alone have been found to be unacceptable because the wafer is not capable of sustaining the high edge pressure without fracture or edge chipping. The usual method is to support the wafer on a somewhat resilient surface, such as a thin polyurethane pad which has sufficient frictional characteristics with the silicon to prevent the wafer from sliding. A template is usually added to assist in positioning the wafers on the mounting plate and to prevent the occasional tendency of wafers to move.

Double side polishing, also necessarily a waxless form of polishing, is performed on a machine essentially like a double-sided lapper. There is considerable interest in double sided processing (both lapping and polishing) because of the potential for improved flatness, thickness and bow. However, double sided polishing has not as yet, received wide acceptance. Double side polishing, since it is performed with counter-rotating pads, tends to equalize lateral polishing forces on the wafers, thereby reducing the probability of edge chipping due to forces from the template.

Pressure: Although processing cost can be reduced by the faster removal rate achievable with higher pressure, there are tradeoffs. Higher pressure can lead to non-uniform removal, increased pad wear, poor temperature control, and breakage.

Temperature: Because of the chemical nature of polishing, an increase of temperature results in increased polishing rate. The optimum temperature which is typical in most polishing facilities is about 40°C. This is usually measured with an optical pyrometer sighting directly on the pad. Higher temperature can cause excessive evaporation of the slurry resulting in non-uniform removal rate. It can also lead to the formation of haze because of the faster chemical reaction, which presents difficulty downstream in the cleaning process.

6.3.2 The Optimum Polishing Process:

Every wafer polishing operation has its own formula for polishing including polishing agent, pad, wafer holding method, polishing temperature, slurry flow rate, pressure, speed, and which has been developed on a particular machine to achieve the desired polished wafer quality. For this reason it is very difficult to describe a standard polishing technique. As the demand for flatter wafers increases, all wafer preparation processes become critcal, especially the lapping and polishing processes. The optimum polishing process for a given facility depends largely upon the interplay of product specification, yields, cost, and quality considerations, and must be developed uniquely for that facility.

6.3.3 Other Methods Of Polishing:

Virtually all wafer production utilizes conventional slurry polishing, yet there is a need for flatter surfaces that are exceptionally clean and damage-free. There are many reported polishing processes but most consist of only minor variances from the general process described above. Three novel processes deserve mention because they depart from the typical processes yet offer the possibility of improvements such as surface quality, speed, or flatness.

Copper ion exchange: A copper ion-containing solution is used in place of the polishing slurry (55). When the solution contacts the silicon wafer, copper ions are reduced to metallic copper. By the action of the polishing pad, the copper is removed along with the oxidized silicon, on a nearly atomic scale. The reported result is a very fast, damage-free polishing process. Although it is being used in production polishing, it has not met wide acceptance, perhaps because of the corrosive nature of the chemicals used, the cost, or concern about the possibility of copper contamination of the wafers.

Float polishing: A recently reported process (56), float polishing closely resembles pitch polishing where the pitch lap is replaced with a diamond-turned tin lap surface. The polishing agent is suspended in a liquid, and by controlling rotational speeds of the lap and the sample holder, the sample wafer is suspended on a film of the slurry. The polishing agent particles must be softer than the silicon, and removal is apparently effected by the bombarding of the silicon by the particles. Reportedly this method, although quite slow, yielded damage-free and scratch-free surfaces of exceptional flatness. "Non-contact" polishing, reported by Magee and Leung (57), appears to be a similar process to float polishing.

Mechano-chemical polishing: This method (58), also performed in a matter similar to conventional polishing processes is unique in that it is typically performed dry. The polishing agent must be a softer material than silicon, for example, calcium carbonate, and removal is apparently brought about by chemical reaction between the polishing agent and the wafer. Flat, scratch and damage-free surfaces have been observed using this method. As with float polishing, the process does not appear to have gained acceptance in production, yet the resulting superior surface may be beneficial for selected applications.

7.0 CLEANING

The topic of wafer cleaning is a complex one, and a comprehensive treatment of it would deserve a complete chapter in itself. There are several steps in the wafer preparation sequence where cleaning may be appropriate, however this discussion includes only post-lap and post-polish cleaning since they are the most important cleaning processes in wafer preparation. For example, a cleaning procedure after slicing to remove mounting epoxy and swarf from wafers is not discussed, although this would be required in a typical facility. Slicing, lapping and polishing are relatively dirty operations compared to downstream wafer fabrication, and the cleaning processes used must be capable of effectively removing the relatively large amounts of particulate materials which are left on the wafers.

In addition to the removal of particles, a post-polish cleaning method should be able to remove organic materials such as oils, waxes and fingerprints, ionic subtances such as sodium and potassium, and trace metals. All of these contaminants are attached to the surface with bonding forces which may be physical, electrostatic or chemical in nature. Those materials that are chemically bonded (ions, atoms, films) are removed by chemical means whereas those that are less strongly attached (particles) may be removed by mechanical methods. The final cleaning step for polished wafers includes both scrubbing and chemical immersions followed by extensive rinsing in deionized water.

7.1 Mechanical Cleaning

Lapping residue consists of abrasive particles, iron particles from the lapping plates, debris from the wear of the carriers plus other contamination from the lapping machine area. Since the lapped wafers are sent to the polishing area after cleaning and etching, it is most important that all particluate contamination is removed to prevent scratching. Simple spray rinsing is not sufficient to assure the removal of all particles of lapping abrasive, however the use of high pressure (2000-3000 psi.) jets is effective in some cases. Scrubbing with high speed revolving brushes is very effective in removing particles. The brushes are made of hydrophilic fibers so that it is believed that the fibers do not actually contact the wafer surface, but ride on a thin film of attached water, providing the severe turbulence required to dislodge particles. Sonic "scrubbing" by the use of high frequency vibration in a water bath causes cavitation, and the collapse of the resulting bubbles effectively removes particles. Before polishing, thin films of ionic or organic contamination which may be on the wafers are of little consequence because the polishing process will adequately remove these from the critical front surface.

In the case of single side processing with wax mounting, the slices are mounted on support plates before lapping, and remain mounted through the polishing step. In this case, the whole mounting plate with wafers attached must be cleaned free of all abrasive and silicon particles, since it passes through to the polishing area.

7.2 Chemical Cleaning

The chemical process that virtually all immersion techniques are based upon is the so-called "RCA Cleaning Method" described by Kern and Puotinen (59). This method utilizes a series of solutions designed to remove any type of contaminating material that is likely to be on the wafers. A summary of the procedure is as follows:

1. Molecular contaminants, such as gross organics, waxes, oils, etc. are removed with organic solvents such as freon with ultrasonic agitation followed by a freon vapor degrease.

2. Ionic contamination and trace organics are removed by immersion in a warm hydrogen peroxide-ammonium hydroxide solution, followed by a dilute hydrofluoric acid dip to remove the oxide film which is formed from the basic peroxide solution.

3. Atomic contamination such as adsorbed heavy metals are removed with a hydrogen peroxide-hydrochloric acid mixture.

Thorough rinsing in deionized water is required after the ionic and atomic removal, and of course the cleaning and rinsing processes must be performed in a clean room. Rinse water must be of the highest quality, otherwise contamination will be redeposited on the wafers. This means that the water must be free of dissolved and particulate contamination including bacteria (60),(61).

7.3 Other Cleaning Methods

Megasonic cleaning (62), is a recent development which appears to clean wafers as effectively as brush scrubbing and chemical cleaning combined. It is a high frequency ultrasonic method using the RCA chemicals at room temperature with the wafers immersed as usual in carriers. Particles down to 0.5 micron are removed from wafers. It is reported to have greater throughput and significantly less chemical waste than chemical immersion cleaning.

Cleaning solutions based upon choline have been reported by Muroka, et. al. (63). This chemical which is similar to ammonia but is a fully ionized strong base, is combined with non-ionic surfactants to form an effective solution for removal of metal contaminants as well as bacteria. It may be used in place of or in addition to the RCA system. It is reported to be more useful in front end applications rather than for initial wafer preparation.

7.4 Equipment

The equipment used for cleaning can be as simple as a manually operated series of baths through which the wafer carriers are sequentially moved, to a fully automatic machine which eliminates the operator entirely. The cleanliness of the wafers will depend upon the ability of the system being used to first remove contamination and second to prevent redeposition, especially of particles. Since the human operator is the primary source of particulates in a clean room, the fully automatic system has a distinct advantage. Commercial equipment which is offered has been described recently in some detail, (64),(65). A wide variety of wafer scrubbers is available. For chemical cleaning, centrifugal spray units offer improvements over immersion baths since each wafer is uniformly exposed to uncontaminated, fresh chemicals.

The importance of high quality cleaning can not be overstressed as it is ultimately a yield determining factor. No cleaning system will be effective that is not protected from contamination from outside sources such as humans, bacteria, and miscellaneous particles as well as from impure chemicals and poor water. Since many contaminants are organic in nature, the use of strong oxidants is effective. Recently the direct and continuous application of ozone to deionized water systems has been reported (66) which significantly reduces bacteria in water, a leading cause of wafer contamination. High quality water is absolutely essential in wafer cleaning operations (67).

8.0 MISCELLANEOUS OPERATIONS

8.1 Heat Treatment

Czochralski crystals have a high level of oxygen impurity which often exceeds substantially the concentration of dopant in the material. When silicon crystals are held at approximately 450°C. for a few hours, the oxygen undergoes a transformation which causes it to behave as an electron donor, much like an n-type dopant. These oxygen donors, or thermal donors, mask the true resistivity of the crystal because they either add carrier electrons to an n-type crystal, or compensate positive holes in a p-type crystal. The concentration of thermal donors varies along the crystal length because the thermal history of the crystal is different from top to bottom. Usually the upper portion of the crystal has been near the 450°C temperature for a sufficient time during crystal growth to create thermal donors, whereas the lower portion of the crystal is cooled quickly through the critical temperature range.

Fortunately, thermal donors can be annihilated by heat treating the material briefly in the range of 500 to 800°C, and then cooling quickly through the 450°C region before donors can re-form. Such a heat treatment is possible for crystal ingots up to about 100 mm. diameter, because they can be cooled quickly in flowing room air. Larger crystals, however, because of their large thermal mass, can not easily be cooled quickly enough to prevent donor re-formation, and therefore heat treatment must be performed on the wafers instead.

Although in principle, donor annihilation can be performed on wafers at any time during their preparation, it is usually best to perform the heat treatment immediately after wafering, since resistivity is often a critical quality control measurement. Reject wafers can then be separated before additional processing costs are expended on them.

Donor annihilation is a bulk effect, and therefore wafer heat treatment can be performed in air, since any surface oxide that may form at the high temperature will be removed in subsequent lapping and polishing steps. Equipment for heat treatment of wafers is relatively simple, and may be either manually operated, or fully automatic and continuous, for example as with a belt conveyor furnace. The process is not a critical one, provided the wafers are heated uniformly. The optimum heat treatment temperature is the lowest one that at which donors are annihilated reasonably fast. Caution is advised when heat treatment is performed at higher temperatures because other effects such as oxygen precipitation and wafer warpage may occur.

8.2 Backside Damage

A "gettering" process is the use of one defect in the crystal lattice to control or annihilate another. Intrinsic gettering utilizes a built-in property of the Czochralski crystal, i.e. its supersaturated oxygen content which can be precipitated by heat treatment, causing damage to the lattice. (Oxygen is removed from the device region of the wafer before precipitation is performed). Extrinsic gettering utilizes external methods to damage the lattice at the backside of the wafer. It is the only method of gettering for float zone material, because of its extremely low oxygen content. Backside abrasion is one of several extrinsic gettering methods.

Silicon wafers produced for semiconductor device and integrated circuit manufacture are prepared from dislocation-free silicon crystals, because dislocations in the vicinity of devices can cause high leakage currents leading to failures. Impurity atoms, especially heavy metals, also have a deleterious effect, because they create traps for the carriers (electrons and holes) which are essential for the operation of transistors and diodes.

Yet it has been known for many years that there are beneficial properties of dislocations, provided they are not in the vicinity of the devices. Generally speaking, the desirable property of a dislocation is its ability to attract other defects, especially point defects like vacancies or other smaller dislocation-related defects, such as stacking faults. Point defects are a naturally occuring phenomenon, and are introduced into the crystal during the high temperature crystal growth process, and essentially frozen in there. These defects become mobile at the high temperatures of device processing (1000 to 1200° C). and they tend to cluster and precipitate causing lattice strain, eventually causing damage sufficient to degrade devices nearby. The combination of point defects and dislocations results in a net decrease in lattice strain, and therefore they are attracted to each other. Thus, point defects will be attracted to regions of greatest lattice damage.

The introduction of a small, but controlled amount of physical damage on the back surface of the wafer causes a network of dislocations extending for a small distance into the wafer (68). This damage becomes a sink for other defects which may either be already in the bulk of the wafer or which may be introduced during device processing. Defects migrate by diffusion toward the damaged backside and away from the device region during the high temperature processing. Heavy metal atoms, which are commonly found in device fab areas, such as chromium, iron, copper, gold, etc. are fast diffusing and therefore backside damage offers protection from these contaminants as well.

Backside damage is a process which usually follows polishing. The wafer is held on a vacuum chuck, good side down, and the exposed side is subject to an abrasive technique such as slurry or sand blasting, lapping, grinding or turning. The process is often performed with simple, home-built equipment. An example of one method for backside gettering (69) uses a jet of abrasive slurry, directed at a spinning wafer to produce a generally circular pattern of surface damage. A more recently developed method uses laser-induced damage (70). Naturally, it is important to avoid damage to the front side of the wafer, and the damage invariably produces dust which must be cleaned off. It is difficult to achieve a constant degree of damage. Alternative backside damage methods which can be performed in the clean room, such as phosphorus diffusion or polysilicon deposition are usually preferred. Since any damage technique introduces stress, it is important that the method used does not introduce unacceptable warpage. Close interaction with device processing is recommended in developing a backside gettering process that is effective, yet avoids warpage of wafers.

8.3 Wafer Marking

The traceability of wafers through both the wafer preparation processes and through the wafer fab line is becoming increasingly important. Identification of key parameters affecting yields is made considerably easier when the past history of the wafer can be retrieved along with the source of the wafer and its material properties. SEMI specification M1.1 (9) defines the type of information that may be included in the numbering system and the dimensions of the alpha numeric characters, so that wafers can be automatically read at various stages of processing. The use of lasers to mark silicon has been shown to be an effective method, and the damage caused by the laser apparently does not degrade the wafer provided it is relieved by a chemical etch. Articles by Singer and Scaroni (71),(72) contain good discussions of the methods and equipment used in wafer marking.

8.4 Packaging

Essential for merchant suppliers and useful for captive facilities, packaging protects the finished wafers from contamination during shipping and storage. Packaging containers must be clean so that the wafers are not contaminated by them, and are sealed until the wafers are used. Careful consideration in the design of packaging methods is important to allow for convenience in transferral of wafers to processing containers without contamination of the wafers at the point of opening and transfer. The fact that most fab lines start with a wafer cleaning operation does not negate the importance of protecting the wafer to avoid contamination between wafer preparation and wafer fabrication lines.

9.0 IN-PROCESS MEASUREMENTS

Referring to the wafer preparation process flow diagram in Figure 1, the third column lists measurements that are made at the corresponding process step. Measurements of the material at selected points throughout the process are an essential part of the preparation of wafers. From crystal and wafer shaping through the final wafer finishing steps, measurements are required for quality control, for meeting customer specifications, to correct problems before they create scrap material, and to avoid further processing of reject material. The types of measurements needed depends to a large extent upon the specifications required by the user, however other considerations relating to cost, product improvement efforts, and reliability of the test will also go into the choice. Provided the selection of measurement techniques is made with care, the cost of the in-process measurement is fully justified by the benefits.

The correlation of process control parameters is an important function of in-process measurement. For example, if a parameter is fluctuating randomly and uncontrollably in a process, it may be possible to correlate this fluctuation with another variable, thereby leading the way to better control and higher yield.

Measurements can be broadly classified into mechanical, electrical, structural, and chemical. Mechanical includes physical dimensions such as thickness, flatness, bow, taper and edge contour. Visual inspection is a more subjective test which may also be included in the mechanical category. Electrical measurements usually include resistivity and lateral resistivity gradient, carrier type and lifetime, and may include more detailed measurements like spreading resistance profiles. Measurements giving information on the crystalline lattice are classified in the structural category, and may include, for example, tests for stacking faults, dislocations, slip, and swirl. Routine chemical measurements are typically limited to the measurement of dissolved oxygen and carbon by fourier transform infrared analysis (FTIR). Trace impurity analyses such as spark source mass spectrometry are time consuming and costly and therefore are not used routinely. There are commercial instruments available for all of these wafer measurements and most can be purchased as manually operated, one wafer at a time, to fully automatic, cassette to cassette systems. The survey papers by Matlock (73) and Logan (74) give current information on measurements and available measurement equipment.

It should be noted that many of the above in-process measurements do not relate directly to wafer preparation, rather they are related to crystal growth or to the starting polysilicon material. Results from these tests are fed back to the crystal growth area where the corresponding characteristics are generated. Measurements of material characteristics are performed during wafer preparation because they cannot easily be performed on the as-grown ingots. The in-process measurements to be discussed here will be limited to those directly related to wafer preparation processes, i.e. the mechanical or dimensional measurements.

9.1 Wafer Specifications And Industry Standards

Wafer specifications define what the customer wants. Industry standards define in detail how measurements are to be made. The standards also, in many cases, determine acceptable ranges for measured values, for example the location and dimensions of identification/orientation flats. Standards also provide reliable information for semiconductor equipment manufacturers, raw material suppliers, and to other related support industries. The use of standards is invaluable in resolving vendor/customer problems for it assures that measurements made in different laboratories are comparable.

The standards to which wafers are prepared and measured have been developed over the last two decades by the American Society of Testing Materials (ASTM) and the Semiconductor Equipment and Materials Institute, (SEMI). By 1970, the ASTM committee F-1 had already developed a set of standards for the electrical and chemical measurement of wafers (75). During the early 1970's SEMI issued the first set of standards for wafer dimensions (76) and these quickly met world-wide acceptance. The ASTM/SEMI standards have evolved since then to a comprehensive collection of documents, including both dimensions and measurement techniques for polished wafers (9),(77). The standards encompass not only semiconductor materials, but also chemicals, equipment automation and interfaces, packaging, and photomasks. They are revised periodically and published annually.

The SEMI/ASTM standards which are of concern to silicon wafer preparation are as follows:

SEMI SPECIFICATIONS

M1 Specifications for Polished Monocrystalline Silicon Wafers.

M1.1 Specification for Alpha Numeric Marking of 100 mm, 125 mm and 150 mm Silicon Slices.

M5 Specifications for Edge Coutours on Silicon Slices.

In addition to the above specifications which define standard dimensions, There are a number of standard test methods which are applicable to wafer preparation:

SEMI/ASTM TEST METHODS

F613. Measuring Diameters of Silicon Slices and Wafers.

F671. Measuring Flat Length on Slices of Electronic Material.

F533. Thickness and Thickness Variation of Silicon Slices.

F928. Method to test for the edge contour of silicon slices.

F534. Bow of Silicon Slices.

F26. Determining the Orientation of Semiconductive Single Crystal.

F21. Hydrophobic Surface Films by the Atomizer Test.

F22. Hydrophobic Surface Films by the Water break test.

F24. Measuring and Counting Particulate Contamination on Surfaces.

F523. Unaided Visual Inspection of Polished Silicon Slices.

F612. Cleaning Surfaces of Silicon slices.

F657. Measuring Warp and total Thickness Variation of Silicon Slices by noncontact Scanning Method.

9.2 Mechanical Measurements

It can be seen from the above list that a great deal of effort has been expended on the measurement of dimensional specifications. The precise registration of multiple high resolution patterns is required for integrated circuit fabrication and the pressure on suppliers to produce flatter wafers with more precise dimensions continues.

9.2.1 Diameter And Flat Length: Standard wafer diameters are specified over a range from 2 inch (50 mm. nominal) to 200 mm. The wafer diameter of course, is controlled initially by the crystal grinding process. Using a manual technique, the diameter is measured by the operator, using a simple micrometer caliper. The ground crystal diameter must be sufficiently large to take into account the subsequent removal due to crystal etching, edge grinding, and wafer etching. The allowed tolerance depends upon the nominal diameter, for example, the acceptable range for a 150 mm. wafer is between 149.5 and 150.5 mm. Final wafer diameter is measured on a

20x optical shadowgraph comparator having a translational stage.

Like wafer diameter, the flat length is initially determined at the crystal grinding step, and controls must be developed which take into account subsequent removal by etching. The final flat length is measured on a finished wafer, using a 20x optical comparator with a translational stage. The wafer is aligned on the stage with the flat parallel to the stage translation movement with one end of the flat positioned at a reference point on the screen. It is then moved so the opposite end of the flat is positioned at the reference point. The distance the wafer was moved between measurements is recorded as the flat length. Care should be taken to properly compensate for the effect of edge rounding on the flat length.

9.2.2 Crystallographic Orientation: This measurement is made at the commencement of slicing the crystal. Once the crystal is properly adjusted to the correct angle for slicing, all wafers will be sliced at the correct orientation. The crystal may be checked when it is mounted on the saw, but the usual technique is to check the first slice, making the appropriate angle corrections as needed on the saw. SEMI specification F26 describes both the optical and the X-ray methods of determining wafer orientation. The optical method does not work as well with <100> as with <111>, and it is a destructive test when performed on wafers, because it requires etching of the sample. It can be set up to operate directly on the slicing equipment which negates the possibility of operator error in transferring angle corrections. The X-ray method is non-destructive and more accurate, but must be used separately, employing transfer fixturing which adapts to both the machine and the saw. Both methods can be adapted to either wafers or crystal boules by appropriate fixturing. the X-ray method is usually favored because of its advantages, even though the equipment is quite costly and there is always the hazard associated with the use of X-rays.

9.2.3 Thickness And Thickness Variation: Wafer thickness is measured several times in the processing sequence. The first measurement is made after wafering to control the slicing process. Wafers may be sliced to a tolerance of +/- 0.0025 mm. (+/- 0.0001 in.) with good slicing equipment and technique. Thickness measurements at the saw may be made with a simple caliper micrometer, (probably not suitable for accurately measuring the above tolerance), with an electromechanical indicator, or with a non-contact electronic instrument. Automatic equipment using the non-contact method is available which is capable of measuring and sorting wafers according to thickness categories. Sorting is used prior to lapping, to assure that the lapping process does not result in tapered wafers. Thickness is also measured routinely after lapping, as well as after polishing. Unspecified thickness is usually intended to mean center point thickness.

As the wafer approaches its finished dimensions, the thickness measurements must be more comprehensive. Five separate measurements are made on the wafer, one at the center and four equally spaced around the perimeter, 6 mm. from the edge to avoid edge effects. Total thickness variation (TTV) is expressed as the difference between the maximum and the minimum values. In wafer processing operations, the wafer is invariably held on a vacuum chuck, with its back side theoretically flat, and thus the TTV measurement is a rough indication of the flatness deviation of the front surface. It can be seen, however, from Figure 16-a, that a tapered wafer may theoretically have two perfectly flat surfaces, and therefore more specific definitions of wafer geometry are required.

9.2.4 Flatness: The flatness of the polished wafer is important to the sucessful processing of wafers. The degree of flatness required is dependent upon the application, however, the trend towards larger wafers, larger chip sizes and smaller geometries is requiring tighter flatness specifications in the industry, in general. Optical imaging processes are especially sensitive to flatness deviations, and the degree of flatness is a good indication of the expected performance of the wafer in lithography steps. One of the more important aspects of the SEMI specifications is the definition of flatness and the various properties of wafers that are related to flatness. Flatness and its related paramaters are affected by several of the wafer preparation steps, such as wafering, heat treatment, lapping, and polishing.

Flatness is the deviation of the front surface, measured relative to a specified reference plane when the back surface of the wafer is ideally flat, as when pulled down by vacuum onto an ideally clean flat chuck Figure 16b. This deviation, when measured over the whole wafer is called total indicator reading (TIR), and maximum focal plane deviation (FPD) when measured over a specified area, or site.

Flatness may be expressed as (1) global flatness (over the whole wafer), (2) the maximum value of site flatness as measured on all sites, or (3) the percentage of sites which have a site flatness equal to or less than a specified value.

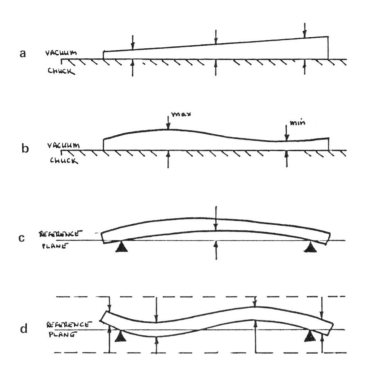

Figure 16: Wafer Mechanical Measurements

(a) Tapered Wafer
(b) Flatness (TIR)
(c) Bow
(d) Warp

9.2.5 Bow and Warp: Figure 16c is a measure of the convex or concave shape of the median surface of the wafer, an imaginary surface equal distant from the front and back surfaces. Bow is therefore a bulk property and not a function of the surface. Although a moderate amount of bow is removed when the wafer is pulled down on a vacuum chuck, excessive amounts of bow cannot be removed in this way, in which case bow will adversely affect wafer performance as if the flatness were degraded. Bow is measured by placing the wafer on a fixture with three points which support the wafer on a circle near the wafer edge. The deviation of the center of the wafer from a theoretical plane defined by the support points is defined as bow.

Since the determination of bow is based on a single center point measurement, it may not detect certain non-planar conditions. Warp is defined as the difference between the maximum and minimum distances of the median surface of the wafer from a defined reference plane Figure 16d. It is used to describe curvature which may have both positive and negative values on the same wafer. Like bow, warp is also a property of the bulk material and not of the surface.

9.2.6 Edge Contour: An image of the edge of the wafers created on a screen with a projection microscope or with an optical comparator, and this image is compared with a template, which has been prepared at the same scale as the image and is mounted on the screen. As long as the wafer contour is within the limits set by the template, the wafer is acceptable as shown in Figure 17.

9.2.7 Surface Inspection: Customarily, this test is performed by unaided visual examination. The wafer is illuminated normal to its surface with a high intensity narrow beam collumnated light or with a diffuse light, and it is viewed at an oblique angle. Defects on the surface can be detected because they scatter the light. It is helpful to inspect the wafers in a black, nonreflective surrounding in order to maximize the inspector's ablity to see defects. Many types of defects have been named, such as pits, orange peel, saw exit marks, crow's feet, edge chips, pits, haze, etc. Although such a test is subjective, it is extremely important, because on the finished wafer, defects can be the result of many causes from the initial sawing process through to the final cleaning and packaging.

Recently, surface inspection instruments based upon laser or light scanning of the wafer have been offered to the industry (78). The sensitivity of these units can be adjusted to display on a video monitor, the number of defects equal to or greater than a certain intensity. Commercial instruments also provide various outputs based upon the collected data. Although it is difficult to obtain agreement between two peices of equipment, the output appears to be quite consistent at a given location, and the method is finding uses in controlling surface quality and wafer cleanliness in individual lines.

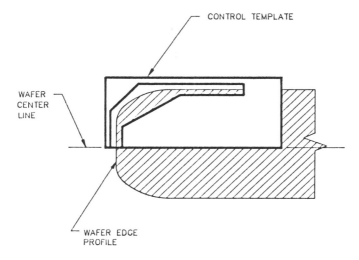

Figure 17: Cross sectional view of typical edge contoured wafer with superimposed control template.

10.0 DISCUSSION

Wafer preparation is often considered one of the more mature technologies in the field of integrated circuit manufacture because it is based upon earlier experience in the abrasive machining and polishing of hard materials such as glass and ceramics. Unique requirements for semiconductor fabrication, however, have resulted in deviations from that experience in a number of ways. If one considers the historical developments over the last two decades, it can be concluded that the technology of wafer preparation continues to evolve, resulting in ever increasing wafer quality. The need for flatter wafers with closer dimensional tolerances continues to be an important factor in wafer preparation methods because circuit geometries have now approached that elusive one micrometer range. Wafer cleanliness and surface defect reduction are becoming more and more important as geometries decrease.

Superimposed on the trends toward closer dimensional control and cleaner, more defect-free surfaces is the continuing trend toward larger diameters. These trends are self contradictory, because the required quality is harder to achieve with larger wafers. Yet 150 mm. production wafers are common now, and 200 mm. diameter wafers are being used in a few pilot lines. The economies of scale in wafer fab are obviously the controlling factor even though the cost per square inch of wafer surface is increases substantially with the larger wafers.

In order to improve dimensional tolerances on large wafers, there are some needs with regard to equipment and control that should be mentioned. In the wafering area, the present measurement of blade deflection is a useful tool for determining that the blade has suffered some degradation or damage, but an advance measure of a blade degradation is the actual cutting force, which begins to increase when the blade needs to be reconditioned by dressing. Commercial equipment suppliers do not offer cutting force measuring devices. Another potential improvement may be the elimination of the cutting beam which is cemented to crystals for wafering. The beam serves two purposes; to prevent breakout edge chipping and to hold the wafer until automatic removal devices can attach to it. Wafers cannot be easily handled with standard automatic handling equipment until the beam has been removed by manual means. Chipping can be prevented by a thin layer as easily as with a large beam, in fact a strippable coating over the whole crystal could serve as a protective layer during both crystal and wafer handling to prevent the type of damage that has been described by Dyer (6), and would simplify automatic handling of wafers directly from the saw.

In the polishing area, wafer handling is still largely manual. Automatic wafer loading and unloading from polishers has not as yet been successfully demonstrated by equipment suppliers. Improved yields would be expected if such equipment could be developed.

Cleaning continues to be an extremely important area, with clean room and water purification technologies making steady improvements.

A quantitative evaluation of surface quality after polishing and cleaning is needed. Surface damage in the first few atomic layers is suspected to be important in device processing, but is difficult to measure. A newly announced method (57) which uses scattered light from a laser is claimed to be able to detect very small amounts of damage such as damage from low energy-low dose ion implantatioin on GaAs and silicon.

The wafer preparation equipment suppliers have always anticipated the needs of the market in their ability to produce equipment to handle the diameters required. They have responded quite well to the needs of the industry in terms of wafer quality improvements. The future requirements are formidable, but the probability that the wafer suppliers will be able to meet the industry needs in the future is high.

ACKNOWLEDGEMENTS

The author would like to sincerely thank Larry Dyer and Sam Rea of Texas Instruments, and Howard Norman of Motorola for valuable comments and assistance.

REFERENCES

1. 1988 *Master Buying Guide*, Semiconductor International, Denver, CO (1988)

2. *Electronics Desk Manual*, Lake Publ. Co., Libertyville, IL (1988)

3. *Microelectronic Manufacturing and Testing 1988 Desk Manual*, Lake Publ. Co., Libertyville, IL

4. *Solid State Technology 1988 Processing and Production Buyers Guide*, Technical Publishing Co., Port Washington, NY

5. Gielisse, P.J., and Stanislao, J., In: The Science Of Ceramic Machining and Finishing, (S.J. Schneider and R.W. Rice, eds.) pp. 5-35, *NBS Special Publication* 348, (1972)

6. Dyer, L.D., "Damage aspects of ingot-to wafer processing" In: Emerging Semiconductor Technology, ASTM 960, D.C. Gupta and P.H. Langer, eds., pp. 297-312, *Am. Soc. for Testing Materials*, (1986) See also: Proceedings of the flat-plate solar array project workshop on crystal growth for high-efficiency silicon solar cells, 3-4 Dec. 1984, pp. 259-277, *DOE/JPL* 1012-109, Jet Propulsion Laboratory, Pasadena CA. (1985)

7. Herring, R.B., Personal communication, August, (1987)

8. Dyer, L.D., In: Proceedings of the low-cost solar array wafering workshop, 8-10 June, 1981, pp. 269-275, *DOE/-JPL*-1012-66, Jet Propulsion Laboratory, Pasadena CA (1982)

9. *Book of Semi Standards* 1987 Vol. 3, Materials, Semiconductor Equipment and Materials Institute, Inc., Mountain View, CA (1987)

10. Gorton, G., "Cutting-Off Machine", *U.S. Patent* No. 1,073,600 (1913)

11. Jansen, M.T., "An improved method for the preparation of serial sections of undecalcified dental tissues", *J. Dental Res.* Vol. 29 pp. 401-6 (1950)

12. Roberts, E. and Johnson, C.M., "Current economic and sensitivity analysis for ID slicing of 4" and 6" diameter silicon ingots", In: Proceedings of the Low-Cost Solar Array Wafering Workshop, *DOE/JPL* 1012-66, pp. 69-77, Jet Propulsion Laboratory, Pasadena CA (1981)

13. Gillis, K., "Plating with General Electric man-made diamond", *Am. Machinist* (1966)

14. Brenner, B., "Industrial diamond tool and method of producing same", *U.S. Patent* No. 2,411,867 (1946)

15. Weiss, S.I., "Methods and apparatus for making annular cutting wheels", *U.S. Patent* No. 3,356,599 (1967)

16. Lane, R.L., "Abrasive cutting element", *U.S. Patent* No. 3,626,921 (1971)

17. Lane, R.L., "ID slicing technology for large diameters", *Solid State Tech.* Vol. 28, pp. 199-123 (1985)

18. Kachajian, G.S., "A systems approach to semiconductor slicing to improve wafer quality and productivity", *Solid State Tech.* pp. 59-64 (1972)

19. Heinrich, R.G., "Cutting apparatus", *U.S. Patent* No. 3,039,235 (1962)

20. Nelke, C.J., and Brucker, J.P., "Hydraulically tensioned saw assembly", *U.S. Patent* No. 3,556,074 (1971)

21. Lynah, F.P. and Ross, J.B., "Design and use of multiple-blade slurry sawing in a production atmosphere" In: Proceedings of the Low Cost Solar Array Wafering Workshop, 8-10 June, 1981, pp. 133-139, *DOE/JPL*-1012-66, Jet Propulsion Laboratory, Pasadena, CA (1982)

22. Wells, R.C., "Wire Saw", *U.S. Patent* No. 4,494,523 (1985)

23. Anderson, J.R., Second Quarterly Progress Report, (Solarex Corp.) *DOE/JPL* 955077-78/3, Jet Propulsion Laboratory, Pasadena, CA (1979)

24. Barry, J.F., et. al., "Method of shaping semiconductive bodies", *U.S. Patent* No. 2,827,427 (1958)

25. Sheff, S., "Electrolytic shaping of germanium, characteristics of the <111>, <110>, and <100> n- and p-type surfaces", *Electrochem Tech.* Vol. 5, No. 1-2, pp. 47-52 (1967)

26. Lane, R.L. "Method and apparatus for slicing semiconductor ingots", *U.S. Patent* No. 4,465,550 (1984)

27. Khattak, C.P., Schmid, F., and Smith, M.B., "Wire-blade development for fixed abrasive slicing technique (FAST) slicing", In: Proceedings Of The Low-Cost Solar Array Wafering Workshop, 8-10 June, 1981, pp. 111-118, *DOE/-JPL*-1012-66, Jet Propulsion Laboratory, Pasadena CA (1982)

28. Raia, E., "Cold cuts", *High Technology* Vol. 6, pp. 58-59 (1985)

29. Christ, M.H., "Edge rounding of silicon slices", *Electronic Pkg. and Prod.* Vol. 18, pp. 160-163 (1978)

30. Herring, R.B., "Silicon wafer technology-State of art 1976", *Solid State Techn.* Vol. 19, pp. 37-42,54 (1976)

31. Guidici, D.C., "Fracture strength of silicon wafers", *Microelectronics* Vol. 9, No. 1, pp. 14-17 (1978)

32. McKee, R.L., "The elusive art of lapping", *Grinding and Finishing*, pp. 21-33 (1967)

33. Murray, C., "Techniques of wafer lapping and polishing", *Semicond. International* Vol. 8, pp. 94-103 (1985)

34. Dudley, J.A., "Abrasive technology for wafer lapping", *Microelect. Manuf. and Testing* Vol. 9, No. 4, pp. 18-20 (1986)

35. Anon. "Direct thickness measurement in lapping process", *Semicond. International* Vol. 7, pp. 258-9 (1984)

36. Newman, D.R., "Ultrasonic thickness measurement for control of machine tools", *Brit. J. NDT* Vol. 5, pp. 73-79 (1972)

37. Katagiri, K., and Honda, M., *U.S. Patent* No. 4,433,510 (1984)

38. Brun, J. and Cazcarra, V., "Lapping method to adapt the silicon wafer bow", *IBM Techn. Disclosure Bull.* Vol. 23, No. 4, pp. 1467-8 (1980)

39. Samuels, L.E., "the mechanisms of abrasive machining", *Sci. American* Vol. 228, pp. 132-142 (1978)

40. Herschel, J.F.W., *Encyclopedia Metropolitana*, article "Light" (1830)

41. Beilby, G., "Surface flow in crystalline solids under mechanical disturbance", *Proc. Roy. Soc.* Vol. 72A, pp. 218-225 (1903), "Flow Theory", *Trans. Opt. Soc.* Vol. 9, p. 22 (1907), See also Aggregation and Flow In Solids, Macmillan and Co. Ltd., London (1921)

42. Rayleigh, Lord, "Polish", *Trans. Opt. Soc.*, London Vol. 19, pp. 38-47 (1917-18)

43. Klemm, W, and Smekal, A., "Uber den Grundvorgang des Polierens von glasern", *Naturwiss.* Vol. 29, pp. 688-690 (1941)

44. French, J.W., "The principles and practice of polishing", *Opt. and Sci. Inst. Mak.* Vol. 62, pp. 253-259 (1921)

45. Bruche, E. and Poppa, H., "The polishing of glass", *J. Soc. Glass Tech.* Vol. 40, pp. 513T-519T (1956)

46. Kaller, A., "Zur poliertheorie des glases", *Silikat Tech.* Vol. 7, pp. 380-390, 399 (1956). See also Jenaer Jahrbuch, s. 145-167 (1957).

47. Schnegg, A., and Lampert, I., "On the polishing of silicon", Abs. No. 271 presented at the *Electrochem. Soc.* meeting, Toronto (1985)

48. Sirtl, E. and Adler, A., Z., *Metallkunde* Vol. 52, p. 529 (1961)

49. Gutsche, H.W., "Surface damage in silicon", Presented at the 6th annual *IEEE Microelectronics Symposium* in Clayton, MO (1967)

50. Lachapelle, R.L., "Process for polishing crystalline silicon", *U.S. Patent* No. 3,328,141 (1967)

51. Payne, C.C., "Silica sol compositions for polishing silicon wafers", *U.S. Patent* No. 4,462,188 (1984)

52. Mandle, R.M., "Process for preparing a polishing compound and product", *U.S. Patent* No. 3,298,807 (1967)

53. Basi, J.S., "Silicon Wafer Polishing", *U.S. Patent* No. 4,057,939 (1977)

54. Walsh, R.J., "Method and apparatus for wax mounting of thin wafers for polishing", *U.S. Patent* No. 4,316,757 (1982)

55. Mendel, E., and Yang, K., "Polishing of silicon by the cupric ion process", *Proc. IEEE* Vol. 57, pp. 1476-1489 (1969)

56. Namba, Y., and Tsuwa, H., In: Laser Induced Damage In Optical Materials: 1980 *NBS Special Publ.* 620, Bennett, H.E., Glass, A.J., Guenther, A.H., and Newman, B.E., eds. p. 171 (1981)

57. Magee, T.J., and Leung, C., "Elimination of excessive surface defects in GaAs IC material using non-contact polishing (NCP)", *Technical report-SBIR Phase 1, ONR* Contract No. N00014-84-C-0057 (1984)

58. Yasunaga, N., Tarumi, N., Obara, A., and Imanaka, O., In The Science of Ceramic Machining and Surface Finishing II, *NBS Special Publ.* 562, Hockey, B.J., and Rice, R.W., eds., p. 171 (1979)

59. Kern, W., and Puotinen, D.A., "Cleaning solutions based on hydrogen peroxide for use in silicon semiconductor technology", *RCA Review* Vol. 31, pp. 187-206 (1970)

60. Balazs, M.K., and Porier, J., "Those confusing pure water specifications: Setting the record straight", *Microelect. Manuf. and Testing* Vol. 8, pp. 22-23 (1984)

61. Lefevre, L.J., "Ion exchange - Crucial mechanism for production of ultrapure water", *Microelect. Manuf. and Testing* Vol. 8, pp. 26-27 (1984)

62. Mayer, A., and Shwartzman, S., "Megasonic cleaning: A new cleaning and drying system for use in semiconductor processing", *J. Elect. Materials* Vol. 8, No. 6, pp. 855-864 (1979)

63. Muraoka, H., Kurosawa, K., Hiratsuka, H., and Usami, T., "Cleaning solutions based on Choline", Abs. No. 238, presented at *Electrochem. Soc.* Meeting, Denver 11-16 (1981)

64. Anon., "Mask and wafer cleaning equipment roundup", *Microelect. Manuf. and Testing* Vol. 7, pp. 24-25 (1984)

65. Burggraaf, P.S., "Wafer cleaning: Part 1, Brush and high pressure scrubbers, Part 2, State-of-the-art chemical technology, "Part 3, Sonic scrubbing", *Semicond. International* Vol. 4, pp. 71-100 (1981)

66. Zoccolante, G., "Innovations in water purification", *Semicond. International* Vol. 10, No. 2, pp. 86-89 (1987)

67. Burggraaf, P., "Ultrapure water focus", *Semicond. International* Vol. 10, No. 6, pp. 128-135 (1987)

68. Lawrence, J.E., "Behavior of dislocations in silicon semiconductor devices: diffusion, electrical", *J. Electrochem. Soc.* Vol. 115, No. 8, pp. 860-865 (1968)

69. Lawrence, J.E., and Santoro, J.C., "Method of preparing high yield semiconductor wafer", *U.S. Patent* 3,905,162 (1975)

70. Eggermont, G.E.J., Falster, R.J., and Hahn, S.K., "Laser induced backside damage gettering", *Solid State Techn.* Vol. 30, No. 11, pp. 171-178 (1983)

71. Singer, P., "Wafer marking and reading", *Semicond. International* Vol. 5, p. 35 (1982)

72. Scaroni, J.H., "Using lasers to mark indentification data on silicon wafers", *Microelect. Manuf. and Testing* Vol. 6, p. 6 (1982)

73. Matlock, J.H., "Current methods for silicon wafer characterization", *Solid State Tech.* Vol. 30, No. 11, pp. 111-116 (1983)

74. Logan, C., "Analyzing semiconductor wafer contamination," *Microelect. Manuf. and Testing* Vol. 8, pp. 9-10 (1984)

75. Bullis, W.M., and Scace, R.I. "Measurement standards for integrated circuit processing", *Proc. IEEE* Vol. 57, No. 9, pp. 1639-1646 (1969)

76. Carlson, J.W., "Industry standardization of silicon substrates", *Solid State Tech.* Vol. 16, No. 9, pp. 49-53 (1973)

77. *Annual Book Of ASTM Standards*, American Society for Testing and Materials, Philadelphia, PA

78. Iscoff, R., *Semicond. International* Vol. 5, No. 2, pp. 77-92 (1982)

5

Silicon Epitaxy

Robert B. Herring

1.0 INTRODUCTION

The term "epitaxial" is applied to a film grown over a crystalline substrate in such a way that the atomic arrangement of the film bears a defined crystallographic relationship to the atomic arrangement of the substrate. The word "epitaxy" is a transliteration of the two Greek words "epi", meaning "upon" and "taxis", meaning "ordered". The epitaxial process is similar to that of crystal growth with the substrate serving as the seed. However, epitaxial processes are differentiated from the crystal growth processes described by Liaw in Chapter 3 of this book in that epitaxy is performed at temperatures well below the melting point of either the film or substrate. Most silicon epitaxial layers used today are prepared by chemical vapor deposition (CVD). Those layers are commonly 1 to 20 microns thick. However, an alternative evaporation/condensation process, molecular beam epitaxy (MBE), widely used for compound semiconductors, has growing applications in silicon technology especially for layers thinner than about 2 micrometers. Three other alternative processes, liquid phase epitaxy, solid phase regrowth, and recrystallization of amorphous or polysilicon layer over insulators are also described in this chapter. However, the emphasis of this chapter is on the technology and equipment for chemical vapor deposition (CVD) (1-5) of silicon epitaxial layers.

1.1 Homoepitaxy and Heteroepitaxy

When a material is grown epitaxially upon a substrate of the same material, the process is called homoepitaxy. That is the case for growth of silicon upon a silicon substrate. If, however, the layer is grown upon a chemically different substrate the process is termed heteroepitaxy. An example of heteroepitaxy is the epitaxial deposition of silicon on sapphire (SOS). That technology has been recently reviewed by Vasudev (6).

Even when the crystal structures of the layer and substrate are basically the same, shifts of composition result in differences in lattice parameters. The resultant mismatch of lattice parameters at the film/substrate interface limit the technologist's ability to produce epitaxial layers of dissimilar materials.

1.2 Applications of Epitaxial Layers

1.2.1 Discrete and Power Devices: Silicon epitaxy was developed to enhance the electrical performance of discrete bipolar transistors (7). The technology change from junction transistors to diffused planar structures in the early 1960's, produced a need for a material structure not achieved by diffusion of dopants from the surface. The breakdown voltage of the discrete transistors was limited by the field avalanche breakdown of the substrate material. Use of higher resistivity substrates produced higher breakdown voltages, but with the performance penalty of increased collector series resistance. The structure needed was a thin, lightly doped but single crystal layer of high perfection upon a more heavily doped silicon substrate. In that structure, the more heavily doped substrate material reduces the collector series resistance while the base-collector breakdown voltage is governed by the lighter doping in the near surface region. Epitaxial deposition of a lightly doped P^+ epitaxial layer on an N^+ substrate is a way of achieving that type of material structure. After additional patterning and diffusion steps to prepare base and emitter regions, the individual transistors were separated, packaged. The backside of the substrate bonded to the package forming the collector contact, while fine gold wires bonded to the emitter and base regions provided the additional contacts as shown schematically in Figure 1.

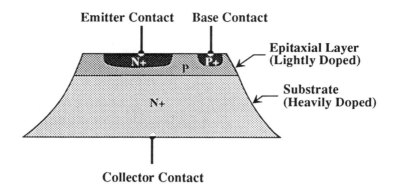

Figure 1. Mesa discrete transistor fabricated in an epitaxial layer on a heavily doped N+ substrate.

Other advantages of fabricating devices in epitaxial layers are the accurate control of doping levels and advantages which arise from the generally low oxygen and carbon levels in epitaxial layers.

The use of epitaxial techniques for discrete transistors was extended to two and three layer epitaxial structures in which not only the lightly doped area of the collector but the base region was grown epitaxially. Today such multi--layer epitaxial structures are used for SCR, Triac, and high voltage or high power discrete products.

1.2.2 Integrated Circuits: The development of planar bipolar integrated circuits introduced the requirement for devices built on the same substrate to be electrically isolated. The use of an oppositely typed substrate and epitaxial layer met part of that requirement. Device isolation was completed by the diffusion of "isolation" regions through the epi layer to contact the substrate between active areas. That structure, shown in Figure 2, is basic to junction isolated bipolar circuits.

In planar bipolar circuits, it is common to employ a heavily doped diffused (or implanted) region under the transistor. That heavily doped subcollector area under the epitaxial layer is usually called a "buried layer" or the "DUF" for diffusion under film. The buried layer serves to lower the lateral series resistance between the collector area below the emitter and the collector contact. In addition the buried layer is needed to produce uniform planar operation of the emitter, avoiding current crowding which leads to hot spots near edges of the emitter.

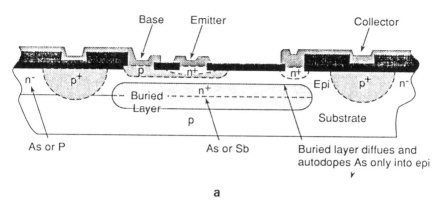

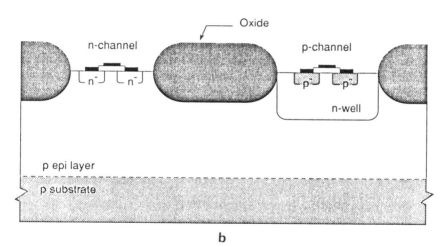

Figure 2. (a) A junction isolated bipolar device fabricated as part of an integrated circuit using a buried layer subcollector and a lightly doped n-epitaxial layer.

(b) An N-Well CMOS structure fabricated in a lightly doped p-epitaxial layer.

1.2.3 Epitaxy for MOS Devices: Unipolar devices such as junction field-effect transistors (JFETs) and VMOS technology also use epitaxial structures (8). Since about 1980, there has been growing use of epitaxial structures to improve the performance of dynamic random-access memories (DRAMs) (9). More recently, VLSI CMOS (complimentary metal-oxide-semiconductor) devices have been built in thin (3 to 8 microns) lightly doped epitaxial layers upon heavily doped substrates of the same type (N or P). That epitaxial structure reduces the "latch-up" of high density CMOS integrated circuits by reducing the unwanted interaction of closely spaced devices.

1.3 Epitaxy as the Complement to Ion Implantation

The techniques of diffusion reviewed by Tsai (10) and ion implantation reviewed by Seidel (11) are available for adding a more heavily doped layer over a lightly doped one. Epitaxy provides the device designer the complementary ability to place a lightly or oppositely doped region over a heavily doped one. In addition, epitaxy provides the ability to contour and tailor the doping profile in ways not possible using diffusion or implantation alone.

However, in addition to control of the dopant profile, epitaxial layers provide a layer of oxygen free material which is also generally low in carbon. That difference can be utilized to prepare "internally gettered" substrates with well developed oxygen precipitates while active devices are built in the precipitate free epitaxial layer (12).

2.0 TECHNIQUES FOR SILICON EPITAXY

The techniques of silicon epitaxy can be classified according to the phase (solid, liquid, or vapor) of the material used in forming the epitaxial layer. Some of the techniques are more commonly applied to epitaxial layers on compound semiconductors than for silicon epitaxy.

2.1 Chemical Vapor Deposition

The most common technique for silicon epitaxy is chemical vapor deposition, usually abbreviated as CVD. This technique first reported by Sangster in 1957 (13) has become the dominant commercial technique for silicon epitaxial film preparation. The technology has been reviewed by Pearce

(1), Liaw and Rose (2), Bollen (3) and others. In the CVD technique, a silicon substrate is heated in a chamber into which reactive silicon containing gaseous compounds are introduced. the gases react on the hot surfaces of the substrate and holder to deposit a silicon layer. Provided the substrate is atomically clean and the temperature is sufficient for the atoms to have surface mobility, the deposit will take on the crystal structure of the silicon substrate. The basic horizontal gas flow reactor is shown schematically in Figure 3.

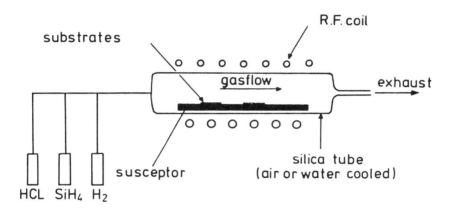

Figure 3. Schematic of a simple horizontal flow, cold wall, CVD reactor.

Subsequent sections of this chapter will deal in more detail with the CVD method for silicon epitaxial film preparation.

2.2 Molecular Beam Epitaxy

Molecular beam epitaxy (MBE) is a non-CVD vapor phase technique that uses an evaporation method. Although the technique was developed in the early 1960's, it has only recently gained major importance for silicon device fabrication. The general principles of MBE have been reviewed by Pearce (1) and Bean (14),(15), and silicon MBE epitaxy has been reviewed recently by Iyer (16).

The principal advantage of MBE for VLSI silicon devices is that the process is carried out at lower temperatures than the 1000-1200 degree C range typical for silicon epitaxy by

CVD. The lower deposition temperature reduces outdiffusion of local areas of dopant diffused into the substrate (i.e. buried layers) and reduces autodoping which is an unintentional transfer of dopant into the epitaxial layer. Thus MBE is the favored technique for preparation of sub-micron thickness epi layers or high frequency devices requiring hyper--abrupt transitions in the doping concentration between the epi layer and the substrate.

2.2.1 MBE - Process Description: In the MBE process, silicon and one or more dopants are evaporated in an ultra--high vacuum (UHV) chamber as depicted in Figure 4. The evaporated atoms are transported at relatively high velocity in a straight line from the source to the substrate. They condense on the lower temperature substrate. Under the high vacuum conditions at pressures of 10^{-8} to 10^{-10} Torr, the mean free path of the atoms (14) is given by:

$$L = \frac{5 \times 10^{-3}}{p} \qquad (1)$$

where L is the mean free path in cm, and p the system pressure in Torr.

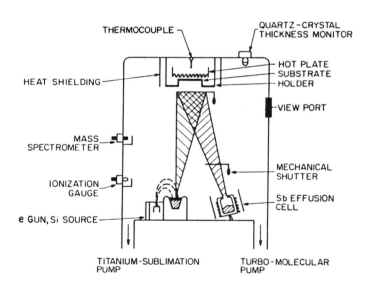

Figure 4. Schematic of a molecular beam epitaxial system (Reference 1).

Provided the substrate is held at a sufficient temperature, (400-800°C), the condensed atoms of silicon or dopant will diffuse about on the surface until they reach a low energy site at which they fit well the atomic structure of the surface. The "adatom" then bonds in that low energy site, extending the underlying crystal by a vapor to solid phase crystal growth.

The usual temperature range for the substrate during deposition is up to 800°C. Higher temperature processes could be used but the increased outdiffusion or lateral diffusion of dopants in the substrate would reduce the advantage of MBE over CVD.

Insitu cleaning of the substrate is accomplished by high temperature bakes at 1000 to 1250°C for several minutes under high vacuum to decompose the native surface oxide and to remove other surface contaminants (17). An alternative clean is the use of a low energy beam of inert gas to sputter clean the substrate. Carbon is difficult to remove but decreases at the surface by diffusion into the substrate during a short anneal at 800-900°C used to reorder the surface damaged by the sputter clean (1).

A wider range of dopants is available for MBE than in CVD epitaxy since the dopant beam is generated by a Knudson effusion cell (18). The preferred operating temperature range for the dopant cells is 200-1300 degrees C. The common N-type silicon dopants such as As and P evaporate so rapidly even at 200 degrees C that they are difficult to control. The common P-type dopant, boron, even at a cell temperature of 1300 degrees, evaporates too slowly for practical use. As a result most MBE uses Antimony (Sb) for N-type layers and either aluminum (Al) or gallium (Ga) for p-type layers. Dopant uniformity with 1% radial uniformity on substrates has been reported by Pearce (1).

An alternative doping technique is use of insitu ion implantation during deposition (19). This technique makes use of a low current (1 micro-amp), low voltage (less than 3 KeV), implant beam while the layer is being grown. Figure 5 illustrates the theoretical and actual profile generated by such an insitu implant doped process.

2.2.2 MBE Equipment: The heart of an MBE system is the ultra-high vacuum chamber. The materials and surface finish of the walls are selected to minimize sticking of gas molecules to ease the maintenance of the 10^{-8} to 10^{-10} Torr vacuum levels. To minimize exposure of the chamber to

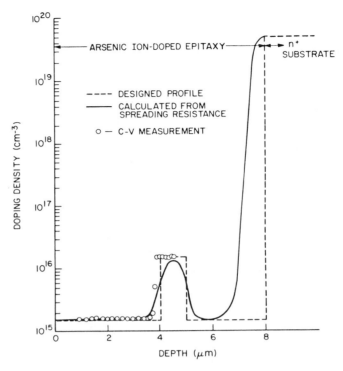

Figure 5. Designed and measured arsenic doping profiles generated by a molecular beam epitaxial system (Reference 19).

atmosphere, the substrates are inserted and withdrawn through a vacuum load lock. The main chamber is commonly pumped by cryopumps.

The components of a practical multi-chamber MBE system (16) are shown schematically in Figure 6. Several ports of the vacuum chamber are devoted to surface analysis equipment used to monitor surface cleanliness and analyze surface atomic structure. The substrate is generally resistance heated while the 400 to 1100 degree C substrate temperature is monitored by thermocouples, optical pyrometry, or infrared detectors.

The perfection of MBE films requires complete surface cleaning and good vacuum levels during deposition. Improvement of vacuum levels and the consequent decrease of impurity flux to the surface results in lowering of contamination on the growth surface. Accumulation of contamination obstructs defect free single crystal growth leading to nucleation of dislocations or stacking faults. Dislocation density,

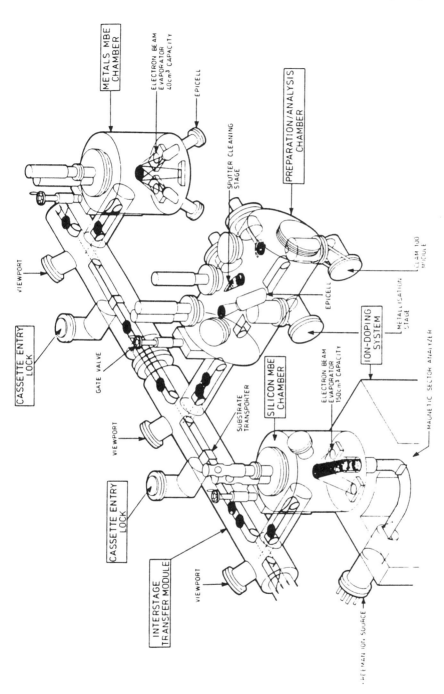

Figure 6. Schematic of a multiple chamber MBE system. (Reference 16)

another measure of film perfection, has been shown to decrease with increased preheat time which improves surface cleanliness. The preheat allows carbon on the surface to diffuse into the bulk, reducing that source of dislocation generation. In addition, operation at higher substrate temperature is also used to reduce dislocation density by increasing the surface diffusivity of added atoms to good fit, low energy sites (17),(20).

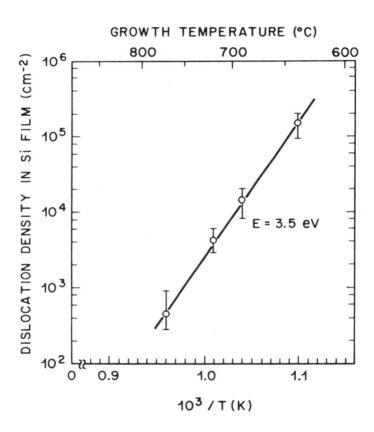

Figure 7. Dislocation density in silicon MBE layers as a function of substrate temperature during film growth (Reference 20).

Silicon Epitaxy 269

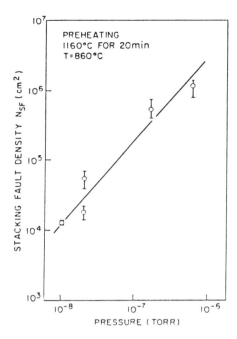

Figure 8. Stacking fault density in silicon MBE layers as a function of the background chamber pressure during epitaxial growth (Reference 20).

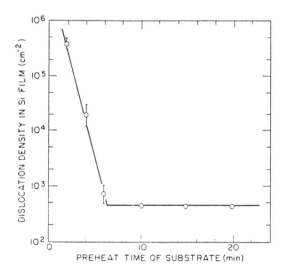

Figure 9. Dislocation density in silicon MBE layers showing the improvement with increased preheat time (for reduction and removal of the surface oxide) (Reference 20).

2.3 Liquid Phase Epitaxy (LPE)

Another non-CVD technique uses the substrate in contact with a liquid melt to freeze an epitaxial layer onto the surface in a manner very much related to the crystal growth techniques described in Chapter 3. The growth of epitaxial layers on silicon from the liquid phase (LPE) can be traced back to the early 1950s to work by Wartenberg (21), Keck and Broder (22), and Goss (23). In these early experiments the silicon was dissolved in a variety of solvents such as aluminum, indium, silver, and zinc. When a seed was introduced into the solution and the temperature lowered to produce supersaturation, layers were grown epitaxially on the seed. This early work which sought to produce bulk crystals of silicon was abandoned as the Czochralski technique was developed.

The first successful liquid phase growth of epitaxial thin films on silicon was reported in 1969 by D'Asaro et. al.(24). This work used either tin or tin-lead as the solvent. A schematic of an LPE system by Baliga (25),(26) is shown in Figure 10.

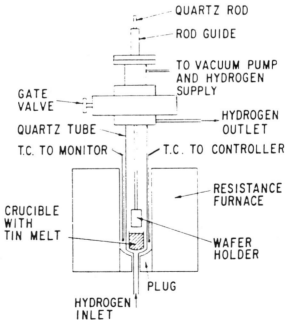

Figure 10. Schematic of a typical silicon liquid phase epi (LPE) apparatus (Reference 26).

This LPE technique is widely used for preparation of epitaxial layers on compound semiconductors and for magnetic bubble memory films on garnet substrates. There the control of a multi component system makes CVD nearly impossible. In film growth by LPE from solution melts, at low cooling rates, when the surface reaction (growth) kinetics are rapid compared with the mass transport of silicon to the seed, the epitaxial layer thickness will vary in proportion to the temperature drop, independently of the cooling rate. With increasing cooling rates, the mass transport rate will increase and the growth rate will increase with cooling rate up to a point at which the growth rate becomes limited by surface reaction kinetics. In Figure 11, it can be seen that the growth rate increases with cooling rate up to about 1 degree per minute while growth at cooling rates above 2 degrees per minute occurred under kinetically limited conditions (25),(26).

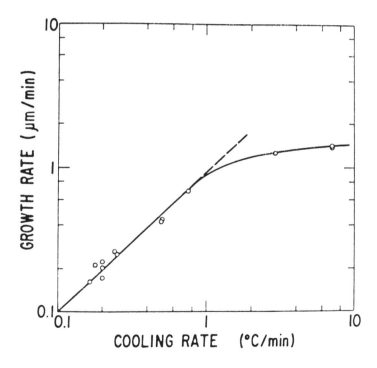

Figure 11. LPE growth rate increasing with cooling rate up to about 1 micron per minute (Reference 26).

Figure 12 shows the fabrication sequence for a vertical channel field effect transistor. The N on N+ epitaxial structure might be built using liquid or vapor phase epitaxial growth. Preferential etching is used to open areas part way through the N type epitaxial layer. In this example by Baliga (26) liquid phase epitaxial growth is used to fill the etched out gate areas which control current flow vertically from the top side source to the N+ substrate drain region.

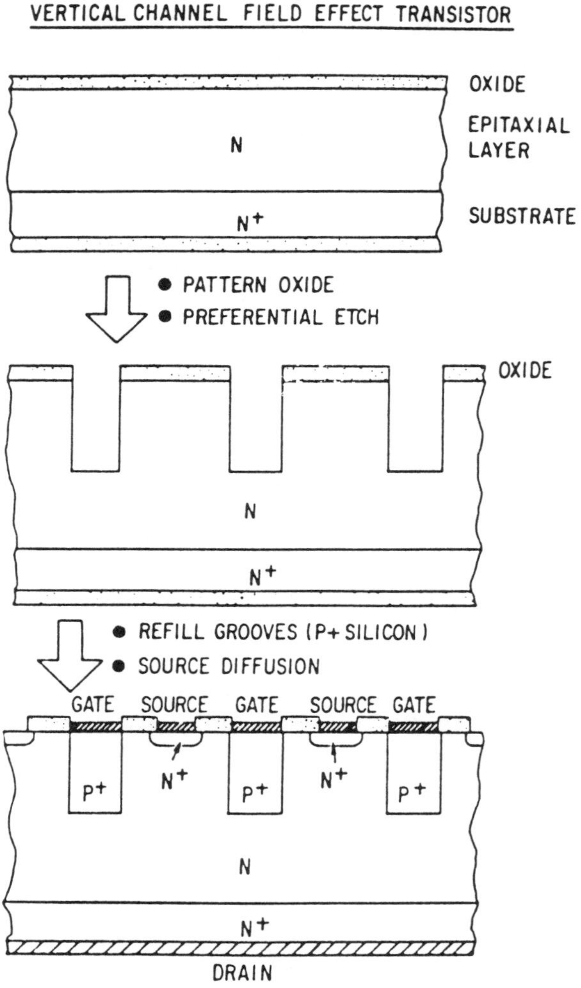

Figure 12. Schematic of fabrication steps in the fabrication vertical field effect transistors by etch and LPE refill techniques (Reference 26).

2.4 Solid Phase Regrowth

2.4.1 Regrowth of Amorphous Layers: Surface layers subjected to high dose ion implants are driven amorphous by the heavy damage inflicted on the lattice as the energetic ions are absorbed. Upon annealing above 600 degrees C, the amorphous layer recrystallizes. The recrystallization proceeds from the interface with the underlying crystal. The reordered interface moves toward the surface resulting in solid phase epitaxial regrowth. There is continuing interest in application of this technique to preparation of thin controlled epitaxial layers.

2.4.2 Recrystallization of Thin Films: Attempts have been made to utilize solid phase regrowth for preparation of controlled thin film epitaxial layers especially for films over an insulating substrate. The process makes use of an amorphous or polysilicon layer deposited by sputtering or CVD upon a substrate of silicon covered by an oxide or nitride. High dose ion implantation with an inert gas such as argon or implantation with silicon may be used to disorder the layer. The layer may be capped by an oxide or nitride to prevent evaporation of dopant and to control heat during the recrystallization. Subsequent heating using a resistance strip heater (27) shown schematically in Figure 13, or heating by a CW (continuous) laser (28) scanned over the surface are used to regrow the layer. Small holes in the oxide or nitride provide contacts between the film and the substrate to serve as seed points for epitaxial growth.

The solid film regrowth technique appears restricted to preparation of thin layers of one micron or less due to the implant energy required to disorder the material to greater depth. In addition, in the anneal of implants, grain nucleation and growth near the surface, has been observed as a competing process to the upward movement of the substrate interface. In layers over one micron such surface nuclei have time to grow to stable size ahead of the epitaxial regrowth interface.

A related solid phase technique relies on the recrystallization of a deposited amorphous or polysilicon film. For this technique, a silicon film is deposited upon a silicon substrate or more commonly upon an oxide over silicon and then heated generally by a strip heater passed over the surface or by a scanned pulsed laser to recrystallize the film to single crystal or large grain polysilicon. A schematic illustration (1) of this technique is shown in Figure 13.

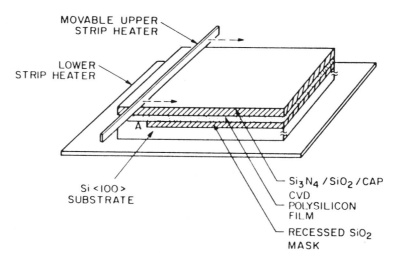

Figure 13. Recrystallization solid phase epitaxy using a moving strip heater.

Figure 14 shows schematically the use of this fabrication technique to produce a stacked n-channel device in recrystallized polysilicon (29) over a thermally grown or deposited oxide. Oriented epitaxial growth may be accomplished by making a series of holes in the oxide to allow points of contact between the underlying substrate and the deposited polysilicon film. The contact points become "seed" areas for establishing the regrowth orientation.

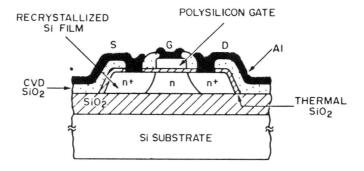

Figure 14. A stacked MOS structure over an insulating oxide fabricated in a recrystallized poly-silicon layer.

The solid phase process is today a laboratory technique and much additional work remains before large scale use, but solid phase techniques have considerable commercial potential. With additional development it could rival MBE as a technique for preparation of sub-micron epitaxial layers.

3.0 SURFACE PREPARATION FOR SILICON EPITAXIAL GROWTH

3.1 Surface Cleaning and Oxide Removal

Contamination on the substrate surface will interfere with the registry of the atoms of the deposited layer. Therefore a basic requirement of epitaxy is an atomically clean surface just prior to beginning deposition. This is accomplished through some combination of precleans outside the reactor and insitu cleans done in the reactor.

3.1.1 Surface Precleans: Chemical precleans such as hot sulfuric acid plus hydrogen peroxide are required if heavy organic contamination such as residual photoresist is expected on substrates coming to the epitaxial process step. If heavy organic residues are not expected, the "RCA-1 clean" (30) using ammonium hydroxide, hydrogen peroxide, and water may be sufficient. This solution is generally used heated to about 50-85°C or at lower temperatures in agitated "megasonic" cleaning tanks especially effective for removal of small micron sized silicon dust particles from the surface (31).

Thick oxides such as those used to mask the patterned diffusions known as the "buried layer" are commonly stripped in HF solutions. This step may also be carried out using HF vapor etching of the silicon dioxide. There is some evidence that metallic contaminants plate out of aqueous HF solutions onto silicon. Hence, the HF vapor stripping of the masking oxides is to be preferred for minimizing the metallic contamination of the substrate surface. If an aqueous HF solution is used, it may be followed by an "RCA-2" type clean (30) using HCl-hydrogen peroxide solutions to complex and remove the surface metallics. The chemical cleaning steps employ D.I. water rinses between cleaning steps.

In the past, it has been common to use a dilute HF solution plus a D.I. water rinse as the last wet cleaning step to minimize the native oxide thickness. However, since the native oxide will rapidly reform and grow to a few monolayers upon exposure to air, some insitu oxide removal is always required. There is evidence that the aqueous HF solutions are a major source of small micron sized particles on surfaces which can produce defects in the epi layer. Hence the growing practice in epi precleaning employs a "megasonic" agitated cleaning bath containing hydrogen peroxide to lightly oxidize the silicon surface. Small micron and submicron sized particles such as silicon dust, held on the surface by electrostatic forces, are efficiently removed by the "megasonic" agitation of a surfactant solution (31).

3.1.2 Drying the Wafers: After the last wet cleaning step followed by a D.I. water rinse, the material must be dried without introduction of contamination to the surface. This is commonly done by use of a centrifugal rinser/dryer. The types which spin two or more carriers about an axis off the wafer surface are preferred over the single cassette type which spin the wafers about an axis through the center of the wafers. Rinser/dryers of most designs accumulate bacteria in the water feed lines. That residue will produce streaks of organic material on the silicon surface which will not rinse away. The organic material will break down during wafer heating but leave high carbon levels in the track of the streak. The high carbon levels can nucleate epi defects and must be avoided. Therefore, it is better practice to operate these dryers in the "dry only" mode. The design of the dryer must be such that particle generation is minimized. The rotor bearing design, materials of construction, filtration of the inlet gases, and prevention of suck-back of residue from the exhaust line are critical to obtaining dry wafers free of contamination.

An alternative to the centrifugal type dryer is the alcohol vapor dryer. These units began as an offshoot of industrial vapor degreasers. Recently improved designs specifically dedicated to silicon wafer drying have been introduced. The cool wet wafers in carriers are lowered into a warm alcohol vapor zone. The cool surfaces condense alcohol vapors which dissolve and remove the water from the surfaces. As the parts are warmed by the vapors, condensation slows. Then as the parts are lifted slowly from the vapor zone the residual alcohol evaporates leaving the parts dry. This vapor drying process when properly operated will dry the substrates with little chance of introduction of small particles to the surface. Older dryers of this type were abandoned in the early 1970's due to concerns about safety of the flammable vapors. However, the newly designed units are enclosed, fully automated, and employ several sensors and interlocks for improved safety.

3.2 Insitu Gas Phase Cleans

Silicon surfaces exposed to air even in cleanrooms used for device fabrication can be expected to have a thin native oxide some 0.5-1.5 nm thick plus some level of carbon at the surface. The native surface oxides which appear to be a hydrated silicon dioxide, must be removed by cleans done "insitu" inside the reactor after the air has been removed.

3.2.1 Removal of the Surface Oxide: It would be possible thermodynamically to remove the surface oxide using a hard vacuum environment and heating the substrates. However, that removal approach is not realistic due to the very high vacuum levels required. In molecular beam epitaxy the oxide is frequently removed by an argon sputter clean (16), while CVD reactors normally employ hydrogen reduction of the native oxide.

Figure 15 shows the calculated line of stability calculated from thermodynamic data for silicon dioxide heated at atmospheric pressure in a hydrogen ambient with low levels of oxidizer (32). For oxidizer levels above the line, the oxide is stable while in the region below the line, silicon dioxide will be reduced. In vapor phase epitaxial growth it is necessary to hold substrates in the region below the line of Figure 15 to reduce the surface oxide prior to beginning epitaxial growth.

Experimental results reported by Ghidini and Smith (33) for the growth of oxide on (111) and (100) oriented silicon from traces of water vapor in an inert ambient suggest the line of Figure 15 may be better represented as a transition band rather than a sharp line separating the regions of clean silicon from oxide covered silicon.

The phase stability data of Figure 15 is shown replotted in Figure 16 in terms of hydrogen to oxidizer ratio versus temperature. When replotted in this way, the stable oxide region is below the line. Above the transition band the surface oxide is reduced and clean silicon surfaces are produced. Principal sources of oxidizer are air or water vapor from leaks in seals of the reactor chamber or in the gas delivery lines or gases adsorbed on surfaces during loading of the reactor. Two cases are shown in Figure 16: one being a leak tight reactor in which the hydrogen to oxygen ratio rises rapidly during heating to plateau at a 0.5 ppm oxidizer level. In this case the process crosses into the oxide reduction region around 950°C. In that type reactor, clean surfaces and epitaxial growth could be carried out at any temperature

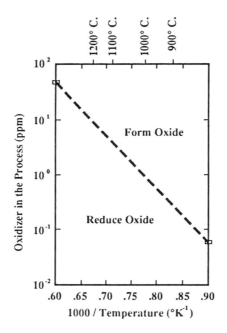

Figure 15. Boundary of the thermodynamic stability region for SiO_2 versus reciprical absolute temperature. Surface oxide is stable above the line but reduced in the region below the line.

above that point. In the second case, the reactor has some greater leak rate and plateaus at 5 ppm oxidizer. In that case, the process will not cross into the oxide reduction region until temperature is raised above about 1125°C. This makes clear the increasingly stringent hydrogen purity requirements for low temperature epitaxial processes.

Figure 17 shows experimental results for oxide removal performed in an AMC-7800 cylinder reactor by rapid heating in hydrogen to temperatures of 1100, 1150, and 1200°C. Following a 3 minute hold at temperature, the power was turned off and the samples rapidly cooled. Extrapolation of the data suggests the oxide removal proceeds above 1060°C under these conditions. This test was performed using thermal oxide grown by dry oxidation at 1000°C. The native oxide formed in pre-epitaxial cleaning may be less well bonded and therefore removed at lower temperature, but Figure 16 shows a method of testing leak integrity and gas purity for an epitaxial reactor and illustrates the application of concepts derived from Figures 15 and 16.

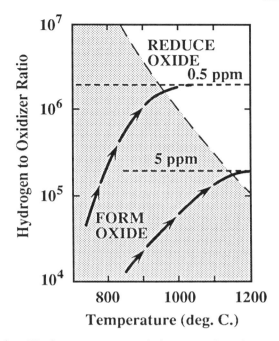

Figure 16. Hydrogen to oxidizer ratio in the reactor increases as water vapor is driven off the interior surfaces. Hydrogen purity becomes limited by the purity of the incoming gases.

        (a)   a tightly sealed system with 0.5 ppm oxidizer and

        (b)   a less well sealed system with 5 ppm oxidizer in the inlet gases.

As will be discussed in a later section, epitaxy using silane is generally performed at temperatures around 950 to 1000°C. As shown by Figure 16, this places more stringent leak rate constraints on the reactor seal design and maintenance than for processes using the chlorinated silanes at temperatures of 1100-1200°C.

3.2.2 Removal of Adsorbed Water Vapor: Since most CVD epitaxial reactors are not loadlocked, the inner surfaces adsorb water vapor during unloading and loading operations. The CVD epitaxial reactor is commonly purged with nitrogen to displace the air and then purged with hydrogen during heating to remove most of the water vapor adsorbed on surfaces during loading.

Figure 18 shows the typical process sequence used for epitaxial depositions. The substrates are heated in a hydrogen

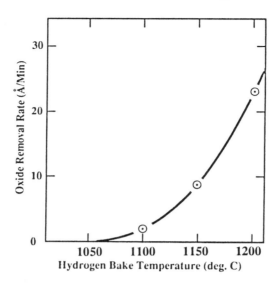

Figure 17. Rate of oxide removal versus temperature in an AMC-7800 cylinder reactor during a hydrogen bake (3 minutes) at atmospheric pressure. Extrapolation shows the onset of oxide reduction at about 1060°C.

ambient from room temperature to about 700-800°C. During that heat-up, water vapor and other oxidizer is driven off from the reactor surfaces. The wafer holders are frequently made of coated graphite which will adsorb and hold water vapor if allowed to cool during the loading operation. Another area which adsorbs water vapor is water cooled metal surfaces in the cooler inlet and outlet regions of the reactor. The hydrogen to oxidizer ratio rises rapidly as the reactor is purged and heated, since the water vapor is removed rapidly from the heated surfaces. However, the hydrogen purity becomes limited for several minutes by the slower release of water vapor from cooled surfaces such as water cooled metallic parts near chamber seals. A very common process step, (Step D in Figure 18), is a dwell or bake at a temperature below the oxide stability line. The purpose of the dwell is to allow time for the water vapor to be desorbed and for the hydrogen purity to rise to the level limited only by chamber leaks and inlet gas purity.

A recent experimental study by Roberge et. al. (34) of the water vapor level in an epitaxial reactor showed a strong dependence on the time the reactor was open to the atmo-

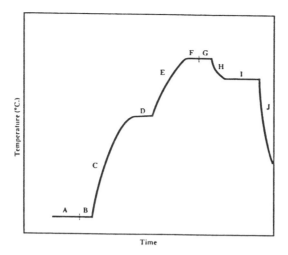

Figure 18. Temperature vs time for a typical CVD silicon epitaxial cycle:

Step	Purpose
A	Reactor unload/load
B	Nitrogen/Hydrogen purge
C	Heat to bake temperature
D	Bake to remove water vapor
E	Heat to oxide reduction region
F	Bake to remove surface oxide
G	Gas phase etch (HCl)
H	Cool to epitaxial temperature
I	Epitaxial film growth
J	Cool in hydrogen then nitrogen

sphere and the temperature of the surfaces during the loading of substrates. That study found peak water vapor levels from a cold start over 15 ppm compared to peak levels under 5 ppm for cycles started with the reactor surfaces heated. The difference is due to the decreased water adsorption on the heated surfaces during loading. That study (34) also quantified the well known impact of water vapor and other gas contaminants on surface texture of the epitaxial layer.

At the end of the bake step, the oxidizer level in the reactor is largely determined by the purity of the inlet gases. The hydrogen used for epitaxial deposition is generally obtained

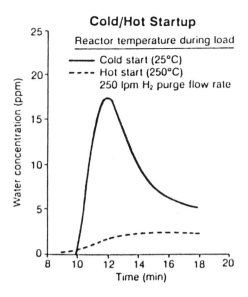

Figure 19. Water vapor concentration measured during heating in a modified vertical (pancake) reactor comparing two different starting conditions (Reference 34).

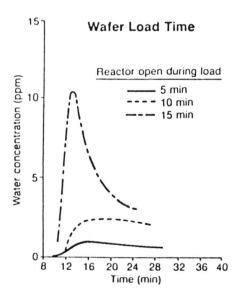

Figure 20. Water vapor concentration measured inside a vertical reactor for varying reactor open periods during reactor unload/load (Reference 34).

by vapor boil-off from a liquid hydrogen tank with delivery through high integrity stainless steel lines with welded connections. With proper attention to details of the piping installation, it is common to deliver hydrogen into the epitaxial reactor with a purity of 0.1 to 0.5 ppm oxidizer. Delivery systems not meeting these purity requirements may require use of a point of use gas dryer system to remove oxidizer to levels below 0.5 ppm.

3.2.3 Oxide Removal: Step E of the process in Figure 18 is further heating of the substrates to cross the oxide stability line into the region where the surface oxide is reduced and removed. It is essential that the stability line be crossed rapidly and the substrates carried to a condition well into the reduction region. In the transition zone in the vicinity of the stability line, the native oxide will be reduced in local regions, reoxidized, and re-reduced. The local repeated oxidation and removal leads to roughening or pitting of the surface which persists even after epitaxial film growth. Results of the study by Roberge et. al. (34) also showed the haze level, or degree of off-axis light scattering produced by surface texture of the epi, was well correlated with the oxidizer levels intentionally introduced in the reactor during heat-up. Those results plotted in Figure 21 show higher haze levels for N-type (111) substrates than for P-type (100) for equivalent oxidizer levels.

Once the oxide stability line of Figure 15 is crossed, the native surface oxide is reduced by the hydrogen such that within a minute or two an oxide free silicon surface results. Silicon gas phase etching using hydrogen chloride or other etchants may be used if damaged silicon is present but etching should not begin before the oxide is removed or pitting will result.

3.2.4 Carbon on the Surface: Another substrate surface contaminant is carbon which may come from organic compounds left from wet chemical cleaning or the carbon may be deposited on the surface during the reactor heating. In general, the carbon will diffuse into the surface at 1050 to 1100°C at a rate sufficient to deplete the surface to levels low enough that high quality epitaxial films can be deposited. That is one of the purposes of a higher temperature bake cycle such as shown in Figure 18 prior to the deposition step.

Carbon at the substrate/epi interface has been reported by Reif (35) as a principal limitation to good quality low temperature (<900°C) epitaxial growth. Insitu removal of the carbon is difficult except by diffusing it into the substrate.

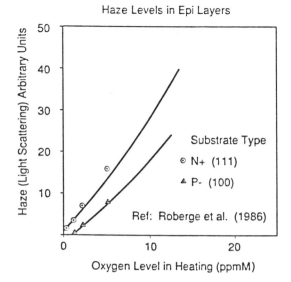

Figure 21. Haze levels as measured by off-axis light scattering for epi grown under conditions of varying levels of oxidizer in the CVD reactor ambient during heat-up prior to growth (Reference 34).

Hence, the control of carbon on the substrate surface is a key requirement of lowering epitaxial temperatures into the range below 1000°C.

3.3 Insitu Etching

When the epitaxial process for silicon was introduced in the early 1960's, wafers were polished using fine diamond paste which left a surface damaged layer up to 2 micrometers deep. That damaged layer had to be removed in order to grow defect free epitaxial layers. Earliest practice used a wet chemical etch to remove that damaged layer, but the use of gaseous anhydrous HCl vapor etching (36),(37) was developed by 1970. The improvement in wafer polishing with the chem-mechanical method by the mid-1970's removed the original need for surface removal. However, reliance on HCl vapor etching as a pre-epi insitu clean has persisted.

Most epitaxial processes which use HCl etching today for high speed bipolar devices remove only 100 to 300 nm of surface material. This is especially true for bipolar processes where etching removes material from the diffused "buried layer" or subcollector regions of the devices.

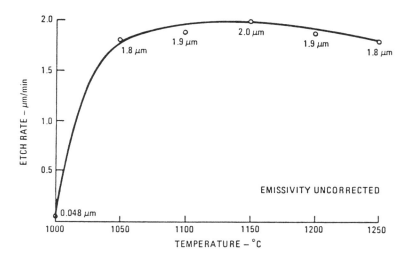

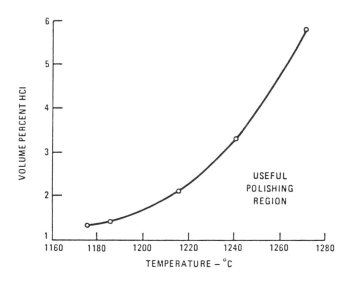

Figure 22. (a) Influence of temperature on HCl gas etch rates.
(b) Limits of the useful polishing etch region. HCl concentrations above the line tend to pit the surface.

Because of a tendency of the HCl vapor etch to pit the surface when removal of the native oxide is incomplete, it is common to perform the HCl etch at 1150-1200°C, then decrease the process temperature during deposition. This high-low cycle also serves to deplete the surface of carbon which diffuses into the substrate and to deplete the surface of dopant through evaporation to reduce autodoping (38).

In cases in which the surface of the "buried layer" has been doped to such high levels that the surface exceeds solid solubility for the epitaxial temperature, HCl etching removes damaged substrate layers and results in improved epitaxial quality. However, for cases in which the substrate surface is clean and of good crystallographic perfection, HCl etching may not be required.

Water vapor from air leaks in stainless steel lines, regulators, and valves of HCl delivery systems result in wet HCl corrosion of those components. The resulting metallic chlorides are volatile enough to be carried into the reactor and to contaminate the substrates. High metallic contamination levels result in pitting or haze on the epi surface, however, lower levels which are hard to identify may still degrade devices and produce junction leakage and lower minority carrier lifetime.

The typical CVD epitaxial cycle is summarized in Figure 23 as a plot of hydrogen/oxidizer ratio versus temperature. Loading, heat-up, and outgassing the reactor all occur in the region of surface oxide stability. After a dwell (A-B) to allow time for water vapor desorbtion, further heating carries the process across the stability line into the region where surface oxide is reduced and removed by the hydrogen. Once atomically clean substrate surfaces are obtained, gas phase etching (if used) is carried out at a higher temperature to avoid pitting the surface. Finally the temperature may be lowered to near the oxide formation boundary and silicon bearing gases introduced for epitaxial film growth as discussed in the following section.

4.0 GROWTH OF SILICON EPITAXY BY CVD

4.1 Growth Chemistries

The chemical vapor deposition of silicon can be carried out by three basic types of reactions: (1) disproportionation, (2) pyrolysis, and (3) hydrogen reduction. Each of these methods has certain advantages and disadvantages.

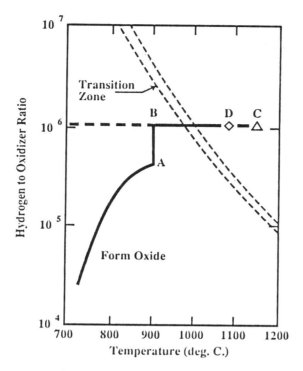

Figure 23. A typical CVD epi cycle employs a bake (heat to 800-900°C and hold 3-5 minutes) at a temperature below the oxide reduction boundary to complete the drive off of adsorbed water vapor (A to B), heating to beyond the oxide reduction boundary (B to C), an optional bake or HCl etch (C), and cooling to the epi growth temperature (D), generally done below the gas etch/bake temperature.

4.1.1 Disproportionation: The disproportionation reaction involves dissociation of divalent silicon halides, SiX_2, into solid silicon and a four-valent silicon halide, SiX_4, which remains in the vapor phase according to the reaction:

$$2\ SiX_2(v) \rightarrow Si(s) + SiX_4(v) \qquad (2)$$

The above reaction has been carried out using SiI_2, $SiCl_2$, $SiBr_2$, and SiF_2 with silicon iodide, SiI_2, the most commonly used reactant. The SiI_2 is generated by passing iodine vapor over a charge of polysilicon at about 1150°C upstream of the substrate upon which deposition of the epi layer is to occur. When the SiI_2 vapors are transported to the vicinity of the substrate held at a lower temperature around 900°C, the reaction is reversed and a solid deposition of silicon results. If a single crystal substrate is placed in the disproportionation region, a single crystal film may be obtained.

This technique has been reported by Glang and Wajda (39), May (40), Taft (41), and Hanssen et al (42). The principal advantages are that this technique requires lower temperatures leading to the possibility of abrupt doping transition from the substrate to the epitaxial film. In addition, the growth takes place under near equilibrium conditions which reduces the nucleation of silicon on some foreign substrates. Thus the technique is ideal for growth of selective epitaxial deposits on silicon or within patterned areas of silicon substrates. The principal disadvantage of the technique is that it requires a closed tube system operated at low pressure. That makes introduction of dopants difficult and the technique has not been widely used for production of silicon epitaxial films.

4.1.2 Pyrolytic Decomposition: Epitaxy can also be obtained by pyrolytic decomposition of SiI_4 or SiH_4. The pyrolysis of SiI_4 was studied in 1960 by Herrick and Krielbe (43) for production of semiconductor grade polysilicon. The early work on the pyrolytic decomposition of silane for epitaxial films was by Joyce and Bradley in 1963 (44). A straight pyrolysis of dichlorosilane, SiH_2Cl_2, using an inert carrier gas is possible but rarely used. The deposition reaction using dichlorosilane will be discussed below under hydrogen reduction reactions.

The pyrolysis reaction produces solid silicon and hydrogen:

$$SiH_4(v) \rightarrow Si(s) + 2H_2(v) \qquad (3)$$

The reaction occurs readily at temperatures above 600°C for polysilicon, with single crystal films obtained at temperatures above 850-900°C. For polysilicon, this reaction is generally carried out at low pressure in pure silane or using an inert carrier gas such as argon or nitrogen.

When epitaxial films are desired, the reaction is generally carried out at atmospheric pressure (or reduced pressures around 1/10 atmospheric) using a carrier gas which is generally hydrogen to obtain a reducing ambient to remove and avoid regrowth of surface oxide on the substrate. The higher gas velocity obtained in a pumped system at reduced pressure is helpful in reducing autodoping and in promoting thickness and resistivity uniformity.

The hydride, silane, is passed over the heated substrate and decomposes on contact, depositing the silicon film. The hydrogen released by the pyrolytic decomposition of silane or dichlorosilane desorbs from the surface and is transported away in the exhaust of the reactor.

The reaction is not reversible, operates far from equilibrium with a high negative enthalpy. As a result this reaction has the potential for high rates of growth at high temperatures if reactant supply can be sustained. However, one common problem is the formation of gas phase nuclei which fall as spurious seeds upon the epitaxial layer leading to defects when the silane partial pressure and temperature are raised for high rates of growth. However, Bloem (45) has shown that the introduction of some HCl with the silane suppresses the gas phase nucleation and using that technique, single crystal films were grown at rates up to 40 micrometers per minute using silane in a jet impinged against a heated surface.

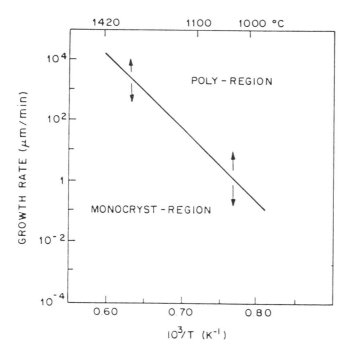

Figure 24. Growth rate boundary of the monocrystalline/polycrystalline transition for silane growth versus reciprocal absolute temperature as determined by Bloem (Reference 45).

Figure 24 shows the limits of epitaxial growth rate versus temperature determined by Bloem. Deposits formed at rates above the line were clearly polycrystalline. Just below the line, films were epitaxial but with significant defect density.

In most commercial batch reactors, the growth rates used are about 0.1 to 1.0 micrometer/min. Beyond that rate, the defects from gas phase nuclei become troublesome. In addition, since the reaction proceeds slowly above 400°C, powdery deposits of silicon form in lower temperature zones of the reactor and this requires frequent cleaning of the reactor to avoid defects in the films. These cleanings reduce the productivity of the reactor and are a disadvantage for this technique. An additional disadvantage is that the silane source material is more costly than chlorosilanes per gram of silicon deposited. However, very high purity silane is available for growth of very high resistivity films (> 500 ohm-cm). An additional disadvantage of silane is the requirement for a more leak tight reactor system. Operating at the lower temperature requires that the oxidizer be reduced to avoid surface oxide formation and defects in the epitaxial layer as discussed in section 3 above. This places higher requirements on reactor design, maintenance, and gas delivery systems.

The lower temperature of deposition and freedom from chlorine transport of the substrate atoms are advantages for heteroepitaxy such as silicon on sapphire where the lower temperature is needed to suppress the outdiffusion and auto doping of aluminum from the substrate. An additional advantage of silane for CVD epitaxial growth of silicon is the simplicity of delivery and control for a gaseous source, and the small amount of pattern shift or distortion for films grown over patterned "buried layers" (46).

For homoepitaxy of silicon the chlorosilanes are the preferred source because they lead to higher reactor productivity. The disadvantages of silane for commercial use are the higher costs and the requirement for higher leak integrity systems which are a consequence of the lower operating temperature.

4.1.3 Reduction of Chlorosilanes: The most common CVD processes for silicon epitaxy are based on the hydrogen reduction of chlorosilanes. Among the chlorinated silanes, the monochlor, SiH_3Cl is unstable and not available. Of the higher chlorinated forms, dichlorosilane, SiH_2Cl_2 is an easily liquefied gas. Since becoming available in commercial quantities in the mid-1970s, dichlorosilane has been growing in importance as the source material for CVD silicon epitaxy. The decomposition of dichlorosilane takes place by pyrolysis without requiring hydrogen reduction:

$$SiH_2Cl_2(v) \rightarrow Si(s) + 2HCl(v) \qquad (4)$$

However, the common practice is to heat silicon substrates in hydrogen to 1050-1150°C to desorb water vapor from the reactor surfaces and reduce the native oxide prior to growth of the expitaxial film. The hydrogen is continued during deposition as a carrier or diluent gas and to maintain a strongly reducing atmosphere to prevent the formation of surface oxide which would disrupt the single crystal growth or introduce defects.

Silicon epitaxial growth using dichlorosilane is generally carried out in the 1050-1150°C range. The growth rates of silicon epitaxy which depend on the substrate temperature and the partial pressure of dichlorosilane are generally in the 0.1 to 1.5 micrometers per minute range.

Trichlorosilane, $SiHCl_3$, and silicon tetrachloride, $SiCl_4$, are both volatile liquids at room temperature. They are generally used with a carrier gas, generally hydrogen, saturated with the source material by bubbling the carrier gas through the liquid. The carrier gas, saturated with the source material vapor, is introduced into the reactor chamber along with a main stream of hydrogen. The higher chlorinated silanes undergo hydrogen reduction with HCl byproduct desorbed and transported away in the exhaust gas stream.

$$SiH_2Cl_2(v) \rightarrow Si(s) + 2HCl(v) \qquad (5)$$

$$SiHCl_3(v) + 3/2\ H_2 \rightarrow Si(s) + 3HCl(v) \qquad (6)$$

$$SiCl_4(v) + 2\ H_2 \rightarrow Si(s) + 4HCl(v) \qquad (7)$$

Trichlorosilane is the silicon based chemical produced in largest commercial quantity for use in making silicone rubber products and for production of semiconductor grade bulk polysilicon (see Chapter 2). Because of those other uses, trichlorosilane is the lowest cost source material for silicon epitaxy. Because of the lower cost, trichlorosilane has been the most common source material for the production of thick (30 - 100 micrometers) epitaxial layers for discrete power devices. The higher temperatures of 1100-1250°C required

Table 1

Boiling Points and Melting Points of Silanes and Chlorosilanes

Compound	Phase	bp (°C)	mp (°C)
SiH_4	gas	-112	-185
SiH_2Cl_2	gas	8.3	-122
$SiHCl_3$	liquid	31.8	-127
$SiCl_4$	liquid	57.6	-68

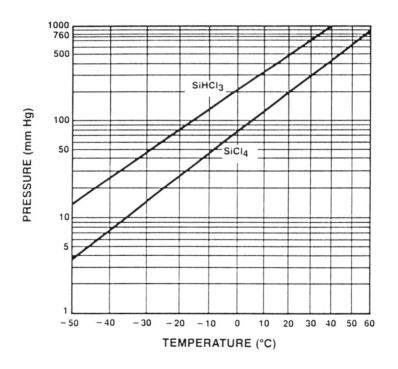

Figure 25. Vapor pressure for trichlorosilane ($SiHCl_3$) and silicon tetrachloride ($SiCl_4$) versus temperature. Due to the subatmospheric vapor pressure, these source chemicals are usually delivered by a bubbler system using an inert carrier gas.

for the hydrogen reduction of this more stable compound limit its use for thin epitaxial layers where narrow transition widths are required but don't present a problem for the thick power device epitaxial films.

Silicon tetrachloride, $SiCl_4$, is obtained as a by-product of the production of semiconductor grade polysilicon from trichlorosilane. In large vertically integrated silicon plants, the byproduct silicon tetrachloride is separated from the polysilicon plant's recycle stream by distillation for use in growth of silicon epitaxial layers. The greater thermal stability of the compound requires that temperatures of 1125 to 1225°C be used to achieve decomposition rates needed to produce growth rates of 0.2 to 1.0 micrometer/min. Those high temperatures are incompatible with the abrupt, narrow transitions of dopant level from substrate to epitaxial layer required for high speed VLSI structures. As a consequence, use of silicon tetrachloride is rapidly declining in silicon epitaxy for integrated circuits.

4.2 Growth Kinetics and Mechanisms

The most common model for addition of atoms to the growing epitaxial layer is the so-called plateau-ledge-kink model shown in Figure 26. In this model, the incoming silicon atom reacts on the plateau region. However, in that type site it forms only one or two bonds, and will surface diffuse to a "better fitting site". The atom surface diffuses to a growth ledge which is a site of higher binding and then along the ledge to a kink site. There the adatom makes several near neighbor bonds, which reduces its ability to diffuse along the surface. Adatoms which fail to reach such low energy sites may react with species in the gas and leave the surface. In molecular beam epitaxy, the probability of an atom returning to the vapor phase is modeled as a sticking coefficient which is temperature and species dependent. A similar concept of a surface reaction probability can be used in vapor phase deposition.

Impurity atoms contained in surface precipitates may react with components of the gas (especially HCl) to remove impurity from the surface locally, resulting in a pit. Or alternatively, growth may be accelerated by the impurity resulting in a mound on the surface. Particles which are on the surface disrupt the crystal growth and result in various types of growth spikes.

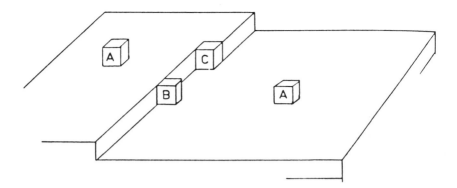

Figure 26. Schematic of the ledge-kink growth model showing adatoms surface diffusing from positions A to B to C. Lower energy, more stable sites being characterized by an increasing number of near neighbor bonds.

4.2.1 Kinetics of Growth from Silane: Early models of growth from silane included only a single step of surface reaction for the pyrolysis reaction. More recently, however, Lee (47) has suggested a more complex model. In his model, the species adsorbed is SiH_2 instead of the SiH_4. He suggested the following sequence: (a) diffusion of silane through a boundary layer to near the surface, (b) dissociation to SiH_2 plus an H_2 molecule at or near the surface, (c) adsorption of the SiH_2, (d) surface diffusion to a kink site, (e) incorporation of silicon into the crystal lattice, and finally (f) desorption of the H_2. At temperatures above 1000°C, steps (b)-(f) occur rapidly and the overall process becomes limited by step (a) which is a diffusion limited supply of reactant to the surface.

Bloem and Giling (48) have shown that for higher temperatures, the reaction rate is directly proportional to silane partial pressure until limited by supply kinetics (see Figure 27). When the silane is diluted with hydrogen, the growth rate remains proportional to silane partial pressure but inversely proportional to the square root of hydrogen partial pressure.

Studies at atmospheric pressure at lower temperatures by Bloem and Classen (49) and by Lee (47) suggest that at the lower temperatures the silicon surface is covered by adsorbed hydrogen. The reaction then becomes limited by a surface reaction, probably the adsorption of the SiH_2. A recent study of polysilicon kinetics at low pressure by Foster

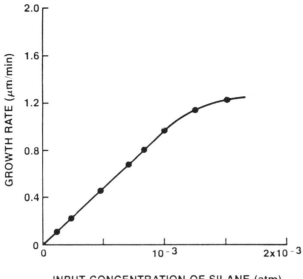

Figure 27. Growth rate for epi increasing linearly with silane concentration until limited by kinetic effects (Reference 48).

et al. (50), however, found no such effect of hydrogen partial pressure at pressures below 1 Torr. Thus the surface site blocking action of adsorbed hydrogen may be dependent on total pressure.

4.2.2 Kinetics of Growth from Dichlorosilane: Bloem and Classen (51) have shown that the process consists of several intermediate steps and that $SiCl_2$ plays an important role in determining the rate of reaction. Analysis of gas species by Sedgwick and Smith (52) using Raman spectroscopy and by Ban and Gilbert (53) have shown that the decomposition of dichlorosilane to $SiCl_2$ plus hydrogen occurs in the gas phase as the gas becomes heated near the surface. With both the hydrogen and $SiCl_2$ gases adsorbed, the surface mobility of the $SiCl_2$ moving to a kink site is limited by the hydrogen covering most of the surface sites. Finally the incorporation of the silicon adatom into the silicon lattice requires reaction with hydrogen and desorption of the HCl byproduct from the surface.

Classen and Bloem (49) found the deposition rate linearly proportional to input dichlorosilane concentration at

1000°C but only parabolically related at the lower temperatures of 800 and 900°C. Other studies by Herring (5, 54) and Liaw and Rose (2) have also found near linearity at higher temperature but strongly non-linear relationships below 1000°C.

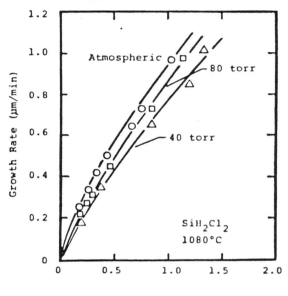

Figure 28. Growth rate as a function of dichlorosilane concentration in hydrogen carrier gas for three total pressures of 40, 80 and 760 Torr. (Reference 5).

4.2.3 Kinetics of Growth from Trichlorosilane and Silicon Tetrachloride: Ban (55) analysed the vapor phase species and identified the same components as from dichlorosilane reactions. They were H_2, HCl, $SiCl_2$, SiH_2Cl_2, $SiHCl_3$ and $SiCl_4$. However the concentration of $SiCl_2$ was much lower than that predicted by equilibrium calculations by Bloem et al.(56) shown in Figure 29.

A more complete model by Van der Putte et al.(57) considered effects of the existence of temperature gradients within the boundary layer. The predictions of Van der Putte's model shown in Figure 30 are in much better agreement with the experimental measurements including the recent work of Aoyama et al.(58).

The experimental growth rate data for silicon tetrachloride by Herring (5) and Liaw and Rose (2) for low concentrations of $SiCl_4$ suggest a parabolic dependence on input reactant concentration as shown in Figures 31 (a) and (b).

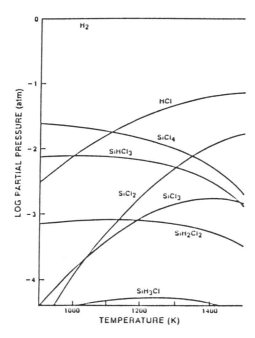

Figure 29. Calculated equilibrium species partial pressures for the Si-H-Cl system versus absolute temperature for a total pressure of 1 atmosphere and a Cl/H ratio of 0.06 (from Bloem et. al, Reference 56).

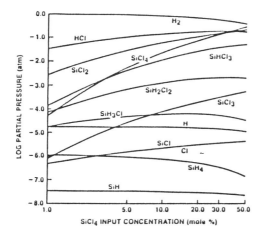

Figure 30. Calculated species partial pressures for the Si-H-Cl system versus $SiCl_4$ input concentration at 1500 degrees K. and 1 atm. (from van den Putte et al. Reference 57).

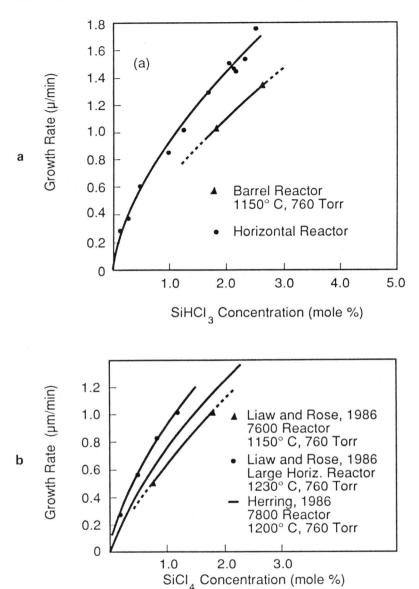

Figure 31. Silicon epi growth rate versus source concentration for

(a) trichlorosilane and
(b) silicon tetrachloride (Refs. 2 and 5).

4.3 Nucleation

4.3.1 Homogeneous Nucleation:

Nucleation of solid silicon from the silicon source gas can occur either in the gas phase (homonucleation) or upon a solid surface (heteronucleation). Nuclei forming in the gas phase are undesirable since they may fall upon the surface and nucleate defects in the growing layer. Since homonucleation requires a higher degree of supersaturation, prevention of gas phase nucleation consists of keeping the input source concentrations below critical levels and by careful ramping of the source gases during start-up of the reaction.

In homogeneous nucleation theory, a particle will only continue to grow if a nuclei reaches a critical size. When particles are below the critical size, the energy for creation of additional surface is greater than the negative energy of formation of additional solid and the particle is unstable. Fluctuations may, however, allow the particle to attain the critical size beyond which is grows spontaneously with a decrease in net energy. Low supersaturation increases the size of the critical radius and makes gas phase nucleation less likely. The rate of nucleation is related to temperature and the change in free energy for formation of the critical radius particle, ΔF. It has been shown by Brice (61) that the nucleation rate, dN/dt, for homogeneous nucleation can be described by:

$$dN/dt = C \exp(-\Delta F/kT) \qquad (8)$$

where C is a constant, k Boltzman's constant, and T the absolute temperature.

Eversteijn (62),(63) evaluated the critical concentration of silane for gas phase (homogeneous) nucleation with the results shown in Figure 32. The critical concentration to avoid gas phase nuclei for silane decreases with increasing temperature (decreasing values of reciprocal temperature). Studies have also shown that addition of some HCl to the gas stream decreases the gas phase nucleation rate (45).

4.3.2 Heterogeneous Nucleation:

Below the supersaturations leading to homogeneous nucleation, nucleation upon a solid surface may still be energetically favored (heterogeneous nucleation). In that heterogeneous nucleation region, silicon is more likely to grow upon a silicon substrate at high temperature than upon a foreign substrate since no nucleation is involved and growth proceeds by the ledge-kink

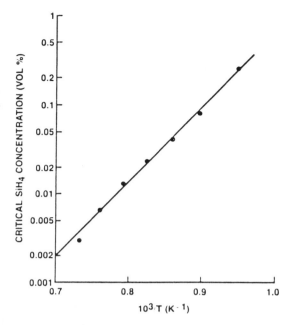

Figure 32. Critical silane concentrations for homogeneous nucleation. Concentrations above the line produce gas phase nucleation (smoke) and highly defective epitaxial growth (after Eversteijn Reference 62, 63).

mechanism. Silicon surfaces near (111) contain few ledges and the growth near the isolated ledge areas develops large growth facets and leads to a rough surface. By cutting the substrate slightly off the (111) plane, (usually 3 degrees), a high density of growth ledges are created and smooth epitaxial growth is obtained. Orientations near (100) have ample surface ledges and those substrates are generally cut on the (100) plane within 0.5 degree.

Liaw et al. (59) and others have found that the heterogeneous nucleation rate is strongly dependent upon the nature of the foreign substrate. At a given temperature and input concentration of the source gas, the nucleation rate decreases in the following order:

> Single crystal silicon
> Polysilicon
> Silicon nitride
> Aluminum oxide
> Silicon oxide

Classen (60) has reported that nucleation is further suppressed by increasing the growth temperature and by additions of HCl to the gas stream as shown by data for the saturation nucleus density on SiO_2 in Figure 33. By taking advantage of those nucleation trends, epitaxial layers may be deposited selectively upon patterned substrates masked with silicon oxide or nitride with little or no deposition upon the mask material (59).

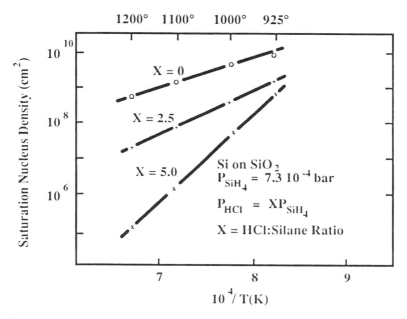

Figure 33. Nucleation density on silicon dioxide versus reciprical temperature for silane growth with and without additions of HCl (after Classen and Bloem Reference 60).

5.0 DOPANT INCORPORATION

5.1 Intentional Dopant Incorporation

In general, epitaxial layers are prepared with a uniform electrical conductivity controlled by the introduction of donor or acceptor atoms in the growing layer. For silicon, this is most commonly done by doping with phosphorus, arsenic, or antimony for N-type layers and by boron for P-type epi layers.

In the early days of silicon epitaxy (1960's), dopants in liquid form were introduced by addition into liquid silicon sources such as silicon tetrachloride. However, gas phase doping developed by Law (64) is so much more easily controlled that today it is the only technique used commercially. Dopants in the form of hydrides of the dopant elements (AsH_3, PH_3, B_2H_6) are used mixed at part per million (ppm) levels dopant gas in hydrogen. Source bottles commonly are prepared with 10 ppm to 0.1 percent dopant hydride mixed in carrier hydrogen. These dopants are delivered through a double dilution control system capable of injecting the dopant gas diluted to levels of parts per billion up to a part per million in the stream of the hydrogen carrier and silicon source gas.

When the dopant hydrides enter the reactor and are heated, they decompose into various vapor species which react with the heated surface of the substrate material. Figure 34 shows schematically the dopant incorporation for arsine doped films (65). Dopant hydrides might be expected to decompose spontaneously, however, due to the high partial pressure of hydrogen, they are relatively stable. The equilibrium partial pressures for the species in the Si-H-P and Si-H-As systems were calculated by Rai-Choudhury and Salkovitz (66) and for the Si-H-P and Si-H-B-Cl systems by Bloem and Giling (48) and Giling (67). Interactions with the silicon deposition chemistry take place, especially for the chlorinated silicon sources, which lead to changes in the growth rate as discussed below.

5.1.1 Measurements of Dopant Incorporation: The incorporation of dopant may be measured by SIMS for direct measurements of the levels of dopant present in the epitaxial layer. However, the level is more commonly measured by spreading resistance profiling which measures the local electrical conductivity of the layer and infers the doping level from the conductivity (68).

Figure 35 shows experimental data by Bloem et al (69) for the incorporation of phosphorus, and Figure 36 shows data of Rai-Chaudhury and Salkovitz (70) for boron incorporation in the epitaxial film versus the hydride gas partial pressures for atmospheric epitaxial growth. For phosphorus the dopant incorporation remains proportional to input concentration up to the 10^{-6} level. Beyond that level the incorporation becomes less efficient even though the incorporation level is well below the lattice solubility. Above the 10^{-6} partial pressure level the incorporation follows a square root dependence on the phosphine partial pressure. This may be

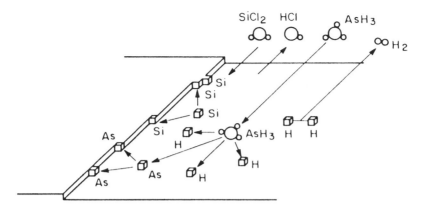

Figure 34. Schematic representation of the arrival of an arsine (AsH$_3$) molecule, dissociation, and the incorporation of an arsenic atom into the growing epitaxial layer (after Reif and Kamins Reference 65).

understood from the thermodynamic calculations by Gilling (67) who found PH$_3$ and PH$_2$ as the principal gas phase components of the Si-H-P system at low input pressures while P$_2$ was a principal gas species at higher input pressures of phosphine.

For boron, the incorporation remains linear up to about 5×10^{-5} beyond which it exhibits a peak near a diborane partial pressure of 5×10^{-4}. The equilibrium partial pressure data suggest that boron-containing species begin to condense at higher diborane partial pressures and do not then contribute to doping the epitaxial layer. That gives rise to the peak in the boron doping curve in Figure 36.

5.1.2 Effect of Temperature: Figure 37 shows data for the dopant incorporation versus temperature for a fixed input partial pressure of 1.4×10^{-6} (66). The incorporation of arsenic and phosphorus decrease with increasing epitaxial temperature while boron shows a reversed trend. This difference must be borne in mind when making reactor temperature adjustments to improve the uniformity of resistivity.

5.1.3 Effect of Growth Rate: Phosphine and arsine slightly depress the silicon deposition rate while diborane enhances the silicon deposition reaction. However, there is also a more complex dependance of doping on the film growth rate. Figure 38 shows the data of Bloem (45) for phosphorus incorporation as a function of the film growth rate. At high

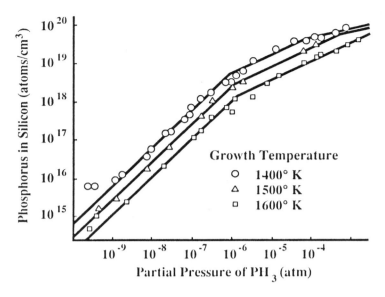

Figure 35. Incorporation of phosphorus in silicon as a function of the input partial pressure of phosphine at 1400, 1500, and 1600 K. (after Bloem et al. Reference 69).

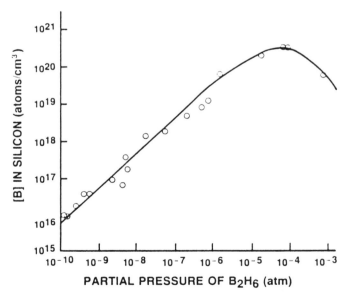

Figure 36. Incorporation of boron in silicon versus the partial pressure of diborane (B_2H_6) as determined by Rai-Chaudhury and Salkovitz (Reference 70).

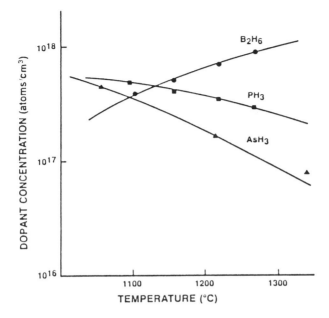

Figure 37. Dopant incorporation versus temperature for a fixed input of dopant partial pressure of 1.4×10^{-6}.

temperature (1500 K, 1227°C), there is little influence of film growth rate on phosphorus incorporation, however, at lower temperatures (1350 K or 1077°C) the incorporation peaks in the vicinity of a growth rate of 1 micrometer per minute.

5.1.4 Effect of Pressure: Figure 39 shows experimental data (5) for incorporation of arsenic at both atmospheric and reduced pressure. Over this low range of arsine partial pressure the incorporation was approximately linear but decreased by about 20-30 percent at reduced pressures of 40 and 80 Torr compared to atmospheric pressure.

5.2 Unintentionally Added Dopants (Autodoping)

5.2.1 Sources of Autodoping: In addition to the intentionally added dopants, unintentional doping may be contributed by the front or backside of the substrate, patterned diffusions in the front side of the substrates, or test or dummy load substrates in the load. This unintentional doping is referred to as autodoping. In addition, autodoping may come from dopant released from the reactor walls as a result of prior doped cycles. This component is referred to as "reactor memory".

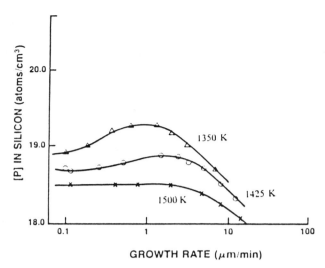

Figure 38. Phosphorus concentration in silicon versus the growth rate for films grown from silane using phosphine dopant at a concentration of 1.2×10^{-6} atm for growth at 1350 K, 1425 K, and 1500 K. (Figure originally presented by Bloem at the Spring 1973 meeting ECS in Chicago, Ill.) (Reference 2).

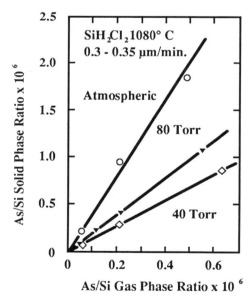

Figure 39. Arsenic incorporation in silicon versus the arsenic to silicon ration in the gas phase for total pressures of 40, 80, and 760 Torr. (Reference 5).

When the doping profile of an epitaxial layer is measured for a lightly doped (or intentionally undoped) film grown over a heavily doped substrate, the profile shows a transition region instead of the expected step change in dopant concentration. The transition behavior may be understood in terms of at least three components. The first component, operating near the substrate/epitaxial interface is due to solid state diffusion of the substrate dopant into the growing film during the time of the film growth process (or a subsequent high temperature cycle). That impurity redistribution was modeled by Rice (71) and by Grove et. al. (72). Price and Goldman (73) reformulated the model of the outdiffusion in terms of more easily measured parameters. The profile near the interface over a uniformly doped substrate generally follows a complementary error function expected for a step change of concentration after redistribution by solid state diffusion.

Because of the lower diffusion coefficient of antimony in silicon compared to phosphorus or arsenic, heavily doped antimony substrates are frequently used as test wafers for

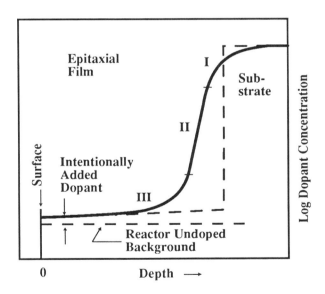

Figure 40. Schematic of the ideal step change in dopant concentration and actual profile with region I autodoping by out-diffusion, II autodoping by on wafer patterned diffusions, and III the autodoping tail from reactor walls and wafer backside sources.

measurement of epitaxial growth rate. However, with the reduced antimony outdiffusion, this may overestimate the epi layer thickness useable for device fabrication.

The second component of autodoping which is seen in the film beyond the outdiffusion region is the gas phase autodoping region. For that region, Thomas et. al. (74) suggested a model in which dopant atoms evaporate from the surface, mix with the gases above the substrate and then reincorporate in the growing film at a rate determined by their concentration in the gas stream. Grossman (75) assumed that the atoms were reincorporated at a rate proportional to the surface concentration and the growth rate of the film. Since most of the evaporated dopant is carried away by the gas stream, both models predicted a exponential decrease of the autodoping concentration with film thickness. More recently Wong and Reif (76) and Wong et al. (77) proposed a model based on trapping of surface adsorbed dopant species by the growing film. These models also predicted an exponential decrease in the autodoping concentration with thickness away from the substrate interface.

A third component of autodoping is contributed by other surfaces in the reactor. This system autodoping tail may also contain the effect of dopant released from the backside of substrates or from other wafers in the load. These sources generally decrease with time during the growth cycle leading to a slow decay of the autodoping tail with increasing film thickness (Figure 40).

Langer and Goldstein (78),(79) studied the autodoping from heavily doped boron substrates and formulated a computer modeling program (Casper) to predict doping profiles in the epitaxial layer.

5.2.2 Lateral Autodoping: Srinivasan (80) developed a model for the autodoping effect by considering evaporation from a point source. The evaporated dopant was considered moving by diffusion but influenced by gas flow near the surface. Contours of dopant concentration which would be circular under zero flow are distorted to ellipses elongated in the direction of the flow (Figure 40).

Srinivasan's model then considered the combined effect of multiple sources representing the diffused buried layer regions on a typical bipolar device product wafer. When the point sources are small and widely spaced, the autodoping concentration will vary laterally around each source and extend over relatively long distances. Srinivasan (38, 80) showed that for arsenic the decay length or distance at

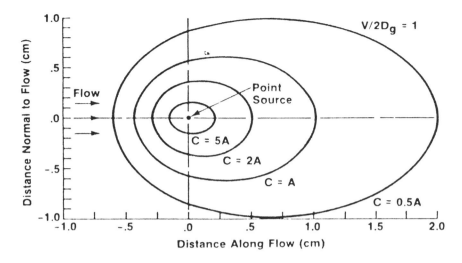

Figure 41. Autodoping concentration contours calculated by Srinivasan (Reference 80). Circular contours are distorted into ellipses in the direction of gas flow.

which the autodoping decreased to 1/e of the peak value was of the order of 6 mm in the downstream and 1 mm in the upstream direction. As the spacing of the autodoping sources is reduced, the lateral effects overlap and beyond 500 local sources per wafer (typical of most VLSI substrates), the autodoping concentration varies less than 5 percent in the space between local source areas.

The model further predicted that the mean autodoping concentration would vary as the square root of the number of the number of point sources on the wafer. Subsequent experiments (81) confirmed that square-root dependency of mean autodoping level on the buried layer density.

5.2.3 Suppression of Autodoping: One technique for reducing this outdiffusion component for phosphorus or arsenic relies on evaporation of substrate dopant from the surface during a high temperature bake just prior to the initiation of epitaxial growth. Longer or higher temperature bake cycles deplete the surface and subsurface concentration prior to the start of film growth. During film growth, the substrate dopant out-diffuses to fill the depletion region but dopant encroachment beyond the original substrate/epitaxial interface is reduced as shown by the profiles of Figure 42.

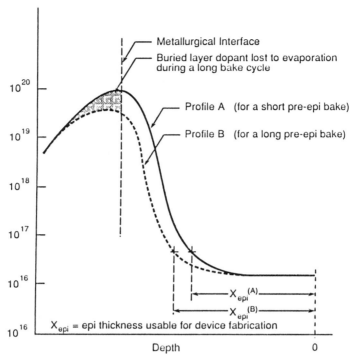

Figure 42. Schematic profile showing the effect of a long bake to evaporate dopant from the surface prior to initiation of epitaxial growth. Surface depletion results in a decrease in outdiffusion into the grown layer.

The component of autodoping contributed from the backsides of substrates may be reduced or eliminated by use of a backside seal consisting of a layer of barrier material deposited on the backs of the wafers. The common backseal layers in order of increasing effectiveness are polysilicon, epitaxial silicon, silicon oxide, and silicon nitride. Backside seals are essential for the growth of high resistivity P-epi layers on heavily doped boron substrates for use in N-well CMOS processes. When an RF heated reactor is used, it is possible to obtain some reduction of backside autodoping from a transfer backseal produced insitu from a polysilicon coating applied to the susceptor. In the RF heated case, the susceptor heated by induction runs hotter than the wafer. When chlorine is present during pre-epitaxial HCl etching, some of the polysilicon substrate coating will be converted to a gaseous species, transported across the narrow gap behind the wafer, and redeposited on the back of the slightly cooler wafer. When epitaxy is done using a chlorinated silicon source the transport will continue during film growth and help to reduce the concentration of substrate dopant reaching the back surface. This mechanism, however, is not available in radiantly heated reactors in which the wafer heated from the front side runs slightly hotter than the wafer support.

Increased velocity of gas flow past the substrates will thin the boundary layer and promote the escape of dopant evolved from the substrate into the exhaust stream of the reactor. One effective way of increasing the gas velocity is reduction of reactor pressure (5),(83),(84). Figure 43 shows profiles measured by capacitance-voltage profiling (5) of thin epi layers deposited on high concentration arsenic doped substrates (oxide backsealed). By decreasing the reactor pressure from atmospheric while maintaining the same mass flow, the transition width was markedly reduced.

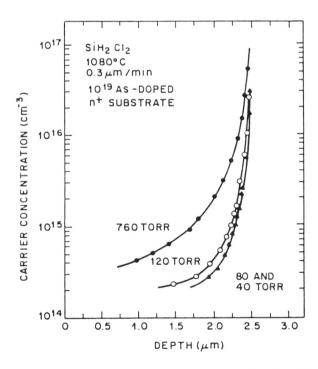

Figure 43. Capacitance-voltage profiles of the dopant profile showing decreased arsenic autodoping in epi layers grown at reduced pressure (Reference 5)

Increased velocity of the gases past the substrate under conditions of reduced pressure epitaxy is also effective in suppressing lateral autodoping measured between the buried layer regions. Figure 44 shows results of spreading resistance profiles taken in the lateral region 0.050 inch from the edge of an arsenic diffused region. While the concentration peak measured about 10^{19} atoms/cm^3 for a profile through the diffused region, the lateral peak of over 5×10^{15} for atmospheric pressure growth was reduced to below 1×10^{15} under conditions of 40-80 Torr pressure during growth of the epitaxial layer (5).

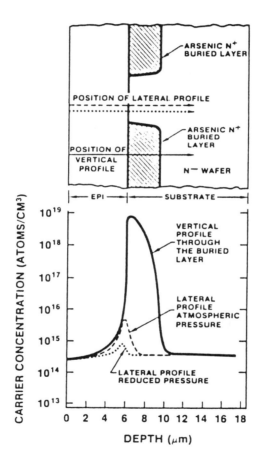

Figure 44. Suppression of lateral autodoping in the region between buried layer diffused regions by use of reduced pressure during epitaxial layer growth.

6.0 SURFACE MORPHOLOGY AND EPITAXIAL DEFECTS

The surface morphology of silicon epitaxial deposits is affected by growth parameters as well as parameters of the substrate. The growth parameters include the temperature, pressure, growth rate, as well as the concentration of the silicon containing gas and the chlorine to hydrogen ratio. The substrate parameters include the substrate orientation, defects in the substrate, and contaminants on the surface of the substrate. Several of these effects will be discussed in this section.

6.1 Substrate Orientation Effects

The crystallographic orientation of the substrate affects both the morphology of the deposited film and the appearance of certain crystallographic defects where they intersect the surface. The growth of smooth epitaxial films can be obtained on (100) or (110) oriented silicon substrates. However, epitaxial growth on substrate surfaces oriented near the (111) plane results in faceted "alligator skin" surfaces. The (111) surfaces contain no atomic steps to provide a density of growth sites. In the absence of atomic steps, the growth produces pyramids and terraces. However, misorientation of the surface by 0.5 degree or more introduces a sufficient density of steps for growth of smooth planar films (84).

6.2 Spikes and Epitaxial Stacking Faults

6.2.1 Growth Spikes: Surface contaminants can also influence the surface smoothness of epitaxial deposits or result in crystallographic defects in the film. Spikes or epitaxial stacking faults are the two most commonly observed defects resulting from surface contaminants. Epitaxial spikes generally originate from a silicon particle on the surface not removed by the pre-epitaxial cleaning process. The chip of silicon may expose faster growing crystal planes than the plane of the substrate. Deposits on the surface of the silicon chip may be epitaxial but in most cases several grains nucleate and a polysilicon nodule results. The chip of material protrudes above the substrate surface into a region of richer supply of gaseous reactants. The result is that the nodule grows at two to ten times the rate of the epitaxial film on the substrate. In thicker epitaxial deposits (20-100 microns) used for bipolar linear or power devices, such spikes may extend 100 micrometers or more above the surface and have a base of 20 to 50 micrometers in diameter. Such large defects will almost certainly interfere with device processing and are generally removed mechanically before the next processing step. Removal generally leaves a region unusable for functional devices. Improvement of pre-epitaxial cleaning, especially by use of megasonic cleans (31) for removal of small particles of silicon dust generally can eliminate epitaxial spikes.

6.2.2 Epitaxial Stacking Faults: Epitaxial stacking faults are crystallographic in nature and arise from defects in atomic arrangement during film growth. This may result from an extra atomic layer (extrinsic fault) or a missing atomic layer (intrinsic fault) along a {111} type plane. The stacking faults extend from the nucleating site upward through the film along one or more of the {111} type planes inclined at 54.7 degrees until they intersect the surface. For (100) oriented substrates there are four such (111) planes along which faults may form. Generally, faults form on all four {111} type planes and the surface etch figure is a square. For substrates oriented with the surface near (111), there are three other {111} type planes inclined at 70.5 degrees from the surface, and the surface intersection figure is triangular. The intersection of the stacking faults forming the edges of the inverted pyramid contain a partial dislocation extending from the nucleation site to the surface.

The intersection of the epitaxial stacking faults with the surface produces a line on the surface observable using a high quality microscope with attachments for Normalski interference contrast. A short (1 to 10 second) chemical etch may be used to enhance the contrast of the surface figure. After the chemical etch which attacks the fault area, the intersection figures may be easily counted or measured. Since the sides of the inverted pyramid lie at known angles to the surface, the size of the figure may be used to measure the thickness of the epitaxial film thickness (85).

The nature of the surface contamination which causes stacking faults has been studied extensively. A good review of this subject was presented by Stowell (86). The contamination may originate from either the substrate surface or internal to the substrate. Outdiffusion of impurities from the substrate is the main internal source. Examples of internal sources include oxygen (87), as well as microprecipitates of metallic impurities (88),(89).

Crystallographic defects from the substrate may also

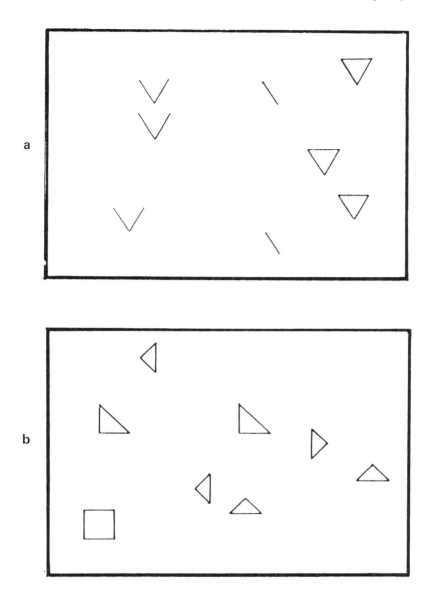

Figure 45. Typical surface etch figures produced by short preferential etch of (a) {111} oriented substrate or (b) {100} oriented substrate containing growth stacking faults intersecting the surface along the trace of one or more inclined {111} type planes.

propagate into the growing epitaxial layer. Pomerantz (88) observed a higher density of epitaxial stacking faults for substrates which had been oxidized prior to epitaxy. Liaw (90) showed that some epitaxial stacking faults originate from oxidation induced stacking faults (OSF's) present in the substrate.

Surface contaminants left by incomplete cleaning are the principle surface source. Other external sources are particulates from the susceptor, reactor quartzware, or the reactor loading mechanism. Static electric charges may attract particles to the substrates during loading. Static charges may also be generated by high velocity gas flow during reactor purging operations.

6.3 Hillocks and Pyramids in Epitaxial Layers

Hillocks are small oval mounds on the surface of the epi while pyramids are faceted regions on the epi surface. Their density is influenced primarily by the growth parameters including the type and concentration of silicon source and the deposition temperature. Joyce (91) reported that high temperature (1200°C) growth using trichlorosilane resulted in smooth featureless surfaces but lower temperatures resulted in rough surfaces. Chang and Baliga (92) have studied hillock formation and have shown that the hillock density increases exponentially with growth rate or inversely with absolute temperature (1/T).

In general, a progressive lowering the growth temperature results in loss of the original wafer mirror finish, replaced by appearance of an orange peel texture on the surface, and at even lower temperature, a matte grey polycrystalline surface finish. The higher the growth rate, determined by the source supersaturation and the temperature, the higher the temperature required for maintaining a good surface quality.

Chernov (3) suggested that the formation of hillocks or pyramids on the surface is the result of a two dimensional nucleation and growth mechanism. He developed a model based on this growth mechanism which predicts surface morphology based on the chemical potential at the growth interface. His model in agreement with observations predicts that the slope of the hillock or pyramid is increased by a decrease in temperature, an increase in source concentration, or an increase in the Si/Cl ratio.

The planarity of selective epitaxial deposits is also

affected by the growth parameters. Tanno et al (94) and Voss and Kurten (95), and Liaw (96) found improved planarity of selective epi deposits at lower temperature and lower pressures where the reaction becomes more limited by surface reactions than by transport of reactants to the surface.

6.4 Dislocations and Slip

Nonuniform heating of the substrate (whether by R.F. or other heating methods) results in nonuniform thermal expansion in the substrate which produces elastic stresses. The thermal stress may produce bowing which may lift the edge of the substrate away from the support in response to the thermal stress. At lower temperatures (generally below 900°C) the yield point of the silicon lattice is sufficiently high that the substrate behaves elastically. Upon cooling, the thermal stress is removed and the substrate returns to its original shape.

However, if the stress exceeds a critical value, the substrate will yield plastically. This occurs by the generation and motion of dislocations which are atomic level line defects which glide along slip planes of the crystal. The passage of one dislocation offsets the material above and below the slip plane by a unit known as the "Berger's vector" of the dislocation.

Figure 46. Typical wafer edge slip as a result of excessive within wafer temperature gradients during heating or during epitaxial film growth.

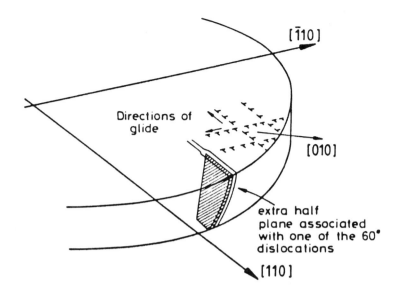

Figure 47. Schematic showing the extra half plane associated with 60° dislocations gliding in [110] type directions.

The dislocations generally propagate from near the edge of the substrate where the stress is highest. They glide toward the center of the substrate producing plastic deformation of the substrate which relieves the thermal stress. However, the dislocation motion slows as the dislocation moves to a region of lower shear stress. The continued slow motion of the dislocations produces "creep" deformation of the crystal, a well studied phenomenon of high temperature materials.

Akiyama et al. (97) have measured the critical radial temperature gradient for inducing dislocations within 150 mm diameter wafers at between 5 and 10 degrees C per centimeter, decreasing with temperature as shown in Figure 48. From those measurements, they inferred a critical shear stress for inducing large scale dislocation generation and motion. The level of that critical shear stress decreased by about a factor of two between 900 and 1200°C as shown in the Figure 49.

Once the crystal lattice yields, the dislocations move along the slip planes for distances determined by the stress. However, even after the substrate is cooled to room temperature, the dislocations remain, frozen in place where they stopped. The passage of large numbers of dislocations from

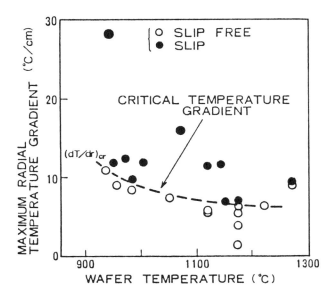

Figure 48. Critical temperature gradients within wafer for slip in 150 mm wafers (Reference 97).

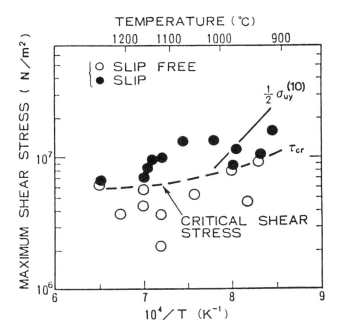

Figure 49. The critical shear stress for slip of 150 mm wafers versus reciprocal absolute temperature (Reference 97).

a single source will produce a step on the surface large enough to be visible by eye or by inspection of the wafer edge under a low power microscope.

The surface steps at the slip lines and the plastic distortion of the substrate might produce alignment problems for fine geometry devices. However, the device impact of slip generally comes from rapid "pipe diffusion" of dopant along the core of the dislocations. When a p-n junction of a device is built by diffusion in a region threaded by dislocations, dopant can diffuse more rapidly along the core region of the dislocation than through the bulk crystal. The result is a non-planar diffusion front. In bipolar transistors, the emitter dopant diffusion faster along the dislocation produces a deeper emitter locally narrowing the base region or producing a shorting of the emitter to the collector region of the device. The effect of dislocations in the epitaxial layer of narrow base bipolar devices is an increase in the density of devices with high leakage or emitter-collector shorts.

X-ray topography is a technique used to image the dislocations within a bulk crystal sample and can be used to check for slip in epitaxial material.

6.5 Microprecipitates (S-pits)

Many common metallic elements such as copper, nickel, iron, and chromium have significant solubility in silicon at high temperatures and fast diffusion rates through the silicon. Such metals may be present in the starting substrates or be picked up during handling in the loading operation or from metal parts or the susceptors within the epitaxial reactor itself.

A detailed study of microprecipitates in silicon epitaxy by Werkoven (98) used preferential etching, transmission electron microscopy, and neutron activation analysis to identify the impurities dominating defect generation. After epitaxial layers were grown, preferential Wright etching revealed a density of 10^4 to 10^7 cm^{-2} of shallow saucer pits. At high density these shallow pits produce a visible "haze" pattern. TEM studies showed small (25-100 nm diameter) metallic precipitates concentrated within a micrometer of the surface as the origin of the shallow pits (or S-pits). Energy-dispersive X-ray analysis showed mainly Fe and Ti as the major impurities and Cr and Ni found occasionally.

He considered several sources but concluded that most iron and titanium contamination was introduced from metallic contamination of the graphite susceptors on which the wafers were placed during epitaxial processing. Werkoven reported pyrolytically coated graphite susceptors somewhat more pure than silicon carbide coated graphite. However, the uncoated and pyrolytically coated graphite had a higher efficiency of transfer of contaminants into the substrates compared to the silicon carbide coatings.

Werkoven found relatively high levels of nickel in the silicon starting material. Schmidt and Pearce (99) reported high levels of nickel contamination of substrates at certain beveling and polishing operations of silicon wafer preparation. Another source may be the nickel electroplate used to bond diamond to the slicing blades used to cut the single crystal silicon ingot.

Werkoven also reported a significant increase in nickel contamination of substrates handled with stainless steel tweezers during wafer loading. By avoiding the use of the stainless steel handling tools, Werkoven produced epitaxial films with no significant increase in nickel over the starting substrate silicon.

At lower temperature, the metal contaminants precipitate near the surface or at defect sites within the substrate. In growing an epitaxial layer, it is common practice to use controlled high temperature dwell steps or "bakes" prior to introduction of the reactants. One purpose of those "bake" steps is to allow microprecipitates near the surface to dissolve and disperse. Failure to provide time for that dissolution will increase the density of stacking faults and other defects in the film.

7.0 PATTERN SHIFT AND DISTORTION

7.1 Patterned Diffusions (Buried Layers)

The substrates used for fabrication of bipolar integrated circuits usually are diffused or implanted prior to epitaxy to create an array of heavily doped regions (subcollector) under the locations where the vertical transistors will be made in the epi layer. This is usually done by thermally oxidizing the substrate, patterning the oxide, and diffusing or implanting arsenic or antimony into the opened part of the pattern. During the drive-in of the diffusion or implant, some oxygen is used to grow a thin (100-400 nm) oxide on the exposed silicon. This helps seal in the dopant reducing evaporation from the surface during the drive-in, accelerates the drive-in, and produces a surface step depressing the patterned area 50-200 nm below the substrate surface plane. When the surface masking oxide is removed in the pre-epitaxial cleaning process, the pattern of the diffused area can be seen using interference contrast optical microscopy which produces light/dark contrast lines at the small surface steps.

An alternative terminology for the "buried layer" is "diffusion under film" or "DUF". The steps in the processing sequence are shown schematically in Figure 50.

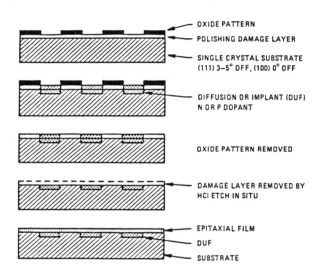

Figure 50. Schematic of the steps in fabrication of a diffusion under film (DUF) or "buried layer".

7.2 Pattern Distortion and Pattern Shift

After epitaxial growth, the surface steps generated in the patterned diffusion step can still be seen as surface steps on the epitaxial film. Those steps are then used as a marker to permit alignment of the next device mask layer relative to the patterned diffusion (buried layer) under the epitaxial film. While the steps at the edge of the patterned areas propagate generally upward during the epitaxial film growth, lateral shifts of the step relative to the original position do occur. These shifts lead to shift of the pattern and distortion of the pattern on the top of the epitaxial film. The degree of the shift and/or distortion is influenced by the chemistry and process parameters of the epitaxial process.

7.3 Pattern Shift Definitions

When the left and right parallel edges of the pattern shift in the same direction by an equal amount, a simple shift is produced as shown in Figure 51. However, the atomic ledge structure of the left and right sides of the depression is different and that may lead to differing amounts of shift of the left and right hand side of the depression. That difference leads to a change in the spacing between the sides of the depression. That change of size is known as distortion. Liaw and Rose (2) have presented photographs of pattern distortion and accompanying surface profile measurements.

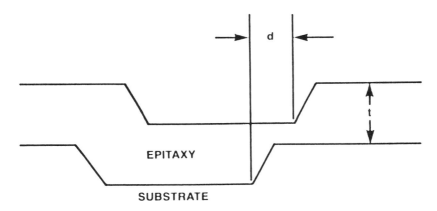

Figure 51. Schematic epitaxial pattern shift. The pattern shift ratio (d/t) is used to characterize the impact of process variables on shift or distortion.

7.4 Role of Crystallography of Surface

Townley (100) related the direction of the pattern shift to crystallography of the substrate surface. For (100) oriented substrates, the edges of rectangular patterned buried layer regions are generally oriented parallel to the surface traces of the {111} planes. The four edges of a rectangular depression are crystallographically equivalent and move together. Hence for wafers accurately cut on the (100) plane, shifts will be small but the pattern may expand or contract.

For substrates oriented near (111), Townley argued that shift was a consequence of imposing square or rectangular depressions on a surface having basically threefold symmetry.

Two edges of a rectangular pattern parallel to the wafer flat lie parallel to a mirror image symmetry plane and will move by similar amounts but in opposite directions. This produces a distortion or size change but low shift for the two sides of a rectangle parallel to the wafer major flat. However, for the two sides of a rectangular depression perpendicular to the wafer flat, the sidewall ledges will be different in character. The two edges under atmospheric pressure epitaxial growth will shift by similar amounts in the same direction leading to a shift.

7.5 Summary of Shift and Distortion Effects

Pattern shift and distortion was first studied by Drum and Clark (101),(102) who developed terminology for describing the motion of the surface steps relative to the underlying diffused area. Others have contributed observations of the effect of silane versus chlorinated sources (46), the effects of temperature (103), reactor type, growth rate and pressure (5),(54),(103).

Figure 52 shows pattern shift data plotted as the ratio of the shift of the center line of a pattern area divided by the epitaxial layer thickness versus the pressure during film growth. Data is shown for two reduced reactor types, one a modified vertical reactor (R.F. heated) and the other a radiantly heated cylinder reactor.

Studies by Weeks (46), and Boydston (104) presented more detailed models of the shift and distortion process. The following summarizes the effects:

1. Pattern shift is generally small for (100) accurately oriented substrates but shift can be significant when the substrate is slightly mis-oriented and growth is done at low rates and low temperatures.

2. Pattern shift for (111) substrates is parallel to the wafer major flat (for material cut following SEMI standards (107), in a direction opposite to the approximately 3 degree tilt off the exact (111) plane used to provide an adequate density of growth ledges.

3. For (111) substrates, the edges of patterns parallel to the wafer major flat move in opposite directions contributing a size change (distortion) but little of no shift.

4. Pattern shift and distortion is smallest for silane based growth chemistry and increases as HCl is added to the silane. Shift and distortion also increase in the order of increasing chlorine in the family of chlorinated silanes: dichlorosilane

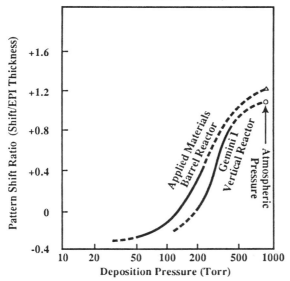

Figure 52. Pattern shift ratio versus epitaxial growth pressure for two types of commercial reduced pressure epitaxial reactors.

(lower), trichlorosilane (moderate), and silicon tetrachloride (the most).

5. Pattern shift and faceting of the edges is reduced for higher temperature growth. Lowering the growth temperature to 1150°C or below for $SiCl_4$, leads to large distortions and in some cases flattening of the steps (pattern washout) which obliterates the pattern.

6. At atmospheric pressure lower growth rates decrease shift and distortion.

7. Use of reduced pressure can decrease pattern shift for (111) substrates but at the tradeoff of enlarged size of the depression part of patterns.

8.0 EQUIPMENT FOR EPITAXY BY CVD

The geometry of the reaction chamber of an epitaxial reactor controls the gas flow patterns and through that impacts properties of the deposited layers. The abruptness of layer to layer changes, degree and pattern of autodoping, as well as the uniformity of thickness and doping are all influenced by the gas flow patterns.

Modeling of chemical reactors makes use of two limiting cases: the plug or displacement flow reactor and the ideally mixed reactor. In the case of a displacement reactor, the gases entering do not mix with the gas in the reactor but displace the gas ahead, pushing it into the exhaust line. This analysis treats a given volume of entering gas as a plug which remains together moving through the reactor. In the plug flow or displacement reactor, when a change of gas composition is introduced at the inlet, the concentration in the reactor is changed completely by one change of gas in the reactor.

At the other extreme, gases entering may be thoroughly mixed with the gas already in the reactor and the gases exiting out the exhaust have average properties for the mixture. In this type reactor, when a step change of concentration of reactant or dopant is applied at the inlet, that change mixes with the gas already in the reactor. The average properties of the gases will change to exponentially approach the concentrations of the inlet gas stream.

Generally speaking, the film properties such as film thickness and resistivity may be expected to be more uniform for the mixed flow reactor even when there is significant autodoping from the substrates. Displacement type reactors, however, can be expected to exhibit sharper layer to layer transitions, with autodoping effects increasing along the process zone in the direction of gas flow. The above generalizations are, however, influenced by high mass flow rates, use of reduced pressures, and other design elements which impact the performance of the reactor.

Commercial epitaxial reactors may not exactly fit either of these two idealized limiting cases of the displacement or fully mixed reactor flow used for analytical models, but it is still useful to compare them to these models.

8.1 Classification of Commercial Reactors by Flow Geometry

Most commercial epitaxial reactors are designed in one of three flow geometries:
- horizontal flow reactors
- vertical flow reactors
- cylinder reactors

The most common types of silicon epitaxial reactors are shown schematically in Figure 53.

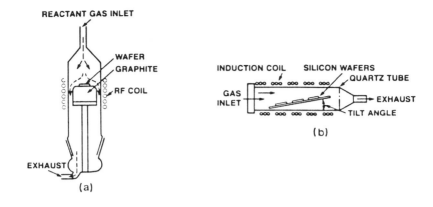
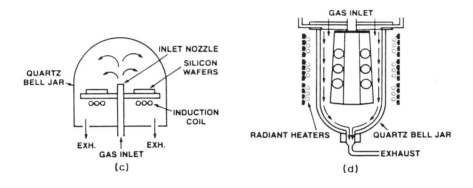

Figure 53. Schematic CVD reactor geometries for (a) true vertical reactor, (b) classic horizontal flow reactor, (c) modified vertical (or pancake) reactor, and (d) downflow cylinder reactor.

8.1.1 Horizontal Reactors: The horizontal flow epitaxial reactor is a displacement type reactor in which gases enter at one end of a tubular chamber and exit from the other end of the reactor chamber. As the gases pass over substrates placed on a heated wafer holder, the gases are heated and react to deposit silicon films on the substrates. The earliest horizontal reactors were mounted within a resistance heated annular furnace. The time to heat and cool the furnace limited productivity and deposits on the hot walls of the tube increased the frequency of cleaning plus consuming valuable reactants. Resistance heating was replaced by R.F. induction heating from a coil wrapped around the process chamber area. When R.F. induction heating is used, the wafers are held on a coated graphite carrier called the susceptor. The graphite susceptor couples inductively to the R.F. field of the coil outside the tube.

A horizontal reactor, like any displacement flow reactor, exhibits strong gas depletion effects. Reactant concentrations change along the direction of flow as reactants are consumed by reactions at the heated surfaces and the gas stream is diluted by the by-products of those reactions. One way of compensating for the depletion is to tilt the susceptor, narrowing the flow channel along the direction of flow. This causes the gas to speed up as it passes through the narrowing section. Higher gas velocity thins the boundary layer near the substrate surfaces increasing the transport of reactant to the surfaces and thereby increasing the growth rate. Overall gas velocities of horizontal flow reactors are generally in the 30-70 cm/sec range.

The reaction tube of horizontal reactors was initially round in cross section but were changed to a rectangular cross section to produce more uniform gas flow velocity over the wider susceptors required for larger loads and larger wafer diameters. Although the horizontal reactor is generally called a displacement reactor, studies of the gas flow patterns show that the flow is more complex. Viscous drag of the reactor walls produces a slow moving boundary layer near the walls and susceptor with a faster moving core of the gas away from the walls. In addition, the heated gas near the susceptor expands and is less dense than the cooler gas moving in the main gas stream above the susceptor. This gives rise to upwelling of "thermals" of hot gas from the susceptor upsetting the stability of the flow. Ban (105) presented a model for flow in a horizontal reactor which included a twist to the flow imparted by the heated gases from the susceptor. This type of thermally driven secondary effects of the flow pattern depend critically on the location of the susceptor relative to the walls contributing some difficulty in reproducing flow conditions run to run.

R.F. heated horizontal flow reactors, the main workhorse of commercial silicon epitaxy in the early 1970's used for about 90% of the commercial work. However, horizontal reactors were prone to temperature gradients during the rapid heating and cooling portions of the cycle leading to slip in the wafers. The temperature of the susceptor in the R.F. heated horizontal reactors is adjusted by coil changes which adjust the coupling to the susceptor. Wafer slip, difficulty maintaining thickness uniformity plus the time consuming difficulty of adjusting and maintaining the precise coil to susceptor coupling led to displacement of horizontal reactors by vertical and cylinder type reactors. By 1989, less than 5% of commercial silicon epitaxy utilizes horizontal reactors.

8.1.2 Vertical Flow Reactors: The original vertical flow reactors first used for silicon epitaxy by Theuerer in 1961 (106) were single wafer reactors in which displacement gas flow moved vertically from an inlet at the top to an exhaust at the bottom. A single wafer was supported on a pedestal holder with the gas flow impinging against the substrate. In that form the vertical flow reactor is a displacement type reactor with flow perpendicular to the substrate in contrast to the flow parallel to the substrate in the horizontal reactor designs.

Commercial modified vertical reactors, introduced in the 1960's, enlarged the wafer support to a large plate capable of holding several substrates and covered the reaction area with a dome shaped bell jar. The gas inlet of these modified vertical reactors is through a central gas injection directed upward to impinge on the dome above the substrate holder. This modified vertical (or pancake) reactor is generally heated by a flat spiral wound R.F. induction coil within the reaction chamber below the wafer holder (coated graphite susceptor). Gases injected upward from near the center of the wafer holder, induce a convection flow producing a radial gas flow estimated at 5-10 cm/sec over the substrates. The fresh gas mixes with the gases in the reactor and some of the mixed gases, flow over the edge of the wafer holder to exit through ports in the baseplate of the reaction chamber. In this modified form, the vertical reactor is a good example of a mixed flow reactor.

By rotating the wafer holder, the heating of the susceptor and flow non-uniformity is time averaged, improving the wafer to wafer uniformity of the film properties. The design feature of rotation of the susceptor during deposition produced about a factor of two improvement in uniformity of thickness wafer to wafer when compared to horizontal reactors.

During the 1980's, larger vertical reactors capable of processing several 150 mm substrates per load have been developed and designs for reduced pressure processing have been introduced. The modified vertical reactor continues in popularity with a reputation for reliability and uniformity of the deposited layers.

330 Handbook of Semiconductor Silicon Technology

8.1.3 Cylinder Reactor Geometry: The commercial cylinder reactors produced by Applied Materials as models 7000, 7600, 7800, and 7810 are an intermediate flow system between the fully mixed flow of modified verticals and the displacement flow of horizontal reactors. The radiantly heated cylinders use an inverted bell jar chamber in which gases are admitted through two inlet jets, swirl and mix in a complex flow pattern to exit at the bell jar neck at the bottom of the chamber. The wafers are held in depressions (pockets) on flat sides of a five to seven sided truncated pyramidal wafer holder. The wafer holder is tapered about 3 to 5 degrees, slightly smaller at the top. The angle of the holder with respect to the chamber wall helps keep the wafers in the pockets and also provides a tapered gas flow passage which helps make up for depletion along the flow path. The two inlet gas jets give the gas a strong downward velocity along the back of the reaction chamber. The wafer holder is rotated at about 2 rpm to time average the non-uniformity of the heating and flow patterns.

An R.F. heated cylinder reactor recently introduced by Applied Materials as the model 7010 employs two wafer holders and an elevated, bottom loaded process chamber. While one wafer holder is in process the other is being unloaded and reloaded by a robotic handling system.

Cylinder reactors introduced about 1970, have gained wide acceptance for commercial production of slip-free epitaxy and for the good uniformity of the deposited films. Due to the higher complexity of the systems, they have not matched the high levels of reliability reputation of modified vertical designs.

8.1.4 Other Reactor Types: The Tetron reactor (107) (Figure 54) introduced in 1986 by the Tetron Division of Gemini, Inc. (now a division of Lam Research), employs a circular array of wafer carriers which form tapered cavities. Each cavity has two wafers facing each other from walls of the tapered cavity. Gas introduced from the outside of the carrier array flows by displacement flow across the wafers through the tapered cavity to exit into the center of the array and downward into an exhaust. Depletion is compensated by the taper of the cavity. The wafer carrier array is rotated to time average flow and heating non-uniformity for improved uniformity. The radial gas flows parallel to the wafers is typically 15-30 cm/sec. while the array is rotated with a tangential velocity of 2-3 cm/sec. Because of the lack of gas mixing between tapered cavities, it is reported that this reactor produces high intrinsic (background) resistivity and low autodoping even from heavily doped substrates in the load.

A single wafer reactor has recently been introduced by the Epsilon Division of ASM which employs R.F. power for an insitu clean and employs rapid deposition rates (3-5 microns/min.) at temperatures in the 1000°C range.

Other reactors including a continuous feed reactor built by Texas Instruments, and a multiple disk reactor built at RCA (108) never have reached commercial manufacture or widespread use.

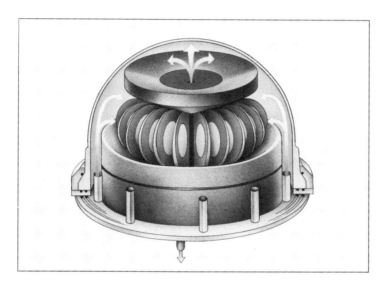

Figure 54. Inward flow gas path in the Tetron reactor.

8.2 Heating Techniques

Silicon epitaxy requires high temperatures in the range of 900-1300°C. Most reactors operate with cold wall quartz chambers to minimize reaction on the walls and to reduce reactions with the walls which can release oxidizer into the process. Elastomeric seals, at the inlet and outlet require combinations of air and water cooling. To keep processing cycle times to a minimum, relatively rapid heating and cooling is required. The principal heating methods are:

a. R.F. induction heating using frequencies of 50 to 450 kHz with water cooled coils internal or external to the chamber,

b. Radiant heating using high intensity quartz halogen resistance heated lamps from outside the reaction chamber,

c. Radiant heating using resistance heaters located within the reaction chamber, or

d. A combination of the above techniques.

8.2.1 Resistance Heating: Resistance heating from outside the chamber was used in some early reactors but was abandoned to reduce cycle time. More recent development of an externally resistance heated reactor has failed to achieve commercial release. The use of resistance heating internal to the chamber in combination with other heating techniques is used for the Tetron tapered cavity reactor.

8.2.2 R.F. Induction Heating: R.F. induction heating in which power from a coil connected to a power supply is coupled inductively to a conductive susceptor (usually coated graphite) was the first technique to supply the power density required for rapid heating required for shortening the cycle time. The term "susceptor", correctly applied in this application is frequently carried over to other reactor types in which the term "wafer holder" is more correct for identifying the wafer support structure.

In R.F. heated reactors where the heat is applied to the susceptor, uniformity of heat transfer to the wafer is a difficult design problem, especially for larger diameter wafers. Figure 55 shows schematically a wafer upon a flat susceptor with temperatures decreasing from T1 on the susceptor, to T2 on the backside of the wafer to T3 which is lower due to heat loss by radiation from the polished front side. The temperature gradient through the wafer produces a bowing which results in wafer edges being raised off the susceptor. Once the edges lose contact, heat transfer across the gap becomes inefficient and the rim of the wafer cools and contracts. The hoop stress developed by the radial temperature gradient may be relieved by crystal slip around the edges of the wafer.

Three susceptor design modifications which have been used to reduce slip are shown schematically in Figure 56. In the first case, the susceptor thickness under the wafer center is reduced to decrease coupling and produce more heating toward the edges of the wafer. The recessed wafer pocket with a recess under the wafer is also directed at maintaining edge contact and reducing the efficiency of

Silicon Epitaxy 333

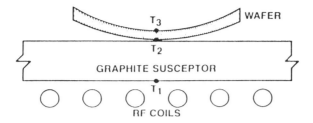

Figure 55. Wafer bow in response to temperature gradients from the highest, T1 in the R.F. heated susceptor, to an intermediate T2 at the wafer contact surface, to the lowest T3, at the free surface radiating to the cold walls of the reactor.

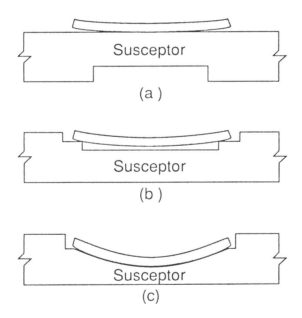

Figure 56. Susceptor designs to reduce within wafer temperature gradients during R.F. heating:

    (a) depressions in the opposite side,
    (b) stepped pockets (contact at rim), and
    (c) pockets contoured to fit the curve of the wafer at processing temperatures.

heating at the wafer center. The third pocket design uses a contoured pocket which can be shaped to produce good uniformity of heating but must be changed for different wafer thickness and different operating temperatures for optimum results.

Liaw and Rose (2) have discussed other susceptor modifications for improved heating uniformity.

8.2.3 Radiant Heating: Radiant heating from outside the chamber employs quartz halogen heating lamps which put out a broad spectrum of energy from the infrared well into the visible range. Quartz walls of the chamber must be as transparent as possible to that broad spectrum of wavelengths or excessive heating of the reactor walls and deposition on the walls will result. In order to prevent overheating of the chamber wall, cylinder reactors generally employ forced air cooling impinged against the chamber walls. The difficulty in keeping the walls cool enough to prevent wall deposits limits the use of silane in the air cooled radiantly heated cylinder reactors.

8.2.4 Combined Mode Heating: Most reactors employ polished metal surfaces to reflect energy back into the processing chamber. In some cases, this passive reflection of heat may be supplemented by addition of radiant heating as a secondary heat source. McDiarmid et al. (109) used combined heating and shaped cavity susceptors to produce slip-free epitaxial deposits on large diameter wafers. Resistance heating within the chamber supplementing another primary heat source has also been used for the Tetron tapered cavity reactor.

8.3 Operating Pressure of Reactors

Although experimental reactors have been operated in the range below 10 Torr. (82),(83), growth rates were slow due to the lower surface arrival rate of reactants at the low pressure. Commercial reactors operating in a reduced pressure range of 40-150 Torr. were introduced in 1978 (5),(54) and the technique was later extended to large vertical reactors (110),(111).

Those reduced pressure reactors use a Roots blower backed by an oil sealed mechanical vacuum pump pumping flows of 100-200 standard liters per minute of hydrogen and deposition source gases.

Reduction of the pressure increases the gas velocity and cuts the average gas residence time by about a factor of ten.

That improves the abruptness of junctions between layers and improves uniformity and influences film properties related to pattern shift and autodoping. Figure 51 shows pattern shift data for two reactor types as a function of pressure. The difference probably is due to the cyclic growth rate during wafer holder rotation in the radiantly heated cylinder compared to a more uniform growth rate in the R.F. heated modified vertical reactor.

9.0 TRENDS FOR THE FUTURE IN SILICON EPITAXY

Silicon epitaxy, long a required process for bipolar processing, is now being widely used for MOS processing, especially for prevention of latch-up in high density CMOS devices. Some of the forces which will shape equipment and process developments for silicon epitaxy were reviewed by Burggraaf (112). Among his projections were the growing application of silicon epitaxy to production of CMOS circuits where pressures to adopt larger size wafers is driven by the large size of the die (over 1 cm^2 area). With 150 mm diameter wafers common in production today, users of equipment are now specifying equipment have capability of processing 200 mm wafers. Advanced technology groups by 1990 will be considering the processing changes required for use of even larger substrates. Equipment for processing 150 mm wafers will dominate equipment purchases between now and 1992, however, equipment for 200 mm wafers will become common by 1992 and will dominate equipment purchases by 1996.

The high density of advanced devices dictates use of progressively lower processing temperatures to reduce lateral diffusion in the devices. The epitaxial deposition which occurs early in the processing sequence remains one of the highest temperature processes. Process development attention is being directed toward lowering those temperatures or at least reducing the integrated time-temperature product. This places emphasis on use of silane source chemistry and more leak tight reactor designs for processing at 900°C or below while remaining in the reduction mode for surface oxides.

The recognition of the importance of controlling water vapor during loading and unloading the reactor, will lead to increased interest in load locks to restrict introduction of air and water vapor between loads. Reactor load locks are probably more easily implemented for single wafer reactors or small loads than for large load reactors. Hence, there is growing pressure in favor of reactors with smaller load size.

Increased attention to insitu cleaning means that new gas cleaning chemistries possibly enhanced by plasma energy can be expected as a part of future commercial processes. Those techniques will probably employ reduced pressures at least for the insitu cleaning portion of the cycle.

While larger load sizes are favored for lowering the cost of capital equipment and per wafer costs, the demands of better uniformity on larger (150, 200, and 250 mm) diameter wafers favor the single wafer or small batch reactor designs. Rapid heating techniques developed initially for implant anneals can be expected to find application in the single wafer or small batch reactor designs for the 1990's.

REFERENCES

1. Pearce, C.W., Epitaxy, Ch. 2 in *VLSI Technology*, (S. M. Sze, ed.), McGraw-Hill, New York (1983)

2. Liaw, H.M. and J.W. Rose, Silicon Vapor-Phase Epitaxy. Chapter 1 in *Epitaxial Silicon Technology* (B.J. Baliga, ed.), pp. 1-89, Academic Press, New York (1986)

3. Bollen, L.J.M., Epitaxial State of the Art. Acta Electronica, **21** #3(Mar.):185-199 (1978)

4. Hammond, M.L., Silicon Epitaxy. *Solid State Technology* **21** #11(Nov.):68-75 (1978)

5. Herring, R.B., Silicon Epitaxy at Reduced Pressure, in *Proc. 7th Intl Conf. on Chemical Vapor Deposition* (T.O. Sedgewick, ed.), Proc. Vol. 79-3, pp. 126-139, Electrochem. Soc., Pennington, NJ (1979)

6. Vasudev, P.K., Silicon-on-Sapphire Heteroepitaxy. Chapter 4 in *Epitaxial Silicon Technology*, (B.J. Baliga, ed.), Academic Press,Inc., New York (1986)

7. Theuerer, H.C., Kleinack, J.J., Roar, H.H., and Christensen, H., Epitaxial Diffused Transistors. *Proc. IRE* **48**:1642-1649 (1960)

8. Yaney, D.S. and Pearce, C.W., The Use of Thin Epitaxial Silicon Layers for MOS VLSI, *Proc. of the 1981 International Electron Devices Meeting*, pp. 236-239, Piscataway, NJ: IEEE (1981)

9. Strzempa-Depre, M., Harter, J., Werner, C., Skapa, H., and Kassing, R., Improved Physical Design Model for Latchup Analysis of VLSI-CMOS Circuits in *Semiconductor Silicon* 1986 (H.R. Huff et. al., eds.) pp. 1053-1064, Electrochem. Soc., Pennington, NJ (1986)

10. Tsai, J.C.C., Diffusion. Chapter 5 in *VLSI Technology* (S.M. Sze, ed.), McGraw-Hill, New York (1983)

11. Seidel, T.E., Ion Implantation, Chapter 6 in *VLSI Technology* (S.M. Sze, ed.), McGraw-Hill, New York (1983)

12. Jastrzebski, L., Soydan, R., Goldsmith, B., and McGinn, J.T., Internal Gettering in Bipolar Processes. *J. Electrochem. Soc.* 131:2944-2953 (1984)

13. Sangster, R.C., Maverick, E.F., and Croutch, M.L., Growth of Silicon Crystals by a Vapor Phase Pyrolitic Deposition Method. *J. Electrochem. Soc.* 104:317-319 (1957)

14. Bean, J.C., *MBE for Devices, in Impurity Doping Processes in Silicon* (F. Wang, ed.), North-Holland, Amsterdam (1981)

15. Bean, J.C., Silicon Molecular Beam Epitaxy as a VLSI Processing Technique, in *Proceedings International Electron Device Meeting*, pp. 6-13, IEEE, Piscataway, NJ (1981)

16. Iyer, S.S., Silicon Molecular-Beam Epitaxy, in *Epitaxial Silicon Technology* (B.J. Baliga, ed.), pp. 91-175, Academic Press, New York (1986)

17. Sugiura, H., and Yamaguchi, M., Crystal Defects in Silicon Films Formed by Molecular Beam Epitaxy. *Japan J. Appl. Physics* 19:583-589 (1980)

18. Luscher, P.E., and Collins, D. M., in *Design Considerations for Molecular Beam Epitaxy Systems* (B.R. Pamplin, ed.), Pergamon Press, London (1981)

19. Ota, Y., N-type Doping Technique in Silicon Molecular Beam Epitaxy by Simultaneous Arsenic Ion Implantation and by Antimony Evaporation. *J. Electrochem. Soc.* **126**:1761-1765 (1979)

20. Sugiura, H. and M. Yamaguchi, Growth of Dislocation-Free Silicon Films by Molecular Beam Epitaxy. *J. Vac. Sci. Technol.* **19**:157-160 (1981)

21. Watenberg, H., *Z. Anorg. Allg. Chem.* **265**:186 (1951)

22. Keck, P.H., and Broder, J., The Solubility of Silicon and Germanium in Gallium and Indium, *Phys. Rev.* **90**: 521-522 (1953)

23. Goss, A.J., Crystallization of Silicon from Solution in Sn, *J. of Metals.* **40**:1085 (1953)

24. D'Asaro, L.A., Langdorf, R.W., and Furnanage, R.A., in *Semiconductor Silicon* (R.R. Haberecht et al., eds.), pp. 233-242, Electrochem. Soc., Princeton, NJ (1969)

25. Baliga, B.J., Kinetics of Epitaxial Growth of Silicon from a Tin Melt. *J. Electrochem. Soc.* **124**:1627-1631 (1977)

26. Baliga, B.J., Silicon Liquid Phase Epitaxy, Chapter 3 in *Epitaxial Silicon Technology* (B.J. Baliga, ed.), Academic Press, New York (1986)

27. Tsaur, B-Y., Fan, J.C.C., Geis, M. W., Silversmith, D.J., and Mountain, R.W., Improved Techniques for Growth of Large Area Single Crystal Si Sheets over SiO_2 Using Lateral Epitaxy by Seeded Solidification. *Appl. Phys. Lett.* **39**:561-563 (1981)

28. Kamins, T.I., and Pianetta, P.A., MOSFETs in Laser Recrystallized Poly-silicon on Quartz. *IEEE Electron. Device Lett.* **EDL-1**:214-216 (198).

29. Tsaur, B-Y., Geis, M.W., Fan, J.C.C., Silversmith, D.J., and Mountain, R. W., N-Channnel Deep Depletion Metal-

Oxide Transistors Fabricated in Zone Melting-Recrystallized Polysilicon Si Films in SiO_2. *Appl. Phys. Lett.* **39**:909-911 (1981)

30. Kern, W. and Puotinen, D.A., Cleaning Solutions Based on Hydrogen Peroxide for use in Silicon Semiconductor Technology, *RCA Review* **31**:187-206 (1970)

31. Shwartzman, S., Mayer, A., and Kern, W., Particle Removal from Solid State Wafers, *RCA Review* **46**:81-105 (1985)

32. Swalin, R.A., *Thermodynamics of Solids*, John Wiley, New York (1962)

33. Ghidini, G. and Smith, F.W., Interaction of H_2O with Si (111) and (100). *J. Electrochem. Soc.* **131**: 2924-2928 (1984)

34. Roberge, R.P., Francis, A.W. Jr., Fisher, S.M., and Schmitz, S.C., Gaseous Impurity Effects in Silicon Epitaxy. *Semiconductor International* **10** #1:77-81 (1981)

35a. Donahue, T.J. and Reif, R., *J. Appl. Phys.* **57**:275 (1985)

35b. Donahue, T.J. and Reif, R., Low Temperature Silicon Epitaxy Deposited by Very Low Pressure Chemical Vapor Deposition Parts I-III. *J. Electrochem. Soc.* **133**:1691-1697, 1697-1701, 1701-1705 (1986)

36. Classen, W.A.P., Rate Determining Reactions and Surface Species in CVD of Silicon, Ch. 5 in *Kinetic Studies on the Nucleation and Growth of Silicon via Chemical Vapor Deposition, Univ. Doctoral Dissertation*, Eindhoven, Netherlands (1981)

37. Bean, K., Chemical Vapor Deposition of Silicon and its Compounds, in *Semiconductor Materials and Process Technology Handbook* (G.E. McGuire, ed.) pp. 80-125, Noyes Publications, Park Ridge, NJ (1987)

38. Srinivasan, G.R., Modeling and Applications of Silicon Epitaxy, in *Silicon Processing* (D.C. Gupta, ed.), ASTM STP 804, American Society for Testing and Materials, pp. 151-173 (1983)

39a. Glang, R. and Wajda, E.S., in *The Art and Science of Growing Crystals* (J.J. Gilman, ed.), pp. 88-92, Wiley, New York (1963)

39b. Wajda, E.S., and Glang, R., in *Metallurgy of Elemental and Compound Semiconductors* (R.O. Grnbel, ed.) p. 229, Interscience, New York (1961)

40. May, J.E., Kinetics of Epitaxial Silicon Deposition by a Low Pressure Iodide Process. *J. Electrochem. Soc.* **112**: 710-713 (1965)

41. Taft, E.A., Epitaxial Growth of Doped Silicon Using an Iodine Cycle, *J. Electrochem. Soc.* **118**:1535-1538 (1971)

42. Hanssen, J.H.L., Saaman, A.A., de Moor, H.H.C., Giling, L.J., and Bloem, J., *J. Cryst. Growth* **65**: 406-410 (1983)

43. Herrick, C.S., and Krieble, J.G., *J. Electrochem. Soc.* **107**:111-117 (1960)

44. Joyce, B.A., and Bradley, R.R., Epitaxial Growth of Silicon from Pyrolysis of Mono-silane on Silicon Substrates. *J. Electrochem. Soc.* **110**:1235-1240 (1963)

45. Bloem, J., High Chemical Vapor Deposition Rates of Epitaxial Silicon Layers. *J. Cryst. Growth* **18**:70-76 (1973)

46. Weeks, S.P., Pattern Shift and Pattern Distortion During CVD Epitaxy on (111) and (100) Silicon. *Solid State Technology* **24**#11(Nov.):111-117 (1981)

47. Lee, H.H., Silicon Growth at Low Temperatures: SiH_4-HCl-H_2 System *J. Cryst. Growth* **69**:82-90 (1984)

48a. Bloem, J. and Giling, L.J., in *Cur. Topics Mat. Sci.* **1**: 147-342 (1978)

48b. Bloem, J., and Giling, L.J., Epitaxial Growth of Silicon by Chemical Vapor Deposition, Ch. 3 in VLSI Electronics Microstructure Science, V. 12 Silicon Materials (N.G. Einspruch, ed.), pp. 89-139, Academic Press, New York (1985)

49. Bloem, J., and Classen, W.A.P., Rate-determining Reactions and Surface Species in CVD Silicon. *J. Cryst. Growth* **49**:435-444 (1980)

50. Foster, D., Learn, A. and Kamins, T.I., Deposition Properties of Silicon Films Formed from Silane in a Vertical Reactor. *Vac. Science and Technology* **4**: 1182-1186 (1986)

51. Bloem, J., and Classen, W.A.P., Nucleation and Growth of Silicon Films by Chemical Vapor Deposition. *Philips Tech. Rev.* **41**:60-69 (1983)

52. Sedgwick, T.O., and Smith, J.E., Raman Scattering Spectroscopy Applied to the Study of Chemical Vapor Deposition Systems. *J. Electrochem. Soc.* **123**:254-258 (1976)

53. Ban, V.S. and Gilbert, S.L., Chemical Processes in Vapor Deposition of Silicon (Part I). *J. Electrochem. Soc.* **122**:1382-1388 (1975)

54. Herring, R.B., Advances in Reduced Pressure Silicon Epitaxy. *Solid State Technology* **22**#11(Nov.):75-80 (1979)

55. Ban, V.S. and Gilbert, S.L., Chemical Processes in Vapor Deposition of Silicon (Part II), *J. Electrochem. Soc.* **122**:1389-1391 (1975)

56. Bloem, J., Oei, Y.S., de Moor, H.H.C., Hanssen, J.H.L., and Giling, L.J., Near Equilibrium Growth of Silicon by CVD - I. The Si-Cl-H System. *J. Cryst. Growth* **65**:399-405 (1983)

57. Van der Putte, P., Giling, L.J., and Bloem, J., Growth and Etching of Silicon in Chemical Vapor Deposition Systems: The Influence of Thermal Diffusion and Temperature Gradients. *J. Cryst. Growth* **31**:299-307 (1975)

58. Aoyama, T., Inoue, Y., and Suzuki, T., Growth Reaction Mechanisms of Vapor Phase Silicon Epitaxy, in *Semiconductor Silicon 1981* (H.R. Huff, R.J. Kriegler, and T. Takeishi, eds.), pp. 379-389, Electrochem. Soc., Pennington, NJ (1981)

59. Liaw, H.M., Weston, D., Reuss, B., Birritella, M., and Rose, J., in *Proc. 9th Intl. Conf. on Chemical Vapor Deposition* (McD. Robinson, et al., eds.), pp. 463-475, Electrochem. Soc., Pennington, NJ (1984)

60. Classen, W.A.P., and Bloem, J., The Nucleation and Growth of Silicon via Chemical Vapor Deposition, in *Semiconductor Silicon 1981*, (H.R. Huff, R.J. Kriegler, and T. Takeishi, eds.), pp. 365-376, Electrochem. Soc., Pennington, NJ (1981)

61. Brice, J.C., *The Growth of Crystals from Liquids*, North-Holland Pub., Amsterdam (1973)

62. Eversteijn, F.C., Gas-Phase Decomposition of Silane in a Horizontal Epitaxial Reactor. *Philips Res. Rep.* **26**:134-144 (1971)

63. Eversteijn, F.C., Chemical Reaction Engineering in the Semiconductor Industry. *Philips Res. Rep.* **29**:45-66 (1974)

64. Law, T.J., *U.S. Patent* No. 3,173,814

65. Reif, R., Kamins, T.I., and Saraswat, K.C., A Model for Dopant Incorporation into Growing Silicon Epitaxial Films (Parts I and II), *J. Electrochem. Soc.* **126**: 644-652 and 653-660 (1979)

66. Rai-Chaudhury, P., and Salkovitz, E.I., Doping of Epitaxial Silicon: Effect of Dopant Partial Pressure. *J. Cryst. Growth* **7**: 361-367 (1970)

67. Giling, L.J., *Mater. Chem. Phys.* **9**: 117-138 (1983)

68a. Mazur, R.G., and Dickey, D.H., A Spreading Resistance Technique for Resistivity Measurements on Silicon. *J. Electrochem. Soc.* **113**:255-259 (1966)

68b. Schumann, P.A., and Gardner, E.E., *J. Electrochem. Soc.* **116**:87-91 (1969)

69. Bloem, J., Giling, L.J., and Graef, M.W.M., The Incorporation of Phosphorous in Silicon Epitaxial Layer Growth. *J. Electrochem. Soc.* **121**:1354-1357 (1974)

70. Rai-Chaudhury, P., and Salkovitz, E.I., Doping of Epitaxial Silicon: Equilibrium Gas Phase and Doping Mechanism. *J. Cryst. Growth* 7: 353-360 (1970)

71. Rice, W., Diffusion of Impurities During Epitaxy. *Proc. IEEE* **802**:284-295 (1964)

72. Grove, A.S., Roder, A., and Sah, C.T., Impurity Distribution in Epitaxial Growth. *J. Appl. Phys.* **36**:802-810 (1965)

73. Price, J.B., and Goldman, J., Silicon Epitaxial Interface Migration. *J. Electrochem. Soc.* **126**:2033-2034 (1979)

74. Thomas, C.O., Kahng, D., and Manz, R.C., Impurity Distribution in Epitaxial Si Films. *J. Electrochem. Soc.* **109**:1055-1061 (1962)

75. Grossman, J.J., Kinetic Theory for Autodoping for Vapor Phase Epitaxial Growth of Germanium. *J. Electrochem. Soc.* **110**:1065-1068 (1963)

76. Wong, M. and Reif, R., A Trapping Mechanism for Autodoping in Silicon Epitaxy - Part I: Theory. *IEEE Trans. Electron Devices* **ED-32**:83-88 (1985)

77. Wang, M., Reif, R., and Srinivasan, R., A Trapping Mechanism for Autodoping in Silicon Epitaxy - Part II: Parameter Extraction and Simulations. *IEEE Trans. Electron Devices* **ED-32**:89-94 (1985)

78. Langer, P.H., and Goldstein, J.I., Impurity Redistribution During Silicon Epitaxial Growth and Semiconductor Device Processing. *J. Electrochem. Soc.* **121**:563-571 (1974)

79. Langer, P.H., and Goldstein, J.I., Boron Autodoping During Silane Epitaxy. *J. Electrochem. Soc.* **124**:591-598 (1977)

80. Srinivasan, G.R., Autodoping Effects in Silicon Epitaxy. *J. Electrochem. Soc.* **127**:1334-1342 (1980)

81. Srinivasan, G.R., A Flow Model for Autodoping in VLSI Substrates. *J. Electrochem. Soc.* **127**:2305-2306 (1980)

82. Ogirima, M., Saida, H., Suzuki, M., and Maki, M., Low Pressure Silicon Epitaxy. *J. Electrochem. Soc.* **124**:903-908 (1977)

83. Ogirima, M., Saida, H., Suzuki, M., and Maki, M., A Multiwafer Growth System for Low Pressure Silicon Epitaxy. *J. Electrochem. Soc.* **125**:1879-1883 (1978)

84. Rode, D.L., Wagner, W.R., and Schumaker, N.E., Singular Instabilities on LPE GaAs, CVD Si, and MBE InP Growth Surfaces. *Appl. Phys. Lett.* **30**:75-78 (1977)

85. ASTM, Standard Test Method for Crystallographic Perfection of Epitaxial Deposits of Silicon by Etching Techniques. Designation F80 - 85, Annual Book of ASTM Standards, Am. Soc. for Testing and Matls., Philadelphia, PA (1987)

86. Stowell, M.J., In *Epitaxial Growth, Part B* (J.W. Matthews, eds.), pp. 437-492 Academic Press, New York (1975)

87. Finch, R.H, Queisser, H.J., Thomas, G., and Washburn, J., *J. Appl. Phys.* **34**:406-415 (1963)

88. Pomerantz, D., A Cause and Cure of Stacking Faults in Silicon Epitaxial Layers. *J. Appl. Phy.* **38**:5020-5026 (1967)

89. Thomas, R.N., and Francombe, M.H., Influence of Impurities on the Surface Structures and Fault Generation in Homo-Epitaxial Silicon (111) Films. Surf. Sci. **25**:357-378 (1971)

90. Liaw, H.M., Rose, J. and Fejes, P.L., Epitaxial Silicon for Bipolar Integrated Circuits. *Solid State Technology* **27** #5(May):135-143 (1984)

91. Joyce B.A., Growth and Perfection of Chemically Deposited Epitaxial Layers of Si and GaAs. *J. Cryst. Growth* **3/4**:43-59 (1968)

92. Chang, H.R., and Baliga, B.J., in *Proc. 9th Intl Conf. on Chemical Vapor Deposition* (McD. Robinson et al., eds.), pp. 315-323, Electrochem. Soc., Pennington, NJ (1984)

93. Chernov, A.A., Growth Kinetics and Capture of Impurities During Gas Phase Crystallization. *J. Cryst. Growth* **42**:55-76 (1977)

94. Tanno, K., Endo, N., Kitajima, H., Kurogi, Y., and Tsuya, H., Selective Silicon Epitaxy Using Reduced Pressure Technique. *Japanese J. Appl. Phys.* **21**:L564-L566 (1982)

95. Voss, H.J. and Kurten, H., *Tech. Dig. International Electron Devices Meeting (IEDM)*, pp. 35-38 (1983)

96. Liaw, H.M., Reuss, R.H., Nguyen, H.T., and Woods, G.P., Surface Morphology of Selective Epitaxial Silicon, in *Semiconductor Silicon 1986* (H.R. Huff, T. Abe, and B. Kolbesen, eds.), pp. 260-273, Electrochem. Soc., Pennington, NJ (1986)

97. Akiyama, N., Inoue, Y., and Suzuki, T., Critical Radial Temperature Gradient Inducing Slip Dislocations in Silicon Epitaxy Using Dual Heating of Two Surfaces of a Wafer. *Japanese J. of Appl. Physics* **25**:1619-1622 (1986)

98. Werkoven, C.J., Source, Transport and Precipitation of Metallic Impurities in Si-Epitaxy, in *Aggregation Phenomena of Point Defects in Silicon* (E. Sirtl and J. Goorissen, eds.), pp. 144-154, Electrochem. Soc., Pennington, NJ (1982)

99. Schmidt, P.F., and C.W. Pearce, A Neutron Activation Analysis Study of the Sources of Transition Group Metal Contamination in the Silicon Devise Manufacturing Process. *J. Electrochem. Soc.* **128**:630-637 (1981)

100. Townley, D.O., Optimum Crystallographic Orientation for Silicon Device Fabrication. *Solid State Technology* **16**#1(Jan.):43-47 (1973)

101. Drum, C.M., and Clark, C.A., Geometrical Stability of Shallow Surface Depressions During Growth of (111) and (100) Epitaxial Silicon. *J. Electrochem. Soc.* **115**:664-669 (1968)

102. Drum, C.M., and Clark, C.A., Anisotropy of Macrostep Motion and Pattern Edge Displacements During Growth of Epitaxial Silicon on Silicon Near {100}. *J. Electrochem. Soc.* 117:1401-1405 (1970)

103. Lee, P.H., Wauk, M.T., Rosler, R. S., and Benzing, W.C., Epitaxial Pattern Shift Comparison in Vertical, Horizontal, and Cylinder Reactor Geometries. *J. Electrochem. Soc.* 124:1824-1826 (1977)

104. Boydston, M.R., Gruber, G.A., and Gupta, D.C., in *ASTM Spec. Tech. Publ.*, **804**, pp. 174-189, American Soc. for Testing and Matls., Philadelphia, PA (1983)

105. Ban, V.S., Transport Phenomena Measurements in Epitaxial Reactors. *J. Electrochem. Soc.* **125**:317-320 (1978)

106. Theuerer, H.C., Epitaxial Silicon Films by the Hydrogen Reduction of $SiCl_4$. *J. Electrochem. Soc.* **108**:644-653 (1961)

107. Burggraaf, P., The Dawn of Epitaxy's New Era. *Semiconductor International* 9#5(May):68-74 (1986)

108. Ban, V.S., Novel Reactor for High Volume Low Cost Silicon Epitaxy. *J. Cryst. Growth* **45**:97-107 (1978)

109. McDairmid, J., Lawrence, L.H., and Hammond, M.L., in *Proceeding of 9th Intl. Conf. on Chemical Vapor Deposition* (McD. Robinson, et al., eds.), pp. 351-355, Electrochem. Soc., Pennington, NJ (1984)

110. Fisher, S.M., Hammond, M.L., and Sandler, N.P., Reduced Pressure Epitaxy in an Induction-Heated Vertical Reactor. *Solid State Technology* 28#1 (Jan.):107-112 (1986)

111. Lawrence, L.H., McDiarmid, and Hammond, M.L., Reduced Pressure Epitaxy in an Induction-Heated Vertical Reactor, in *Proceedings of 9th Intl. Conf. on Chemical Vapor Deposition*, (McD. Robinson et al., eds.), p. 454, Electrochem. Soc., Pennington, NJ (1984)

112. Burggraaf, P.S., The Forces Behind Epitaxial Silicon Trends. *Semiconductor International* 6#10(Oct.):44-51 (1983)

6

Silicon Material Properties

W. Murray Bullis

1.0 INTRODUCTION

Silicon is the second most abundant element on earth. It always occurs in the form of a compound, most often as the oxide (silica) or as a silicate. These compounds form a wide variety of rocks and minerals. The common mineral quartz is a form of silica. Until late in the 18th century, silica was considered to be an element, but it was correctly proposed to be an oxide by Lavoisier in 1787 (1). By the middle of the next century, silicon had been successfully obtained in crystalline form.

Also during this century, the properties now associated with semiconductors were first observed. These properties include the negative temperature dependence of resistivity due to the exponential variation of the free carrier density with temperature; photoconductivity, a decrease in resistance on exposure to light; various thermoelectric effects; the Hall and other magnetoelectric effects; and rectification between the semiconductor and a metal contact. This last property was widely used for detectors in crystal radio receivers in the first half of the 20th century; a patent for a silicon crystal detector was issued in 1906 (2).

The purity of the silicon samples available for study during the first four decades of the 20th century was inadequate to provide sufficiently reproducible properties to establish accurate models for the electrical and optical properties. However, work with other materials demonstrated that the classical theories of the 19th century were not fully adequate. It was not until quantum mechanics was developed in the late 1920's and early 1930's that all the various experimental observations could be properly interpreted.

The concept that minute amounts of impurities control the conductivity of semiconductors such as silicon was first recognized in the early 1930's by Gudden (3). The intensive development of silicon crystal detectors for radar sets during World War II (4) led to improved purification techniques and increased understanding of conduction phenomena. These advances led in turn to the junction diode, the point contact transistor, and the junction transistor (5).

In the following decade, the basic properties of both germanium and silicon were solidly established. The many refinements of both theory and experiment provided the basic understanding essential for the technological application of these materials on a wide scale for the next two decades. During this period, silicon became the material of choice for integrated circuits, both because it could operate at higher temperatures than germanium and because it, like aluminum, has a thin native oxide "skin" which serves as a protection to the silicon surface as well as acting as a diffusion barrier.

Advances in control of crystal perfection and purity continued throughout the next several decades; during this period silicon materials technology more than kept pace with the device technology requirements. In the mid-1970's, however, there was a resurgence of interest and activity in the fundamental properties of silicon. This came about because, as the sizes of the features in the device structures shrank and die and wafer sizes increased, material properties once again had a major impact on device and circuit yields and characteristics.

This recent activity is concerned not so much with the basic properties of silicon but rather with more complex phenomena related to transport and aggregation of the impurities and defect structures in the crystal and their interactions with electrons and holes. Successful studies of these phenomena are possible only because of the widespread availability of very high quality silicon crystals and because of the development and application of very powerful and sensitive analytical techniques.

2.0 CRYSTALLOGRAPHIC PROPERTIES

The atoms in crystalline materials are arranged in a periodic pattern which is characteristic of the material. This pattern can be visualized as a number of parallelepipeds with atoms at the corners. These parallelepipeds, or unit cells, repeat themselves throughout the structure. There are seven different crystal systems; some of these have variants with extra atoms between the corner atoms so that 14 different atomic structures are possible (6). Cubic crystals are characterized by unit cells with three equal lengths along three mutually perpendicular axes. The unit cell of a cubic crystal is described by the lattice constant, a, which is the length of a side of the cube.

2.1 Silicon Crystal Structure

Silicon, like other group IV insulators and semiconductors including diamond, germanium, and gray tin, crystallizes in the diamond cubic lattice structure. Each silicon atom has four bonds, one to each of its four nearest neighbors, as shown in Figure 1. This lattice structure can be constructed from two interpenetrating face-centered cubic lattices displaced from each other along the body diagonal by a distance equal to one-fourth of its length. Group III-V compounds and some II-VI compounds crystallize in a similar arrangement, called the zinc-blende lattice, where one of the constituents occupies sites on one of the face-centered cubic lattices and the other constituent occupies sites on the other.

The lattice constant, a, is 0.5430710 nm in pure, float-zoned silicon (7). The diamond cubic lattice is much more open than most crystal configurations. Five large interatomic voids in which interstitial atoms are easily accommodated are located along the body diagonals as shown in Figure 2. However, inclusion of atoms in these voids may change a; for example, if about 10^{18} interstitial oxygen atoms/cm^3 are present, a = 0.5430747 nm (7). Though it is quite small, this difference becomes very important in the utilization of silicon samples for accurate determination of the Avogadro constant by the X ray/crystal density method (8). Although oxygen-free, semiconductor-grade silicon represents the most perfect material available for this purpose, refinements in the determination are still being made as material with fewer and fewer crystal defects becomes available.

It is convenient to describe the various planes which

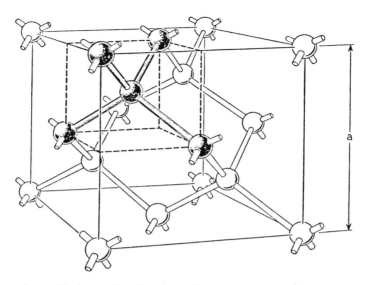

Figure 1: Unit cell of the diamond cubic lattice structure showing the lattice constant, a, and the locations of the silicon-silicon bonds. [From Shockley, Reference 5.]

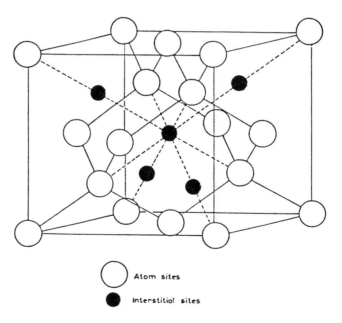

Figure 2: Unit cell of the diamond cubic lattice structure showing the five large interatomic voids along the body diagonals. [From Rhodes, Reference 58.]

pass through the atomic sites in a crystal by their Miller indices. These indices, h, k, and l, are the reciprocals of the intercepts of the plane with the x, y, and z axes, respectively. A plane is denoted by the symbol (hkl); negative intercepts are denoted by a bar over the index. A family of planes, e.g., (100), (010),(001), (00$\bar{1}$), (0$\bar{1}$0), and ($\bar{1}$00), is denoted by a bracketed symbol: {100}. In general, the indices are expressed as the smallest integers having the same ratio. Selected low index planes are shown in Figure 3.

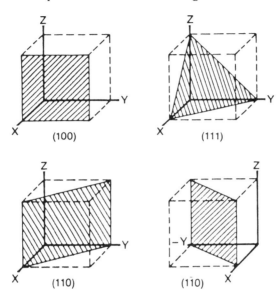

Figure 3: Examples of low-index planes.

The Miller indices may also be used to indicate a direction. In a cubic crystal, the x, y, and z axes form an orthogonal set; therefore the direction perpendicular to a particular plane has the same Miller indices as the plane. A direction is denoted by a square bracketed symbol: [hkl]; a family of directions is denoted by angular brackets: <hkl>. For example, the x axis is the [100] direction and the three coordinate axes are the <100> directions. The body diagonals are the <111> directions.

In a simple cubic lattice, the spacing between planes, d_{hkl}, is given by:

$$d_{hkl} = \frac{a}{[(h^2 + k^2 + l^2)]^{1/2}} \qquad (1)$$

The distance between the {100} planes (which pass through the corner atoms) is a. However, because there are atoms between the corner atoms in the diamond cubic structure, planes which pass through the atoms in the crystal are actually closer together than this if the Miller indices are expressed in the usual form as the smallest integers. There are eight atoms in the unit cell of the diamond lattice. These are located at the following points: (0,0,0), (0,1/2,1/2), (1/2, 0,1/2), (1/2,1/2,0), (1/4,1/4,1/4), (1/4,1/4,3/4), (3/4,1/4,3/4), and (3/4,3/4,1/4). Therefore the {100} planes which include all the atoms are the {400} planes, and so the actual distance between the atomic planes along any of the cubic axes is a/4 or 0.1358 nm.

Because there is one atomic plane between each of the {110} planes, the planes of this type which include all the atoms are {220} planes and the distance between them is $a/\sqrt{8}$ or 0.1920 nm.

The four nearest neighbors of each atom in the diamond lattice are located at the points of a tetrahedron which has the atom at its center. Lines joining nearest neighbor pairs are along <111> directions. If the lattice is drawn with the [111] direction vertically as in Figure 4, it can be seen that the (111) planes are not uniformly spaced. Each pair of closely spaced planes is separated from the adjacent pairs by a much larger distance. The set consisting of one plane from each pair of closely spaced planes (labelled A, B, C, in the Figure) belongs to one of the interpenetrating face-centered cubic lattices and the other (labelled α, β, γ) belongs to the other. The distance between adjacent (111) planes of the same set is 1/3 of the unit cell diagonal, or $a/\sqrt{3}$) (= 0.3135 nm). The distance between the closely spaced planes is 1/4 of this distance or 0.0784 nm.

2.2 Crystal Habit

The preferred growth habit of silicon is octahedral, bounded by slow growing {111} planes. In Czochralski growth (the most common growth procedure used for commercial silicon crystals), externally imposed temperature gradients cause the crystal to grow with radial symmetry, but the various growth orientations have characteristic markings.

Commercial silicon crystals are most frequently grown in the [100] or the [111] direction. The top of a [100] crystal is usually somewhat squarish with four ribs extending radially from the seed. These ribs occur where {110} planes intersect

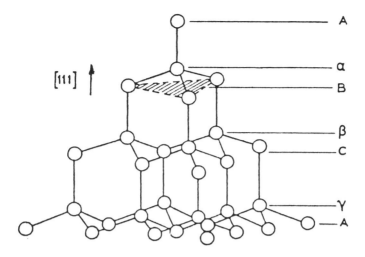

Figure 4: Schematic view of diamond cubic lattice structure showing the arrangement of pairs of (111)-planes. [From Rhodes, Reference 58.]

the surface of the crystal as illustrated in Figure 5. If the shoulder is quite flat, the top surface approximates a (001) plane. As shown in Figure 5c, well defined flats occur in place of the ribs if the shoulder is conical with an angle of about 54°, which is the angle a {111} plane makes with the (001) plane. During body growth, flats may sometimes form on the sides of the crystal, extending down from the ribs. Usually, the growth conditions are such as to make these flats very small.

On the other hand, six ribs usually extend radially from the seed on the top of a [111] crystal with a relatively flat shoulder. These ribs occur where the three {110} planes intersect the top of the crystal as shown in Figure 6. The top of such a crystal approximates a (111) plane and flats corresponding with one face of an octahedron (see Figure 7) may occur. Three equally spaced flats extend down the sides of a [111] crystal. These flats are actually successions of {111} planes (faces 2, 3, and 4 of the octahedron) which are inclined slightly (about 20°) to the growth axis and so have a somewhat rough or shingled appearance. Occasionally, the crystal forms a "dog-leg," and the flats appear to shift by 60°, being formed from the other set of {111} planes (faces 5, 6, and 7 of the octahedron).

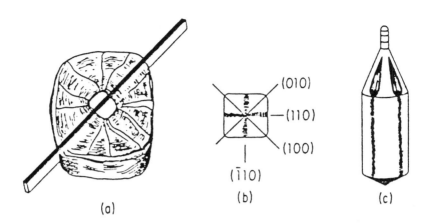

Figure 5: Rib structure at the top of a [100] silicon crystal. (a) Top of crystal grown with a flat shoulder. (b) Identification of the low index planes which cut the crystal at and between the ribs. (c) Crystal grown with a conical shoulder. [From Runyan, Reference 1.]

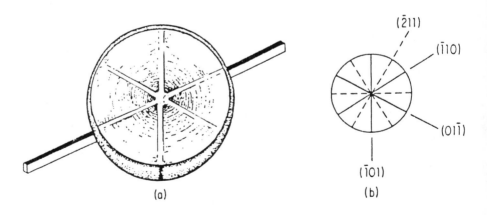

Figure 6: Rib structure at the top of a [111] silicon crystal. (a) Pictorial view of top of crystal. (b) Identification of the low index planes which cut the crystal at and between the ribs. [From Runyan, Reference 1.]

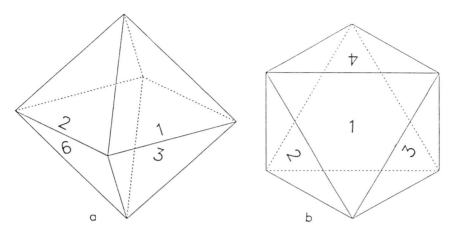

Figure 7: Octahedra formed by the {111} planes of a silicon crystal. In view (a), the median square is a {100} plane. View (b) is looking down on a {100} plane. In this view plane 5 is between planes 2 and 4, plane 6 is between planes 2 and 3, plane 7 is between planes 3 and 4, and plane 8 is opposite plane 1.

2.3 Crystal Orientation

Although cubic crystals are isotropic to first order, higher order effects are frequently significant in device processing. In particular, mechanical properties (including fracture), crystal growth, diffusion, ion implantation, and etching may exhibit observable directional characteristics. Therefore, it is desirable to control the directions of the atomic planes of the crystal in relation to the surfaces of the wafer.

Rough crystal orientation can be determined from the growth marks discussed above. However, the requirements of modern device technology are such as to require more accurate determination of the crystal and wafer orientations than can be done in this way. To establish the relationships between the geometrical surfaces of crystals and wafers and the structure of the crystal with sufficient accuracy, X-ray or optical methods must be used. X-ray methods are based on the principles of the diffraction of x rays from the atomic planes in a crystal. When a parallel, monochromatic beam of x rays strikes a crystal, it is scattered by the various atoms it hits. If the incoming beam has certain specific relationships with the atomic planes, coherent scattering or diffraction occurs. In particular, when the wavelength, λ, and the angle of incidence, ϑ, obey Bragg's law:

$$n\lambda = 2d \sin \vartheta \tag{2}$$

where n is an integer and d is the spacing of the atomic planes, the scattering is in phase and the diffracted beam has a maximum intensity. This is known as a Bragg reflection. By combining Equations (1) and (2), it can be seen that, for cubic structures, the Bragg angle is given by:

$$\sin \vartheta = \frac{n\lambda(h^2 + k^2 + l^2)^{1/2}}{2a}. \tag{3}$$

Not all values of h, k, and l result in Bragg reflections. For diamond cubic structures, h, k, and l must be either all even (including 0) or all odd, with the further condition that if they are all even the sum $h + k + l$ must be a multiple of 4. These conditions may be summarized by the following relations

$$h^2 + k^2 + l^2 = 8m + 3, \text{ or} \tag{4a}$$

$$h^2 + k^2 + l^2 = 8(m + 1), \tag{4b}$$

where m is any integer. Values of $h^2 + k^2 + l^2$ are listed in Table 1 for selected low index planes. Occasionally, different planes have the same value of $h^2 + k^2 + l_2$ and hence the same Bragg angle. In such cases it is desirable to check a higher order reflection (larger value of n in Equation (3)) to assure that the desired plane has been found.

Details of the procedure for determining the orientation of a silicon crystal by X-ray diffraction and goniometer techniques are given as Method A of ASTM Standard Test Methods F 26 (9). Additional refinements of the technique and many practical suggestions for carrying it out are given in the manual by Wood (10). Method A of ASTM Standard Test Methods F 847 (11) describes the application of X-ray diffraction techniques to determination of the orientation of flats on silicon wafers.

Another technique for orienting crystals also involves x rays, but in this case broad band, or white, radiation is used rather than monochromatic radiation. The preferred technique is to direct a collimated beam of white x radiation at

Table 1

Low Index Planes From Which

X-Ray Reflections Are Allowed

Allowable Planes	m	$h^2 + k^2 + l^2$
{111}	0	3
{220}	0	8
{311}	1	11
{400}	1	16
{311}	2	19
{422}	2	24
{333}	3	27
{511}	3	27
{440}	3	32
{531}	4	35
{620}	4	40
{533}	5	43
{444}	5	48

the crystal surface through a small circular hole in a photographic plate or film and observe the pattern of spots, known as a Laue pattern, made by the backscattered radiation on the film.

When white radiation is used, the Bragg Equation (Equation (2)) is satisfied for many different planes (different values of d) at the same time; each plane will utilize the correct value of wavelength to satisfy Equation (2) for the fixed value of ϑ. In each case the beam diffracted from a particular atomic plane forms a spot on the film. The identity of the various planes can be determined from the angular positions of the spots. Details of the method are given in ASTM Standard Method E 82 (12) and in Wood (10). Method B of ASTM Standard Test Methods F 847 (11) covers the application of the Laue Back Reflection Method to determination of the orientation of wafer flats.

A third method for orienting both wafers and crystals involves the use of optical reflections from etch pits in a plane surface. Caustic etches (up to a 50% solution of KOH or NaOH) may be used. Details of the technique, which is best applied to low index planes such as the commonly used

{100} or {111} surfaces, are given as Method B of ASTM Standard Test Methods F 26 (9). This method is not as widely used in present day technology as are X-ray methods.

Townley (13) discusses the requirements and procedures for wafer orientation in detail from the points of view of wafer scribing and pattern shift during epitaxial growth. This paper summarizes the salient properties as follows:

1. Crystal growth perpendicular to a {111} plane proceeds more slowly than growth in other directions.

2. Fracture occurs most readily along {111} planes.

3. Preferential etching exposes {111} planes.

By studying (100) and (111) stereographic projections, reproduced in Figures 8 and 9, respectively, he determined optimum locations for the reference flats and directions of slice orientation.

Crystals grown in the [100] direction have four-fold symmetry; therefore any {110} plane is equivalent to any other. These {110} planes are in the same directions as the inclined {111}, easy-break, planes so the preferred scribe lines are parallel with and perpendicular to the intersections of any {110} plane with the wafer surface. Drum and Clark (14) studied pattern shift on {100} surfaces and concluded that {100} wafers should be oriented as closely as possible to an exact {100} crystal plane to minimize the pattern shift.

For crystals grown in the [111] direction, the symmetry is six-fold rather than four-fold, and the preferred orientation planes are less obvious. There are three inclined {111} planes, 120° apart, the ($1\bar{1}1$), ($1\bar{1}1$), and ($11\bar{1}$). Wafers cut perpendicularly to the crystal axis prefer to break along these inclined planes. Preferential etching of the surface facing toward the tail, or tang end, of the crystal (as for optical orientation) results in triangular etch pits pointed in these directions as shown in Figure 9a. If the reference flat is on the ($1\bar{1}0$) plane (or the equivalent ($\bar{1}01$) or ($01\bar{1}$) planes) and if the wafer is placed so the reference flat is placed toward the operator, the etch figure points from left to right; this is the preferred direction for scribe tool travel along the scribe line parallel with the reference flat. Along the scribe line perpendicular to the reference flat, either direction of scribe tool travel is equivalent. If the other set of {110} planes (the ($1\bar{1}0$), ($\bar{1}01$) or ($01\bar{1}$) planes) is chosen for

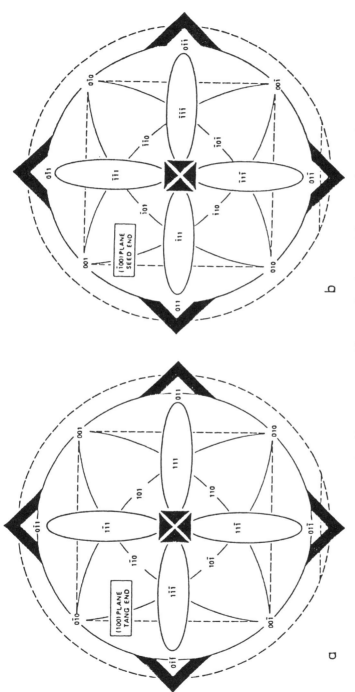

Figure 8: (100) stereographic projection with superimposed growth lines, etch figure, optical reflection lobes, and break planes. (a) Tang end of crystal: (100) plane. (b) Seed end of crystal: ($\bar{1}$00) plane. [From Townley, Reference 13. Reprinted with permission of *Solid State Technology*.]

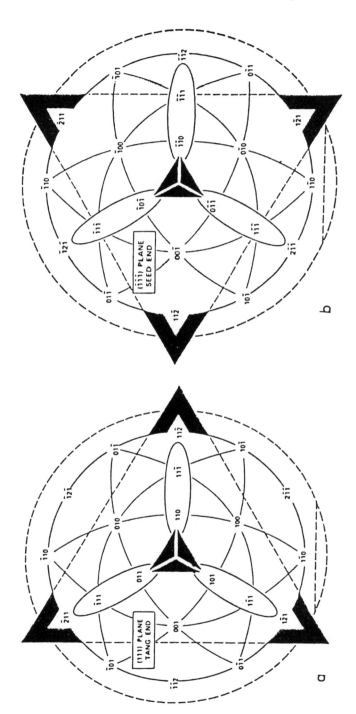

Figure 9: (111) stereographic projection with superimposed growth lines, etch figure, optical reflection lobes, and break planes. (a) Tang end of crystal: (111) plane. (b) Seed end of crystal: ($\bar{1}\bar{1}\bar{1}$) plane. [From Townley, Reference 13. Reprinted with permission of *Solid State Technology*.]

the reference flat, the etch Figure points from right to left along the preferred scribe direction.

It is possible to distinguish these sets of {110} planes either by using a Laue photograph or by combining X-ray diffraction with additional information regarding the growth conditions. It should be noted that although nowadays device dice are no longer separated by scribing, the specifications for flat orientation still in use derive from these scribing requirements.

Wafers from crystals grown in the [111] direction are frequently cut slightly off orientation to minimize pattern distortion during epitaxial layer growth. Drum and Clark (15) have demonstrated that this distortion is a minimum when the normal to the wafer surface is slightly inclined toward a <110> direction. Typically the normal is rotated an angle α from the [111] direction in a plane parallel with the reference flat, as shown in Figure 10. Off orientation angles are most frequently specified to be between 2 and 4.5° with a tolerance of ±0.5°. It should be noted that tilting the normal to the left is not eiquivalent to tilting it to the right since there are two non-equivalent sets of {110} planes.

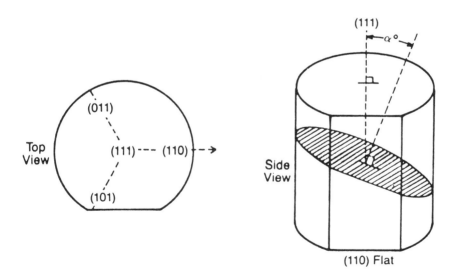

Figure 10: Schematic side view and top view of a [111] crystal showing the intentional misorientation of a wafer by $\alpha°$ from the [111] direction towards the nearest [110] direction.

362 Handbook of Semiconductor Silicon Technology

2.4 Crystal Defects

Reference to Figure 4 shows that the atoms in a silicon crystal are stacked in regular order. Along the [111] direction, adjacent pairs of atomic planes are displaced horizontally from each other. There are three different locations for the atoms; the cycle repeats every third group. Atoms in planes identified by the same symbol lie directly under each other along the [111] direction. Any disturbance in this regular order of atoms is a crystal defect.

A wide variety of defects can form in a silicon crystal during and after growth. Their properties are summarized in the following paragraphs. In addition, it should be noted that defects in silicon have been the object of intensive research efforts in the past decade or so. The review by Newman(16) summarizes some of this work for defects both in as-grown crystals (as considered in this chapter) and in wafers following certain device processing procedures. Compilations of more recent research results appear in several proceedings volumes (17),(18),(19),(20),(21). New results and models continue to appear frequently. For the most up-to-date information, the interested reader is advised to consult the current literature regularly.

2.4.1 Point Defects: Vacancies, silicon self interstitials, and impurity atoms in substitutional or interstitial sites, are defects which occur at a lattice point and do not distort the crystal over an extended distance.

Vacancies and self interstitials: The combination of a self interstitial and an associated vacancy is known as a Frenkel defect while an independent vacancy is known as a Schottky defect. These point defects exist in thermodynamic equilibrium in the crystal in relatively large concentrations at temperatures near the melting point of silicon. Their concentrations at room temperature are governed both by the thermal history of the crystal and by the presence of sources and sinks for the defects.

Vacancies may be quenched in during cooling or they may be generated by the plastic deformation which may occur as a result of thermal stresses during the cool down process. Vacancies which diffuse to the surface of the crystal disappear. The surface is the principal source and sink for vacancies, but extended defects within the crystal can also serve as sinks. A pair of vacancies can also combine to form a divacancy which appears to be a stable defect in silicon.

Originally, vacancies were thought to be the dominant

form of point defect in silicon crystals. However, at the present time, interstitials appear to be assuming a greater significance. Silicon self interstitials are generated during the oxidation of silicon, not only at the surface but also during the formation of oxide precipitates in the bulk.

Both vacancies and self interstitials interact extensively with other defects and play a dominant role in impurity diffusion in silicon. Studies of the effects of surface oxidation on impurity diffusion have led to a considerable increase in our understanding of the characteristics of vacancies and self interstitials in silicon (22),(23).

Atomic impurities: Impurity atoms may exist in the silicon crystal either substitutionally on sites usually occupied by silicon atoms or in interstitial sites. Solubilities of various atomic impurities in silicon as reported in a classic paper by Trumbore (24) are summarized in Figure 11 together with additional data on silver (25), nickel (26), and sodium and potassium (27). More recent solubility data obtained by Weber (28) for some of the transition metals are given in Figure 12.

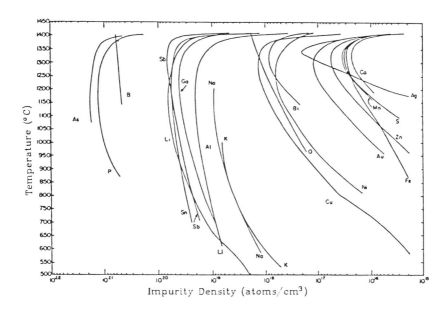

Figure 11: Solid solubilities of impurities in silicon as a function of temperature. [From Neuberger, M., and Welles, S.J., "Silicon," DS-162, October 1969.]

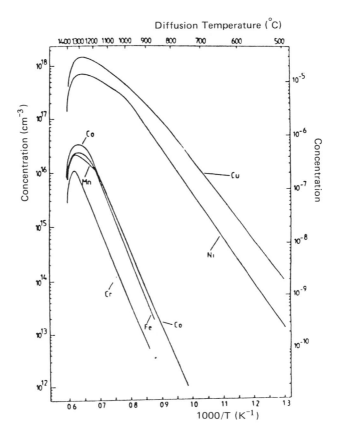

Figure 12: Solubilities of transition metals in silicon. [From Weber, Reference 28.]

Dopant impurities: When introduced into a silicon crystal, impurities from the third column of the periodic table (such as boron, aluminum, or gallium) or impurities from the fifth column (such as phosphorus, antimony, or arsenic) occupy substitutional sites and control the electrical characteristics of silicon. These impurities are intentionally introduced in concentrations ranging from less than 2 parts per billion atomic (ppba) ($<10^{14}$ atoms/cm^3) to more than 200 parts per million atomic (ppma) ($>10^{19}$ atoms/cm^3), depending on the requirements. The electrical effects of these impurities are discussed in detail in section 3 of this chapter.

Oxygen: In silicon pulled from melts contained in quartz crucibles by the modified Czochralski technique, oxygen occurs in concentrations ranging from 15 to 30 ppma, as measured by charged particle activation analysis (29). The oxygen concentration in silicon grown by the float-zone technique is very much lower, of the order of 20 ppba or less; in crystals pulled from the melt in a magnetic field (Magnetic Czochralski, or MCz, growth) the oxygen concentration may be low, <5 ppma, or very high, >35 ppma, depending on the growing conditions.

Oxygen atoms dissolved in silicon predominantly occupy interstitial sites. Two infrared absorption bands, one at 1107 cm^{-1} and the other at 514 cm^{-1}, are associated with the interstitial oxygen impurities in silicon. The first of these is widely used for quantitative analysis of oxygen in silicon as outlined in section 4.4.3 of this chapter.

Oxygen plays three important roles in silicon: thermal donor formation; precipitate formation and intrinsic gettering; and mechanical strength of the crystal. Bourret (30) has recently reviewed the various models which have been proposed to account for the formation of thermal donors and SiO$_2$ precipitates. Tiller et al. (31) provide a theoretical discussion of the various types of precipitates which may be formed. The influence of oxygen on the mechanical properties of silicon is discussed in section 5.3 of this chapter.

The properties of oxygen in silicon are discussed in greater detail in the chapter on Oxygen, Carbon, and Nitrogen in Silicon.

Carbon: Like silicon, carbon has four valence electrons, but it is considerably smaller. It may occur in small concentrations (<0.5 ppma) in all forms of silicon and sometimes occurs in higher concentrations. In substitutional sites, it is electrically neutral, but it may complex with other atoms. An infrared absorption band at about 605 cm^{-1} on the edge of the very large two-phonon silicon lattice absorption band is associated with carbon.

There is general agreement that carbon is involved in the precipitation of oxygen in silicon, but there is considerable controversy concerning the responsible mechanism(s). Barraclough and Wilkes (32) cite 23 papers published between 1979 and 1985 which report various effects of carbon on oxide precipitate formation in Czochralski silicon. They demonstrate that carbon in excess of 0.6 ppma plays a major role in oxygen precipitation at temperatures below 750°C and point out that such precipitates may dissolve at temperatures above 1150°C. Carbon is also shown, in this and other reports, to be involved with "new donor" formation.

Föll et al. (33) demonstrated that carbon has a marked

effect on the formation, density, and behavior of B-swirl defects in float-zone material. More recently, Kolbesen and Mühlbauer (34) reviewed the properties of carbon in considerable detail with emphasis on the behavior of carbon in crystals grown by the float-zone process. They show that carbon interacts strongly with oxygen, which is introduced into float-zone silicon in significant quantities during the long-time, high-temperature diffusion cycles associated with power device fabrication, forming defects in the material which may adversely affect device properties.

On the other hand, they conclude that the presence of relatively large amounts of carbon is not harmful to integrated circuits and other shallow devices which are fabricated for shorter times and at lower temperatures. Even when such devices are fabricated from Czochralski material, there is usually an oxygen-free region (denuded zone) near the surface where the devices exist. However, they recognize that carbon may affect the bulk precipitate formation which in turn may affect device properties.

Carbon has also been identified as the defect which pairs with group III acceptor centers to form complexes called x levels (35), and interactions with metallic impurities have been reported (16).

The properties of carbon in silicon are discussed in greater detail in the chapter on Oxygen, Carbon, and Nitrogen in Silicon.

Hydrogen: For some time, hydrogen has been known to affect the properties of the interface between silicon and silicon dioxide (36). More recently, hydrogen has been shown to passivate donor (37) and acceptor (38) dopant species, oxygen-related thermal donors (39), and chalcogen double donors (40).

Hydrogen is a very rapid diffuser and can penetrate deeply into a silicon wafer even at very low temperatures (41). Dry etching processes can frequently result in introduction of hydrogen (42) which can permeate the surface regions of the wafer (43). Since hydrogen in silicon is the subject of intensive investigation at the present time, greater understanding of its properties in silicon can be expected to be forthcoming.

Nitrogen: Unlike the other group V donor impurities, nitrogen appears to have almost no electrical activity. Its solubility is between 20 and 200 ppba. Several infrared absorption bands are associated with nitrogen in various states (44), (45). Even in very low concentrations, nitrogen serves as an effective means for strengthening the crystal, apparently through dislocation locking of nitrogen-oxygen complexes both in Czochralski crystals grown from silicon nitride crucibles (46) and in nitrogen-doped float-zone crystals (45). Typical concentrations of nitrogen in doped crystals range from 40 to 150 ppba.

The properties of nitrogen in silicon are discussed in greater detail in the chapter on Oxygen, Carbon, and Nitrogen in Silicon.

Transition metals: Another important class of atomic impurities is the transition metal group which includes titanium, vanadium, chromium, manganese, iron, cobalt, nickel, and copper. Their properties in silicon have recently been thoroughly reviewed by Weber (28) with particular emphasis on diffusion, solubility, and electrical activity.

The properties of these impurities tend to vary systematically with increasing atomic number. They tend to form complexes with dopants, especially boron (28),(47). Some also precipitate out as silicides (48). All these impurities are fast interstitial diffusers but some are faster than others. In equilibrium at high temperatures, they are in interstitial sites, but whether they can be quenched into interstitial sites at lower temperatures or whether they precipitate out depends on the relative values of the solubilities and diffusion coefficients.

As is discussed in section 3 of this chapter, if these impurities are present in large enough concentration, they can affect the electrical conductivity, but they are more important as recombination centers as discussed in the chapter on Carrier Lifetime and Recombination Properties. In general, these impurities are considered to be undesirable in silicon; direct impact of the presence of iron on integrated circuit yields has been demonstrated (49). An elegant technique for detecting iron in silicon was developed in connection with this work (50).

Recently, flat band shifts in MOS capacitors with about 800 Å thick oxides were correlated with the presence of iron contamination in the 5% HF solution used for preoxidation clean (51). Subsequent work has confirmed that iron surface contamination degrades the properties of MOS device structures (52), but this work found that while iron ions in 1% HF solutions do not adhere to the wafer surface, the iron concentration on the surface is directly proportional to the iron concentration in HNO_3 solutions. In both studies, the iron was introduced as ferric nitrate; the only apparent difference is the dilution of the HF solution.

More recently, new methods based on total reflection X-ray fluorescence (TXRF) have been reported for analysis of metallic impurities on silicon surfaces (53). The sensitivity of these methods varies with atomic number of the impurity but is of the order of 10^9 to 10^{11} cm^{-2} for a variety of transition metals. The analysis depth is of the order of 3 nm.

Other atomic impurities: Other impurities sometimes appear in silicon in significant quantities. Gold is widely used to control carrier lifetimes in power devices and high speed integrated circuits. Silver and platinum have also been used for this purpose. Lithium is used to form very high resistivity silicon for nuclear particle detection. The group VI series, sulfur, selenium, and tellurium has been investigated extensively because of its scientific interest (54). Germanium is completely miscible in silicon; germanium-silicon films have been used to introduce misfit dislocations to provide an extrinsic gettering mechanism in epitaxial silicon wafers (55). Such alloy films open new doors to defect engineering to control the properties of active multi-layer structures fabricated on silicon substrates (56).

2.4.2 Extended Defects: Dislocations are line defects which extend over many atomic spacings within the crystal. Stacking faults, localized regions in which the stacking order of the atoms in the crystal is disturbed, are bounded by partial dislocations. Precipitates are inclusions of other materials within the crystal structure.

Dislocations and slip: Dislocations arise when a portion of the crystal is plastically deformed and slips across another portion. In silicon, as in all diamond cubic crystals, the slip plane between the two portions of the crystal is a {111} plane. The stresses in the crystal cause atomic planes to move from one equilibrium position to another. In some cases, there is an extra plane of atoms on one side of the slip plane; the line of atoms at the end of this extra half plane is called an edge dislocation, as shown in Figure 13. The dislocation may move through the crystal along the direction of the shearing forces.

Under some conditions, the crystal may twist. In this case, there is no extra half plane of atoms. The dislocation, known as a screw dislocation, occurs along the line where the planes above and below the slip plane return to their original relationship, as shown in Figure 14. Screw dislocations move through the crystal in the direction perpendicular to the shearing forces.

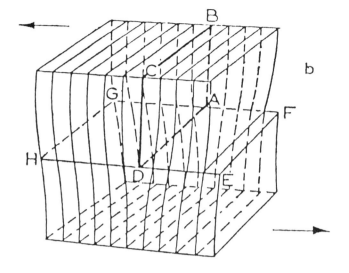

Figure 13: Edge dislocation denoted by AD in the slip plane EFGH. [From Rhodes, Reference 58.]

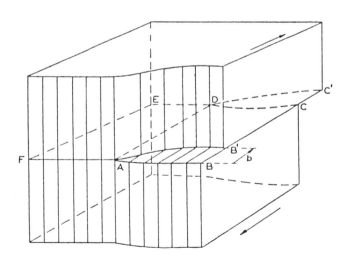

Figure 14: Screw dislocation denoted by AD in the slip plane FBCE. [From Rhodes, Reference 58.]

Dislocations in real crystals may be pure edge, pure screw, or, more commonly, a combination of both. Dislocations are characterized by a vector known as the Burgers vector which is the closure failure between a circuit around the dislocation and a similar circuit in a perfect crystal (57). This vector, denoted by the symbol **b**, is perpendicular to an edge dislocation and parallel with a screw dislocation. A dislocation cannot end within a single crystal except at another dislocation; thus dislocations form closed loops or interconnecting networks. A more complete discussion of the nature of dislocations and their motion within a crystal may be found in standard texts (58).

The correspondence between the intersection of a dislocation line with the surface and the etch pit formed by using a preferential etch has been demonstrated by the use of copper decoration and direct imaging of the dislocations (59). A standard method is available for developing and counting dislocation etch pits (60). In addition a variety of X-ray techniques can also be used in studying dislocation characteristics (58).

In 1958, Dash (61) developed a procedure for growing crystals free from dislocations. This is accomplished by necking down a small seed so that the few dislocations which might remain grow out to a surface and then shouldering the crystal out to the desired diameter very carefully to avoid introducing new dislocations. All commercially available silicon crystals are today grown using this method. It should be noted that in dislocation-free crystals, the vacancy supersaturation may lead to undesirable aggregation in the form of voids or stacking faults (62).

Even in dislocation-free crystals, slip may be introduced during crystal growth, generally at the bottom of the crystal, due to excessive thermal stresses. In addition, slip may occur during thermal processing of wafers, particularly at points where the wafer is confined such as boat resting points. Slip may also be introduced during epitaxial layer growth. ASTM Standard Test Method F 47 (60) covers procedures for identifying slipped regions of crystals and wafers. Slip is characterized by lines of etch pits along <110> directions.

Stacking faults: Stacking faults may occur during epitaxial layer growth, although modern methods have greatly reduced their incidence. Stacking faults are also induced by oxidation, either at the surface during device processing or in the bulk during oxide precipitate growth. Since silicon dioxide occupies more volume than the silicon consumed in its formation, an excess of silicon interstitials forms in the neighborhood, and if suitable nuclei and stress fields are present, stacking faults may form. Rozgonyi and Seidel (63) have discussed the origin and differentiation of surface and bulk oxidation induced stacking faults in considerable detail.

A twin is a portion of a crystal which contains atoms stacked in one order (e.g., ABCAB) on one side of a boundary (the twin plane) and in the mirror image order (e.g., BACBA) on the other. Twins form in dendritic silicon crystals but are rarely seen in commercial crystals pulled from the melt.

Precipitates: If impurities are present in concentrations larger than their solubility at a given temperature, they may precipitate, usually as compounds of silicon, provided that suitable nuclei are present. These nuclei may arise from the presence of other types of defects (heterogeneous nucleation) or from tiny aggregates of the same atoms which form the precipitate (homogeneous nucleation) (64).

Precipitates may form during the cool down cycle after growth or during subsequent processing. Other defects, such as dislocation loops and stacking faults, sometimes form as a result of precipitation. ASTM Standard Test Method F 416 (65) covers detection of precipitates and other oxidation induced crystal defects by means of preferential etching. Identification of the chemical nature of precipitates and of precipitate morphology requires the use of transmission electron microscopy. Further information on SiO_2 precipitate formation and characteristics may be found in the chapter on Oxygen, Carbon, and Nitrogen in Silicon.

3.0 ELECTRICAL PROPERTIES

3.1 Bands and Bonds in Pure Silicon Crystals

A silicon atom has 14 electrons. These are distributed two each in 1s and 2s states, six in 2p states, and three each in 3s and 3p states. When silicon atoms are brought together to form a crystal, the discrete levels are broadened into bands so that all the electrons originally in a given energy state have energies slightly different from each other. Because there are so many atoms per unit volume, these bands may be regarded as continuous. In the crystal, the 3s and 3p states intermingle to form two bands separated by a large energy gap without energy states. There are four quantum states per atom in the lower band, called the valence band, and four quantum states per atom in the upper band, called the conduction band. The gap between these bands is called the forbidden energy gap, or simply the forbidden gap.

As noted previously, each atomic site in silicon is surrounded by four nearest neighbor sites which occur at the points of a tetrahedron. The chemical bonds in silicon are formed between nearest neighbors, each atom of the pair contributing one electron. As a result, each atom is surrounded by eight electrons, four of its own and one from each of its nearest neighbors. Such a configuration, which is similar to that of a noble gas, is chemically very stable. All of the available electrons have places in the valence bonds and none are free to move in the lattice unless some energy is applied to break the bond.

Another way of looking at the situation is that all the electrons occupy all of the energy states in the valence band. For an electron to move, it must gain enough energy to be elevated across the gap to an energy state in the conduction band. The energy required may be supplied from a photon of energy equal to or greater than the forbidden energy gap (photoconductivity) as shown schematically in Figure 15 or from thermal energy if the temperature is sufficiently high. Electrons which leave the broken bonds enter states in the conduction band and are free to move throughout the crystal lattice. The broken bonds left behind are free to accept electrons from neighboring bonds, and as a result the empty states in the broken bonds, known as "holes", also seem to move throughout the lattice. Note that equal numbers of holes and electrons are produced when electrons are excited across the forbidden gap.

3.2 Dopant Impurities

So far, we have considered pure silicon without impurities or defects, and we have assumed that each lattice site is occupied by one and only one silicon atom. Suppose now that a phosphorus atom is substituted for a silicon atom as shown schematically in Figure 16. Phosphorus is in the fifth column of the periodic table and has five valence electrons. Four of these electrons contribute to the valence bonds in the same way as the four valence electrons of the replaced silicon atom.

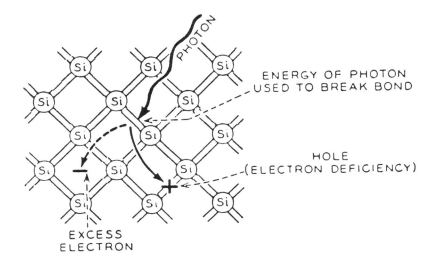

Figure 15: Two dimensional representation of the covalent bonds between adjacent silicon atoms in an intrinsic (pure) silicon crystal showing the generation of a hole-electron pair by a photon of energy greater than the forbidden energy gap. [Adapted from Shockley, Reference 5.]

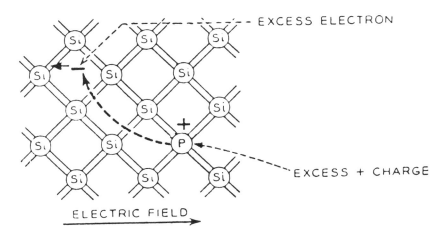

Figure 16: Two dimensional representation of the covalent bonds between adjacent silicon atoms in a silicon crystal doped with phosphorus, a donor impurity, showing the generation of a free electron by thermal ionization of the donor impurity. [Adapted from Shockley, Reference 5.]

The fifth electron is not a part of the valence bond structure. It is attached to the phosphorus atom only by the electrostatic attraction of the positive charge of the nucleus. The energy required to free this electron is much smaller than that required to break a valence bond. Because one of its electrons can easily be freed to move about in the crystal, phosphorus is known as a **donor** impurity. In an energy band diagram, the phosphorus atom is represented by an energy level in the forbidden gap very close to the bottom of the conduction band. The energy difference between this level and the bottom of the conduction band is the energy required to free the electron from the phosphorus atom; this energy is known as the donor activation energy. Because mobile electrons have a negative charge, silicon containing phosphorus or another group V impurity is called n-type.

If on the other hand a silicon atom is replaced by a boron atom, or another atom from column III of the periodic table, the impurity atom contributes only three valence electrons. These three electrons form valence bonds with three of the neighboring silicon atoms. In the fourth bond there is a missing electron, or hole. An electron from a neighboring bond may move into this hole with very little expenditure of energy, leaving a hole behind, as shown schematically in Figure 17. The energies required for this are similar to those required to free an electron from a donor atom. Because they accept electrons from the valence band, impurities with three valence electrons are called **acceptor** impurities. An acceptor impurity is represented by an energy level in the forbidden gap near the top of the valence band; the energy difference is known as the acceptor activation energy. Since mobile holes behave similarly to "electrons" with a positive charge, silicon which contains acceptor impurities is called p-type.

3.3 Statistics

Our qualitative treatment has yielded a physical picture of donor and acceptor impurities in silicon. However, to relate the impurity properties to electrical conductivity in a quantitative way, we must find expressions for the number of free holes and electrons as a function of temperature and impurity concentration. Statistical methods may be used to solve this problem because the bands may be treated as though they are continuous.

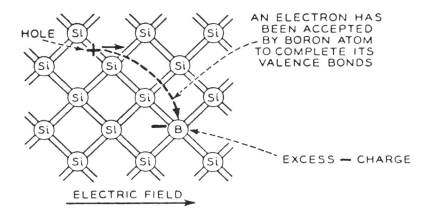

Figure 17: Two dimensional representation of the covalent bonds between adjacent silicon atoms in a silicon crystal doped with boron, an acceptor impurity, showing the generation of a free hole by thermal ionization of the acceptor impurity. [Adapted from Shockley, Reference 5.]

3.3.1 Fermi Function: The probability of finding an electron in a particular energy state, E, is given by the Fermi function, F(E):

$$F(E) = \{\exp[(E-E_F)/kT] + 1\}^{-1}, \tag{5}$$

where k is Boltzmann's constant, T is the absolute temperature, and E_F is the thermodynamic free energy per electron in the crystal. This quantity, which is called the Fermi energy, is continuous across the boundary between two conductors in equilibrium.

The form of the Fermi function is given in Figure 18. At absolute zero, F(E) is unity below E_F (all states are filled with electrons) and zero above it (all states are empty). At higher temperatures, the Fermi function is rounded off in the region around the Fermi energy so that some states below E_F are empty and electrons occupy some states above E_F. The probability of a state at the Fermi energy being filled is one-half. The higher the temperature, the greater the rounding and the more electrons that can occupy states with $E>E_F$. For energies sufficiently far from the Fermi energy that $(E-E_F) \gg kT$, the Fermi function may be approximated by the Boltzmann distribution function, f(E):

$$F(E) \approx f(E) = \exp[-(E-E_F)/kT]. \tag{6}$$

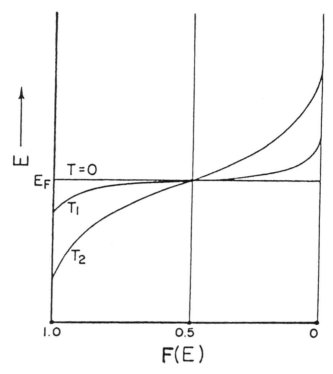

Figure 18: The Fermi function at temperatures, T=0 K, T=T_1 (> 0), and T=T_2 (> T_1).

3.3.2 Density-of-States Function: Having found an expression for the probability of finding an electron in a particular energy state, it is next necessary to determine the distribution of these energy states in the crystal. The density of states function, D(E), has finite values in the valence and conduction bands, is a sharp spike at any discrete donor or acceptor level, and is zero elsewhere in the forbidden gap. The exact calculation of D(E) requires quantum mechanical techniques (66). Useful working relationships can be obtained from the effective mass approximation, which is valid near the band edges.

An electron's energy, E, is related to its momentum, **p**, by:

$$E = (1/2m_0)\mathbf{p}^2, \qquad (7)$$

where the reciprocal of the mass of the electron (m_0) is the second derivative of E with respect to **p**. The graph of Equation (7) is a parabola, centered at the origin and increasing in an upward direction. The number of electronic states per unit volume, for one direction of electronic spin, lying in the range between E and E+dE is given by:

$$D(E)dE = h^{-3} \int d\mathbf{p}, \tag{8}$$

where h is Planck's constant and the integral is taken over the volume of **p**-space between E and E+dE. Since the energy contours are spherical (i.e., $D(E)dE = 4\pi h^{-3} p^2 dp$), substituting for p and pdp from Equation (7) yields:

$$D(E) = 2\pi h^{-3}(2m_0)^{3/2} E^{1/2}. \tag{9}$$

In a solid, the energy-momentum functions are periodic. At the bottom of a band the curvature of the E vs. p curve is upward. In the simplest case, the band would be spherically symmetric and the second derivative would be the same in all directions. This is not the case in silicon (67), but nevertheless it is possible to compute an appropriate average mass, m_e, known as the density-of-states effective mass. This quantity, which is a function of both temperature, T, and electron density, n, can be used in an equation similar to Equation (9) to calculate the density of states in the conduction band:

$$D_C(E) = 2\pi h^{-3}[2m_e(T,n)]^{3/2} E^{1/2}. \tag{10}$$

An analogous situation applies at the top of the valence band but now the band is inverted. The density of states in the valence band may be written:

$$D_V(E) = 2\pi h^{-3}[2m_h(T,p)]^{3/2} E^{1/2}, \tag{11}$$

where m_h is the density-of-states effective mass for holes in the valence band and p is the hole density. Although the curvature of the bands is such that m_h would appear to be a negative number, it is conventional to treat the effective mass as a positive number and assign a positive charge to the holes to account for their behavior under the influence of electromagnetic fields.

3.3.3 Intrinsic Carrier Density:

The number density of electrons, n(E), at a given energy, E, is given by the product of D(E) and F(E). In the conduction band, therefore, the number density of electrons is:

$$n = \int_{E_c}^{E_c'} n(E)dE = \int_{E_c}^{E_c'} D_c(E)F(E)dE, \qquad (12)$$

where E_c and E_c' are the values of energy at the bottom and top of the conduction band, respectively. Because the Fermi function goes rapidly to zero at energies far removed from E_F, the shape of D(E) well into the conduction band need not be taken into account and the upper integration limit may be replaced by infinity. Also, in semiconductors which do not have too many dopant impurities, F(E) may be replaced by f(E), Equation (6). With these approximations, Equation (12) may be integrated to yield:

$$n = (2\pi m_e kT/h^2)^{3/2} \exp[-(E_c-E_F)/kT]. \qquad (13)$$

Similarly, the hole density in the valence band may be calculated as:

$$p = (2\pi m_h kT/h^2)^{3/2} \exp[-(E_F-E_v)/kT]. \qquad (14)$$

Both of these quantities depend on the position of the Fermi energy which we have not yet determined. However, their product, np, does not since it is a function only of the width of the forbidden energy gap, $E_c - E_v$:

$$np = (2\pi kT/h^2)^3 (m_e m_h)^{3/2} \exp[-(E_c-E_v)/kT]. \qquad (15)$$

The square root of this product is a fundamental property of a semiconductor called the intrinsic carrier density, n_i.

Barber (68) has carried out an extensive study of the temperature and dopant impurity dependencies of the density-of-states effective masses and the intrinsic carrier density in silicon. Thurber and Buehler (69) have calculated polynomial fits of the temperature variation of the forbidden energy gap, based on the work of MacFarlane et al. (70), and the density-of-states effective masses, based on Barber's analysis. These functions are of the form:

$$F = \sum_{n=0}^{N} a_n T^n, \qquad (16)$$

where T is the absolute temperature in kelvin, a_n is the coefficient of the nth term in the polynomial, and N is the order of the polynomial. Coefficients for the forbidden energy gap in units of electron volts and the effective masses in units of the free electron mass are listed in Table 2. Results are presented in Figures 19 through 21. Table 3 lists values of these quantities over the temperature range 0 to 600 K.

Table 2

Coefficients of the Polynomial Fits to the

Energy Band Gap, Electron Mass, and Hole Mass of Silicon

	$E_g = E_c - E_v$	m_e/m_0	m_h/m_0
a_0	1.15556	1.06270	0.590525
a_1	3.23741×10^{-5}	-1.61708×10^{-4}	-5.23548×10^{-4}
a_2	-8.70110×10^{-7}	6.83008×10^{-6}	1.85678×10^{-5}
a_3	9.95402×10^{-10}	-3.32013×10^{-8}	-9.67212×10^{-8}
a_4	$-3.809777 \times 10^{-13}$	8.04032×10^{-11}	2.30049×10^{-10}
a_5		-9.66067×10^{-14}	-2.59673×10^{-13}
a_6		4.54649×10^{-17}	1.11997×10^{-16}
residual std. dev.	0.00032	0.00087	0.0019

With these parameters, n_i can be calculated as the square root of Equation (15). This parameter is also listed for the temperature range 100 to 600 K in Table 3. At 300 K, n_i in silicon is equal to 1.13×10^{10} cm^{-3}, about 2000 times less than the value of n_i in germanium.

3.3.4 Qualitative Description of the Energy Structure of Silicon: It is now possible to examine qualitatively the dependence of the Fermi energy on impurity concentrations and temperature; for a quantitative treatment, the reader is referred to Blakemore (66) or other standard text. We know that the Fermi energy is the energy where the probability of an energy state being occupied is one-half. This is true even if there is no state at that particular value of energy.

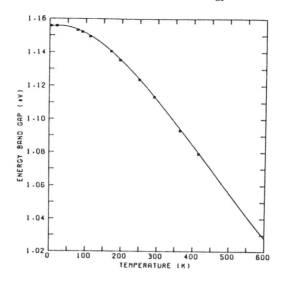

Figure 19: Energy band gap of silicon as a function of temperature. The points are experimental data from MacFarlane et al., Reference 70, and the solid line is from the polynomial fit. [From Thurber and Buehler, Ref 69.]

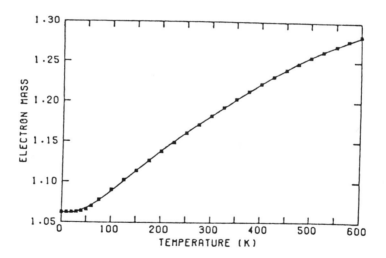

Figure 20: Electron density-of-states effective mass in silicon with donor density less than 5×10^{17} cm^{-3} as a function of temperature. The points are experimental data from Barber, Reference 68, and the solid line is from the polynomial fit. [From Thurber and Buehler, Reference 69.]

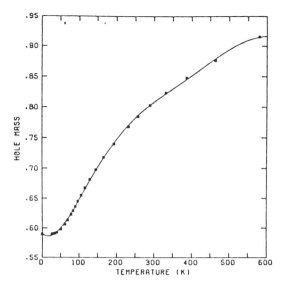

Figure 21: Hole density-of-states effective mass in silicon with acceptor density less than 5×10^{17} cm^{-3} as a function of temperature. The points are experimental data from Barber, Reference 68, and the solid line is from the polynomial fit. [From Thurber and Buehler, Reference 69.]

Table 3

Band Gap, Electron Mass, Hole Mass, and Intrinsic Carrier Density of Silicon Calculated from Polynomial Fits

T (K)	E_g (eV)	m_e/m_0	m_h/m_0	n_i (cm^{-3})
0	1.1556	1.063	0.591	
50	1.1551	1.068	0.600	
100	1.1511	1.089	0.648	3.67×10^{-11}
150	1.1440	1.114	0.701	4.45×10^{-1}
200	1.1346	1.139	0.747	6.13×10^{4}
250	1.1233	1.161	0.781	8.43×10^{7}
300	1.1108	1.182	0.807	1.13×10^{10}
350	1.0973	1.202	0.829	3.97×10^{11}
400	1.0832	1.221	0.851	5.96×10^{12}
450	1.0690	1.239	0.873	5.05×10^{13}
500	1.0548	1.256	0.894	2.84×10^{14}
550	1.0409	1.270	0.910	1.18×10^{15}
600	1.0274	1.282	0.917	3.88×10^{15}

Moreover, the probability function is symmetrical about this energy. In the case of pure, or intrinsic, silicon, the hole and electron densities are equal at all temperatures. Therefore, since $m_e > m_h$, the Fermi energy lies slightly below the middle of the forbidden energy gap:

$$E_F = (1/2)(E_c - E_v) - (3/4)\ln(m_e/m_h). \tag{17}$$

When the silicon contains only a single donor impurity species, additional energy levels are introduced into the forbidden energy gap slightly below the bottom of the conduction band. At absolute zero, all the electrons associated with the donors fill these additional energy levels. The Fermi energy must therefore lie between the donor levels and the bottom of the conduction band.

As the temperature increases, some electrons from the donors are thermally excited into the conduction band. In this "extrinsic" or "exhaustion," region, the electron density varies exponentially with $1/T$ with a slope of $-(E_c - E_d)/2k$, where E_d is the energy level of the donor impurity and the difference is the donor activation energy. Values of $E_c - E_d$ for typical donor impurities are listed in Table 4. Note that these energies are given in units of electron volts; in these units, k has the value 8.61735×10^{-5} eV/K.

Table 4

Activation Energies of Donor and Acceptor Impurities in Silicon

Donor Impurities	$E_c - E_d$ (eV)
Phosphorus	0.044
Arsenic	0.049
Antimony	0.039
Acceptor Impurities	$E_a - E_v$ (eV)
Boron	0.045
Aluminum	0.057
Gallium	0.065
Indium	0.160

At a temperature somewhat below room temperature, the donor impurity levels are nearly empty. In this "saturation" region, the electron density is nearly constant with a value equal to the donor impurity density to within a few percent.

At some higher temperature, electrons begin to be excited from the valence band until, at temperatures well above room temperature, electrons from this source predominate. At these temperatures, the properties of the silicon approach those of intrinsic silicon, and the electron density varies exponentially with $1/T$ with a slope of $-(E_c-E_v)/2k$. In this "intrinsic" region, the electron and hole densities are nearly equal.

Analogous arguments apply if the only impurities are a single acceptor species. In this case, the carriers are holes with a density, p, the impurity levels are located at E_a, near the top of the valence band, and the acceptor activation energy is given by E_a-E_v. Therefore, the slope of the curve of ln p vs. $1/T$ is $-(E_a-E_v)/2k$. Values of E_a-E_v for typical acceptor impurities are also given in Table 4.

In the more usual case, both donor and acceptor impurities are present in the silicon, although one (called the majority impurity) is usually present in significantly greater number and dominates the properties of the silicon in the extrinsic and saturation regions.

For purposes of illustration, assume that donors are the majority impurities ($N_d \gg N_a$). Even at T = 0 K, electrons from some of the donors fill all available acceptors, so the Fermi energy must lie at E_d. In this case, the slope of the curve of ln n vs. $1/T$ is $-(E_c-E_d)/k$ in the extrinsic region, and the value of n in the saturation region is approximately equal to N_d-N_a. Again, analogous arguments apply if acceptors are the majority impurity.

3.3.5 Actual Band Structure of Silicon:

The actual band structure of silicon is quite complex. Both experimental and theoretical work have established the shapes of the conduction and valence bands. Detailed calculations have been made by several authors (71),(72). These are in good agreement with the results of cyclotron resonance experiments (67). There are two qualitative aspects of the band structure which are generally important. First, the minimum of the conduction band is displaced from the maximum of the valence band so that a momentum change is required when electrons move across the gap from the valence band to the conduction band. Second, the structure of the bands is such that certain properties exhibit anisotropies.

The band structure of silicon is shown schematically in

Figure 22. This Figure shows the lowest energy of the bands above the forbidden gap and the highest energy of the bands below the gap as a function of k (= $2\pi p/h$) along two low index directions. The minima at the bottom of the lowest conduction band (C2) occur along the six principal cubic axes, between the center and the edge of the Brillouin zone. The surfaces of constant energy are ellipsoids of revolution rather than spheres so that in a given minimum the effective mass is different in different directions. The cubic symmetry is such, however, that the effective mass associated with the density of states can be represented by a scalar quantity as discussed above. There is also a third conduction band (C3) which has its minimum at the zone center located about 2.5 eV above the band edge. Direct transitions to and from this band can be observed under certain conditions.

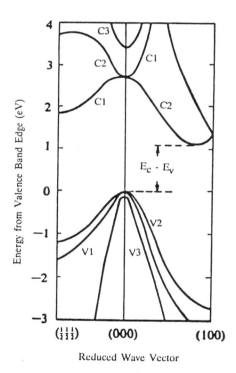

Figure 22: Schematic of the energy bands of silicon in the energy range near the forbidden energy gap. [After Blakemore, J.S., Solid State Physics, Cambridge: Cambridge University Press (1985).]

The valence bands have a single extremum at the center of the zone. The two upper valence bands are degenerate at the maximum and the constant energy surfaces are warped spheres. One of these bands is referred to as the heavy hole band (V1) and the other as the light hole band (V2). The maximum of the third (split-off) valence band (V3) occurs 0.04 eV below the maximum of the light and heavy hole bands. The effective mass of the holes in this band, which is spherical, is intermediate between the effective masses at the maxima of the heavy and light hole bands. The heavy hole band dominates the density-of-states hole effective mass. As with the electron effective mass, the cubic symmetry requires that the hole effective mass be a scalar, but it is a rather complex function of the temperature and hole density as discussed above.

3.4 Electronic Conduction

The current density, J, of a stream of electrons of density n is given by the product:

$$J = nev, \qquad (18)$$

where e is the electronic charge and v is the average velocity of the electrons in the stream. If an electric field, E, is applied to a crystalline solid, v is proportional to the field: $v = \mu E$, provided that the field is not too large (in which case the velocity approaches a saturation value). The constant of proportionality, μ, is called the electron mobility. From these relations and Ohm's law, $J = \sigma E$, it is clear that the electrical conductivity, σ, is given by the product:

$$\sigma = ne\mu. \qquad (19)$$

If more than one type of carrier is present, the contributions are additive; for example, if electrons and holes are both present in comparable number, the conductivity becomes:

$$\sigma = ne\mu_n + pe\mu_p. \qquad (20)$$

The resistivity, ρ, is the reciprocal of the conductivity.

The motion of electrons and holes in a crystal is affected by collisions which change their speed or direction. These collisions occur at places in the crystal where the lattice periodicity is disturbed. We should expect that a more perfect lattice at very low temperatures would result in fewer collisions and greater mobility. Factors such as increased thermal agitation of the silicon atoms, replacement of silicon atoms in the lattice by impurities, and the existence of crystal defects all cause disturbances which can "scatter" the carriers.

The two most important scattering mechanisms in silicon are lattice scattering (due to the thermal motion of atoms in the lattice) and ionized impurity scattering. The mobility of carriers limited by lattice scattering is approximately proportional to $T^{-3/2}$, and thus this mechanism is dominant at higher temperatures. On the other hand, the mobility of carriers limited by ionized impurity scattering is approximately proportional to $T^{3/2}$, and thus this mechanism dominates at lower temperatures. Ionized impurity scattering, of course, increases with impurity density. Carrier-carrier and neutral impurity scattering also may become important at high impurity densities.

The relationship between resistivity and dopant density was measured initially by Irvin (73); from these data, Caughey and Thomas (74) developed empirical formulas to fit the carrier mobilities. Efforts to fit these data with calculated curves based on first principles met with mixed success. Later Wagner (75) discovered a serious discrepancy between the Irvin curves and experimental data derived from boron implantation. These difficulties led to a detailed experimental and theoretical study of the issue at the National Bureau of Standards. In this work it was found that under some conditions it is essential to consider the effects of carrier-carrier and neutral impurity scattering (76),(77). Curves of electron mobility as a function of total donor density and temperature are given in Figures 23 and 24, respectively. Curves of hole mobility as a function of hole density and temperature are given in Figures 25 and 26, respectively.

By combining the lattice scattering mobilities for electrons and holes with the intrinsic carrier density [Equation (15)], the intrinsic resistivity of undoped silicon can be calculated for various temperatures. The results of this calculation are given in Figure 27 in which carrier density and intrinsic resistivity are plotted in the usual way (as a function of 1000/T) to show the dominant exponential dependence and the intrinsic resistivity is also plotted linearly in T to facilitate comparison with the resistivity against temperature curves for doped silicon.

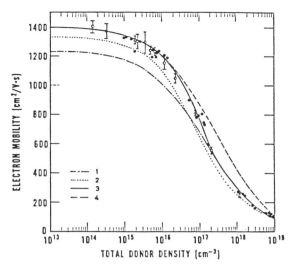

Figure 23: Electron mobility in n-type silicon at 300 K as a function of total donor density. Curve 3 is the theoretical calculation, curves 1 and 4 are calculations made with and without the inclusion of electron-electron scattering over the entire range, respectively; and curve 2 is the Caughey and Thomas (74) calculation based on Irvin's data (73). The points are more recent experimental data reported in the literature. [From Li, NBS Spec. Publ. 400-33, Reference 76.]

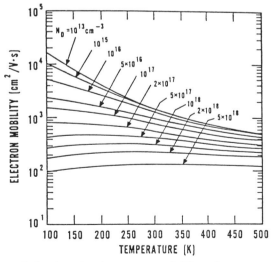

Figure 24: Calculated electron mobility in n-type silicon as a function of temperature for various dopant densities. [From Li, NBS Spec. Publ. 400-33, Reference 76.]

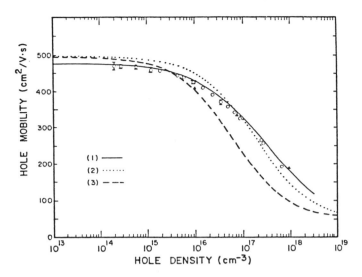

Figure 25: Hole mobility in p-type (boron-doped) silicon at 300 K as a function of hole density. Curve 1 is the theoretical calculation; curve 2 is based on the work of Wagner (75); and curve 3 is the Caughey and Thomas (74) calculation based on Irvin's data (73). The points are more recent experimental data reported in the literature. [After Li, NBS Spec. Publ. 400-47, Reference 77.]

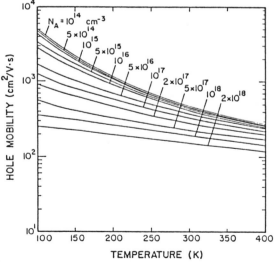

Figure 26: Calculated hole mobility in p-type silicon as a function of temperature for various boron densities. [From Li, NBS Spec. Publ. 400-47, Reference 77.]

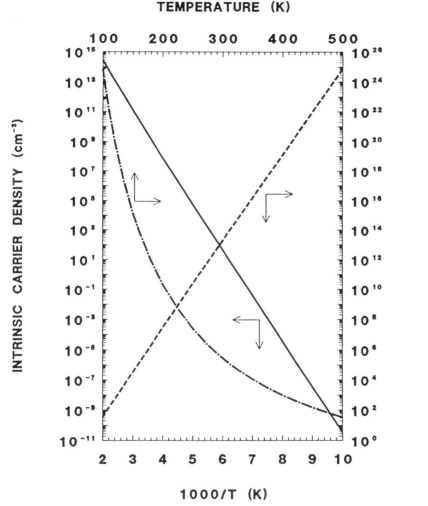

Figure 27: Intrinsic carrier density (solid line) and resistivity of undoped silicon (dashed line) as a function of reciprocal temperature and resistivity of undoped silicon as a function of temperature (chain dot line) over the temperature range from 100 to 500 K.

These curves are shown in Figures 28 and 29 for n- and p-type silicon, respectively. In the high and low temperature regions beyond the range of these plots, the exponential behavior of the carrier density dominates the form of the resistivity curve. In the intermediate temperature range shown, the resistivity varies much more slowly with temperature; in this region, most of the temperature dependence comes from the mobility term. However, because the dopant atoms are not fully ionized even at room temperature, a small amount of the temperature dependence comes from changes in carrier density with temperature. Both because the acceptor activation energies are more widely separated and because the degeneracy factor (66) is such as to cause the percentage of acceptors which are ionized to be less than that for donors of the same activation energy, the influence of carrier density changes is more pronounced in p-type silicon. The hole mobility as a function of total acceptor density is measurably different for the various dopant species (78). Curves of the fraction of phosphorus and boron atoms ionized at selected temperatures (including 300 K) are given as a function of the total dopant density in Figures 30 and 31.

It must be noted that the results discussed above relate to "uncompensated" silicon, i.e., silicon in which the majority impurity density is greater than about 1000 times the minority impurity density. If the donor and acceptor densities are equal to within a factor of 1000, the silicon is said to be "compensated." In this case, significant numbers of the majority impurity atoms are always ionized and the carrier mobilities are reduced below the values in uncompensated material with the same majority carrier density because of the increased ionized impurity scattering.

3.5 Electrical Characterization

Resistivity is the most widely used electrical parameter for characterizing silicon. There are three standard methods for measuring resistivity: ASTM Standard Method F 84 (79), which covers application of the four-probe technique to measurements on wafers and serves as a referee method; ASTM Method F 673 (80), which covers a contactless method (which is the method most widely used in production) and its calibration by the four-probe method; and ASTM Test Methods F 43 (81), which cover routine applications of the two- and four-probe methods.

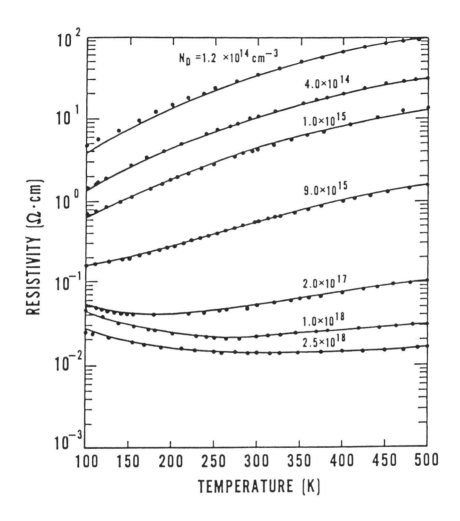

Figure 28: Resistivity as a function of temperature for seven phosphorus-doped silicon wafers. Solid lines are the theoretical calculations and the dots are experimentally measured values. [From Li, NBS Spec. Publ. 400-33, Reference 76.]

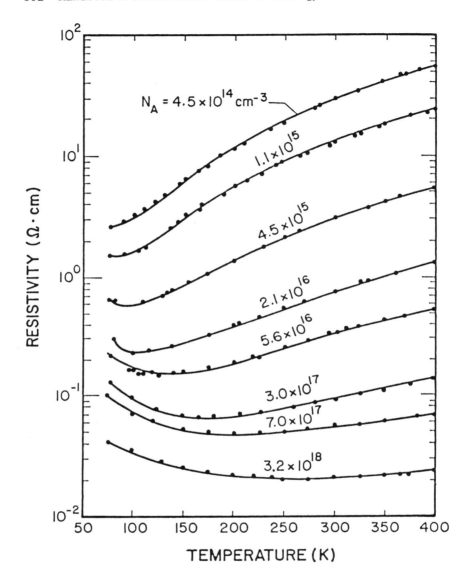

Figure 29: Resistivity as a function of temperature for eight boron-doped silicon wafers. Solid lines are the theoretical calculations and the dots are experimentally measured values. [From Li, NBS Spec. Publ. 400-47, Reference 77.]

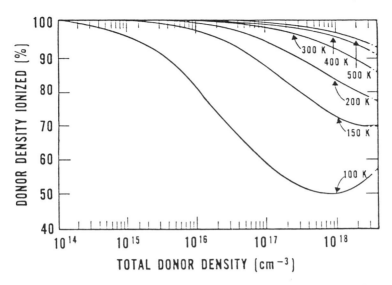

Figure 30: Fraction of ionized phosphorus atoms in n-type silicon as a function of total phosphorus density for various temperatures. [From Li, NBS Spec. Publ. 400-33, Reference 76.]

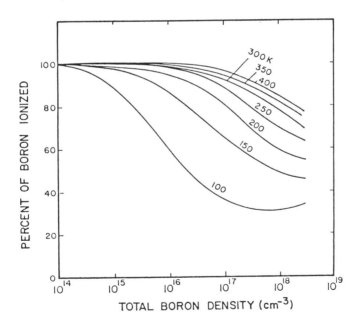

Figure 31: Fraction of ionized boron atoms in p-type silicon as a function of the total boron density for various temperatures. [From Li, NBS Spec. Publ. 400-47, Reference 77.]

Bullis et al. (82) established the temperature variation of the resistivity of silicon in the vicinity of room temperature for both boron- and phosphorus-doped silicon. This variation, though small, is important for accurate resistivity measurements. The necessary correction factors have been incorporated into the standard methods (79),(80),(81). The correction factors given for phosphorus-doped silicon are generally suitable for use with silicon doped with other donor impurities. However, because of the differences in ionization fraction discussed above, the correction factors for boron-doped silicon do not apply to silicon doped with other acceptor impurities.

Resistivity measurements are widely used to establish the carrier or dopant densities in silicon. Thurber and coworkers have carefully determined the relationships between resistivity and dopant density for uncompensated boron-doped (83) and phosphorus-doped (84) silicon. The results in the form of a graph are given in Figure 32. Additional details of the measurements and analytical procedures are available (85); the results have been incorporated into ASTM Practice F 723 (86). In addition, the Practice contains a chart which shows the differences between these results and the sometimes still used results of Irvin (73); it should be noted that Irvin's results for p-type silicon were based largely on data from gallium-doped samples.

Determination of the relationship between resistivity and electron or hole density is straightforward in the low dopant density regime (up to about 10^{17} impurity atoms/cm^3) where the carrier and dopant densities are quite closely equal. In moderate density regions (10^{17} to 10^{19} impurity atoms/cm^3), the dopant impurities are not fully ionized at room temperature and so the carrier density is less than the dopant density. Li and coworkers have made detailed theoretical analyses of the electron and hole mobilities which can be used to determine the relationship between resistivity and carrier density (76),(77),(78),(87).

3.6 Conduction at High Electric Fields

When the electric field becomes very large, the average velocity of the carriers approaches a saturation value. Caughey and Thomas (74) have developed empirical curves to fit the dependence of the hole and electron velocities, v_h and v_e, respectively, on field strength, E:

Silicon Material Properties 395

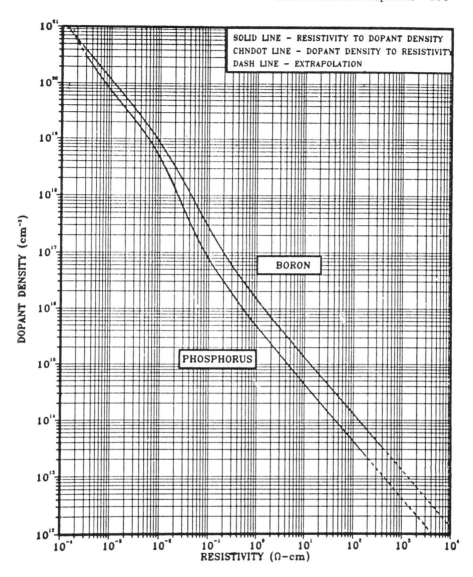

Figure 32: Dopant density as a function of resistivity at 23°C for silicon doped with phosphorus or boron. On the scale of this graph, the solid and chain dot curves are not distinguishable; however, in regions where the self-consistency error is appreciable, the two curves can be distinguished on the wall chart available from ASTM as Adjunct PCN 12-607230-46. [From ASTM Standard Practice F 723, Reference 86. Copyright ASTM. Reprinted with permission.]

$$v_h = v_m \frac{(E/E_c)}{1 + (E/E_c)} \quad \text{and} \tag{21}$$

$$v_e = v_m \frac{(E/E_c)}{[1 + (E/E_c)^2]^{1/2}}, \tag{22}$$

where E_c is the critical field strength (1.95×10^4 V/cm for holes and 8×10^3 V/cm for electrons) and v_m is the saturation velocity (9.5×10^6 cm/s for holes and 1.1×10^7 cm/s for electrons). These relations provide a good fit to the experimental data as summarized by Gibbons (88) as shown in Figure 33. Because of this saturation of the carrier velocity, the mobilities decrease from their constant low field values at fields above about 1000 V/cm.

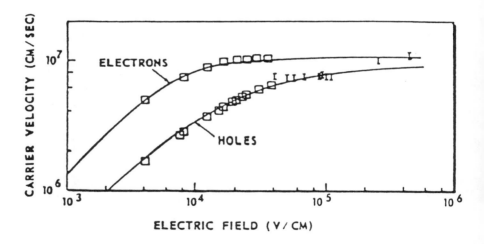

Figure 33: Hole and electron velocities as a function of the electric field applied to high-purity silicon. Solid curves are empirical functions [Equations (21) and (22), respectively] and the data points are from Gibbons' review [Reference 88]. [From Caughey and Thomas, Reference 74.]

At fields above about 10^5 V/cm the carriers can pick up enough energy from the field between successive collisions to excite carriers across the forbidden gap, thus increasing the number of carriers available for conduction. This process, known as avalanche breakdown, most frequently occurs in the space charge region of a reverse biased *p-n* junction where high fields can be easily generated. These high field effects are particularly important in short channel MOS devices. Caughey and Thomas (74) provide a summary of available data on high field mobilities for electrons and holes together with empirical relationships for the dependence of the mobilities on electric field.

3.7 Conduction in a Magnetic Field

The force on a charged particle of charge q and mass m moving under the influence of electric field **E** and magnetic induction **B** is given by:

$$\mathbf{F} = (q/m)[\mathbf{E} + (\mathbf{v} \times \mathbf{B})], \qquad (23)$$

where **v** is the velocity of the particle. If the magnetic induction is in the z-direction and the current is in the x-direction, both electrons and holes are deflected in the -y-direction. If the geometry is such that the net current in the y-direction must be zero, as for example if the semiconductor specimen is a rectangular bar, a field is set up to oppose the motion of the charged particles in this direction. This field is shown in Figure 34 for conduction by either electrons or holes. It is given by:

$$E_y = R_H J_x B_z, \qquad (24)$$

where the coefficient of proportionality, R_H, is called the Hall coefficient after E. H. Hall who discovered this effect in 1879.

The importance of this effect arises from the fact that the Hall coefficient is proportional to $-1/ne$ in the extrinsic region of *n*-type silicon and to $1/pe$ in the extrinsic region of *p*-type silicon. This quantity can therefore be used to determine the carrier concentration from which the impurity density can be deduced. The slope of a plot of $-\ln(R_H)$ vs. $1/T$ is proportional to the impurity activation energy in the extrinsic region; much of the early information about activation energies of donor and acceptor impurities was derived from

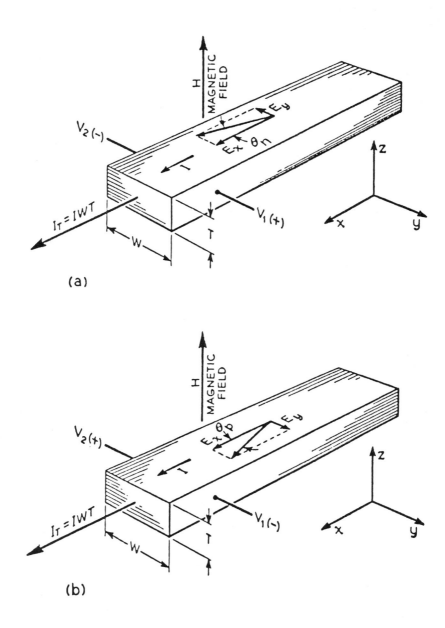

Figure 34: Voltage components for conduction by electrons (a) and holes (b) in a rectangular parallelepiped in a magnetic field, showing the transverse Hall field. [From Shockley, Reference 5.]

Hall measurements. Standardized methods for measuring the Hall coefficient have been developed (89).

Referring to Equation (19), it can be seen that in extrinsic silicon the product $R_H\sigma$ is proportional to the carrier mobility. Because the constant of proportionality is usually close to unity, this product has come to be known as the Hall mobility, μ_H. The difference between μ_H and μ results from the fact that different velocity averages are used in deriving the two quantities (87).

In a magnetic field, the resistance of silicon and other materials increases as the field is increased. This second order effect is known as magnetoresistance. The increase in resistance occurs because the mean free path in the direction of the applied electric field is shortened due to the transverse motion of the charge carriers in the magnetic field. Since all the carriers do not have the same velocity, there are transverse currents in both directions, even though the net transverse current is zero. This effect depends on the square of the magnetic induction whereas the Hall effect depends on the first power. For silicon, the resistance change is only a few per cent at reasonable values of magnetic induction (0.1 to 1 T).

3.8 Deep-level Impurities

Many non-dopant impurities and some crystal defects in silicon are also electrically active. Such impurities and defects also have energy levels in the forbidden energy gap. Because these levels are generally quite far removed from the band edges, they are known as deep levels (90). More recently, a quantitative criterion for determining when a level is deep has been proposed (91).

By analogy with group III and group V impurities which can be treated as hydrogenic and have a single energy state, group II and group VI impurities in silicon may be roughly treated as helium-like with two energy states, one nearer to the band edge than the other. In this model, group II impurities, such as zinc, cadmium and mercury, would be double acceptors and group VI impurities, such as sulfur, selenium and tellurium, would be double donors.

There is a trend toward deeper levels (further removed from the band edges) as the mass (or size) of the impurity atom increases. In fact, even the heaviest group III atoms, indium and thallium, have deep lying energy levels; these are the only known deep-level impurities with a single energy state.

Detailed calculations of the energy states of deep impurities are much more difficult than calculation of the energy states associated with dopant impurities which lie near the band edges. It is only recently that these calculations have been carried out successfully (92).

The earliest experimental investigations of the electrical properties of deep-level impurities in silicon involved Hall effect and resistivity measurements at temperatures below room temperature. Such measurements require relatively large concentrations of the impurity because the electrical properties are dominated by dopant impurities unless the concentration of the deep-level impurity exceeds that of the dopant impurity.

In n-type silicon which contains a deep-level impurity with a partially filled acceptor-like state at energy E_A, the electron density is given by:

$$n = \frac{N_d - N_a}{N_A - (N_d - N_a)} \; g_A \; N_c \; \exp[-(E_c-E_A)/kT], \qquad (25)$$

where N_d and N_a are the donor and acceptor (dopant) impurity densities, respectively, N_A is the deep acceptor impurity density, g_A is the degeneracy of the deep acceptor level, N_c is the effective density of states for the conduction band $[=2(2\pi m_e kT/h^2)^{3/2}]$, E_c is the energy of the conduction band edge, k is Boltzmann's constant, and T is the absolute temperature, provided the following conditions apply:

$N_d \gg N_a$, $N_A > (N_d - N_a)$, $n \ll N_A - (N_d - N_a)$, and $n \ll N_d - N_a$.

On the other hand, if the following conditions apply:

$N_A < (N_d - N_a)$, $N_d > N_A + N_a$, $n \ll N_d - (N_A + N_a)$, and $n \ll N_A + N_a$,

the electron density is given by:

$$n = \frac{N_d - (N_A + N_a)}{N_A + N_a} \; g_d \; N_c \; \exp[-(E_c-E_d)/kT], \qquad (26)$$

where g_d and E_d are the degeneracy and activation energy of the donor (dopant) impurity, respectively. Under these conditions, the deep-level impurity does not directly affect the electrical properties except to provide additional compensating states and thus reduce the electron density for a given donor dopant density. Similar arguments apply for donor-like deep-level impurities in p-type silicon.

Detailed calculations of the resistivity of gold in silicon have been compared with experimental measurements (93). In addition to the two gold levels, the calculations take into account the gold-related shallow level reported by Brückner (94) which causes the resistivity of silicon very heavily doped with gold to fall to much lower values than would be expected on the basis of the gold levels only. Results of these calculations are shown in Figures 35 to 37. Although the results are qualitatively satisfactory, there are several quantitative features which do not agree with the experimental results. First, the sharp rise in resistivity of n-type silicon does not occur when the gold and shallow donor concentrations are equal as would be expected on the basis of the model. Second, in p-type silicon, no value of the degeneracy factor for the gold donor, g_D, could be found for which the calculations agree with experimental results over the full range of shallow acceptor concentrations. Because of these discrepancies, the model used for these calculations must be applied with some reservations.

Although they do not affect the resistivity at the very low densities usually found in silicon, deep-level impurities may have a significant effect on the recombination properties of the silicon even at densities as low as 10^{10} atoms/cm^3. These properties of deep levels are discussed in the chapter on Carrier Lifetime in Silicon. In addition, the activation energies themselves can be measured by means of deep-level transient spectroscopy (DLTS) or by other nonequilibrium techniques.

In many cases, there is considerable disagreement in the values of the energy levels and other characteristics of deep-level impurities reported in the literature. This scatter is greater than would be expected simply from the experimental uncertainties associated with the measurement techniques used. In the case of gold, this situation is thought to arise because of the complex nature of the deep levels associated with the gold in silicon (95). However, Sah and coworkers have recently provided additional insight into this question (96).

Other deep-level impurities, particularly the transition

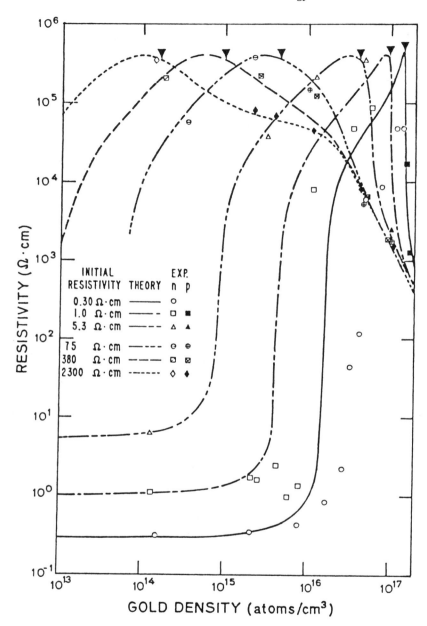

Figure 35: Resistivity as a function of gold density in silicon doped with phosphorus and gold. To the left of the ▽, the Hall coefficient is negative; to the right it is positive. [From Thurber et al., Reference 93.]

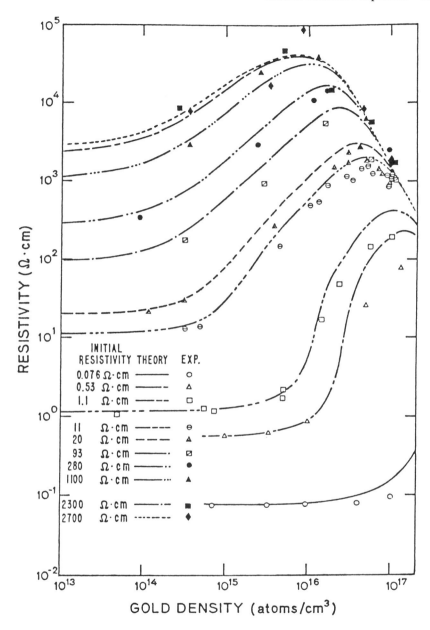

Figure 36: Resistivity as a function of gold density in silicon doped with boron and gold with the assumption that the degeneracy factor of the gold donor level is 8. [From Thurber et al., Reference 93.]

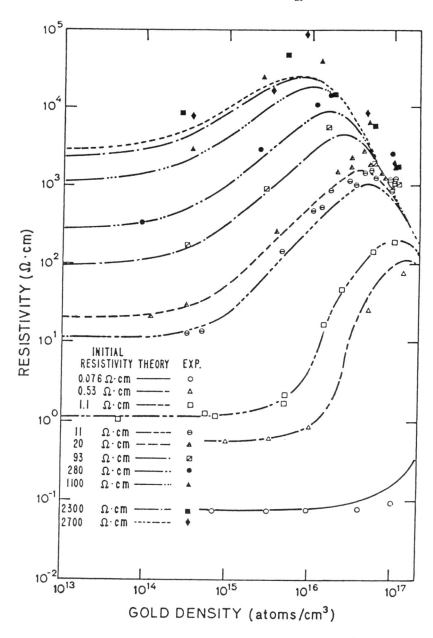

Figure 37: Resistivity as a function of gold density in silicon doped with boron and gold with the assumption that the degeneracy factor of the gold donor level is 16. [From Thurber et al., Reference 93.]

metals, also complex easily with crystal defects or with shallow acceptor impurities. This results in the formation of a multiplicity of energy states associated with the deep-level impurity. Different complexes may be formed for different concentrations of the deep-level impurity, further adding to the confusion. The reader is referred to the current literature for additional information on this subject.

3.9 Rectification

Much of the usefulness of silicon as an electronic material derives from the fact that rectifying junctions can be formed between silicon and a metal and between n- and p-type silicon. Consider first the latter case, the p-n junction. Figure 38 shows the energy band structure for isolated regions of p-and n-type silicon. In p-type silicon, holes are the majority carriers and the Fermi energy is relatively close to the valence band edge. In n-type silicon, electrons are the majority carriers and the Fermi energy is close to the conduction band edge. If the n- and p-type regions are brought together, there is a concentration gradient of both electrons and holes at the boundary. Electrons would like to flow into the p-type region, and holes would like to flow into the n-type region. If the circuit is not closed, or if it is closed by a very high impedance voltmeter, there can be no current, and a potential barrier is set up to counteract the charge flow. This is illustrated in Figure 39; in equilibrium, the Fermi energy is the same on both sides of the junction, and the junction region is depleted of both holes and electrons.

For the case of zero applied bias voltage, there is no net current, and the potential barrier prevents the flow of electrons and holes across the junction. If the p-type side is connected to a positive potential and the n-type side to a negative potential (forward bias), the Fermi energy is tilted such that the potential barrier is reduced. Because the junction region is depleted of free carriers, nearly all of the voltage drop occurs across the junction region. The barrier reduction permits the electrons and holes to diffuse across the junction into the opposite type material and a large forward current results. If the opposite potentials are applied, the potential barrier is increased. In this case, any electrons in the p-type region or holes in the n-type region are attracted to the opposite side of the barrier. Since these minority carriers are present only in very small quantities, the current

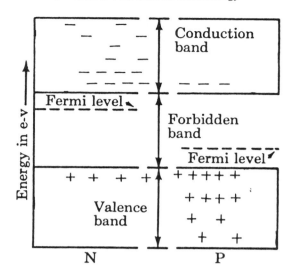

Figure 38: Energy band structure for separate isolated *p*- and *n*-type silicon regions showing the different values of the electrochemical potential (Fermi energy or Fermi level) in the two regions. [After Pierce, J.F., *Semiconductor Junction Devices*, Columbus: Charles E. Merrill (1967).]

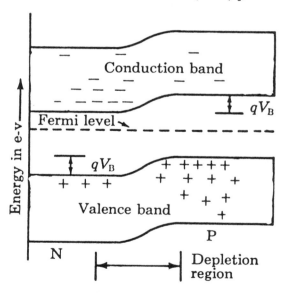

Figure 39: Energy band structure of an isolated *p-n* junction showing the common electrochemical potential, the depletion region, and the built-in potential barrier, qV_b. [After Pierce, J.F., *Semiconductor Junction Devices*, Columbus: Charles E. Merrill (1967).]

in this case is much smaller. This reverse current can provide useful information about the generation and recombination characteristics of the minority carriers; see the chapter on Carrier Lifetime in Silicon for more details.

Rectification can also occur at a metal-silicon junction. In this case there is only a single minority carrier, holes if the silicon is n-type and electrons if the silicon is p-type. Rectifying metal-silicon junctions may be readily formed with point contacts. Such junctions formed the detectors used in early crystal radio receivers and in World War II radars (4). Such junctions may also be used in determining the conductivity type of a silicon sample. If an alternating voltage source is connected to the silicon through a point contact (rectifying) and a large area contact (non-rectifying or "Ohmic"), there is a positive or negative current according to whether the silicon is p- or n-type, respectively (97).

3.10 Thermoelectric Effects

The **Seebeck effect** occurs when two different materials are joined together. If the junctions are maintained at different temperatures, an electromotive force known as the Seebeck emf is developed. If the hot and cold junctions are interchanged, the direction of the emf is reversed. This is the familiar thermocouple voltage and is given by:

$$(\alpha_1 - \alpha_2) \Delta T, \tag{27}$$

where α_1 is the Seebeck coefficient for one material, α_2 is the Seebeck coefficient for the second material, and ΔT is the temperature difference.

Silicon is not a practical thermoelectric material; its thermoelectric figure of merit ($Z = \alpha^2 \sigma / \kappa$, where σ is the electrical conductivity and κ is the thermal conductivity) is not very large because of its relatively low electrical conductivity and Seebeck coefficient and its relatively large thermal conductivity. The practical application of the Seebeck effect for silicon comes in determination of its conductivity type. Use of a thermoelectric probe is a standard technique for this test (97). The method can be understood by reference to Figure 40 which shows a thermoelectric circuit consisting of a metal and p-type silicon.

If it is assumed that the voltmeter in the circuit has infinite impedance so that there is no electric current, it

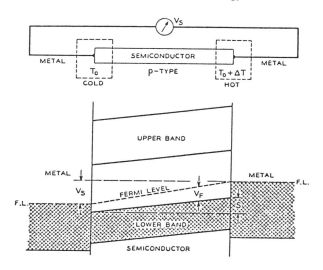

Figure 40: Illustration of a thermoelectric couple consisting of a metal and a *p*-type semiconductor. [From Shive, J.W., *The Properties, Physics, and Design of Semiconductor Devices*, Princeton: Van Nostrand (1959).]

indicates the Seebeck voltage, V_S. There are three contributions to this voltage. First, the Fermi energy in extrinsic silicon moves further away from the band edge as the temperature is increased. This change in Fermi energy relative to the band edge is reflected in the external voltage.

Second, holes at the warmer side are more energetic than those at the cooler side; therefore they have higher random velocity. Because there can be no net current in the circuit, a potential is set up such that the colder holes are attracted to the warmer side in order to counteract the natural flow of the warmer holes to the colder side.

Finally, in some temperature ranges, the acceptor impurities are more completely ionized at a higher temperature; this results in an increased hole density at the warmer side. A potential is set up to counteract the diffusion of holes from the region of higher concentration. All of these potentials cause the warmer side to be negative and the cooler side to be positive. The Seebeck effect in the metal is negligibly small compared to that in the silicon.

For *n*-type silicon, the same effects occur but the sign of the potentials is reversed; the warmer side becomes positive and the cooler side negative. For silicon and other semiconducting materials the sign of the Seebeck coefficient is positive if the majority carriers are holes (p-type) and negative if the majority carriers are electrons (n-type). If holes and electrons are present in approximately equal amounts, the sign of the Seebeck coefficient in silicon is negative because the electron mobility is greater than the hole mobility; under such circumstances, the thermoelectric test for conductivity type fails.

If there is a current in the circuit, two other thermoelectric effects come into play. The **Peltier effect** is the evolution or absorption of heat when an electric current is passed across the junction, or couple, between two dissimilar materials at the same temperature. The effect is reversible; if heat is evolved when the current is in one direction, it is absorbed when the current is reversed. The amount of heat absorbed or evolved is given by $\pi \cdot I$, where π is the Peltier coefficient and I is the current. The Peltier coefficient of a couple is related to the Seebeck coefficients of the two materials which comprise the couple:

$$\pi_{1,2} = T \cdot (\alpha_2 - \alpha_1). \tag{28}$$

The **Thomson effect** is the evolution or absorption of heat when an electric current is passed through material across which there is a temperature difference, ΔT. Thomson heating, which is reversible, occurs in addition to irreversible Joule (RI^2) heating. The amount of heat evolved or absorbed is given by $\vartheta \cdot \Delta T \cdot I$, where ϑ is the Thomson coefficient. The Seebeck coefficient of a material is related to its Thomson coefficient as follows:

$$\alpha = \int_0^T (\vartheta/T)dT. \tag{29}$$

The subject of thermoelectric properties is actually much more complex than the above discussion would imply. The interested reader is referred to the literature (98) for a more complete and rigorous discussion of this topic.

4.0 OPTICAL PROPERTIES

Although silicon is opaque in the visible range of the optical spectrum, it is transparent in the near infrared. It is highly reflective and has a large index of refraction. Because of these properties, optical components, such as lenses and windows, for use in the near infrared region of the spectrum are frequently made of silicon. Silicon also can be used as a detector of infrared and gamma radiation and is widely used for solar cells. The optical properties of silicon are also frequently employed for identifying and counting impurities, especially those which are inactive electrically, and for measuring recombination characteristics.

4.1 Index of Refraction and Reflectivity

The index of refraction, n, of silicon is shown as a function of wavenumber in Figure 41 over the range 800 to 8000 cm^{-1} (99). The reflectivity, R, is given by:

$$R = [(n - 1)^2 + k^2]/[(n + 1)^2 + k^2], \qquad (30)$$

where k, the absorption index at wavelength λ (cm), = $\alpha\lambda/4\pi$, and α is the absorption coefficient (cm^{-1}). In the near infrared region, k can be neglected in comparison with n, and Equation (30) reduces to:

$$R = (n - 1)^2/(n + 1)^2. \qquad (31)$$

The reflectivity of silicon is very nearly 0.3 over the wavenumber range 400 to 5000 cm^{-1}, increasing gradually to about 0.34 at 15,000 cm^{-1} (100). Phillips (101) has developed a quadratic expression to fit the reflectivity data over the wavelength range from 0.70 to 1.05 μm (14300 to 9500 cm^{-1}):

$$R = 0.3214 - 3.565 \times 10^{-6} \nu + 3.149 \times 10^{-10} \nu^2, \qquad (32)$$

where ν is the wavenumber in cm^{-1}. This relation is particularly useful in surface photovoltage measurements of the carrier diffusion length. The reflectivity of silicon is plotted over the range from about 1000 to 15,000 cm^{-1} in Figure 42. In this plot, the left hand branch (1000 to 7500 cm^{-1}) is calculated from Equation (31), and the right hand branch is calculated from Equation (32).

4.2 Antireflection Coatings

Because the reflection coefficient of silicon is so large, considerable energy is lost when the material is used for optical components or for solar cells. The transmission of

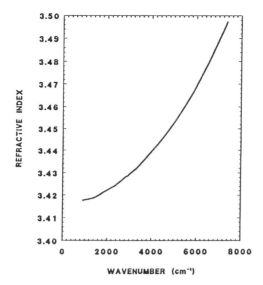

Figure 41: Refractive index of 5 Ω·cm n-type silicon over the wavenumber range 950 to 7500 cm^{-1} at a refracting angle of 15.8°. [Based on data from Salzberg and Villa, Reference 99.]

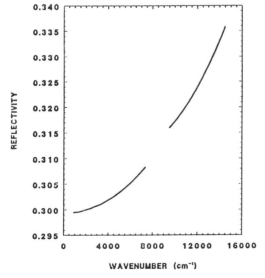

Figure 42: Reflectivity of silicon in the wavenumber range 950 to 14500 cm^{-1}. The left branch of the curve is calculated from Equation (31) using the refractive index data in Figure 41 and the right branch is calculated from Equation (32).

energy through the silicon in the first case or absorption of energy in the second case can be increased through the use of a thin coating of a dielectric with an index of refraction intermediate between that of silicon (n_{Si} = 3.42) and that of air (n_{air} = 1.00).

For one-layer dielectric films the reflectance is reduced to zero at the wavelength where the film thickness is equal to one quarter of the wavelength. For silicon, the optimum dielectric has an index of refraction between 1.8 and 1.9 (102). For thicker films, the reflectance at both interfaces must be considered, but for intermediate wavelengths the total reflectance is reduced significantly below that of the bare silicon.

An alternative technique involving preferential etching of tetrahedra into the surface of the silicon and so capturing the radiation by multiple reflections can increase the amount of radiation absorbed at a silicon-air surface from 70% to 91% (103). This technique has more recently been applied on a larger scale to solar cell fabrication.

4.3 Relationships between Wavelength, Wavenumber, and Photon Energy

Early spectrophotometers produced optical spectra which were linear in wavelength, λ, given in Ångstroms (Å), micrometers (μm), or nanometers (nm). The frequency of the radiation, ν, is found by dividing the velocity of light, c, by the wavelength. The photon energy is given by $h\nu$, where h is Planck's constant. Although the proper unit for energy is joules (J), electron-volts (eV) are more customary in dealing with semiconductors. Modern spectrophotometers produce spectra which are linear in wavenumber, which is the reciprocal of the wavelength, and is usually given in reciprocal centimeters (cm^{-1}). Relationships between these various quantities are summarized in Table 5.

4.4 Absorption

The absorption of electromagnetic radiation is described in terms of an absorption coefficient, α, which is the distance in centimeters required for the intensity of the incident radiation to fall to 1/e of its initial value. For normal incidence, the intensity, $I(\nu)$, of radiation of wavenumber ν at a distance x below the incident surface is given by:

$$I(\nu) = I_0(\nu)\exp[-\alpha(\nu)x], \tag{33}$$

Table 5

Unit Conversions

To convert from:	to:	multiply:	by:
wavelength (Å)	wavelength (μm)	Å	10^{-4}
wavelength (Å)	wavelength (nm)	Å	0.1
wavelength (cm)	wavenumber (cm^{-1})	1/wavelength	1.0
wavelength (μm)	wavenumber (cm^{-1})	1/wavelength	10^4
wavelength (μm)	energy (eV)	1/wavelength	1.240
wavenumber (cm^{-1})	energy (eV)	wavenumber	1.240×10^{-4}
wavelength (μm)	frequency (GHz)	1/wavelength	2.998×10^5
energy (J)	energy (eV)	J	6.241×10^{18}

where $I_0(\nu)$ is the intensity of the radiation incident on the surface of the sample.

4.4.1 Photoconductivity: One way in which electromagnetic radiation is absorbed in a semiconductor is by transferring energy to an electron in an atom thus elevating the electron to a higher energy state. If the electron is in a silicon atom, the electron must be excited from a state in the valence band to a state in the conduction band. Therefore, the energy must be greater than the forbidden energy gap. If the electron is in a donor-type impurity atom with energy states in the forbidden gap, less energy is required to excite the electron to the conduction band. Similarly, the energy required to excite an electron in the valence band to an acceptor-type impurity atom with an empty state in the forbidden gap is also smaller than the band gap energy. This type of absorption results in photoconductivity, or an increase in conductivity resulting from absorption of electromagnetic radiation. Photoconductivity is a transient effect; when the irradiation is removed, the conductivity decays back to its steady state value. In addition to its use in radiation detection and solar energy conversion, this effect is the basis for a method for measuring the minority carrier lifetime (104).

A band-to-band transition is a direct one if the momentum of both states is the same; the transition is indirect if a momentum change is required. As discussed previously, the lowest energy states in the conduction band of silicon do not occur at the same momentum as the highest states in the valence band. Therefore, band-to-band transitions in silicon are indirect until the energy is sufficient to excite the electrons into higher states in the conduction band which are located at the same momentum as the highest valence band states. As a consequence, silicon is not as efficient an energy converter as are direct-gap semiconductors such as gallium arsenide and other binary, ternary, and quaternary III-V compounds.

In silicon, radiation with wavelength less than about 1 μm is required to excite electrons across the forbidden gap from the valence band to the conduction band. Because there are large numbers of electrons in the valence band, the energy is used up very quickly as the radiation enters the silicon, and the penetration depth is short. Therefore α is large as illustrated in Figure 43, which shows the room temperature absorption coefficient as a function of wavelength for radiation energies greater than the band gap (100),(105). The significant increase in absorption coefficient below about 0.4 μm is due to the onset of the direct transition to band C3 (see Figure 22).

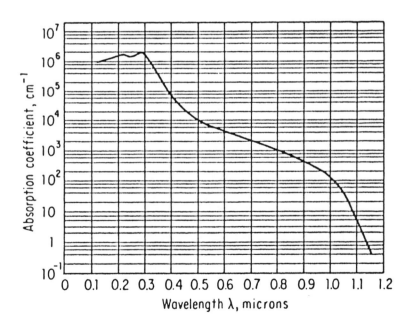

Figure 43. Room temperature absorption of silicon in the wavelength range 0.1 to 1.15 μm (100,000 to 8700 cm^{-1}). [From Runyan, Reference 1.]

Although this Figure shows the general form of the absorption coefficient, it is not sufficiently accurate for stress-relieved surfaces typical of those found in modern, chem-mechanically polished silicon wafers. Nartowitz and Goodman (106) have critically evaluated more recent published data for the room-temperature absorption coefficient over the range 0.70 to 1.04 μm. They used as a criterion the results obtained when applying these data in measurements of the minority carier diffusion length by the surface photovoltage method which is very sensitive to the accuracy of the absorption coefficient used. They conclude that the most reliable results can be obtained when using a "compromise" or composite set of data given by the following relationship:

$$\alpha = (8.4732 \times 10^{-3} \nu - 76.417)^2, \tag{34}$$

where α is the absorption coefficient in cm^{-1} and ν is the wavenumber in cm^{-1}. This expression is plotted in Figure 44.

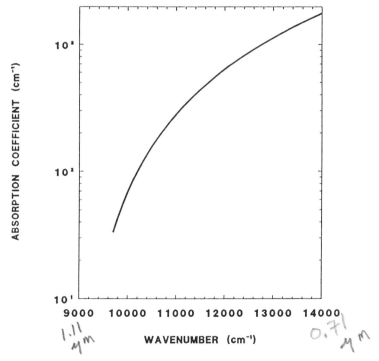

Figure 44: Room temperature absorption coefficient of silicon in the range from 9700 to 13000 cm^{-1}, calculated from Equation (34).

4.4.2 Lattice Absorption: In the longer wavelength (lower energy) region of the spectrum, the photon energy may be absorbed by the electrons in a bond arrangement. Because only the vibrational energy is increased without removing the electron from the bond, this type of absorption does not result in photoconductivity. Absorption by electrons in a bond generally occurs at a localized frequency resulting in absorption peaks; these may be due either to lattice absorption (absorption of radiative energy by the silicon-silicon bonds) or impurity absorption (absorption of radiative energy by silicon-impurity states).

The lattice absorbance spectrum of a 2-mm thick double-side polished silicon slice is shown over the wavenumber range from 1800 to 450 cm^{-1} in Figure 45; there are nine strong lattice peaks of which the strongest by far is located at about 610 cm^{-1}. The position of the lattice peaks at room temperature is given in Table 6 together with suggested assignments of the phonons responsible for the transitions (107). Additional details about the silicon lattice spectrum may also be found in this paper.

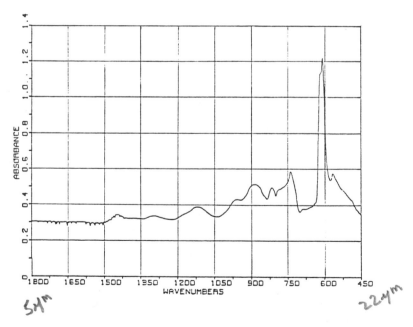

Figure 45: Lattice spectrum of oxygen and carbon free silicon.

Table 6

Lattice Absorption Peaks in Silicon (107)

Peak $\nu(cm^{-1})$	Location $\frac{hc\nu}{k}(K)$	Suggested Phonon Assignments[a] and Their Characteristic Temperatures
566	815	LO(595) + TA(220) or
		LO(575) + TA(200)
610	880	TO(695) + TA(185) or
		TO(680) + TA(200)
689	992	LO(595) + TA(198) + TA(198)
740	1065	LO(595) + LA(470)
766	1103	TO(695) + TA(204) + TA(204)
819	1180	TO(695) + LA(485)
896	1290	TO(695) + LO(595)
964	1388	TO(695) + TO(695)
1302	1875	TO(695) + LO(595) + LO(595) or
		TO(695) + TO(695) + LA(485)
1378	1985	TO(695) + TO(695) + LO(595)
1448	2085	TO(695) + TO(695) + TO(695)

[a] T = transverse, L = longitudinal, O = optical, A = acoustical

4.4.3 Impurity Absorption: Oxygen, carbon, and nitrogen have absorption peaks in the same region of the infrared spectrum that contains the lattice peaks. Oxygen has peaks at 1107 and 515 cm^{-1} (108); carbon has a peak at 605 cm^{-1}, on the edge of the very strong lattice band at 610 cm^{-1} (109), and nitrogen has strong peaks at 963 and 766 cm^{-1} (44). These bands are due to absorption of energy in the silicon-impurity bonds.

There are standard test methods for both carbon(110) and oxygen (111). The latter is based on the use of the 1107 cm^{-1} peak for the quantitative determination of the oxygen content. However, there has been considerable uncertainty as to the calibration factors which relate the impurity concentration to the absorption coefficient in both cases (112). Table 7 summarizes the variety of calibration factors for oxygen and carbon which have been adopted in recognized standards. Recently an international interlaboratory experiment was completed which determined the calibration factor for the 1107 cm^{-1} oxygen band at room temperature to be 6.28 ±

0.18 ppma·cm (113); this factor has been designated the IOC-88, for International Oxygen Coefficient, 1988. At present, 2.0 ppma·cm is the standard calibration factor for carbon (110), but a decrease of about 15%, based on recent interlaboratory testing (112), is being considered by ASTM. The calibration factor for the 966 cm^{-1} nitrogen peak is reported to be 3.64 ppma·cm^{-1} (114).

Table 7

Standardized Values for the Calibration Factors

for Oxygen and Carbon in Silicon

Oxygen

Factor	Location cited	Name	To obtain IOC-88 multiply by
4.9 ppma·cm	DIN 50 438/1 and ASTM 121-80[a]	DIN or New ASTM	1.282
5.5 ppma·cm	ASTM F 45[b]	Kaiser & Keck	1.142
6.06 ppma·cm	JEIDA experiment[c]	JEIDA	1.047
6.28 ppma·cm	ASTM F 1188, F 1189 [d]	IOC-88	1.000
9.63 ppma·cm	ASTM F121-79[e]	Old ASTM	0.652

Carbon

Factor	Location cited	Name
2.2 ppma·cm	ASTM F123-74[f]	Old ASTM
2.0 ppma·cm	DIN 50 438/2 and ASTM F 123-81[g]	New ASTM
1.7 ppma·cm	JEIDA experiment[h]	JEIDA

[a] This value is based on the work of Graff et al. [Graff, K., Grallath, E., Ades, S., Goldbach, G., and Tolg, G., *Determination of Parts per Billion of Oxygen in Silicon by Measurement of the IR Absorption at 77 K*, Solid-State Electron. **16**: 887-893 (1973)]; ASTM method F 121 was withdrawn in 1989.

[b] This was the original calibration curve reported in the literature [Kaiser, W., and Keck, P.H., *Oxygen Content of Silicon Single Crystals*, J. Appl. Phys. **28**: 882-887 (1957)]; the ASTM method which cited it was withdrawn in 1970

[c] This value was found in an interlaboratory experiment conducted by the Japan Electronic Industries Development Association; see Iizuka, T., Takasu, S., Tajima, M., Arai, T., Nozaki, T., Inoue, N., and Watanabe, M., *Determination of Conversion Factor for Infrared Measurement of Oxygen in Silicon*, J. Electrochem. Soc. **132**: 1707-1713 (1985)

[d] This value is based on an international interlaboratory experiment, Ref. 113.

[e] This value is based on the work of J. A. Baker [*Determination of Parts per Billion of Oxygen in Silicon*, Soild-State Electron. **13**: 1431-1434 (1970)] and was cited in the ASTM method between 1970 and 1979

[f] This value is based on the work of Newman and Willis, Ref. 107

[g] This value is based on the work of E. Haas and B. O. Kolbesen (1980), cited in DIN 50 438/2

[h] This value was found in an interlaboratory experiment conducted by the Japan Elecronic Industries Development Association; see Inoue et al., Ref. 112

The standard methods for measuring oxygen and carbon by infrared absorption specify that the measurements be made on samples with parallel, polished sides. The absorption coefficient, α, is found from measurements of the transmittance, T:

$$T = (I/I_0) = [(1-R)^2 \exp(-\alpha d)]/[1-R^2 \exp(-2\alpha d)], \qquad (35)$$

where I is the relative transmitted intensity at the wavelength of the absorption peak, I_0 is the relative baseline intensity at the same wavelength, R is the reflectance (= 0.3 for silicon in the wavelength range of interest), and d is the sample thickness. This equation takes account of the multiple reflection between the faces of the sample (115).

4.4.4 Free Carrier Absorption. Long wavelength radiation can also be absorbed by free carriers. This type of absorption becomes significant at large carrier densities. For *n*-type silicon with resistivity less than about 0.2 Ω·cm and for *p*-type silicon with resistivity less than about 0.5 Ω·cm, the dominant absorption is due to free carriers. The absorption due to both holes (116) and electrons (117) is proportional to the carrier density and to the square of the wavelength; it is therefore greater at longer wavelengths. Consequently, the transmittance of heavily doped silicon in the infrared is reduced so that the material is unsuitable for use in optical components. In addition, absorption measurements cannot be used to determine the oxygen and carbon concentrations and other, chemical, techniques must be employed (118) unless the free carriers are removed by some treatment such as electron irradiation (119).

4.5 Optical Methods for Detecting Dopant Impurities

Dopant impurities have absorption peaks at very long wavelengths (in the range from 600 to 200 cm^{-1}) which can be seen at low temperatures (<25 K). It is possible to determine the densities of both majority and minority dopant impurities at the same time (120). Locations of the absorption peaks for selected impurities at cryogenic temperatures are listed in Table 8 (121).

Dopant impurities may also be detected in very pure silicon by photoluminescence (122). This technique also requires the use of liquid helium to reduce the temperature of the sample to <10 K. Calibration curves for boron, phosphorus, and aluminum have been published. Colley and

Table 8

Absorption Peaks for Selected Impurities in Silicon at Cryogenic Temperatures (122)

Impurity	Wavenumber of Most Intense Peak
Boron	317.7 cm^{-1}
Aluminum	471.7 cm^{-1}
Gallium	548.0 cm^{-1}
Indium	1175.9 cm^{-1}
Phosphorus	315.9 cm^{-1}
Arsenic	382.2 cm^{-1}
Antimony	293.6 cm^{-1}

Lightowlers (122) discuss the various factors which affect the accuracy of the calibration curve. A standard method for carrying out photoluminescence measurements is under development by cooperating task groups from ASTM Committee F-1 on Electronics and the Silicon Committee of the Japan Electronic Industries Development Association (123).

Plasma resonance may also be used to detect the doping density in silicon and other semiconductors (124). Unlike the two methods described above, this method is not impurity specific and relies on the change in reflectivity caused by the free carriers. The location of the minimum as a function of wavenumber has been established empirically for both n- and p-type silicon (124). Plots of these relationships are given in Figure 46.

4.6 Emissivity

The emissivity of silicon has been determined at a wavelength of 0.65 μm (a common wavelength for optical pyrometers) in the temperature range from 1000 K to the melting point (125). The emissivity for a polished silicon surface is given in Figure 47; for a sandblasted surface the emissivity is somewhat higher, about 0.75 over the temperature range from 1000 to 1600 K.

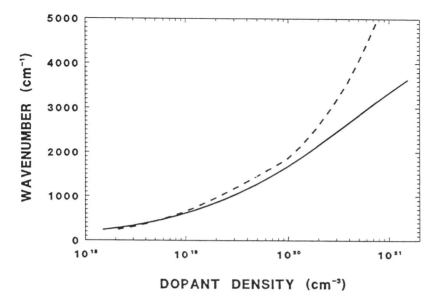

Figure 46: Location of plasma resonance minimum for heavily doped *n*- (solid line) and *p*- type (dashed line) silicon as a function of dopant density.

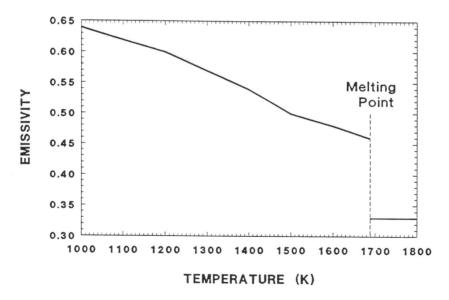

Figure 47: Spectral emissivity of etched and molten silicon surfaces at 0.65 μm as a function of true sample temperature.

Emissivity is also an important consideration for molten silicon when the temperature of the melt is monitored by optical pyrometry during crystal growth. The emissivity of molten silicon in the vicinity of the melting point is 0.33 (126).

5.0 THERMAL AND MECHANICAL PROPERTIES

At room temperature, single crystal silicon is a brittle material. Although some plastic flow may occur at temperatures above 800°C, silicon can be stressed to the fracture point without significant creep or flow at room temperature. Consequently, it is a good material for use in piezoresistive transducer elements because it can cover a wide range without zero point shift.

During integrated circuit processing, silicon is subjected to large thermal and mechanical stresses. Similar stresses may also be encountered during device operation. Such stresses can result in wafer distortion (warp), slip, and fracture. In both cases, the extent of the deformation depends on the nature and severity of the stress and the mechanical strength of the silicon. The thermal conductivity of silicon is also an important factor in dissipation of heat generated in device junctions.

5.1 Elastic Constants

Herring (127) provides a thorough introduction to the elastic constants of silicon. The stress-strain relationship for a cubic crystal is given by:

$$|T_1, T_2, T_3, T_4, T_5, T_6| = \begin{vmatrix} c_{11} & c_{12} & c_{12} & 0 & 0 & 0 \\ c_{12} & c_{11} & c_{12} & 0 & 0 & 0 \\ c_{12} & c_{12} & c_{11} & 0 & 0 & 0 \\ 0 & 0 & 0 & c_{44} & 0 & 0 \\ 0 & 0 & 0 & 0 & c_{44} & 0 \\ 0 & 0 & 0 & 0 & 0 & c_{44} \end{vmatrix} \cdot \begin{vmatrix} S_1 \\ S_2 \\ S_3 \\ S_4 \\ S_5 \\ S_6 \end{vmatrix}, \quad (36)$$

where T_1, T_2, and T_3 are compressive stresses applied in the direction of the x, y, and z axes, respectively; T_4, T_5, and T_6 are shear couples acting parallel to the cube faces and tending to twist; S_1 through S_6 are the components of strain; and c_{11}, c_{12}, and c_{44} are the elastic constants. Values of the elastic constants at 298 K (128) and their temperature coefficients (129) are listed in Table 9.

Table 9

Elastic Constants of Silicon and Their Temperature Coefficients

Elastic Constants (125)		
	c_{11}	1.6564×10^{11} Pa
	c_{12}	0.6394×10^{11} Pa
	c_{44}	0.7951×10^{11} Pa
Temperature Coefficients of Elastic Constants (126)		
	K_{11}	$-75 \times 10^{-6}/°C$
	K_{12}	$-24.5 \times 10^{-6}/°C$
	K_{44}	$-55.5 \times 10^{-6}/°C$

This relationship is frequently written as the inverse:

$$|S_1, S_2, S_3, S_4, S_5, S_6| = \begin{vmatrix} s_{11} & s_{12} & s_{12} & 0 & 0 & 0 \\ s_{12} & s_{11} & s_{12} & 0 & 0 & 0 \\ s_{12} & s_{12} & s_{11} & 0 & 0 & 0 \\ 0 & 0 & 0 & s_{44} & 0 & 0 \\ 0 & 0 & 0 & 0 & s_{44} & 0 \\ 0 & 0 & 0 & 0 & 0 & s_{44} \end{vmatrix} \cdot \begin{vmatrix} T_1 \\ T_2 \\ T_3 \\ T_4 \\ T_5 \\ T_6 \end{vmatrix}, \text{ Eq.(37)}$$

where the s_{ij} are known as elastic moduli or compliance coefficients. For cubic crystals the relationships between the elastic constants and the elastic moduli are as follows:

$$s_{11} = \frac{(c_{11} + c_{12})}{(c_{11} - c_{12})(c_{11} + 2c_{12})}, \tag{38}$$

$$s_{12} = \frac{-c_{12}}{(c_{11} - c_{12})(c_{11} + 2c_{12})}, \text{ and} \tag{39}$$

$$s_{44} = \frac{1}{c_{44}}. \tag{40}$$

In these equations, the c's and the s's may be interchanged to express the c_{ij} in terms of the s_{ij}.

5.1.1 Young's Modulus. Young's modulus, E, is the coefficient which relates the strain to the stress when the stress (tension or compression) is applied in one direction only and the sides of the body in other directions are not constrained:

$$S_i = (1/E)T_i. \tag{41}$$

The value of E depends on the direction of the applied stress. In general, $(1/E)$ can be expressed in terms of the elastic moduli and the direction cosines of the applied force with respect to the crystallographic axes:

$$(1/E) = s_{11} - 2(s_{11}-s_{12}-\tfrac{1}{2}s_{44})(\alpha^2\beta^2+\alpha^2\gamma^2+\beta^2\gamma^2), \tag{42}$$

where α, β, and γ are the direction cosines.

For a force along a [100] direction, Young's modulus is given by:

$$E_{[100]} = 1/s_{11} = 1.31 \times 10^{11} \text{ Pa}. \tag{43}$$

If the force is along a [110] direction, Young's modulus becomes:

$$E_{[110]} = 4/(2s_{11} + 2s_{12} + s_{44}) = 1.69 \times 10^{11} \text{ Pa}. \tag{44}$$

For forces lying along other directions in a (100) plane, the value of Young's modulus is between the extremes given in eqs (43) and (44). For forces lying along any direction in a [111] plane it has been shown that Young's modulus is the same as $E_{[110]}$, given above (130). For a force along a [111] direction, Young's modulus becomes:

$$E_{[111]} = 3/(s_{11} + 2s_{12} + s_{44}) = 1.87 \times 10^{11} \text{ Pa}. \tag{45}$$

Because most silicon integrated circuits and other devices are fabricated on {100} or {111} wafers, the values of Young's modulus in this section are appropriate for analysis of the stress induced in wafers during circuit or device fabrication and use.

5.1.2 Modulus of Compression: When a sample of silicon is subjected to hydrostatic pressure, $T_1 = T_2 = T_3$ and the other stresses are zero. The modulus of compression, K, is the coefficient which relates the pressure, P, to the volume reduction, δ: $P = K\delta$. The modulus of compression is given by:

$$K = \sum_{i,j=1}^{3} s_{ij} = 3(s_{11} + 2s_{12}). \tag{46}$$

The longitudinal linear compressibility is given by K/3.

5.1.3 Shear Modulus: If a single couple acts on a piece of silicon, the angular displacement, S_i (where i=4, 5, or 6), is related to the torsional stress, T_i, by the shear modulus, G: $T_i = GS_i$. For a specimen with a circular cross section, the mean value of the shear modulus is:

$$G = [s_{44} + 4(s_{11} - s_{12} - \tfrac{1}{2}s_{44})(\alpha^2\beta^2 + \alpha^2\gamma^2 + \beta^2\gamma^2)]^{-1}, \tag{47}$$

where α, β, and γ are the direction cosines of the axis of the sample perpendicular to the plane containing the couple.

5.1.4 Poisson's Ratio: Because Poisson's ratio is the ratio of the strain perpendicular to the applied stress to the strain in the direction of the applied stress, it depends on both the direction of the applied stress and the perpendicular direction. On a {111} plane Poisson's ratio is given by:

$$\sigma_{\{111\}} = (1/6)(5s_{12} + s_{11} - s_{44}/2)E_{\{111\}} = 0.29. \tag{48}$$

It turns out that in silicon σ varies only slightly with direction and usually has a value a little over 1/4.

5.1.5 Other Relationships: For additional details on the relationships between the various elastic constants and moduli, the interested reader is referred to the literature (131).

5.2 Piezoresistivity

Uniaxial stresses result in resistivity changes in both n- and p-type silicon. The ratio of the change of resistivity, $\delta\rho_{ij}$, measured in the j-th direction for a stress applied in the i-th direction to the resistivity, ρ_0, in the absence of stress is given by $\Delta_{ij} = \delta\rho_{ij}/\rho_0$. This ratio is related to the applied stress, T, as follows:

$$[\Delta] = [\pi][T], \tag{49}$$

where the piezoresistance tensor $[\pi]$ has the same form as that of the elastic moduli [Equation (37)] with three independent components: π_{11}, π_{12}, and π_{44}. The longitudinal piezoresistance coefficient, π_l, is given by:

$$\pi_l = \pi_{11} - 2(\pi_{11} - \pi_{12} - \pi_{44})(\alpha^2\beta^2 + \alpha^2\gamma^2 + \beta^2\gamma^2), \tag{50}$$

where α, β, and γ are the direction cosines with respect to the crystallographic axes. The directions which are most important for silicon piezoresistive devices are the [100], [111], and [110] directions. The value of the piezoresistance coefficients depends on impurity density and temperature. Results of measurements of π_{11} on n-type silicon are shown in Figure 48 (132) and results of measurements of π_l for the [110] direction on p-type silicon are shown in Figure 49 (133). More recently, Kondo (134) has provided a graphical description of the piezoresistance coefficients in silicon.

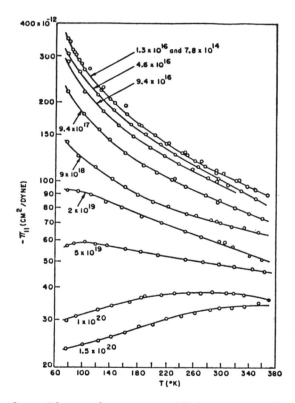

Figure 48: Piezoresistance coefficient π_{11} of n-type silicon as a function of temperature for various impurity concentrations. [From Tufte and Stelzer, Reference 132.]

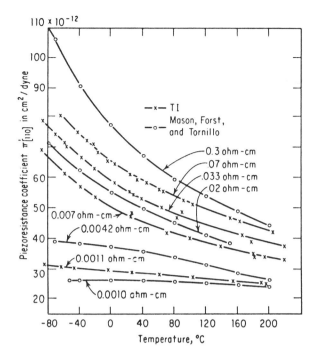

Figure 49: Piezoresistance coefficient $\pi_{[110]}$ of p-type silicon as a function of temperature for various resistivities as reported by Sobey (133). [From Runyan, Reference 1.]

5.3 Mechanical Strength and Plastic Deformation

Although it is brittle at room temperature, silicon is a very strong material. It is possible to hang very large crystals of nearly 100 kg mass on seed crystals necked down to a cross section of a few square millimeters.

The strength depends on the crystal perfection and in some cases on the oxygen content of the silicon. Figures 50 and 51 show stress-strain curves for dislocation-free and dislocated silicon crystals, respectively, which had a range of oxygen content (135). These results show that the oxygen content has a significant effect on the strength of dislocated crystals but very little effect on the strength of dislocation-free crystals. Consequently, the authors conclude that interstitial oxygen atoms dispersed throughout the crystal have no influence on the dynamic processes of dislocations generated as a result of deformation of silicon crystals at elevated temperatures. On the other hand, oxygen atoms in a dislocated crystal congregate on the dislocations at rest and lock them effectively; therefore, crystals with higher oxygen content have a higher mechanical strength. These authors also studied the effect of oxygen precipitation on the mechanical strength and found that drastic softening occurs in crystals with high precipitate density.

These results are consistent with earlier work which concluded that oxygen in the crystal does not significantly affect the ultimate strength of the material, but does affect the upper yield point (136). This work also showed that over the temperature range from 950° to 1400°C, the ultimate tensile strength of silicon decreases from about 3.5×10^8 to about 1×10^8 Pa.

5.3.1 Plastic Deformation: Plastic deformation of silicon wafers can occur at the high temperatures required for device and circuit fabrication. It is most readily observed at temperatures above 600°C but can also be observed at room temperature during microhardness measurements or scribing. The deformation occurs by slip between {111} planes along a <110> direction (137). This is the mechanism by which slip can be propagated from very small surface damage deeply into the wafer during high temperature processing (62). Plastic flow due to excessive stress from thermal shock is responsible for the star pattern of dislocations which can be seen on wafers cut from improperly annealed crystals. Similar patterns can be seen radiating out from pinch points where wafers are supported during diffusion or oxidation if the thermal stresses associated with these process steps are too great.

More recent work using 10-mm wide bars cut along a <110> direction from double-side polished, high-resistivity p-type (100) wafers, 3 in. in diameter and 800 μm thick, has demonstrated that wafers from Czochralski-grown crystals with about 21.5 ppma (IOC-88) dissolved oxygen are less susceptible to plastic deformation than are wafers from float-zoned crystals with oxygen below the detection limit (138).

Following thermal preheating at 1050°C, Czochralski wafers could be deformed at lower temperatures than float-zoned wafers. This is attributed to the influence of bulk defects due to oxygen precipitation in the Czochralski crystal. Following preheating at 800°C, the interesting result was obtained that under low stress (2×10^7 N/m^2) the deformation temperature increased while for higher stress (5 or 10×10^7 N/m^2) the deformation temperature decreased. These results

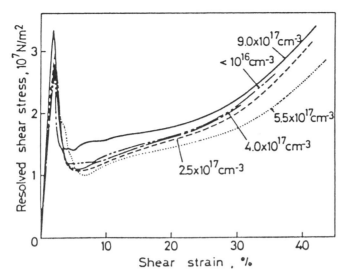

Figure 50: Stress-strain curves of dislocation-free silicon crystals containing various concentrations of oxygen as shown by the figures on the curves (Kaiser-Keck calibration). [123] tensile deformation at 900°C under a shear strain rate of 1.1×10^{-4} s^{-1}. [From Yonenaga et al., Reference 135.]

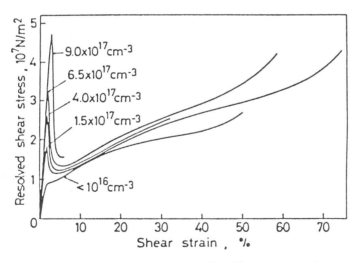

Figure 51: Stress-strain curves of silicon crystals containing dislocations at densities of about 1×10^6 cm^{-2} and various concentrations of oxygen as shown by the figures on the curves (Kaiser-Keck calibration). [123] tensile deformation at 800°C under a shear strain rate of 1.1×10^{-4} s^{-1}. [From Yonenaga et al., Reference 135.]

are consistent with earlier reports of hardening after heat treatment at 700° or 900°C (139) and indicate that plastic deformation of Czochralski crystals depends on the nature as well as the number of bulk defects. They emphasize the necessity for controlling the stress induced during processing when intentionally introducing the bulk defects necessary for intrinsic gettering.

5.3.2 **Warp:** With the introduction of larger diameter crystal, the problem of warp has become more significant, particularly as the ratio of diameter to thickness increases. Leroy and Plougonven (140) explain this result on the basis of a model which fits experimental data of the critical stress in silicon, temperature gradients induced by nonuniform cooling, and initial wafer curvature. This model includes a complete stress analysis previously carried out by Dyer et al. (141) which takes into account deviations from a simple thin membrane model. They observe that the area affected by slip is much greater on the concave side of the wafer than it is on the convex side. They also note that different heat treatment cycles result in different dependencies of the results on oxygen precipitation characteristics.

Lee and Tobin (142) have carried out experiments of the effect of oxygen concentration on warpage as wafers are passed through a typical CMOS processing cycle. They found that precipitation occurred only for oxygen concentrations greater than 19.6 ppma (IOC-88), and that severe warpage was introduced during an implant anneal step for oxygen concentrations in excess of 20.9 ppma. However they found that process modifications including reduction of both the push/pull temperature for the critical step and the temperature of a subsequent annealing step from 900° to 650°C resulted in immunity to process induced distortion for initial oxygen concentrations up to 24.8 ppma. Although the details of the effects of various process conditions may change with the crystalline properties of the starting material, these results again emphasize the importance of considering both the electrical and mechanical aspects of the changes in silicon wafers which are induced during device and circuit processing.

5.3.3 **Fracture:** Dyer has studied the formation of fractures in silicon wafers during processing in some detail (143),(144). He shows that small burrs on vacuum chucks against which the wafer is pulled down can serve as initiation points for fracture. The easiest direction for fracture is along <111> directions, or, in {100} planes, along <110> directions. Fracture considerations loomed large in relation to the scribing operations associated with dicing as noted in section 2.3 of this chapter.

5.4 Thermal Expansion

Differential thermal expansion between the silicon substrate and various films such as silicon dioxide and silicon nitride, which are used in device fabrication is responsible for stresses which can generate and propagate dislocations. These effects are particularly important at corners in windows in the films (145). Swenson (146) has complied data on the thermal expansion coefficient of silicon over the temperature range 4 to 1000 K. In relatively pure silicon the linear thermal expansion coefficient is very small and positive below 18 K. It is negative between 18 and about 120 K, reaching a minimum of -0.7×10^{-6}/K about 80 K. Above 120 K, it increases monotonically reaching a value of about 4.2×10^{-6}/K at 850 K as shown in Figure 52. Values above this temperature are extrapolated. In heavily doped material, the thermal expansion is expected to be larger (147).

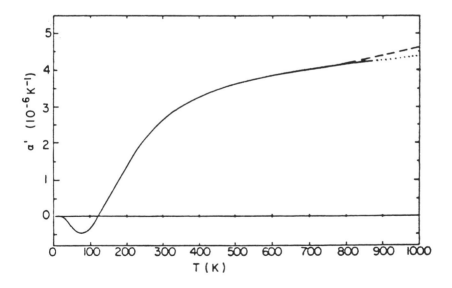

Figure 52: Linear coefficient of thermal expansion of silicon. [Reprinted with permission from Swenson, Reference 146. Copyright 1983 American Chemical Society.]

5.5 Thermal Conductivity

At room temperature, the thermal conductivity of silicon is essentially independent of impurity concentration except for concentrations greater than $10^{18}/cm^3$. Below room temperature the thermal conductivity increases until it reaches a maximum in the vicinity of 20 to 30 K, below which temperature it decreases. In this region, the thermal conductivity is strongly affected by the impurity concentration (148). As temperatures increase above room temperature, the thermal conductivity decreases until it reaches a value of about 0.22 W/cm^2·K at the melting point of silicon as shown in Figure 53 (149).

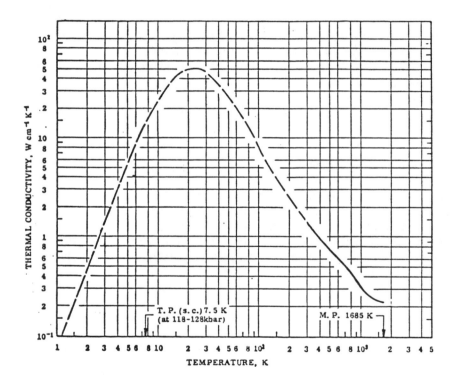

Figure 53: Thermal conductivity of high purity single crystal silicon as a function of temperature. Values for temperatures above 300 K are recommended values; below 300 K, typical values are shown merely to indicate the general trend. [Reprinted with permission from Ho et al., Reference 149. Copyright 1972 American Chemical Society.]

5.6 Hardness

Silicon has a hardness of 7 on the Moh scale. It can be scratched by diamond, alumina, and silicon carbide. Generally diamond is used to grind or cut silicon; diamond is also used if it is desired to mechanically polish silicon wafers. However, silicon today is most often polished by a chemical-mechanical process described more completely in the chapter on Silicon Wafer Processing.

Tables 10 and 11 show data on the effects of orientation on the hardness of silicon at room temperature as measured by a Knoop Indenter with a 100 g load (150). The hardness of silicon can be reduced by increasing the temperature, by passing a current through the sample (151), or by shining a strong source of infrared radiation on the sample (152).

5.7 Other Physical and Thermodynamic Properties

Yaws et al. have reviewed and analyzed many of the reported physical and thermodynamic properties of silicon (153). More recently, Desai has critically evaluated the thermodynamic properties of silicon (154). Table 12 provides data on a variety of physical and thermodynamic constants of silicon from these and other sources (155),(156).

Table 10

Effect of Crystal Orientation on

Hardness of Silicon (150)

(Knoop Indenter, 100-g load)

Face	Average	Range (depending on direction of long axis of indenter)
(111)	948	935-970
(110)	964	940-980
(100)	964	950-980

Table 11

Effect of Grinding Direction on Comparative Hardness of Silicon (150)

Grinding on Plane	In the direction toward	Comparative hardness
(101)	(001)	2
(001)	(101)	1.95
(112)	(101)	1.55
(101)	(112)	1.50
(001)	(111)	1.50
(111)	(001)	1.5
(101)	(111)	1.2
(111)	(101)	1.05

Table 12

Selected Physical and Thermodynamic Constants of Silicon

Property	Value
Atomic Weight	28.0855
Density (155)	2.3290730 g/cm^3
Melting Point (154)	1687 ± 2 K (1414°C)
Boiling Point (154)	3490 K (3217°C)
Enthalpy of Fusion (154)	50.250 kJ/mol
Enthalpy of Vaporization (154)	109.380 kJ/mol [at m.p.]
Enthalpy of Sublimation (154)	450.000 kJ/mol [at 25°C]
Critical Temperature (153)	4886°C
Critical Pressure (153)	537 Pa
Critical Volume (153)	232.6 cm^3/mol
Vapor Pressure (154)	
@ 1000 K	1.57 x 10^{-10} Pa
@ 1300 K	4.34 x 10^{-6} Pa
@ 1687 K	0.055 Pa
@ 2000 K	6.41 Pa
Liquid Heat Capacity (154)	27.200 J/mol•K [at m. p.]
Solid Heat Capacity (154)	20.007 J/mol•K [at 25°C]
Surface Tension (153)	0.736 N/m [at m. p.]
Liquid Viscosity (153)	0.88 mPa•s [at m.p.]
Expansion on Freezing (156)	(9 ± 1)% volume increase

6.0 REFERENCES

1. Runyan, W.R., *Silicon Semiconductor Technology*, McGraw Hill Book Co., New York (1965)

2. Packard, G.W., "Means for Receiving Intelligence by Electrical Waves," *U.S. Patent* 836,531 (1906)

3. Gudden, B., "Electrical Conductivity in Semiconductors," *S. B. Phys.-Med. Soz. Erlangen* 62:289 (1930); for a lucid discussion of the early development of conduction theories, see Wilson, A.H., *Theory of Metals*, 2nd Ed., Cambridge University Press, New York (1953)

4. Torrey, H.C., and Whitmer, C.A., *Crystal Rectifiers*, McGraw Hill, New York (1948)

5. See, for example, Shockley, W., *Electrons and Holes in Semiconductors*, Van Nostrand, Princeton, N.J. (1950)

6. For a general treatment of crystal structure, see Hannay, N.B., *Treatise on Solid State Chemistry*, Vol 1, *The Chemical Structure of Solids*, Plenum Press, New York (1973)

7. Bond, W.L., and Kaiser, W., "Interstitial versus Substitutional Oxygen in Silicon," *J. Phys. Chem. Solids* 16:44-7.45 (1960)

8. Deslattes, R.D., "The Avogadro Constant," *Ann. Rev. Phys. Chem.* 31:435-461 (1980)

9. ASTM Standard Test Methods F 26 for Orientation of a Semiconductive Single Crystal, *Annual Book of ASTM Standards* Vol. 10.05, American Society for Testing and Materials, Philadelphia, PA

10. Wood, E.A., *Crystal Orientation Manual*, Columbia Univ. Press, New York (1963)

11. ASTM Standard Methods F 847 for Measuring Crystallographic Orientation of Flats on Single Crystal Silicon Slices and Wafers by X-Ray Techniques, *Annual Book of ASTM Standards*, Vol. 10.05, American Society for Testing and Materials, Philadelphia, PA

12. ASTM Standard Test Methods E 82 for Determining the Orientation of a Metal Crystal, *Annual Book of ASTM Standards*, Vol. 03.03, American Society for Testing and Materials, Philadelphia, PA

13. Townley, D.O., "Optimum Crystallographic Orientation for Silicon Device Fabrication," *Solid State Technology* 16(1):43-47 (January 1973)

14. Drum, C.M., and Clark, C.A, "Anisotropy of Macrostep Motion and Pattern Edge Displacement during Growth of Epitaxial Silicon on Silicon near [100]," *J. Electrochem. Soc.* 117:1401-1405 (1970)

15. Drum, C.M., and Clark, C.A., "Geometrical Stability of Shallow Surface Depressions during Growth of (111) and (100) Epitaxial Silicon," *J. Electrochem. Soc.* 115:664-669 (1968)

16. Newman, R.C., "Defects in Silicon," *Rep. Prog. Phys.* 45:1163-1210 (1982)

17. *Aggregation Phenomena of Point Defects in Silicon*, E. Sirtl and J. Goorissen, eds, The Electrochemical Society, Pennington, NJ (1983)

18. *Defects in Silicon*, W.M. Bullis and L.C. Kimerling, eds, The Electrochemical Society, Pennington, NJ (1983)

19. *Thirteenth International Conference on Defects in Semiconductors*, L.C. Kimerling and J.M. Parsey, Jr., eds, The Metallurgical Society of the AIME, Warrendale, PA (1984)

20. *Defects in Semiconductors - II*, S. Mahajan and J.W. Corbett, eds, North-Holland, New York (1983)

21. *Semiconductor Silicon 1986*, H.R. Huff, T. Abe, and B. Kolbesen, eds, The Electrochemical Society, Pennington, NJ (1986)

22. Antoniadis, D.A., and Moskowitz, I., "Diffusion of Substitutional Impurities in Silicon at Short Oxidation Times- An Insight into Point Defect Kinetics," *J. Appl. Phys.* 53:6788-6796 (1982)

23. Gösele, U., and Tan, T.Y., "The Nature of Point Defects and Their Influence on Diffusion Processes in Silicon at High Temperatures," in Ref. 20, pp. 45-59

24. Trumbore, F.A., "Solid Solubilities of Impurity Elements in Germanium and Silicon," *Bell Syst. Tech. J.* 39:205-234 (1960)

25. Boltaks, B.I., and Hsüeh, S.-Y., "Diffusion, Solubility and the Effect of Silver Impurities on Electrical Properties of Silicon," *Sov. Phys. - Solid State* 2:2383-2388 (1960-61)

26. Yoshida, M., and Furusho, K., "Behavior of Nickel as an Impurity in Silicon," *Japan J. Appl. Phys.* 3:521-529 (1964)

27. Svob, L., "Solubility and Diffusion Coefficient of Sodium and Potassium in Silicon," *Solid-State Electron.* 10:991-996 (1967)

28. Weber, E.R., "Transition Metals in Silicon," *Appl. Phys.* A 30:1-22 (1983)

29. Nozaki, T., Yatsurugi, Y., Akiyama, N., Endo, Y., and Makida, Y., "Behavior of Light Impurity Elements in the Production of Semiconductor Silicon," *J. Radioanal. Chem.* 19:109-128 (1974)

30. Bourret, A., "Oxygen Aggregation in Silicon," in Ref. 19, pp. 129-146

31. Tiller, W.A., Hahn, S., and Ponce, F.A., "Thermodynamic and Kinetic Considerations on the Equilibrium Shape for Thermally Induced Microdefects in Czochralski Silicon," *J. Appl. Phys.* 59:3255-3266 (1986)

32. Barraclough, K.G., and Wilkes, J.G., "The Incorporation and Disposition of Carbon in Czochralski Silicon," in Ref. 21, pp. 889-902

33. Föll, H., Gösele, U., and Kolbesen, B.O., "The Formation of Swirl Defects in Silicon by Agglomeration of Self-Interstitials," *J. Cryst. Growth* 49: 90-108 (1977); see also Föll, H., Gösele, U., and Kolbesen, B.O., "Swirl-Defects in Silicon" in *Semiconductor Silicon 1977*, H.R. Huff and E. Sirtl, eds, The Electrochemical Society, Pennington, NJ (1977)

34. Kolbesen, B.O. and Mühlbauer, "Carbon in Silicon: Properties and Impact on Devices," *Solid-State Electron.* 25:759-775 (1982)

35. Scott, W., "Infrared Spectra of New Acceptor Levels in Indium- or Aluminum-Doped Silicon," *Appl. Phys. Lett.* 32:540-542 (1978)

36. Poindexter, E.H., Gerardi, G.J., Rueckel, M.-E., Caplan, P.J., Johnson, N.M., and Biegelsen, D.K., "Electronic Traps and P_b Centers at the Si/SiO_2 Interface: Band-Gap Energy Distribution," *J. Appl. Phys.* 56:2844-2849 (1984); Johnson, N.M., and Biegelsen, D.K., "Identification of Deep-Gap States in a-Si:H by Photodepopulation-Induced Electron-Spin Resonance," *Phys. Rev.* B 31:4066-4069 (1985)

37. Johnson, N.M., Herring, C., and Chadi, D.J., "Interstitial Hydrogen and Neutralization of Shallow-Donor Impurities in Single-Crystal Silicon," *Phys. Rev. Lett.* 56:769-772 (1986)

38. Pankove, J.I., Carlson, D.E., Berkeyheiser, J.E., and Wance, R.O., "Neutralization of Shallow Acceptor Levels in Silicon by Atomic Hydrogen," *Phys. Rev. Lett.*, 51:2224-2225 (1983); Johnson, N.M., "Mechanism for Hydrogen Compensation of Shallow-Acceptor Impurities in Single-Crystal Silicon," *Phys. Rev.*, B 31:5525-5528 (1985); DeLeo, G.G., and Fowler, W.G., "Hydrogen-Acceptor Pairs in Silicon: Pairing Effect on the Hydrogen Vibrational Frequency," *Phys. Rev.* B 31:6861-6864 (1985)

39. Johnson, N.M., and Hahn, S.K., "Hydrogen Passivation of the Oxygen-Related Thermal-Donor Defect in Silicon," *Appl. Phys. Lett.* 48:709-711 (1986)

40. Pensl, G., Roos, G., Holm, C., Sirtl, E., and Johnson, N.M., "Hydrogen Neutralization of Chalcogen Double Donors in Silicon," *Appl. Phys. Lett.* 51:451-453 (1987)

41. Schnegg, A., Grunder, M., and Jacob, H., "On the Polishing of Silicon: Acceptor Compensation by Atomic Hydrogen," in Ref. 21, pp. 198-205

42. Fonash, S.J., and Rohaatgi, A., "RIE Damage and its Control in Silicon Processing," in *Emerging Semiconductor Technology*, ASTM STP 960, D.C. Gupta and P.H. Langer,

eds, American Society for Testing and Materials, Philadelphia, (1986)

43. Horn, M.W., Heddleson, J.M., and Fonash, S.J., "Permeation of Hydrogen into Silicon during Low-Energy Hydrogen Ion Beam Bombardment," *Appl. Phys. Lett.* 51:490-492 (1987)

44. Stein, H.J., "Infrared Absorption Band for Substitutional Nitrogen in Silicon," *Appl. Phys. Lett.* 47:1339-1341 (1985)

45. Abe, T., Masui, T., Harada, and H., Chikawa, J., "The Characteristics of Nitrogen in Silicon Crystals," in *VLSI Science and Technology/1985*, W.M. Bullis and S. Broydo, eds, The Electrochemical Society, Pennington, NJ (1985)

46. Watanabe, M., Usami,T. Muraoka, H., Matsuo, S., Imanishi. Y., and Nagashima, H., "Oxygen-Free Silicon Crystal Grown from Silicon Nitride Crucible," in *Semiconductor Silicon 1981*, H.R. Huff, R.J. Kriegler, Y. Takeishi, eds, The Electrochemical Society, Pennington, NJ (1981)

47. Conzelmann, H., Graff, K., and Weber, E.R., "Chromium and Chromium-Boron Pairs in Silicon," *Appl. Phys.* A 30: 169-175 (1983)

48. Ourmazd, A., and Schröter, W., "Gettering of Metallic Impurities in Silicon," in *Impurity Diffusion and Gettering in Silicon*, R.B. Fair, C.W. Pearce, and J. Washburn, eds, Materials Research Society, Pittsburgh (1985)

49. Ward, P.J., "A Survey of Iron Contamination in Silicon Substrates and Its Impact on Circuit Yield," *J. Electrochem. Soc.* 129:2573-2576 (1982)

50. Stewart, D.W., and Newton, D.C., "Determination of Iron in Semiconductor-Grade Silicon by Furnace Atomic Absorption Spectrometry," *Analyst* 108:1450-1458 (1983)

51. McGillivray, I.G., Robertson, J.M., and Walton, A.J., "Effects of Process Chemical Purity on MOS Capacitor Electrical Parameters," in Ref. 21, pp. 999-1010

52. Takizawa, R., Nakanishi, T., and Ohsawa, A., "Degradation of Metal-Oxide-Semiconductor Devices Caused by Iron Impurities on the Silicon Wafer Surface," J. Appl. Phys. 62:4933-4935 (1987)

53. Huber, A., Rath, H.J., and Eichinger, P., "Sub-ppm Monitoring of Transition Metal Contamination on Silicon Wafer Surfaces by VPO-TXRF," in *Diagnostic Techniques for Semiconductor Materials and Devices*, T.J. Schaffner and D.K. Schroder, eds, The Electrochemical Society, Pennington, NJ (1988); Hockett, R.S., Baumann, S., and Schemmel, E., "An Evaluation of Ultra-Surface (<3 nm), Trace ($10^{11}/cm^2$) Impurity Analysis of Silicon Using a New X-Ray Technique," ibid., pp. 113-130

54. Grimmeiss, H.G., Janzen, E., Ennen, H., Schirmer, O., Schneider, J., Wörmer, R., Holm, C., Sirtl, E., and Wagner, P., "Tellurium Donors in Silicon," *Phys. Rev.* B 24:4571-4580 (1981)

55. Salih, A.S.M., Kim, H.J., Davis, R.F., and Rozgonyi, G.A., "Extrinsic Gettering via the Controlled Introduction of Misfit Dislocations," in *Semiconductor Processing*, ASTM STP 850, D.C. Gupta, ed, American Society for Testing and Materials, Philadelphia (1984)

56. Bean, K.E., Lindberg, K., and Rozgonyi, G.A., "Defect Engineering," in *Proc. First Int. Symp. on Advanced Materials for ULSI*. M.P. Scott, Y. Ahasaka, and R. Reif, eds, The Electrochemical Society, Pennington, NJ (1988)

57. Frank, F.C., "Crystal Dislocations - Elementary Concepts and Definitions," *Phil. Mag.* 42:809-819 (1951)

58. Rhodes, R.G., *Imperfections and Active Centers in Semiconductors*, Pergamon Press, Oxford, U.K. (1964)

59. Dash, W.C., "Copper Precipitation on Dislocations in Silicon," *J. Appl. Phys.* 27:1193-1195 (1956)

60. ASTM Standard Test Method F 47 for Crystallographic Perfection of Silicon by Preferential Etch Techniques, *Annual Book of ASTM Standards*, Vol. 10.05, American Society for Testing and Materials, Philadelphia, PA

61. Dash, W.C., "Silicon Crystals Free of Dislocations," *J. Appl. Phys.* 29:736-737 (1958); , "Growth of Silicon Crystals Free from Dislocations," ibid. 30:459-474 (1959)

62. Dash, W.C., "Dislocations in Silicon and Germanium Crys-

tals," in *Properties of Elemental and Compound Semiconductors*, Metallurgical Society of the AIME, Warrendale, PA (1960)

63. Rozgonyi, G.A., and Seidel, T.E., "Surface versus Bulk Nucleated Oxidation-Induced Stacking Faults in Silicon Wafers," *J. Cryst. Growth* 38:359-363 (1977).

64. Inoue, N., Osaka, J., and Wada, K., "Oxide Micro-Precipitates in As-Grown Cz Silicon," *J. Electrochem. Soc.* 129: 2780-2788 (1982)

65. ASTM Standard Test Method F 416 for Detection of Oxidation Induced Defects in Polished Silicon Wafers, *Annual Book of ASTM Standards*, Vol. 10.05, American Society for Testing and Materials, Philadelphia, PA

66. Blakemore, J.S., *Semiconductor Statistics*, Pergamon Press, Oxford, U.K. (1962)

67. Dexter, R.N., Zeiger, H.J., and Lax, B., "Cyclotron Resonance Experiments in Silicon and Germanium," *Phys. Rev.* 104:637-644 (1956)

68. Barber, H.D., "Effective Mass and Intrinsic Concentration in Silicon," *Solid-State Electron.* 10:1039-1051 (1967)

69. Thurber, W.R., and Buehler, M.G., in "Semiconductor Measurement Technology," NBS Special Publication 400-4, Sect. 3.7 (1974)

70. MacFarlane, G.G., McLean, T.P., Quarrington, J.E., and Roberts, V., "Fine Structure in the Absorption-Edge Spectrum of Si," *Phys. Rev.* 111:1245-1254 (1958)

71. Cohen, M.L., and Bergstresser, T.K., "Band Structures and Pseudopotential Form Factors for Fourteen Semiconductors of the Diamond and Zinc-Blende Structures," *Phys. Rev.* 141:789-796 (1966)

72. Herman, F., "Electronic Energy Band Structure of Silicon and Germanium," *Proc. IRE* 43:1703-1732 (1955)

73. Irvin, J.C., "Resistivity of Bulk Silicon and of Diffused Layers in Silicon," *Bell Syst. Tech. J.* 41:387-410 (1960)

74. Caughey, D.M., and Thomas, R.E., "Carrier Mobilities in Silicon Empirically Related to Doping and Field," *Proc. IEEE* 55:2192-2193 (1967)

75. Wagner, S., "Diffusion of Boron from Shallow Ion Implants in Silicon," *J. Electrochem. Soc.* 119:1570-1576 (1972)

76. Li, S.S., and Thurber, W.R., "The Dopant Density and Temperature Dependence of Electron Mobility and Resistivity in n-Type Silicon," *Solid-State Electron.* 20:609-616 (1977); see also Li, S.S., "The Dopant Density and Temperature Dependence of Electron Mobility and Resistivity in N-Type Silicon," NBS Special Publication 400-33 (1977)

77. Li, S.S., "The Dopant Density and Temperature Dependence of Hole Mobility and Resistivity in Boron Doped Silicon," *Solid-State Electron.* 21:1109-1117 (1978); see also Li, S.S., "The Theoretical and Experimental Study of the Temperature and Dopant Density Dependence of Hole Mobility, Effective Mass, and Resistivity in Boron-Doped Silicon," NBS Special Publication 400-47 (1979)

78. Linares, L.C., and Li, S.S., "An Improved Model for Analyzing Hole Mobility and Resistivity in p-Type Silicon Doped with Boron, Gallium, and Indium," *J. Electrochem. Soc.* 128:601-608 (1981)

79. ASTM Standard Method F 84 for Measuring Resistivity of Silicon Slices with a Collinear Four-Probe Array, *Annual Book of ASTM Standards*, Vol. 10.05, American Society for Testing and Materials, Philadelphia, PA

80. ASTM Standard Test Method F 673 for Measuring Resistivity of Semiconductor Slices with a Noncontact Eddy-Current Gage, *Annual Book of ASTM Standards*, Vol. 10.05, American Society for Testing and Materials, Philadelphia, PA

81. ASTM Standard Test Methods F 43 for Resistivity of Semiconductor Materials, *Annual Book of ASTM Standards*, Vol. 10.05, American Society for Testing and Materials, Philadelphia, PA

82. Bullis, W.M., Brewer, F.H., Kolstad, C.D., and Swartzendruber, L.J., "Temperature Coefficient of Resistivity of Silicon and Germanium near Room Temperature," *Solid-State Electron.* 11:639-646 (1968); see also Bullis, W.M., "Standard Measurements of the Resistivity of Silicon by the Four-Probe Method," NBS IR 74-476 (1974) (Available from NTIS as COM 74-11576)

83. Thurber, W.R., Mattis, R.L., Liu, Y.M., and Filliben, J.J., "Resistivity-Dopant Density Relationship for Boron-Doped Silicon," *J. Electrochem. Soc.* 127:2291-2294 (1980)

84. Thurber, W.R., Mattis, R.L., Liu, Y.M., and Filliben, J.J., "Resistivity-Dopant Density Relationship for Phosphorus Doped Silicon," *J. Electrochem. Soc.* 127:1807-1812 (1980)

85. Thurber, W.R., Mattis, R.L., Liu, Y.M., and Filliben, J.J., "The Relationship between Resistivity and Dopant Density for Phosphorus- and Boron-Doped Silicon," NBS Spec. Publ. 400-64 (1981)

86. ASTM Standard Practice F 723 for Conversion between Resistivity and Dopant Density for Boron-Doped and Phosphorus-Doped Silicon, Annual Book of ASTM Standards, Vol. 10.05, American Society for Testing and Materials, Philadelphia, PA

87. Lin, J.F., Li, S.S., Linares, L.C., and Teng, K.W., "Theoretical Analysis of Hall Factor and Hall Mobility in p-Type Silicon," *Solid-State Electron.* 24:827-833 (1981)

88. Gibbons, J.F., "Papers on Carrier Drift Velocities in Silicon at High Electric Field Strengths," *IEEE Trans. Electron Devices* ED-14:37 (1967)

89. ASTM Standard Test Methods F 76 for Measuring Resistivity and Hall Coefficient and Determining Hall Mobility in Single-Crystal Semiconductors, *Annual Book of ASTM Standards*, Vol. 10.05, American Society for Testing and Materials, Philadelphia, PA

90. Milnes, A.G., *Deep Impurities in Semiconductors*, John Wiley & Sons, New York (1973)

91. Jantsch, W., Wünstel, K., Kumagai, O., and Vogl, P., "Deep Levels in Semiconductors: A Quantitative Criterion," *Phys. Rev.* B 25:5515-5518 (1982)

92. Jaros, M., *Deep Levels in Semiconductors*, Adam Hilger, Ltd., Bristol, U.K. (1982)

93. Thurber, W.R., Lewis, D.C., and Bullis, W.M., "Resistivity and Carrier Lifetime in Gold-Doped Silicon," AFCRL-TR-73-0107, 1973, (Available from NTIS as AD 760150)

94. Brückner, B., "Electrical Properties of Gold-Doped Silicon," *Phys. Stat. Sol.* A 4:685-692 (1971)

95. Lang, D.V., Grimmeiss, H.G., Meijer, E., and Jaros, M., "Complex Nature of Gold-Related Deep Levels in Silicon," *Phys. Rev.* B, 22:3917-3934 (1980)

96. Lu, L.S., Nishida, T., and Sah, C.-T., "Thermal Emission and Capture Rates of Holes at the Gold Donor Level in Silicon," *J. Appl. Phys.* 62:4773-4780 (1987)

97. ASTM Standard Test Methods F 42 for Conductivity Type of Extrinsic Semiconducting Materials, *Annual Book of ASTM Standards*, Vol. 10.05, American Society for Testing and Materials, Philadelphia, PA

98. See, for example, Tauc, J., *Photo and Thermoelectric Effects in Semiconductors*, Pergamon Press, New York (1962), which provides extensive citations to earlier literature

99. Salzberg, C.D., and Villa, J.J., "Infrared Refractive Indexes of Silicon, Germanium, and Modified Selenium Glass," *J. Opt. Soc. Am.* 47:244-246 (1957)

100. Philipp, H.R., and Taft. E.A., "Optical Constants of Silicon in the Region 1 to 10 eV," *Phys. Rev.* 120: 37-38 (1960)

101. Phillips, W.E., "Interpretation of Steady-State Surface Photovoltage Measurements in Epitaxial Semiconductor Layers," *Solid-State Electron.* 15:1097-1102 (1972)

102. Runyan, W.R., Herrmann, W., and Jones, L., "Crystal Development Program," Final Report on Contract AF33(600)-33736, Texas Instruments Incorporated (1958)

103. Dale, B., "Research on Efficient Photovoltaic Solar Energy Converters," Final Report, Contract AF10(604)5585, Transitron, Inc. (1960)

104. ASTM Test Method F 28 for Measuring the Minority Carrier Lifetime in Bulk Germanium and Silicon, *Annual Book of ASTM Standards*, Vol. 10.05, American Society for Testing and Materials, Philadelphia, PA

105. Dash, W.C., and Newman, R., "Intrinsic Optical Absorption in Single-Crystal Germanium and Silicon at 77°K and 300°K," *Phys. Rev.* 99:1151-1155 (1955)

106. Nartowitz, E.S., and Goodman, A.M., Evaluation of Silicon Optical Absorption Data for Use in Minority-Carrier-Diffusion Length Measurements by the SPV Method," *J. Electrochem. Soc.* 132:2992-2997 (1985)

107. Johnson, F.A., "Lattice Absorption Bands in Silicon," *Proc. Phys. Soc.* 73:265-272 (1959)

108. Pajot, B., Stein, H.J., Cales, B., and Naud, C., "Quantitative Spectroscopy of Interstitial Oxygen in Silicon," *J. Electrochem. Soc.* 132:3034-3037 (1985)

109. Newman, R.C., and Willis, J.B., "Vibrational Absorption of Carbon in Silicon," *J. Phys. Chem. Solids* 26:373-379 (1965)

110. ASTM Standard Test Method F 123 for Substitutional Carbon Content of Silicon by Infrared Absorption, *Annual Book of ASTM Standards*, Vol. 10.05, American Society for Testing and Materials, Philadelphia

111. ASTM Standard Test Method F 121 for Interstitial Atomic Oxygen Content of Silicon by Infrared Absorption, *Annual Book of ASTM Standards*, Vol. 10.05, American Society for Testing and Materials, Philadelphia, PA (This method was withdrawn in 1989 and replaced by ASTM Standard Test Method F 1188 for Interstitial Atomic Oxygen content of Silicon by Infrared Absorption, *ibid*. A related procedure is ASTM Standard Test Method

F 1189 for using Computer Assisted Infrared Spectrophotometry to measure the interstitial oxygen content of silicon slices polished on both sides, *ibid*.)

112. For oxygen, see Bullis, W.M., Watanabe, M., Baghdadi, A., Li, Y.-Z., Scace, R.I., Series, R.W., and Stallhofer, P., "Calibration of Infrared Absorption Measurements of Interstitial Oxygen Concentration in Silicon," in Ref. 21, pp. 166-180; for carbon, see Inoue, N., Arai, T., Nozaki, T., Endo, R., and Mizuma, K., High Reliability Infrared Measurements of Oxygen and Carbon in Silicon," in *Emerging Semiconductor Technology*, ASTM STP 960, D.C. Gupta, and P.H. Langer, eds, American Society for Testing and Materials, Philadelphia, PA (1986)

113. Baghdadi, A., Bullis, W.M., Croarkin, M.C., Li Y.-Z., Scace, R.I., Series, R.W., Stallhofer, P., and Watanabe, M., "Interlaboratory Determination of the Calibration Factor for the Measurement of the Interstitial Oxygen Content in Silicon by Infrared Absorption," *J. Electrochem. Soc.* 136:2015-2024 (1989); see also Baghdadi, A., Scace, R.I., and Walters, E.J., eds, "Database for and Statistical Analysis of the Calibration Factor for the Measurement of the Interstitial Oxygen Content of Silicon by Infrared Absorption," NIST Spec. Publn. 400-82 (1989)

114. Itoh, Y., Nozaki, T., Masui, T., and Abe, T., "Calibration Curve for Infrared Spectrophotometry of Nitrogen in Silicon," *Appl. Phys. Lett.* 47:486-487 (1985)

115. Pajot, B., "Characterization of Oxygen in Silicon by Infrared Absorption — A Review," *Analusis* 5: 293-303 (1977)

116. Fan, H.Y., and Becker, M., "Infrared Optical Properties of Silicon and Germanium," in *Semiconducting Materials*, H.K. Henisch, ed, Butterworths Scientific Publications, London, U.K. (1951)

117. Spitzer, W., and Fan, H.Y., "Infrared Absorption in n-Type Silicon," *Phys. Rev.* 108:268-271 (1957)

118. Walitzki, J., Rath, H.-J., Reffle, J., Pahlke, S., and Blätte, M., "Control of Oxygen and Precipitation Behavior of Heavily Doped Silicon Substrate Materials," in Ref 21, pp 86-99

119. Tsuya, H., Kanamori, M., Takeda, M., and Yasuda, K., "Infrared Optical Measurement of Interstitial Oxygen Content in Heavily Doped Silicon Crystals," in *VLSI Science and Technology/1985*, W.M. Bullis and S. Broydo, eds, The Electrochemical Society, Pennington, NJ (1985)

120. Kolbesen, B., "Simultaneous Determination of the Total Content of Boron and Phosphorus in High Resistivity Silicon by IR Absorption," *Appl. Phys. Lett.* 27:353-355 (1975)

121. Baber, S.C., "Net and Total Shallow Impurity Analysis of Silicon by FTIR Spectroscopy," *Thin Solid Films* 72:201-210 (1980)

122. Tajima, M., "Quantitative Impurity Analysis in Si by the Photoluminescence Technique," in *Japan Ann. Rev. in Electronics, Computers, and Telecommunications—Semiconductor Technology 1982*, (J. Nishizawa, ed), Tokyo: OHM (1982), pp 1-12; see also Tajima, M., Masui, T., Abe, T., and Iizuka T., "Photoluminescence Analysis of Silicon Crystals," in *Semiconductor Silicon* 1981, H.R. Huff, R.J. Kriegler, Y. Takeishi, eds, The Electrochemical Society, Pennington, NJ (1981), pp 72-89; see also Colley, P. McL., and Lightowlers, E.C., "Calibration of the Photoluminescence Technique for Measuring B, P and Al Concentrations in Si in the Range 10^{12} to 10^{15} cm^{-3} using Fourier Transform Spectroscopy," Semiconductor Sci. Tech. 2:157-166 (1987)

123. Proposed Test Method (p. 213) for Photoluminescence Analysis of Single Crystal Silicon for III-V Impurities, to be published in *Annual Book of ASTM Standards*, Vol. 10.05, American Society for Testing and Materials, Philadelphia, PA

124. ASTM Standard Test Method F 398 for Majority Carrier Concentration in Semiconductors by Measurement of Wavenumber or Wavelength of the Plasma Resonance Minimum, *Annual Book of ASTM Standards*, Vol 10.05, American Society for Testing And Materials, Philadelphia, PA

125. Allen, F.G., Emissivity at 0.65 Micron of Silicon and Germanium at High Temperatures," *J. Appl. Phys.* 101: 1676-1678 (1956)

126. Rea, S.N., "Large Area Czochralski Silicon," *Texas Instruments Report* 03-76-31, ERDA/JPL 9554475-76II (1976)

127. Herring, C., "Elastic Constants," in *Fundamental Formulas of Physics*, D.H. Menzel, ed, Dover, New York (1960), Chapter 25, Section 2, (Originally published in 1955); see also Smith, C.S., "Piezoresistance Effect in Germanium and Silicon," *Phys. Rev.* 94:42-49 (1954)

128. Hall, J.J., "Electronic Effects in the Elastic Constants of Silicon," *Phys. Rev.* 116:756-761 (1967)

129. McSkimin, H.J., Bond, W.L., Buehler, E. and Teal, G.K., "Measurement of the Elastic Constants of Silicon Single Crystals and Their Thermal constants," *Phys. Rev.* 83: 1080 (L) (1951)

130. Riney, T.D., "Residual Thermoelastic Stress in Bonded Silicon Wafers," *J. Appl. Phys.* 32:454-460 (1961)

131. Wortman, J.J., and Evans, R.A., "Young's Modulus, Shear Modulus, and Poisson's Ratio in Silicon and Germanium," *J. Appl. Phys.* 36:153-156 (1965)

132. Tufte, O.N., and Stelzer E.L., "Piezoresistance Properties of Heavily Doped n-Type Silicon," *Phys. Rev.* 133:A1705-A1716 (1964)

133. Sobey, A.E., (unpublished) cited in Ref. 1, p 182

134. Kondo, Y., "A Graphical Representation of the Piezoresistance Coefficients in Silicon," *IEEE Trans. Electron Devices* ED-29, 64-70 (1982)

135. Yonenaga, I., Sumino, K., and Hoshi, K., "Mechanical Strength of Silicon Crystals as a Function of the Oxygen Concentration," *J. Appl. Phys.* 56:2346-2350 (1984)

136. Sylwestrowicz, W.D., "Mechanical Properties of Single Crystals of Silicon," *Phil. Mag.* (ser. 8) 7:1825-1845 (1962)

137. Pearson, G.L., Read W.T., Jr., and Feldman, W.L., "Deformation and Fracture of Small Silicon Crystals," *Acta Met.* 5:181-191 (1957)

138. Kondo, Y., "Plastic Deformation and Preheat Treatment Effects in Cz and FZ Silicon Crystals," in *Semiconductor Silicon/1981*, (H.R. Huff, R.J. Kreigler, and Y. Takeishi, eds), The Electrochemical Society, Pennington, NJ (1981)

139. Hu, S.M., and Patrick, W.J., "Effect of Oxygen on Dislocation Movement in Silicon," *J. Appl. Phys.* 46:1869-1874 (1975)

140. Leroy, B., and Plougonven, C., "Warpage of Silicon Wafers," *J. Electrochem. Soc.* 127:961-970 (1980)

141. Dyer, L.D., Huff, H.R., and Boyd, W.W., "Plastic Deformation in Central Regions of Epitaxial Silicon Slices," *J. Appl. Phys.* 42:5680-5688 (1971)

142. Lee, C.-O., and Tobin, P.J., "The Effect of CMOS Processing on Oxygen Precipitation, Wafer Warpage, and Flatness," *J. Electrochem. Soc.* 133:2147-2152 (1986)

143. Dyer, L.D., and Medders, J.B., "Defects Caused by Vacuum Chuck Burrs in Silicon Wafer Processing," in *VLSI Science and Technology/1984*, K.E. Bean and F.A. Rozgonyi, eds, The Electrochemical Society, Pennington, NJ (1984)

144. Dyer, L.D., "Damage Aspects of Ingot-to-Wafer Processing," in *Emerging Semiconductor Technology*, ASTM STP 960, D.C. Gupta and P.H. Langer, eds., American Society for Testing and Materials, Philadelphia, PA

145. Ahlgren, D.C., Domenicucci, A.G., Karcher, R., Mader, S.R., Popniak, M.R., "Defect Induced Leakage from Bipolar Transistor Isolation Processing," in Ref. 18, pp 472-481

146. Swenson, C.A., "Recommended Values for the Thermal Expansivity of Silicon from 0 to 1000 K," *J. Phys. Chem. Ref. Data* 12:179-182 (1983)

147. Navikova, S.I., "Effect of Impurities on the Thermal Expansion of Silicon," *Sov. Phys. - Solid State* 6:269 (1964-65)

148. Holland, M.G., "The Effect of Oxygen on the Thermal Conductivity of Silicon," in *Proc. 7th Int. Conf. on Low Temp. Phys.*, Univ. of Toronto Press, Toronto (1961)

149. Ho, C.Y., Powell, R.W., and Liley, P.E., "Thermal Conductivity of the Elements," *J. Phys. Chem. Ref. Data* 1: 279-421 (silicon is found on page 394) (1972); for detailed information on the basis of the recommended values, see loc. cit., 3 (Supp. 1): 587-605 (1974)

150. Giardini, A.A., "A Study of the Directional Hardness in Silicon," *Am. Mineralogist* 43: 957-969 (1958)

151. Westbrook, J.H., and Gilman, J.J., "An Electromechanical Effect in Semiconductors," *J. Appl. Phys.* 33:2360-2369 (1962)

152. Kuczynski, G.C., and Hochman, R.F., "Light-Induced Plasticity in Semiconductors," *Phys. Rev.* 108:946-948 (1957)

153. Yaws, C.L., Dickens, L.L., Lutwack, R., and Hsu, G., "Semiconductor Industry Silicon: Physical and Thermodynamic Properties," *Solid State Tech.* 24 (1): 87-92 (Jan. 1981)

154. Desai, P.D., "Thermodynamic Properties of Iron and Silicon," *J. Phys. Chem. Ref. Data* 15:967-983 (1986)

155. Hennis, I., "Precision Density Measurement of Silicon," *J. Res. Nat. Bur. Stds.* 68A: 529-533 (1964)

156. Logan, R.A., and Bond, W.L., "Density Change in Silicon on Melting," *J. Appl. Phys.* 30:322 (1959)

7

Oxygen, Carbon and Nitrogen in Silicon

William C. O'Mara

1.0 INTRODUCTION

The technology for the growth of silicon crystals was developed with a view towards controlled incorporation of electrically active impurities. In Chapter 3, Figure 6, we can see the relationship established between solid solubility and segregation coefficient for the electrically active elements as well as for metals. Most elements fall on a line, originally mentioned by Fischler for both silicon and germanium, that corresponds to $C_S = 0.1\, k_0$ (1).

This same figure shows the special behavior of the elements oxygen, carbon and nitrogen. These electrically inactive impurities have significantly lower solubilities than do other elements. Yatsurugi and coworkers have explained the low solubility of these three elements in terms of the heat of dissociation of the corresponding 'silicides', SiO, SiC, and Si_3N_4 (2). The heats of dissociation of these compounds are as large as the heat of vaporization of refractory metals, being 160, 190, and 160 kcal/mol respectively. Solubility is expressed by the relation

$$M = \exp\left[\frac{\Delta S}{R}\right] \exp\left[\frac{-\Delta H}{RT}\right] \tag{1}$$

where

M = solute solubility in mole fraction
ΔS = entropy of solution
R = gas constant
ΔH = heat of solution
T = absolute temperature

For a solid solution, ΔH includes the breaking of silicon--silicon bonds. The larger the heat of vaporization of the solute element, the higher the corresponding ΔH and the lower the solubility in the solid. In addition, the difference in atomic size between silicon and the light impurity atoms further mitigates against high solubilities.

Oxygen and carbon have played important, if sometimes unsuspected roles in determining the properties of silicon wafers. The normal crystal growth environment for silicon, involving a quartz crucible and graphite heater, ensured incorporation of these impurities. In the past ten years, attention has been focused on control of these impurities and understanding of their effects on the properties of silicon. More recently, nitrogen has been deliberately introduced into the lattice. This chapter is a survey of the properties of these three elements in silicon. Their influence on some properties, such as mechanical strength and metallic impurity removal, is so great that this has become a distinct area for research and development. However, those immortal words of the inveterate scholar still apply - much more investigation is required for a complete understanding.

2.0 PROPERTIES OF DISSOLVED OXYGEN IN SILICON

Beginning with Kaiser and Keck (3), the study of oxygen in silicon has occupied researchers for thirty years. During this time, a number of careful investigations have been made of the basic properties of dissolved or dispersed oxygen in the material, such as solubility and diffusivity as a function of temperature, segregation coefficient at the melting point, and so on. It is a fact that at least two distinct values of each property have been reported! In reviewing the basic properties of oxygen in silicon, it is important to remember that the previous thermal history of the sample and the details of the experimental conditions are part of the explanation for the observed duality. The reason why the influence of thermal history can be so decisive on observed properties is a matter of some debate; in this chapter, we will discuss

2.1 Solubility

The equilibrium value of the solid solubility of oxygen in silicon at the melting point is generally accepted to be ~1×10^{18} atom/cm^3, when differences in quantitative measurement techniques are taken into account. However, the solubility as a function of temperature differs markedly with experimental method, as shown in Figure 1. Curves corresponding to low values of oxygen solubility were reported by Logan and Peters (4), and by Bean and Newman (5). In the former case, oxygen was heated at 450°C to form thermal donors. The total oxygen concentration was deduced using a power law relation between thermal donor and total oxygen concentration. The samples of Bean and Newman contained substantial quantities of dissolved carbon. Oxygen content was measured by infrared absorption after sample heat treatment. It is possible that the presence of carbon enhanced oxygen clustering and precipitation. How this might occur will be discussed in the section on thermal donor formation.

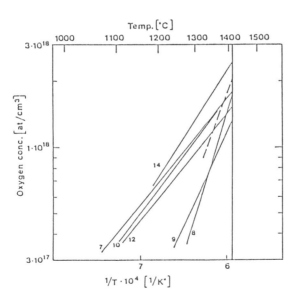

Figure 1. Solubility of oxygen in silicon as a function of temperature (13). (Reprinted by permission of the publisher, The Electrochemical Society, Inc.)

Work of Hrostowski and Kaiser (6), Takano and Maki (7), Mikkelson (8), and Craven (9), give a higher value of solubility as a function of temperature. In addition, Mikkelsen has reported a metastable solubility for oxygen in-diffusion in oxidizing conditions (dashed curve in Figure 1). These samples showed reduced solubility when annealed in vacuum, corresponding to the solid solubility curve. He suggests a second metastable species which exists under oxidizing conditions.

More recent work by Itoh and Nozaki employed charged particle activation to determine oxygen concentration after extended heating of samples in either oxygen or argon ambient (10). They determined the solubility over the temperature range 1000-1375°C to be

$$L = 9.3 \times 10^{21} \exp\left[\frac{-27.6 kcal \cdot mol^{-1}}{RT}\right] atom/cm^3 \quad (2)$$

where L is solubility in atomic fraction. The solubility can also be expressed as

$$L = \exp\left[\frac{\Delta S}{R}\right] \exp\left[\frac{-\Delta H}{RT}\right] \quad (3)$$

ΔS and ΔH are the entropy and enthalpy of solution, R is the gas constant, and T is the absolute temperature. This equation is similar in form to equation 1. They obtained values of

ΔH = 27.6 kcal x mol^{-1}

ΔS = 3.3 cal x mol^{-1} x K^{-3}

An important point made by these authors is that the solubility of oxygen in silicon is determined by an equilibrium with the oxide which is stable at a given temperature. They propose that the stable form of silicon oxide changes from SiO_2 to SiO above about 1200°C. Therefore the solubility versus temperature curve might be expected to bend at this point. The bending could not be clearly observed in the experiment. Similar results for the heat of solution were obtained by Gass and coworkers (11).

Table 1 shows a summary of results obtained by various authors for the heat of solution of oxygen in silicon, corresponding to the slope of the solubility curve. The fact that the curves converge at the melting point indicates that the experiments were all carefully performed, and that the experimental conditions determine the results (12). Figure 1 shows a solubility curve calculated by Carlberg (13). For a discussion of the calculation, see section 2.4.

Table 1

Heat of Solution of Oxygen in Silicon

Author	Method	Temperature Range (°C)	Heat of Solution (kcal/mol)	Ref.
Logan and Peters	SR[1]	1250-1400	53±7	4
Bean and Newman	IR	1100-1200	22±2	5
Hrostowski and Kaiser	IR	1000-1250	38±4	6
Takano and Maki	X-Ray	1100-1200	25	7
Gass et al	CPAA[2]	1000-1280	24.8±0.5	11
Mikkelsen	SIMS[3]			8
oxidizing anneal		600-1240	23.7	
inert anneal		600-1240	15.4	
Craven	IR	950-1350	23.7	9
Itoh and Nozaki	CPAA[4]	1000-1375	27.6±3.0	10

1. SR = Spreading Resistance
2. CPAA = Charged Particle Activation Analysis,(^{18}O)
3. SIMS = Secondary Ion Mass Spectrometry
4. CPAA = Charged Particle Activation Analysis,(^{16}O)

In as-grown crystals, oxygen is typically incorporated at levels of $5-10 \times 10^{17}$ atoms/cm^3. At device processing temperatures in the range of 600-1200°C, this corresponds to substantial supersaturation, as seen from the solubility versus temperature curve. Thus, there is a thermodynamic driving force for precipitation from solid solution at process temperatures. The extent to which precipitation occurs and the approach to the equilibrium solubility value of oxygen in silicon at a given temperature depends on many factors, such as the presence of nucleation sites for precipitation and the kinetics of a particular precipitation process.

2.2 Diffusivity

Oxygen is a very fast diffuser in silicon, compared to impurity dopants commonly used in device manufacture. Figure 2 shows the diffusion coefficient of oxygen in silicon as a function of temperature. Oxygen is intermediate in the magnitude of its diffusion coefficient between such dopants as boron and phosphorus on the one hand, and fast diffusing metals such as copper and lithium. Several investigators have measured the diffusion coefficient of oxygen in silicon over the temperature range 270-1400°C. The line in Figure 2 represents a diffusion coefficient of

$$D = 0.13 \exp\left[\frac{-58.4 kcal/mol}{RT}\right] cm^2/s \qquad (4)$$

where 1 eV=23.1 kcal/mol. The results obtained by various investigators are shown in Table 2. Generally, results at high temperatures are independent of wafer orientation, processing ambient, and the presence of other elements such as dopants. In addition, high temperature outdiffusion gives similar coefficient to in-diffusion results, indicating a single diffusion mechanism (14).

For low temperature experiments, in the 270-400°C range, two distinct diffusion coefficients have been observed (15). These two coefficients were derived from optical dichroism relaxation of stressed silicon samples, and correspond to a hopping motion of oxygen from one site to another. The two values were obtained in silicon crystals with distinctly different thermal histories. For a crystal heated at 1350°C prior to the experiment (referred to as "annealed" in Table 2), the diffusivity measured at low temperature corresponds to the high temperature value previously mentioned above (16). However, if the crystal was preheated at 900°C but not at 1350°C (referred to as "as received" in Table 2), the low temperature diffusion coefficient had a value *one hundred times* the high temperature result. Figure 3 shows the corresponding relaxation times as a function of temperature.

This low temperature anomalous diffusion of oxygen in silicon is important in a number of respects. First of all, the kinetics of thermal donor formation cannot be explained in

Table 2

Diffusion Coefficient of Oxygen in Silicon

Author	Method	Temperature Range (°C)	Activation Energy (kcal/mol)	Pre-factor (cm^2s^{-1})	Ref.
Logan and Peters	SR	1250-1400	80.7	135	4
Corbett, McDonald and Watkins	IR	1100-1280	59.0	0.23	16
Takano and Maki	XRAY	1100-1280	55.3	9.1×10^{-2}	7
Gass et al	CPAA		72.6	22.6	11
Mikkelsen	SIMS	700-1240	56.2	7×10^{-2}	14
Stavola et al	IR	270-400			15
annealed			58.6	0.17	
as received			45.3*	3.2×10^{-4}	
Itoh and Nozaki	CPAA	1150-1375	67.1	3.2	10

* Second value is fast diffusion coefficient at low temperature

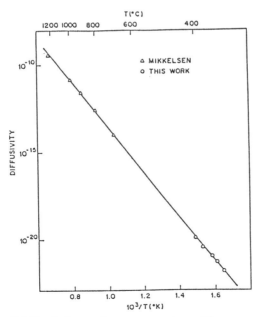

Figure 2. Diffusivity of oxygen in silicon as a function of temperature (15).

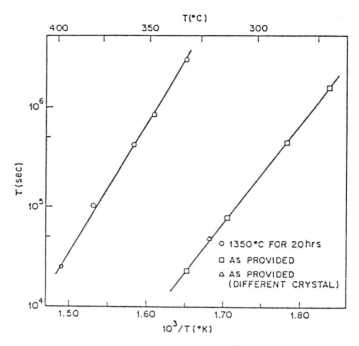

Figure 3. Low temperature reorientation lifetime of oxygen in silicon (15).

terms of the high temperature oxygen diffusivity; a much faster movement of oxygen atoms is required. Second, the existence of two distinct diffusion coefficients implies two distinct diffusing species. Gosele and Tan proposed a molecular species as the fast diffusing entity (17). Alternatively, one may postulate a substitutional oxygen as the fast diffuser. Both its solubility and diffusivity would be expected to be very different from the interstitial species. The diffusion coefficient of either can predominate at low temperature depending on the previous thermal history of the sample.

2.3 Segregation Coefficient

Many attempts have been made to understand the segregation of oxygen between the liquid and solid phase. Most of these attempts have involved the use of crucible grown silicon where variations in oxygen content due to variations in growth rate were used to infer a segregation coefficient. Because oxygen concentration in the melt cannot be obtained in such an experiment, the result is at best an effective segregation coefficient which involves assumptions about oxygen levels in the melt near the solid-liquid interface.

Yatsurugi and coworkers (2) have performed the only direct determination of the segregation coefficient of oxygen in silicon. In this experiment, small diameter silicon crystals were grown by the float zone method, and doped with oxygen from the gas phase. Various growth rates were employed, and oxygen concentration was measured as a function of zone velocity. Figure 4 shows the results of the experiment. Oxygen content was measured in two ways; total oxygen present chemically was measured by charged particle activation analysis (CPAA), while the 9μm infrared absorption (IR) due to interstitial oxygen was also determined.

Oxygen concentration in the solid as determined by CPAA was a maximum for fast growth. The minimum value of oxygen in a solidified sample corresponds to a rapidly quenched sample, and is taken as the concentration in the liquid at the melting point. The ratio of the two concentrations gives a segregation coefficient $k_0 = 1.25$. The 9μm absorption coefficient was constant as a function of growth rate, at a value of $\alpha = 7$ cm^{-1}. This means that for interstitial oxygen, $k = 1.0$, since the concentration of interstitial oxygen in the solid always corresponds to the (quenched) value in the liquid. The identity of the additional oxygen, determined by CPAA, was attributed to small oxygen clusters - termed "infrared inactive" by the investigators.

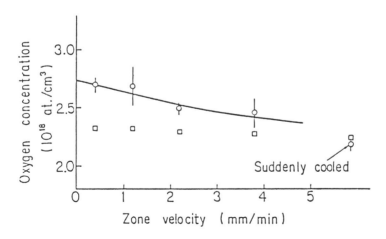

Figure 4. Oxygen content of float zone crystals as a function of growth rate (2). (Reprinted by permission of the publisher, The Electrochemical Society, Inc.)

Other experiments designed to measure the segregation behavior of oxygen in silicon have produced values of k=0.3-2.0. Table 3 summarizes some recent work, including that of Abe et al. (18), and W. Lin (19),(20).

Arguments based on published values of various solubility, and comparison of various segregation coefficients of impurities in silicon are discussed by Jaccodine and Pearce (21), and shown graphically in Chapter 3, Figure 6. Carlberg's calculation, discussed in the following section, results in a value of k=0.995 (13).

Table 3

Segregation Coefficient of Oxygen in Silicon

Author	Method	k_o	Solid Solubility of Oxygen at M.P. (C_s, atom/cm^3)	Liquid Solubility of Oxygen at M.P. (C_l, atom/cm^3)	Ref.
Yatsurugi	FZ growth +CPAA	1.25 ±0.17	2.75±0.15 x10^{18}	2.20±0.15 x10^{18}	2
Abe et al	Crucible growth +IR	1.48	—	—	18
Lin	Crucible growth +IR	0.3	—	—	20
Jaccodine and Pearce	Literature Study	0.65	—	—	21
Carlberg	Calculation	0.995	2.08x10^{18}	2.09x10^{18}	13

2.4 Phase Diagram

The difficulties associated with measuring oxygen concentration in molten silicon have prevented the construction of a detailed phase diagram of the oxygen—silicon system. In addition, the equilibrium between silicon saturated with oxygen in either liquid or solid form and solid SiO_2 and gaseous SiO need to be determined. Figure 5 shows a schematic of the silicon phase diagram constructed by Yatsurugi et al, based on their determination of the segregation coefficient.

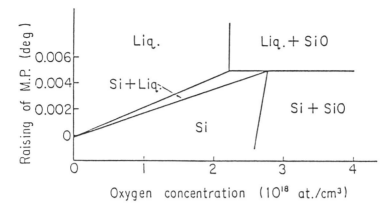

Figure 5. Phase diagram of the silicon-oxygen system according to (2). (Reprinted by permission of the publisher, The Electrochemical Society, Inc.)

Recently, theoretical calculations by Carlberg (12) have provided an alternative to the phase diagram construction.

Thermodynamic data are available to determine free energies for reactions in the silicon-oxygen system. For the reaction

$$\frac{1}{2}SiO_2 = O + \frac{1}{2}Si(l) \tag{5}$$

$$\Delta G^\circ = 18,5970 - 64.9T (J/mol) \tag{6}$$

The solubilities of oxygen in the liquid and solid are given as

$$\ln O^l(mol\%) = \frac{-22370}{T} + 7.8 \tag{7}$$

$$\ln O^s(mol\%) = \frac{-25,414}{T} + 9.6 \tag{8}$$

These correspond to solubilities in equilibrium with SiO_2. When the presence of SiO is considered, we have the additional reaction

$$\frac{1}{2}SiO_2 + \frac{1}{2}Si(l) = SiO(g) \tag{9}$$

for which

$$\ln P_{sio}(101,325 Pa) = \frac{-39,656}{T} + 18.9 \quad (10)$$

The additional reaction between SiO and silicon is

$$SiO(g) = O + Si(l) \quad (11)$$

for which

$$\ln \frac{[O]^l(mol\%)}{P_{sio}(101,325 Pa)} = \frac{17,290}{T} - 11.1. \quad (12)$$

From these relations, Carlberg calculated the phase diagram shown in Figure 6. The diagram is drawn for a total pressure of 2.6×10^3 Pa (solid lines), the normal condition for crucible growth of crystals. Additional, dotted lines are shown for higher and lower temperatures. When the equilibrium between SiO_2 and silicon alone is considered at higher pressures, the phase diagram appears as eutectic; when the SiO equilibrium data is included, and below a pressure of 1.3×10^2 Pa, a peritectic phase diagram results.

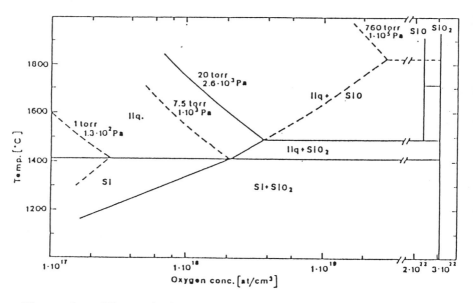

Figure 6. Theoretical phase diagram of the silicon-oxygen system (13). (Reprinted by permission of the publisher, The Electrochemical Society, Inc.)

3.0 OXYGEN CLUSTER AND PRECIPITATE FORMATION

Oxygen is incorporated into the silicon lattice during crystal growth at levels of $5-10 \times 10^{17}$ atoms/cm^3. This amount of oxygen is above the solid solubility at device processing temperatures, which range from 400-1200°C for times which total 24-48 hours. During device fabrication, the dissolved oxygen can change its state to form a variety of complexes, including precipitates, stacking faults, low oxygen surfaces, and thermal donors. The rate at which these are formed as well as the defect density are complicated functions of initial oxygen level, temperature, time, and the thermal history of the sample. The simplest model for these processes assumes homogeneous nucleation, and can apply to silicon which is relatively lightly doped, and free of such impurities as carbon. We will first present the homogeneous nucleation theory developed by Inoue and coworkers (22), and later discuss the influence of high levels of carbon on the precipitation of oxygen from solid solution.

3.1 Precipitate Nucleation

At temperatures below about 900°C, the high supersaturation of oxygen leads to the formation of small oxygen clusters, termed nuclei. Inoue has developed homogeneous nucleation theory to apply to this process. Nucleation rate is

$$J = n^*Zw^* \tag{13}$$

where

n^* = equilibrium critical nucleus density.
Z = Zeldovich factor for nonequilibrium distribution
w^* = oxygen impingement frequency onto a critical nucleus.

Assuming a spherical nucleus and negligible strain in the nucleation process, n^* and w^* can be expressed as

$$n^* = n_1 \exp(-\Delta G^*(r^*)/kT) \tag{14}$$

with

$$\Delta G^*(r^*) = (4/3)\pi(r^*)^3 \Delta G_v + 4\pi(r^*)^2 \sigma \tag{15}$$

$$w^* = 4\pi(r^*)^2 n_1 dp\nu \exp(-E/kT) \tag{16}$$

$$= 4\pi(r^*)^2 n_1 D/d. \tag{17}$$

$$r^* = -2\frac{\sigma}{\Delta G_v} \tag{18}$$

n_1 = nucleation site density

  = dissolved oxygen concentration for homogenous nucleation.

ΔG^* = free energy for critical nucleus formation

k = Boltzmann constant

T = annealing temperature

σ = interface energy between precipitate and silicon lattice.

r^* = critical radius

d = characteristic atomic distance (2.35Å)

p = probability of oxygen impingement onto a nucleus

ν = oxygen vibrational frequency

D = oxygen diffusion coefficient.

E = oxygen diffusion barrier height

(E=2.4 eV is used here, although 1.7 eV can apply in certain situations. See section on thermal donor formation and decay).

Assumptions include that the volume free energy change can be written as

$$\Delta G_\nu = \Delta H \nu (T_E - T)/T_E \tag{19}$$

where

ΔH_ν = dissolution enthalpy

= 6.67 x 10^{10} erg/cm^3

T_E = solid solubility temperature for a given oxygen concentration

= 1300°C for [O] = 11x10^{17} atoms/cm^3

Substituting equations (14) to (19) into (13), the nucleation rate is obtained.

$$J = n_1 4\pi \left(\frac{2\sigma T_E}{\Delta H_\nu (T_E - T)}\right)^2 d\left(\frac{D}{d^2}\right) Z n_1 \times \qquad (20)$$

$$\times \exp\left(\frac{-16\pi\sigma^3 T_E^2}{3\Delta H \nu^2 (T_E - T)^2 kT}\right).$$

Figure 7 shows the nucleation rate as a function of temperature as calculated by this expression. The curve was fit by taking σ=430 erg/cm^3 and Z=0.001. Comparison to observed densities for oxygen content of 11x10^{17} atoms/cm^3 is good. In addition, nucleation rate at 750°C was calculated as a function of oxygen concentration. The results are presented in Figure 8 and compare very favorably with experiment.

Nucleation, the formation of small clusters of oxygen atoms, has a rate which is maximum at about 750°C. The clusters, involving only a few oxygen atoms, have a size distribution which depends on the annealing temperature, The distribution is given by

$$n(r) = n_1 \exp(-\Delta G(r)/kT). \qquad (21)$$

where

$$\Delta G(r) = (4\pi/3) r^3 \Delta G_\nu + 4\pi r^2 \sigma. \qquad (22)$$

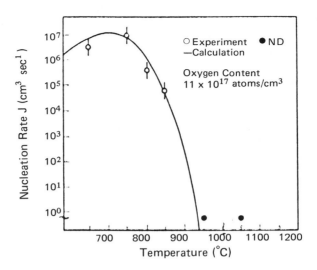

Figure 7. Oxygen precipitate nucleation rate as a function of temperature (22). (This paper was originally presented at the Spring 1981 Meeting of the Electrochemical Society, Inc. held in Minneapolis, MN.)

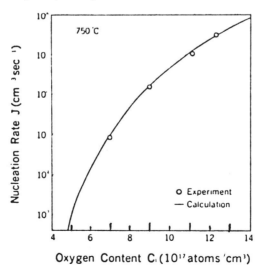

Figure 8. Oxygen precipitate nucleation rate as a function of oxygen content (22). (This paper was originally presented at the Spring 1981 Meeting of the Electrochemical Society, Inc. held in Minneapolis, MN.)

Maximum density is achieved at lower temperatures; size distribution is 2-3Å radius, with 2-5 oxygen atoms per cluster at 500°C, and a density of $10^{14}/cm^3$, for $\sigma=430$ erg/cm^3. On the other end of the nucleation temperature scale, a 900°C nucleation step gives a 2-5Å cluster radius, with 2-20 oxygen atoms per cluster. Cluster density ranges from $10^{14}/cm^3$ for smaller radii to $10^8/cm^3$ for larger clusters. The size of the clusters plays a crucial role in determining subsequent precipitation. Depending on the annealing temperature, clusters of small diameter will dissolve while those of larger diameter will grow and form precipitates. The critical diameter as well as supersaturation phenomena are discussed in the next section.

3.2 Precipitation from Solution

In the temperature range from 1000-1250°C, oxygen precipitates from solution at a rate that depends on initial oxygen content, temperature, time at temperature, and the number and size of the pre-existing nuclei whose formation was discussed above. Figure 9 shows the critical radius as a function of temperature for an oxygen concentration of 11×10^{17} atom/cm^3. When silicon is heated to precipitate formation temperatures, clusters with a radius less than the critical radius at that temperature will dissolve. Those with a size greater than the critical radius will grow into precipitates. In addition, longtime heating at 650-900°C could reduce the oxygen content to a value below the solubility level for the second, higher temperature anneal. In that case, precipitates will actually shrink until the dissolved oxygen level reaches the solubility limit for that temperature.

For annealing temperatures of 900-1000°C, precipitates have a platelet morphology. Higher precipitation temperatures produce an octahedral precipitate, and under certain conditions, such precipitates adopt the β-crystobalite phase of SiO$_2$. Longtime annealing at low temperatures (400-500°C) can produce slender rods of SiOx precipitates, identified in some cases as the coesite form of SiO$_2$ - a high pressure modification of quartz.

Figure 10 shows a transmission electron micrograph (TEM) photo of oxygen precipitates formed after a thermal cycle simulating a device process. The formation of SiO$_2$ is accompanied by a volume increase of a factor of two, which causes strain on the surrounding lattice. After a certain point, the strain can be relieved by punching out dislocations around the precipitate. The upper precipitate of Figure 10 has caused several dislocation loops to form during its growth. These loops appear as crescents on each side of the precipitate where they intersect the plane of the photograph (23).

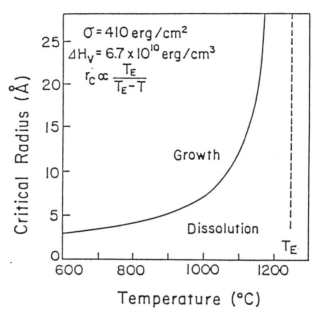

Figure 9. Critical radius as a function of temperature for homogeneous nucleation of oxygen precipitates (22). (Reprinted by permission of the publisher, The Electrochemical Society, Inc.)

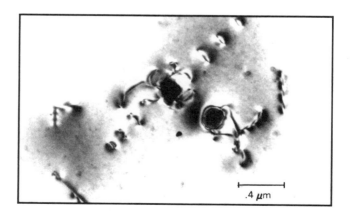

Figure 10. TEM photo of oxide precipitate with an associated punched-out dislocation (23).

3.3 Influence of Thermal History on Precipitate Formation

3.3.1 Ingot Cooling History:
Previously we mentioned that the thermal history of the sample could have a significant effect on the diffusivity of oxygen at the donor formation temperature. Equally dramatic effects due to thermal history are observed in oxygen precipitation at temperatures of 900-1200°C. The thermal history of the crystal begins with the growth of the ingot and is strongly influenced by the cooling of the ingot after growth. Crucible grown ingots experience a thermal history that is different for different portions of the boule. Figure 11 shows a computer simulation of the thermal history of an ingot (24). The furnace is shut off after crystal growth, and the tang or tail of the ingot cools rapidly from temperatures near the melting point to below 400°C. However, the seed end has cooled slowly as the ingot has been withdrawn, and spends considerable time in the temperature range of 500-800°C. This is just the range where homogeneous nucleation is at its maximum, and small nuclei are expected to form.

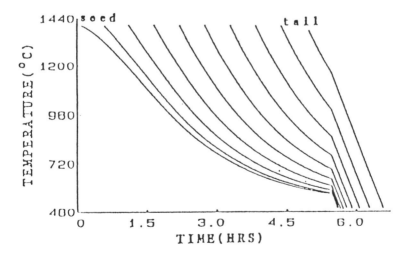

Figure 11. Computer simulated thermal history of a silicon ingot (24). (This paper was originally presented at the Spring 1984 Meeting of the Electrochemical Society, Inc. held in Cincinnati, OH.)

It is worth noting that precipitation inhomogeneities are also observed due to microscopic temperature and growth rate fluctuations during growth itself. These fluctuations cause inhomogeneous distribution of oxygen, resulting in precipitation patterns that resemble heterogeneous nucleation phenomena. These are discussed later in this section.

The distribution of "grown-in" nuclei is shown schematically in Figure 12. The distribution is expected to vary from wafer to wafer, depending on the position in the ingot. Subsequent heating can cause the grown-in distribution to form precipitates whose density depends on the initial, as grown, distribution. Intentional heating of wafers in the temperature range 600-900°C can blur or eliminate differences in ingot thermal history as far as subsequent precipitation is concerned. In addition to the effects of heating in the 600--900°C range, a peculiar additional defect has been seen due to holding silicon at 1200°C. Figure 13 shows the oxygen reduction due to 1050°C annealing along an ingot annealed at a temperature distribution shown on the horizontal axis (25). Significant precipitation occurs both for 600-900°C annealing and for the 1200°C holding temperature.

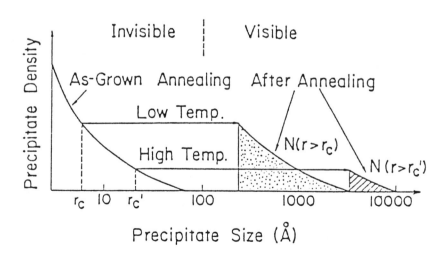

Figure 12. Precipitate nuclei density versus size (22).

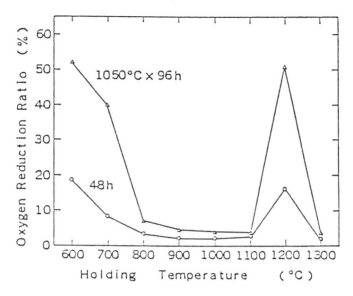

Figure 13. Reduction of dissolved oxygen in an ingot as a function of annealing temperature (25).

Other researchers have begun to model the nucleation and precipitation behavior of oxygen in silicon. Cavitt employed Ham's model to calculate precipitate density after a two step heat treatment (26). Lavine et al. studied previously reported experimental data for precipitation and compared it to a numerical model (27). A model including ingot cooling effects was reported by Hartzell and coworkers (28). Practical application of models is just at its beginning stages.

3.3.2 Microscopic Growth Fluctuations: Temperature fluctuations in the melt can cause local fluctuations in the growth rate of the crystal which lead to inhomogeneous properties of the material. In certain cases, the crystal can be freezing on one side and melting on the other; crystal rotation can result in periodic melting and freezing which maximizes the inhomogeneities on a local scale. These effects are most pronounced when an impurity with a low segregation coefficient is alternatively frozen and then melted and "piled up" in concentration at the solid liquid interface. The result is impurity striations, which lead to resistivity fluctuations if the dopant is electrically active.

In the case of oxygen, local fluctuations in the growth rate lead to variations in precipitate density upon subsequent heating. This is in spite of the fact that the oxygen level, as measured by 9µm infrared absorption, appears uniform across the wafer. An example of this was given by Abe for undoped, crucible grown silicon (18). Two crystals were grown, one with uniform growth rate, and one with fluctuating growth rate. On subsequent heating at 1100°C, the precipitation of oxygen was uniform in the material grown with uniform solidification rate. However, the crystal grown with fluctuations showed inhomogeneous precipitation after heating. Precipitates were distributed in concentric rings from the center to the edge of the wafer. Such rings are sometimes termed "oxygen swirl".

The inhomogeneous precipitation occurred in spite of the fact that impurities such as carbon or electrical dopants were absent from the crystal. In order to account for such behavior, the distribution of oxygen must somehow be nonuniform in the second crystal in spite of the uniformity of interstitial oxygen, as measured by 9µm IR absorption. Work on the distribution coefficient by Yatsurugi and coworkers established the fact that for interstitial oxygen, k=1.0; this means that fluctuations in growth rate by themselves cannot produce variations in local (interstitial) oxygen concentration. A substitutional oxygen species *could* serve as a nucleus for precipitation of the interstitial species. The substitutional atom, entering with a segregation coefficient of k=0.25, would show significant inhomogeneity in distribution in the crystal when the growth rate is varied. The idea that substitutional oxygen can constitute a nucleus for precipitation receives support when the formation of thermal donors is considered. It may be possible that the oxygen clusters formed during the 600-900°C nucleation cycle originate in the association of a substitutional oxygen atom with one or more interstitial species.

3.3.3 Retardation of Nucleation: A phenomenon influencing the precipitation of oxygen at device processing temperatures, reported by Tan and Kung, is the retardation of the nucleation process in the 600-900°C range (29). Figure 14 shows the influence of annealing time at 650°C on subsequent precipitation during a simulated MOS process cycle. Precipitation increases with 650°C annealing time up to 2 hours, then decreases. After 8 hours of low temperature annealing, the precipitate density seen at higher temperatures again increases. There seems to be no explanation for this phenomenon in

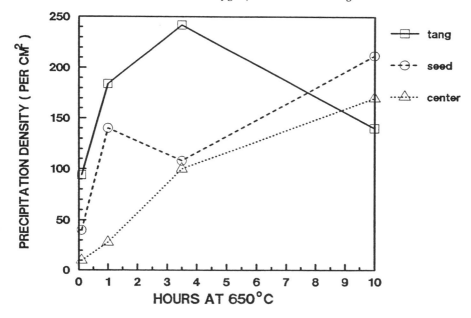

Figure 14. Oxygen precipitate density in silicon as a function of annealing time at 650°C (92).

classical nucleation theory. Tan has suggested that excess silicon interstitials play a role in the retardation phenomenon.

3.3.4 Influence of Substrate Doping on Oxygen Precipitation: When silicon is heavily doped with electronically active impurities, the precipitation of oxygen can be significantly affected. Most studies have concentrated on heavy doping with boron or antimony, since these elements are chosen for substrate dopants in p/p+ and n/n+ epitaxial silicon structures. The presence of boron enhances precipitation while antimony retards it. In the case of heavy doping with antimony, part of the explanation may lie in the oxygen content of the sample.

If crystals are grown under reduced pressure, antimony evaporates during growth. It has been reported that this enhances oxygen evaporation as well, and that the oxygen level in the resulting crystal is below the level at which precipitation will occur. The high electron concentration due to antimony makes it impossible to determine oxygen content by infrared analysis unless special measures are taken. For example, Tsuya has shown how electron irradiation will compensate and neutralize intentionally added dopants (30). Previously, this had also been reported for heavily doped boron samples where the infrared absorption of boron itself was needed, Other, destructive tests for oxygen content in heavily doped substrates include sample fusion and SIMS analysis, discussed in the following section.

Measurements on Sb-doped crystals by Walitzki et al. suggest that "ordinary" crystal growth methods result in low oxygen content (31). Oxygen content and corresponding precipitate density after high temperature can be increased by special ingot growth techniques. Corresponding growth of heavily B-doped crystals by "ordinary" techniques results in higher oxygen concentrations than usual, and increased precipitation of oxygen during device thermal cycles.

Given equal amounts of oxygen in heavily doped n and p-type crystals, there is still the tendency to greater precipitation in p-type material. Wada and Inoue have suggested that the influence of electrically active impurities on the precipitation of oxygen can be understood by examining thermal donor (TD) formation as a function of doping (31). TD formation is enhanced in p-type silicon and reduced in heavily doped n-type silicon. Wada has suggested that the intrinsic electron concentration at the donor formation temperature limits donor formation quantitatively. Therefore, TD population can be larger in p-type but is reduced in n-type silicon. The corresponding effect on oxygen precipitation suggests that the thermal donor cluster is the basic nucleus for oxygen precipitation. That is, in p-type material, increased donor concentration is naturally formed. The donor is the "seed" for nucleation, which means increased precipitate density after heat treatment. On the other hand, donor formation would be suppressed in heavily doped n-type material, which would reduce precipitate density. This latter point is somewhat controversial, and is difficult to verify experimentally.

3.4 Oxidation Induced Stacking Faults

Heating of silicon can produce localized defects termed stacking faults. These are always extrinsic in nature, consisting of an extra plane of atoms which grows as a disc between adjacent (111) planes. The stress resulting from this extra plane of atoms is relieved by a bounding partial dislocation. Stacking fault formation is influenced by the oxygen content of the material, electrical dopant, and the details of the crystal growth process. In addition, the condition of the silicon surface, the presence of thin films such as oxide or nitride, and the ambient for heating can also influence the formation of stacking faults in the bulk. This section is limited to stacking faults related to the presence of oxygen. The formation of oxygen precipitates serves to nucleate stacking faults which grow as a function of annealing time and temperature. If these faults are present near the silicon wafer surface, they can degrade device performance, serving as diffusion pipes, causing electrical shorts between adjacent p and n regions, or nucleate defects in deposited epitaxial layers.

In as-grown silicon, stacking faults (SF) and most other imperfections are usually absent. SF formation can occur after a high temperature step such as oxidation, when oxygen precipitation occurs. SF density is typically measured by performing a defect etch appropriate to the surface orientation. In the case of an oxidation, the oxide is first removed from the surface. Etch figures for bulk stacking faults show the characteristic dumbbell shape shown in Figure 15. This is a (100) surface that was Secco etched after oxidation. Notice the nearby surface stacking fault which lacks the etched-out bounding dislocation of the bulk stacking fault. The surface fault is due to contamination or mechanical abrasion of the surface - not to the presence of oxygen precipitates. Figure 16 shows a TEM photo of a bulk stacking fault, showing the precipitate at the center and the bounding partial dislocation (22). The fault itself is seen as an interference pattern in electron microscopy. This fault is a circular disc which makes an angle with respect to the plane of the photograph.

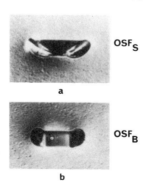

Figure 15. Optical photomicrograph of silicon wafer after Secco etching, showing characteristics of bulk and surface stacking faults. (From G.A. Rozgonyi and T.E. Seidel in *Semiconductor Silicon 1977*, op. cit., p 616-625.)

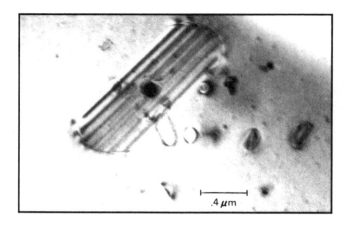

Figure 16. TEM photo of a bulk stacking fault with an oxide precipitate at its center (23).

3.4.1 Influence of Crystal Growth: If wafers from various positions in an ingot are subjected to an oxidation and defect etch, a SF distribution from seed to tail is seen such as shown in Figure 17. This result is for a lightly boron doped ingot; the stacking fault density is high in the seed end of the crystal, and drops abruptly to a low level for the rest of the crystal (33). The oxygen content varies much

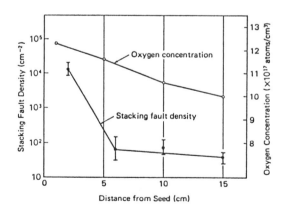

Figure 17. Axial distribution of dissolved oxygen and stacking faults (seen after oxidation and defect etch) in a boron-doped silicon ingot (33).

more slowly from seed to tang and is also shown in the figure. The seed end of the crystal is the region where maximum changes in growth rate occur, from very slow as the crystal is grown laterally to the final diameter, to very fast as the shoulder is grown, to an intermediate growth rate for the body of the crystal.

Daido has shown that varying the crystal rotation rate influences the density and distribution of stacking faults, as seen in Figure 18. Increasing the rotation rate at first decreases the area of high stacking fault density. Further increase of rotation rate merely moves the pattern. The change in SF distribution must be closely related to the change in interface shape produced by changes in rotation.

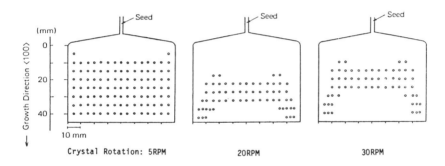

Figure 18. Distribution of oxidation-induced-stacking faults in an ingot as a function of crystal and crucible rotation rate during growth (33).

3.4.2 Influence of Dopant: If a crystal lightly doped with phosphorous is subjected to an oxidation and defect etch, the resulting level of stacking faults is three orders of magnitude higher than for a boron doped crystal. This result is the reverse of precipitate density behavior for heavily doped substrates discussed in the previous section. There is no current explanation for this effect. It is possibly related to microscopic fluctuations in the distribution of phosphorus (low segregation coefficient) as compared to boron. Both n-channel MOS and some bipolar devices tend to start with p-type substrates, and benefit in yield from the reduced stacking fault density in this material compared to n-type silicon. Recent unpublished advances in crucible manufacture which reduce metallic impurity levels, and the crystal growth technology employed for control of oxygen content have reportedly reduced stacking fault levels dramatically, both for p and n-type material.

3.4.3 Influence of Temperature and Time: Stacking fault growth is observed in the temperature range 1000-1200°C, and is enhanced by preannealing at 750°C. The density of the faults depends on many factors, but the length of the faults is a function of temperature and time. Figure 20 shows the length of stacking faults as a function of these variables. The time dependence is accounted for by a $t^{3/4}$ law, whose interpretation is a matter of some dispute (34). Such a $t^{3/4}$ law has also been observed for small precipitates at 750°C, and one interpretation suggests that bulk stacking fault formation is controlled by oxygen diffusion rate (35). The temperature dependence is currently not understood.

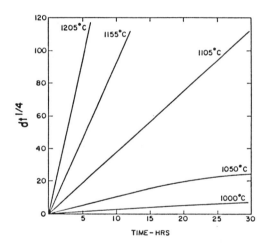

Figure 19. Length of oxidation-induced-stacking faults as a function of oxidation time and temperature (34).

3.4.4 Elimination of Near Surface Stacking Faults: High temperature annealing can eliminate stacking faults formed near the surface of a silicon wafer, The process, described by Shimizu and coworkers (36), is reminiscent of the precipitate dissolution above $T_E=1200°C$ described by Inoue for high oxygen content samples, and also resembles a denuding process, in which oxygen out-diffuses from the surface. Annealing was carried out in a nitrogen ambient after stacking faults had been created by prior heat treatment. For an initial oxygen level of about 1×10^{18} atoms/cm^3, the thickness of a SF-free region grew rapidly at a temperature of 1250°C. An incubation time, corresponding to the elimination of surface stacking faults, was noticed for bulk SF elimination. The thickness of the SF-free surface region followed a power law,

$$d \propto t^n \tag{23}$$

where d is the thickness of the SF-free layer, t is annealing time, and n=0.63. The temperature dependence was fitted with an exponential term, but the activation energy varied from 3.4-4.7 eV, depending on the depth of the stacking fault-free zone.

SF formation can be minimized or prevented in the first place by adding a source of chlorine such as HCl to the ambient used for initial oxidation of the wafer, According to Shiraki, the chlorine causes excess vacancy formation at the wafer surface (37). Vacancies diffuse into the silicon, providing sites for the silicon atoms of which the SF is composed, The silicon atoms move to the vacancies and the stacking fault shrinks. Ultimately, impurity atoms, which provide the nucleus for SF formation, can also be dispersed into lattice sites via the availability of excess vacancies. The mechanism works for both crucible grown silicon, with oxygen as the impurity, and float zone silicon where carbon may provide the stacking fault nucleus.

3.5 Oxygen Out Diffusion and Wafer Surface Denuding

3.5.1 Out Diffusion: Oxygen will diffuse out of a crucible grown substrate, even in an oxidizing ambient. The rate of out-diffusion is equal to that obtained for in-diffusion. This is the high temperature or slower value of diffusivity mentioned in the previous section on that topic. Out diffusion is discussed separately here, since it forms the basis for achieving a defect free silicon wafer surface, which is essential for device fabrication. Together with nucleation and subsequent precipitation, the out-diffusion or denuding cycle constitutes the infant stage of silicon wafer engineering to tailor material properties to device requirements.

The rate at which oxygen will out diffuse depends on the value of the diffusion coefficient at a particular temperature, the initial oxygen content, and the equilibrium with oxygen in the gaseous ambient. The effect of the ambient is somewhat uncertain, but a recent investigation suggests that Henry's Law applies (38). It has the form

$$C_s = \frac{2 p_{O_2} C^{eq}}{p_{SiO}} \tag{24}$$

where

C_s = oxygen concentration in silicon

p_{O2} = partial pressure of oxygen in the atmosphere

C^{eq} = equilibrium pressure of oxygen in the atmosphere

p_{SiO} = partial pressure of SiO in the atmosphere

The factor of two in the above equation enters because two oxygen atoms combine to form one molecule of gas. For extended heating at 1200°C, a surface concentration of 4×10^{17} atom/cm^3 was measured, about a factor of two lower than the solubility of oxygen at this temperature. This means that oxygen precipitates cannot form at the surface for process temperatures above about 1000°C.

3.5.2 Model for Out Diffusion: The out-diffusion of oxygen from silicon has been modeled by Andrews (39). Out diffusion rate is faster the higher the temperature. However, initial oxygen content is usually above the solid solubility at process temperatures below 1300°C, and the lower the temperature, the greater the supersaturation, which provides an additional driving force for out diffusion. These competing effects mean that an optimum temperature will be found for a given initial oxygen concentration, if maximum out diffusion is desired. The one dimensional diffusion equation was used in Andrews' model, subject to the boundary conditions of oxygen concentration equal to the initial value, C_0, at infinite distance into the bulk, but pinned to the solid solubility value at the surface.

The oxygen concentration as a function of distance and time is given by

$$C(x,t) = C_s \text{erfc}(u) + C_0 \text{erf}(u) \text{cm}^{-3} \tag{25}$$

where

$C(x,t)$ = the oxygen concentration as a function of depth and time

C_s = solid solubility of oxygen at the denuding temperature

C_0 = initial oxygen concentration

$\text{erf}(u)$ = error function of u

$\text{erfc}(u)$ = complementary error function of u

The dimensionless parameter u is

$$u = \frac{x}{2}(Dt)^{\frac{1}{2}} \tag{26}$$

where D is the diffusion coefficient of oxygen as a function of temperature.

Figure 20 shows the computation results for the surface oxygen concentration as a function of annealing temperature for an initial oxygen concentration of 1×10^{18} atom/cm^3. Optimum temperature is in the 1050-1150°C range, depending on annealing time. This is the normal temperature range for initial wafer oxidation, and a denuded zone is therefore "automatically" formed for many device processes. In this model, Andrews took the value of the solid solubility as the lower limit for the surface concentration. However, the work of Sugita et al. on the equilibrium between bulk silicon and gaseous ambient suggests that values lower than the solubility value can be achieved, and that this boundary condition may require reevaluation.

Andrews' model predicts a denuded zone depth proportional to the square root of the diffusion time at a given temperature. For 1000°C,

$$DZW = 5.6 + 7.4(t)^{\frac{1}{2}} \tag{27}$$

where DZW is the denuded zone width in μm, and t is the time at temperature. A dry oxygen ambient was used for annealing. Similar denuding results were reported by Huber and Reffle (40). Figure 21 shows results at three temperatures for an initial oxygen content of 7×10^{17} atom/cm^3. These results also show an approximate square root of time dependence.

Figure 20. Oxygen concentration 5 μm below the silicon wafer surface as a function of denuding time and temperature (39). (This paper was originally presented at the Spring 1983 Meeting of The Electrochemical Society, Inc. held in San Francisco, CA.)

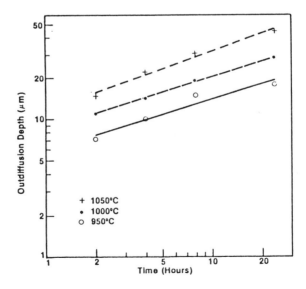

Figure 21. Width of the denuded zone as a function of time and temperature (40).

3.5.3 Measurement of Denuded Zone Depth:

A number of uncertainties arise in the measurement of denuded zone depth. If one is developing a denuding process to add to an existing device process, it is necessary to verify that a given depth of denuding is repeatably achieved. However, after the denuding process, which is often the first step in manufacturing, the denuded zone is typically difficult or impossible to see.

A wafer which has completed the high temperature process sequence can be cleaved and etched to reveal bulk defects. Oxygen precipitates, stacking faults and associated dislocations are seen in the bulk. The denuded zone is free of these and its depth can be measured through the microscope. After the denuding step itself, precipitates have typically not been formed, so the cleave and etch procedure is not useful at this stage. Instead, the oxygen present in the bulk material can be transformed to thermal donors, and the profile due to out-diffusion measured by angle lapping and spreading resistance profiling (SRP). This provides an indirect indication of the denuded zone at that point in the process. Figure 22 shows an outdiffusion profile obtained in this way. The wafer was originally p-type, and the surface remains p-type due to oxygen depletion. The oxygen level increases into the bulk of the wafer, and thermal donors show n-type bulk resistivity.

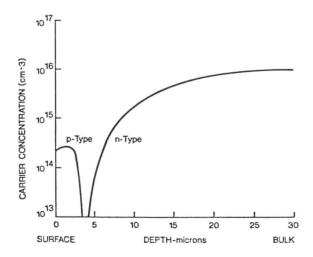

Figure 22. Spreading resistance profile of thermal donor concentration from the surface into the bulk of a silicon wafer after a denuding process.

Although some feeling for the denuding cycle can be obtained by this technique, subsequent thermal cycles will cause the denuded zone to change, and generally to increase in thickness. Sugita and coworkers (37) measured the denuded zone width by X-ray lattice parameter and by charged particle activation as well as SRP, and concluded that the latter technique was less reliable than the former two. In addition to uncertainties with the SRP measurement, additional denuding can occur with added heating, so denuded zone depths are generally deeper than measured by SRP after the denuding step.

Practical problems often arise in specifying a denuding step. The reproducibility of denuding is affected by the precision with which oxygen can be measured by IR absorption. Current practice gives oxygen content with an uncertainty of ±10%. Consequently, denuding must be targeted for larger zone width than is often desired, in order to account for oxygen variations. Even though the active region of device operation may be only 5-10μm, the denuded zone must be targeted for at least 20μm thickness. The active region equals the deepest junction depth plus the associated depletion region when the circuit is operating. Denuding typically is designed for 20-50μm defect free zones. More precise oxygen measurement may allow reduction in targeted depth.

3.5.4 Effect of Ambient: Since the partial pressure of oxygen plays a role in the rate of oxygen out diffusion, one might expect that the ambient would affect the denuding process. However, the partial pressure of oxygen in equilibrium with the silicon surface is only about 9×10^{-5} atm. at 1200°C. Even an inert gas such as nitrogen will have trace amounts of oxygen at this level, and many evaluations show no difference between an oxidizing or inert ambient for denuding. It may be possible that other effects operate since argon ambient seems to produce deeper denuded zones. The concentration of vacancies, reportedly enhanced by the presence of chlorine, may also enhance the out diffusion process as well as the previously noted effect on stacking faults and precipitates. More work is required to determine the role of the gaseous ambient and to establish the denuding process as a repeatable first step in device manufacture.

3.6 Intrinsic Gettering

Gettering is a term left over from the days of vacuum tube manufacturing, when a small amount of titanium was added to the tube before sealing. The metal reacted with residual oxygen, removing it from ambient, thus "gettering" it. Impurity removal in silicon technology is also known as "gettering", but usually refers to the removal of metals which can degrade device performance. Very little is understood about gettering mechanisms, and most theories on this topic invoke some form of action at a distance, the most abhorrent *deus ex machina* of all to a scientist.

Intrinsic gettering refers to the beneficial effects of oxygen precipitates in removing impurities, most likely fast diffusing metals, from the surface region of the silicon wafer where device structures are built. The beneficial effects are seen in increased minority carrier generation lifetime near the surface, reduced leakage of capacitors and p-n junctions, and improved yield of complex devices.

3.6.1 Internal Precipitation: Intrinsic gettering requires the formation of precipitates during the device fabrication process. Nucleation of such precipitates is often added as a separate step at the beginning of the device thermal cycle to enhance the precipitation process. Precipitate growth is different at different process temperatures and is often separated into three temperature ranges. In the range 400--800°C, rod-like agglomerates are seen after extended annealing. These seem to be linear polymers of the small clusters formed during a few hours at these temperatures, and have been identified as the coesite form of SiO_2. From 800-1000°C, platelets are typically seen, while higher temperatures produce octahedral SiO_2 precipitation. An IC manufacturing scheme is a complex combination of process temperatures from 90-1200°C, with ramp cycles which add to nucleation of precipitates.

The improvement in measured device and materials properties due to oxygen precipitation in the bulk has been discussed by Huff and coworkers (41). They suggest that fast diffusing metal impurities, mobile at high temperatures, are entrained in SiO_2 precipitates as they form. There are two consequences of the theory. First, in order to be effective throughout the device manufacturing process, precipitation must continue to occur. Second, a process temperature high enough to dissolve the precipitates will release the trapped impurities and degrade the material properties.

3.6.2 Denuding and Precipitation for Device Processing: The internal precipitation must be confined to the bulk material, and excluded from the device active region in order to provide optimum results. The sequence of processing may include denuding, nucleation and precipitation steps. Figure 23 shows a set of etched wafer cross-sections which show the effects of nucleation and precipitation as a function of time and temperature. The process sequence is typically designed to meld into device manufacturing, and the steps for oxidation, diffusion and so forth may in fact produce the effects of denuding and precipitation. Nucleation can be effected by an extended ramp from 650°C to the oxidation temperature of about 1050°C (42). A recent multistep process combining all cycles has been described by Hirao and Maegawa, with applications to bipolar and MOS device processing (43). Huff and coworkers described its application to DRAM manufacture (40). Even the production of epitaxial silicon layers can benefit in terms of material quality from a denuding and precipitation process (44). Heavily doped substrates, used for epitaxial wafer production, often differ in precipitation rate from the lightly doped material used for mainstream device manufacturing.

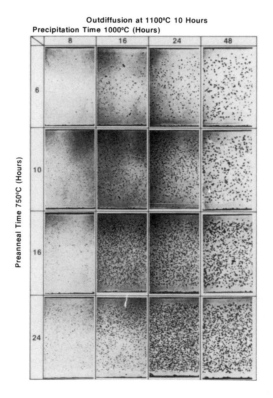

Figure 23. Cross-sections of wafers subjected to various denuding, nucleation, and precipitation cycles (40).

In the case of heavily antimony doped substrates, the oxygen content of the ingot is lower than for a lightly doped ingot grown under the same conditions. This may be due to differences in melt temperature and crucible dissolution rate. Since heavy doping with electrically active impurities makes oxygen measurement by IR impossible, wafer suppliers were not immediately aware of the lower oxygen level in Sb-doped material. Only after SIMS was developed and applied as a quantitative technique was the oxygen level seen to be lower by about a factor of two for antimony material (45). Because of the lower oxygen content, precipitation occurs only with extended nucleation; in some cases, precipitation will not occur at all in a normal process scheme. Lately, advances in the crystal growth process have resulted in antimony doped material with higher oxygen levels (31).

In contrast to Sb-doped material, precipitation is enhanced in substrates heavily doped with boron. The oxygen suppression effect does not occur during crystal growth with boron doping, and oxygen content of heavily B-doped material is slightly higher than in a lightly doped ingot. In addition, the precipitation in heavily B-doped material is enhanced. The mechanism for enhanced precipitation has something to do with the fact that oxygen donor formation is also enhanced in strongly p-type silicon. This enhanced clustering to produce donors can therefore serve to greatly increase the number of nuclei for precipitation.

Precipitated oxygen in the bulk of the wafer can produce performance enhancements in devices beyond the gettering function. A high density of precipitates, coupled with a denuded zone of the appropriate depth, can serve as an epi-like structure as far as the electrical properties of devices are concerned. The precipitates reduce the minority carrier lifetime in the substrate, reducing stray currents which otherwise flow laterally between devices. Lifetime reduction by internal precipitation has been applied to dynamic MOS memories to increase holding time of stored charge (46), and to photosensor arrays and CMOS devices to reduce crosstalk and eliminate latchup (47).

4.0 QUANTITATIVE AND QUALITATIVE MEASUREMENT OF OXYGEN

4.1 Quantitative Analysis by Chemical and Physical Methods

Trace analyses of oxygen in silicon requires exacting

analytical techniques which are useful primarily as ways to calibrate non-destructive infrared analysis. These methods include charged particle and photon activation, gas and vacuum fusion of the sample, and secondary ion mass spectroscopy. These quantitative techniques are reviewed in this section, and infrared absorption spectroscopy is discussed in the following section.

4.1.1 Sample Fusion Analysis: The first determination of the oxygen content of silicon single crystals was made by Kaiser and Keck using vacuum fusion analysis (3). The experimental work of fusion analysis was actually performed by W.H. Smith. The method consisted of melting the silicon sample in a graphite crucible containing liquid iron at 1700°C. Oxygen in the sample reacted with the graphite to form carbon monoxide, which was quantitatively determined in the gaseous state by IR absorption. Oxygen loss in the form of SiO was minimized by inserting a graphite plug into the crucible opening. Accuracy of the determination was reported as $\pm$ 2×10^{16} oxygen atom/cm^3. Vacuum fusion analysis was also performed by Graff and coworkers, with similar reported accuracy (48). However, Their results provided a different calibration than obtained by Kaiser and Keck. Inert gas fusion analysis was reported by Baker, and served as the infrared calibration of choice for several years (48). The accuracy of this work has been called into question by more recent analyses.

4.1.2 Activation Analysis: Oxygen can be transmuted to a radioisotope by means of charged particles or energetic photons. ^{3}He charged particles cause the reaction

$$^{16}O\ (^3He,p)\ ^{18}F. \tag{28}$$

Several other reactions are possible in the silicon matrix and care must be taken in sample preparation. The procedure is described by Kim (50), and by Nozaki et al. (51). Samples with high oxygen content can be counted directly for radioactive fluorine. For low oxygen levels, elaborate chemical separations are required to eliminated interfering α emitters. Recent work has also been reported using this quantitative technique (52). Accuracy is reported as $\pm$ 2×10^{16} atom/cm^3.

Radioactive oxygen can be produced by high energy photons according to the reaction

$$^{16}O\ (\alpha,p)\ ^{15}O. \tag{29}$$

Rath and coworkers combined the activation analysis with gas fusion for separation of ^{15}O as carbon monoxide (53). Stated error of the chemical analysis technique is ± 15%.

Regolini and coworkers recently described the use of tritium activation to covert oxygen to ^{18}F, and obtained a calibration of the 9μm oxygen infrared absorption. No error for the activation measurement was explicitly stated (54).

4.1.3 Secondary Ion Mass Spectrometry: Recent advances in experimental technique using secondary ion mass spectrometry for silicon samples have made possible the determination of oxygen content in the parts per million range. Improvements in vacuum technology have allowed the background concentration of oxygen to be reduced to below one part per million, and Harada et al. reported a linear relation between the SIMS signal and either infrared absorption (lightly doped samples), or charged particle activation analysis (44). As a quantitative technique, SIMS is perhaps easier to employ than activation analysis or fusion methods in samples where electrically active dopant concentrations prevent IR measurements. A comparison of SIMS and gas fusion determination of oxygen in heavily doped silicon was recently reported by Walitzki et al. (31).

4.2 Infrared Absorption

4.2.1 Absorption at 9μm: Undoped silicon is transparent in the infrared, below the band gap absorption edge at about one μm wavelength. Figure 24a shows the spectrum of pure silicon from 1400 to 400 cm^{-1}. The scale of measurement in cm^{-1} is inversely proportional to the wavelength and the figure corresponds to a wavelength range of 7-27 μm. In this region of the spectrum, vibrational absorptions can occur at localized impurity sites in the silicon lattice at room temperature. Figure 24a shows a series of absorptions due to the silicon lattice itself. These absorptions are strictly forbidden, even in second order, since no polarization of charge can occur in the lattice if only silicon is present. Evidently a fourth order interaction leads to these lattice vibrations (55).

When oxygen and carbon are present in silicon, the spectrum is modified as shown in Figure 24b. The prominent oxygen band at 1106 cm^{-1} (9μm) is visible in the single beam absorbance spectrum, but the carbon peak and a second peak due to oxygen are difficult to see. Figure 24c represents the difference in absorption between the traces of a) and b).

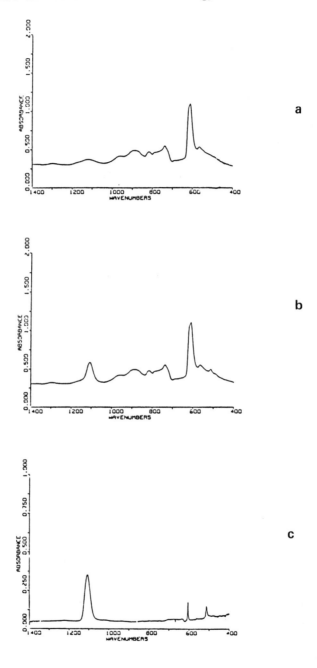

Figure 24. a) Single beam spectrum of pure silicon in the mid-infrared. b) Single beam spectrum of silicon containing oxygen and carbon. c) Difference spectrum of a) and b).

Only the impurity vibrations are seen. The 1106 cm^{-1} peak is symmetrical, with a full width at half maximum (FWHM) of 32 cm^{-1} The intensity of this peak serves as the basis for quantitative measurement of oxygen content in silicon, as the 607 cm^{-1} peak shown in Figure 24c does for carbon. The 513 cm^{-1} peak shown in Figure 24c, which is due to oxygen, will be discussed in a subsequent section.

The concentration of an impurity dissolved in a transparent medium can be determined by optical absorption, if the impurity has a well defined absorption such as the 1106 cm^{-1} vibrational absorption due to oxygen in silicon. For dilute solutions, Beer's Law applies. It can be written

$$\alpha = \frac{1}{t} \log_{10} ac \qquad (30)$$

where

α = absorption coefficient

a = absorptivity

c = concentration of impurity

t = sample thickness

Quantities actually measured in infrared analysis are α, the absorption coefficient, and t, the sample thickness. In order to use Beer's law for measurement of impurity concentration, some relation between α and a, the absorptivity must be determined. This requires an independent chemical analysis of the concentration of oxygen, in this case, for a series of samples used to establish the value of the Beer's law constant, a. The previous section has summarized the relevant chemical methods which have been used for the calibration procedure.

Since the measurement of trace amounts of oxygen is very difficult and subject to error, the calibration procedure has led to widely varying results. Table 4 shows results obtained by a variety of analytical procedures. Intensive work in the last few years has led to the conclusion that the original work of Nozaki et al. represents the most accurate conversion factor. Recent cooperative measurements by the Japan Electronic Industry Development Association (JEIDA), represent a refinement of this work, and a conversion coefficient of 6.06 ppma/cm^{-1} (51). This is also expressed as 3.03x10^{17} atom/cm^3/cm^{-1}. Considerable confusion has arisen due to two other calibration constants adopted by the American Society for Testing and Materials (ASTM). The first conversion factor adopted was that of Baker, and is sometimes referred to as the "Old ASTM" calibration (56). Subsequently, the society adopted the calibration due to Graff. This came to be known as the "New ASTM" calibration (57). International effort has refined and extended the JEIDA effort, which results in yet another calibration for this measurement. A list of experimental results obtained for quantitative calibration of infrared absorption of oxygen in silicon is given in (58) and in the chapter on Materials Properties. See also results of He and coworkers (59), and the review article by Pajot (60).

Table 4

Summary of Calibration of 9 μm Infrared Absorption Coefficient for Determining the Concentration of Oxygen in Silicon

Author	Method	Conversion Factor (ppma*cm)	Stated Error (ppma)	Ref.
Kaiser and Keck	vacuum fusion	5.5	---	3
Baker	gas fusion	9.63	±2.29	48
Nozaki et al	CPAA	6.0	±0.4	50
Graff et al	vacuum fusion	4.9	±0.3	47
Kim	CPAA	7.6	±1.5	49
Rath et al	photon activation	6.0	±0.4	52
Iizuka et al	CPAA	6.06	±0.05	51
Bullis et al	various	6.28	±0.04	58
He et al	fusion	6.2	---	59
Regolini et al	CPAA	6.1	±0.4	54

4.2.2 Interferences in Measurement of Oxygen Content: Accurate determination of the oxygen content of silicon crystals may be impaired by a number of interferences. These include effects due to oxygen precipitation, electrically active impurities, complex formation with carbon, and artifacts of sample preparation which interfere with obtaining the true absorption coefficient of the 9μm absorption.

Infrared absorption at 9μm is due to oxygen in an interstitial position, with no near neighbors except silicon atoms. Any oxygen present as clusters or in different lattice sites may not contribute to the 9μm absorption, so that the total oxygen content of the sample is not actually measured by the infrared procedure. At best, the 9μm IR measurement of concentration will be *proportional* to the true oxygen concentration, the constant of proportionality being set by the previously mentioned chemical experiments. However, Yatsurugi found that at very high oxygen concentrations, corresponding to a 9μm absorption of 7 cm^{-1}, the proportionality between IR and total oxygen was no longer observed (2). More oxygen was detected chemically than was indicated by IR. The excess oxygen was termed "infrared inactive", presumably because no other absorptions were seen in the area of the 9μm peak. For some samples, the excess oxygen amounted to 20-25% of the total oxygen content. This behavior is shown in Figure 4, and led to the determination of the oxygen segregation coefficient with a value of k=1.25.

When oxygen atoms cluster as a result of heat treatments above 400°C, the infrared absorption spectrum can undergo a marked change. At some point, the oxygen atoms are no longer dissolved in the silicon matrix, but form a second phase with silicon. This phase can be termed SiO_x, where x ranges from 1-2, and the material may be amorphous or crystalline, depending on the thermal history of the sample. This results in broad absorption bands which can obscure or distort the dissolved oxygen band. Tempelhoff and Spiegelberg have reported bands centered at 1030 cm^{-1} to 1075 cm^{-1} for heat treatments between 400 and 780°C (61). In addition, higher temperatures, above 870°, produced a band with a broad doublet at 1124 and 1224 cm^{-1}, which they assigned to α crystobalite. Precipitated oxygen which is infrared active invariably makes the measurement of residual dissolved oxygen content more difficult. Also, Freeland reported a broadening of the 9μm absorption itself after extended heat treatment at 650°C, which further complicates quantitative measurement (62).

A second effect which can interfere with the measurement of oxygen concentration by infrared absorption is the absorption due to free carriers. The classical formula for free carrier absorption relates the absorption coefficient to the carrier concentration and the square of the wavelength (55).

$$\alpha = CN\lambda^2 \tag{31}$$

where N is the carrier concentration, lambda is the wavelength, and C is a constant which includes the carrier relaxation time. For carrier concentrations above about $5\times10^{15}/cm^3$, corresponding to resistivities below 1 ohm-cm, the free carrier absorption is strong enough to make quantitative measurement of oxygen impossible unless special measures are taken. For heavily doped silicon samples where $9\mu m$ absorption measurements are required, the free carriers must be trapped in order to eliminate their optical absorption effects. Tsuya has reported a method of trap creation by electron bombardment that allowed oxygen measurements in heavily Sb-doped silicon samples (30). Antimony concentrations were on the order of $5\times10^{18}/cm^3$, and 7MeV electrons with a flux density of $5\times10^{18}/cm^2$ were used for trap creation in silicon wafer samples.

Majority carriers for heavily doped samples can also be trapped by irradiating the sample with fast neutrons (63). Heavily doped samples require doses up to 6×10^{17} neutrons/cm^2 for complete carrier removal. Moreover, fast neutrons can produce "A centers" in silicon. "A centers" consist of a vacancy and an oxygen atom bonded to two next-near-neighbor silicon atoms. These centers are formed at the expense of interstitial oxygen atoms, which complicates the subsequent measurement by infrared.

A third source of interference in quantitative determination of oxygen in silicon is due to the sample surface condition and thickness. A rough silicon surface causes scattering of the incident radiation, leading to apparent absorption coefficients higher than the true values. For polished surfaces, multiple internal reflectance must be accounted for. For very thin samples, ≤ 15 mils in thickness, interference fringes or oscillations disturb the transmission of infrared radiation. These effects must be accounted for in order to obtain an accurate value of the absorption coefficient, and a correspondingly accurate value for oxygen concentration. These interferences and appropriate corrections are described by Stallhofer and Huber (65).

The surface condition of the silicon sample affects the apparent absorptivity because of reflective scattering at the two silicon/air interfaces. The optimum sample for quantitative measurements is $2\mu m$ thick and has a transmission relative to air of 30% or more at 300 cm^{-1}.

Taking reflections into account, the simple Beer's law expression becomes

$$\alpha = -\frac{1}{t}\ln\frac{-B+\sqrt{B^2+4AT}}{2}A \tag{32}$$

where

- t = sample thickness in centimeters
- A = $T R^2 e^{-2\alpha_L t}$
- B = $1 - R^2 e^{-2\alpha_L t}$
- R = reflection index, = 0.3
- T = relative transmission
- α_L = silicon lattice absorption,
  = 1 cm^{-1} at 1106 cm^{-1}

The simple Beer's Law expression is obtained by setting the term

$$R^2 e^{-2\alpha_L t} = 0$$

However, this results in oxygen concentrations which are too high by 5-7% for 2μm thick samples, and by 20% for thinner samples.

Often the standard silicon wafer, which is polished on one side only, is used for the measurement of oxygen concentration. The backside, as-etched surface, causes scattering of the radiation which varies according to the etching process. For a given process, the infrared transmission can be set equal to

$$e^{-\frac{C}{\lambda^2}} e^{-\alpha_l t}$$

where C characterizes the scattering behavior of the backside. If a reference sample of the same thickness and backside characteristics is used, then the simple Beer's Law relation applies. For this relation to be valid, the transmission at 1600 cm^{-1} should be 25% or less.

4.3 Interpretation of the Infrared Spectrum of Oxygen

4.3.1 The Si$_2$O "Molecule": The 9μm absorption used for quantitative analysis arises from the vibration of the oxygen atom with respect to its near neighbor silicon partners. The oxygen atom has inserted itself into a Si-Si bond, forming a nonlinear Si-O-Si structure. This is called "interstitial" oxygen, although the oxygen atom is covalently bonded and is not in a true interstitial position. Depending on the chemical bonding of the structure, other vibrational absorptions should be possible. Figure 25 shows a normal mode analysis of a hypothetical Si_2O "molecule" (66). This molecule is shown with a divalent oxygen at its center in a highly nonlinear form reminiscent of H_2O. This species can exhibit three normal vibrations; ν_1 is termed symmetric stretching, ν_2, angle bending, and ν_3 asymmetric stretching. For a nonlinear molecule, all three modes are infrared active. This is the model originally used by Kaiser, Keck and Lange to interpret the infrared absorption of oxygen in silicon (67). The asymmetric stretching mode, ν_3, should be the most intense absorption, and was identified as the 9μm band. The room temperature band at 513 cm^{-1} (19.5μm) was assigned to ν_2, the angle bending mode. This line can be seen in Figure 24 c). The symmetric stretching mode, ν_1, was assigned to a line that appears only at low temperature at 1205 cm^{-1} (8.3μm).

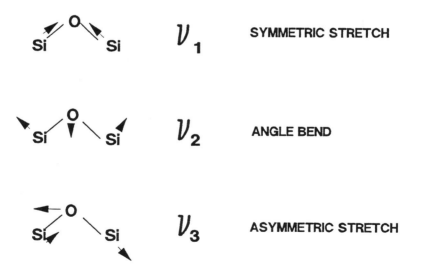

Figure 25. Infrared active normal modes of vibration for the "Si_2O" molecule, by analogy with H_2O (66).

It is amazing that any interpretation of the infrared absorption of oxygen in silicon could successfully ignore the effects of the lattice and could take into account only the near neighbors. But the interpretation of Kaiser et al. explained many of the features of the infrared absorption and allowed development of a picture of the way that oxygen is incorporated into the lattice. As mentioned above, the oxygen atom inserts itself into the silicon-silicon bond, and the resulting Si-O-Si structure is slightly nonlinear. Actually, the structure is more like the linear CO_2 molecule than the H_2O molecule, since it is nearly linear. The insertion of an oxygen between two near-neighbor silicon atoms expands the lattice, which agrees with X-ray measurements of the lattice parameter by Bond and Kaiser (68). They determined the values of the lattice constant of silicon to be

Vacuum grown $a_0 = 5.430710 \pm 0.000010 >$ Å

Oxygen grown $a_0 = 5.430747 \pm 0.00007 >$ Å

A more extensive investigation of the infrared spectrum of oxygen in silicon was undertaken by Bosomworth and coworkers (69). Their results confirm parts of the Kaiser interpretation while casting doubts on others.

4.3.2 Low Temperature and Far Infrared Spectra: Bosomworth et al. investigated oxygen absorptions at low temperatures, and in the far infrared. At low temperatures, the 9μm line splits into as many as four lines, with relative intensity depending on the temperature. At high resolution, satellites are seen due to the presence of silicon isotopes of mass 29 and 30. The splitting of the 9μm line is due to a set of multiple groundstates for the absorption. Additional transitions among these groundstates are seen in the far infrared. Four transitions are seen in the frequency range 29-49 cm^{-1}. These were assigned by Bosomworth to the v_2 or angle bending mode, which had previously been assigned by Kaiser et al. to the 513 cm^{-1} absorption. The low temperature line at 1205 cm^{-1} could be explained as a double quantum absorption. Figure 26 shows the vibrational absorption transitions of the Si_2O species as determined by Bosomworth. Since the Si-O-Si bond angle is nearly linear, at 150-160°, the v_2 or symmetric stretching mode of the species is quenched (70). This implies that the 513 cm^{-1} line is *not* a normal mode of vibration of the "interstitial" Si_2O species.

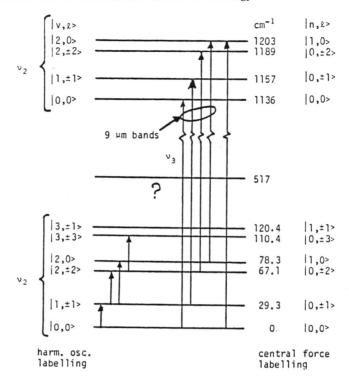

Figure 26. Allowed transitions in the mid and far infrared for the "Si$_2$O" molecule (66).

4.3.3 Assignment of the 513 cm^{-1} Oxygen Absorption: Certainly the 513 cm^{-1} absorption is due to oxygen, since an isotope shift is observed when O^{18} is substituted for O^{16} (71). However, when silicon samples are heated, the absorption coefficient of this line often decays more rapidly than does that of the 9μm line (62). This is additional evidence that the two lines are independent absorptions of two different species, although Shimura explained this effect by an oxygen polymerization reaction.

A close examination of this absorption shows a unique lineshape, unlike the shape of other absorption lines. The 513 cm^{-1} line is asymmetrical, with a broad tail to the low wavenumber side. The lineshape is identical to the prominent oxygen absorption in *germanium*, although the positions and linewidths are different (73). Oxygen in germanium shows a single line at 855 cm^{-1} whose FWHM is 6 cm^{-1} at room temperature (74). This contrasts to the 513 cm^{-1} line in silicon with a FWHM of 11 cm^{-1} and the 1106 cm^{-1} with a FWHM of 32 cm^{-1}. At low temperatures, the 855 cm^{-1} oxygen in germanium line *does not split* into multiple components, but rather decreases in linewidth to less than 1 cm^{-1}. This indicates a very well defined environment on an atomic scale. Taken together with the fact that the segregation coefficient of oxygen in germanium is 0.11 (75), this indicates a species very different from "interstitial" oxygen in silicon. A very weak line at 1260 cm^{-1} also appears due to oxygen in germanium. Anticipating the results of the argument which follows, I assigned the 513 cm^{-1} line in silicon and the 855 cm^{-1} line in germanium to a site which has the full symmetry of the lattice. This is either a true interstitial site or a fully substitutional position. Originally, I thought the substitutional site must be ruled out on chemical grounds, since tetravalent oxygen is a very unusual species. However, evidence from other impurity absorptions in silicon forced me to conclude that the substitutional position is the one actually occupied.

The differences in the incorporation of oxygen in silicon and germanium have implications for the historic choice of silicon as the basic material for electronic devices. Table 5 summarizes the observations on infrared absorption and offers assignments which allow understanding of the key difference between the materials. The substitutional oxygen species is the predominant form in germanium, while the reverse is true for silicon. The interstitial species, easily formed in silicon, may be thought of as a "monomer" of SiO_2. The fact that it forms easily in silicon but not in germanium leads to the ease of native oxide growth on silicon but not on germanium. The ease of SiO_2 layer growth was the basis for choosing silicon as the preferred device material. Substitutional oxygen is much more like a normal dopant, with a low segregation coefficient. In germanium, evidently discrete Ge-O complexes form rather than the infinite Ge-O polymer which would correspond to SiO_2.

4.3.4 Substitutional Oxygen in Silicon: The assignment of the 513 cm^{-1} oxygen band has so far been discussed in terms of what it is *not*, that is it is not a normal mode of the Si_2O interstitial oxygen species. The evidence includes complete analysis of low temperature and far infrared spectra, the relative absorption strength of this band to the 9μm band after heat treatment, and the peculiar lineshape of the 513 cm^{-1} band which is the same as the prominent oxygen band in germanium.

Table 5

Comparison of Infrared Absorption and other Properties of Oxygen in Silicon and Germanium

Species	Silicon	Germanium
Interstitial Oxygen		
Frequency (cm^{-1})	1106	1260
Linewidth at Half Maximum (cm^{-1})	38	20
Lineshape	Symmetric	?
Relation to Donor	3:1	N.A.
Segregation Coefficient	1.0	?
Substitutional Oxygen		
Frequency (cm^{-1})	513	855
Linewidth at Half Maximum (cm^{-1})	11	6
Lineshape	Asymmetric	Asymmetric
Relation to Donor	1:1(?)	4:1
Segregation Coefficient	0.25	0.11
Ratio of Substitutional to Interstitial Oxygen	1:16	16:1

Hrostowski and Kaiser have reported the isotope shift of this band, which definitely established that it is due to oxygen. A look at the absorptions of other impurities in silicon helps to establish its location in the lattice. Table 6 shows the frequencies reported for isotopes of oxygen (68), carbon (76), and boron (77) in silicon. These are low temperature frequencies, to allow direct comparison of the others to the boron lines which were reported only at low temperature. The lines shift to higher frequency in going from room temperature to liquid nitrogen temperature, and then do not shift further at temperatures below about 78K.

If these frequencies are plotted versus the inverse of the square root of isotopic mass, a straight line relationship is seen, as plotted in Figure 27. If omega versus inverse square root of mass results in a linear relation, the absorptions all correspond to the same impurity center, i.e. they all have the same site symmetry in the lattice. Furthermore, the linear frequency shift with different isotopic mass corresponds

Table 6

Low Temperature Absorption Frequencies for Light Elements in Silicon

Isotope	Frequency (cm-1)	Ref.
O^{18}	506.4	71
O^{16}	517.3	71
C^{14}	572.8	76
C^{13}	589.1	76
C^{12}	607.5	76
B^{11}	622.8	77
B^{10}	645.8	77

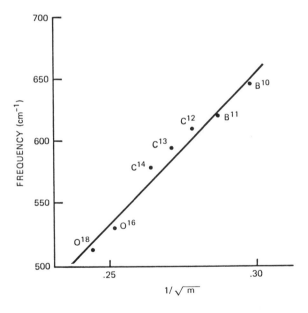

Figure 27. Plot of absorption frequencies of light impurity elements in silicon as a function of inverse square root of isotopic mass (75).

to simple harmonic oscillator behavior. This means that all impurities have the full symmetry of the lattice. That is, they occupy either the fully substitutional or true interstitial lattice position (78).

The choice of which position—substitutional or interstitial—is the correct one cannot be made on the basis of infrared absorption alone. When the observations of carbon absorption in silicon were made, the boron lines had already been reported and this was the original basis for assigning the substitutional site to carbon. Everybody *knows* that boron is substitutional in silicon, therefore the harmonic oscillator relation means that carbon is substitutional also. Actually, there are several objections that can be raised to this interpretation, including the fact that there is only one experiment reported which implies that boron is really substitutional (115). In addition, the boron absorptions were observed in heavily doped samples, and the transition moments of the boron and carbon absorptions, which determine their intensities as a function of concentration, are actually easier to interpret if both are assumed to be interstitial.

On the other hand, Baker et al. observed a contraction in the silicon lattice constant with increasing carbon levels, which indicates a substitutional species if Vegard's Law holds (79). This important piece of evidence means that the infrared absorption at 513 cm^{-1} is due to substitutional oxygen. Objections to this interpretation include one based on observation of the stress-induced dichroism of the band (80). A careful analysis of this experiment shows no inconsistency with the substitutional assignment.

In this experiment, silicon samples were subjected to uniaxial stress along a <110> direction, and the oxygen absorptions at 9 and 19μm were observed while under stress. Optical dichroism can be observed, since one third of the lattice sites are stressed, while the other sites are unaffected by the stress. Site symmetry is reduced, and polarized radiation can distinguish between stressed and unstressed sites. The sign of the dichroism of the 9μm line, previously observed by Bosomworth et al. (69), is opposite to that of the 19μm line. Since the asymmetric character of the 9μm absorption had been established, Stavola termed the 19μm line a symmetric absorption, based on the observed dichroism.

The persistence of the dichroism as a function of temperature was also measured. For both absorptions, the dichroism begins to disappear at temperatures above 350°C.

Stavola concluded that both were therefore due to the same species, interstitial oxygen.

If one examines the effects of <110> stress on the tetrahedral lattice position, it is apparent that such a stress reduces the symmetry of one third of these sites. If an oxygen atom is on such a site (this also would apply to the true interstitial site), then the oxygen complex with its four silicon near neighbors has C_{2v} symmetry. This is the same symmetry as the "interstitial" oxygen species. Thus the observed dichroism could apply to either species.

The fact that the dichroism of both 9 and 19μm lines disappears at 350°C is an indication that thermal donor complexes are forming, causing a reorientation of oxygen atoms which causes all lattice sites to become equivalent. The simultaneous disappearance is due to the fact that both the "interstitial" and substitutional oxygen species are a part of the thermal donor complex. This is additional support for the model of Keller, discussed in the following section.

4.3.5 Properties of Substitutional Oxygen: The relative intensities of the absorptions at 19μm and 9μm can be used to get some idea of the relative concentrations of the two oxygen species. This can be done assuming that the "oscillator strength" of the two transitions is the same. Hrostowski and Kaiser reported an intensity ratio of 10:1 for the 9μm and 19μm bands. Measurements at room temperature on commercially available crucible grown silicon samples show an intensity ratio of 16:1, which varies slightly in the as-grown condition from sample to sample. This means that the 19μm species represents a few percent of the total oxygen in these samples where the 9μm absorption coefficient is in the range of 3-4 cm^{-1}. This explains how the 9μm band can be proportional to the oxygen concentration measured by activation analysis. Yatsurugi and coworkers reported deviations from this proportionality only at higher concentration levels, when the 9μm absorption coefficient approached 7 cm^{-1}. The deviation from proportionality was always towards more oxygen measured by activation than determined by infrared.

Yatsurugi et al. determined the segregation coefficient of oxygen in silicon to be k=1.25. They also showed that k=1.0 for the 9μm species, with the remainder termed "infrared inactive". This additional oxygen may well be the 19μm species, whose concentration drops to a relatively low value in samples normally used for device manufacture, If so, it is possible that this second species has a segregation coefficient of k=0.25 and is incorporated into the lattice independently of the 9μm oxygen atom. This allows the understanding of many of the puzzling aspects of the apparent duality of properties of oxygen in silicon.

In a previous section of this chapter, we discussed the diffusion coefficient of oxygen in silicon, and the fact that two values were observed around 400°C depending on the previous thermal history of the sample. If there are two distinct species of oxygen in the lattice, the possibility of two diffusion coefficients is easily understood.

Some of the peculiar features of oxygen precipitation can also be understood in terms of two types of oxygen in silicon. In a wafer, the oxygen level as measured by infrared absorption is fairly uniform, measured from center to edge. On the other hand, precipitated oxygen can show marked radial gradients, often referred to as "oxygen swirl". A striking example of this was presented by Abe, comparing two wafers of identical oxygen level (18). One was pulled with growth-rate fluctuations, the other without. The wafer from the crystal pulled with growth-rate fluctuations showed inhomogeneous oxygen precipitation, in a banded or "swirl" pattern. There were no impurities present in the crystal which could have served as nucleation sites. This readily explained by a substitutional species which is subject to a fluctuating incorporation when the growth rate varies - an interstitial species would maintain constant concentration. The substitutional oxygen atom then serves as the nucleus for the precipitation of nearby interstitial oxygens, and the precipitation pattern follows the substitutional incorporation pattern. Table 7 summarizes the properties of interstitial oxygen and the proposed substitutional species.

Table 7
Properties of Oxygen in Silicon

Property	Interstitial	Substitutional
Segregation Coefficient	1.0	0.25
Relative Concentration	16	1
Relative Diffusivity	1	100
Relation to Thermal Donor	Satellite	Central Atom
Inhomogeneous Nucleation		+
Precipitate Nucleation Center		+

The implication of this observation is that the 19μm species controls the nucleation of oxygen precipitates in the temperature range 400-900°C. Around 450°C, this results in donor formation by a mechanism discussed below. Between 500-900°C, a nucleation process occurs which leads to precipitate formation when the material is subsequently heated at temperatures between 1000-1200°C. The nucleation and low temperature diffusion effects are reduced if the wafer is first heated to 1250-1350°C and quenched, suggesting that a redistribution of oxygen occurs in this temperature range. The redistribution very likely involves an equilibration between interstitial and substitutional sites, perhaps reducing the quantity of substitutional oxygen or eliminating severe gradients. Most likely, the solubility of substitutional oxygen as a function of temperature has a steeper gradient than does the interstitial species, and heating close to the melting point favors the interstitial species.

This hypothesis is illustrated schematically in Figure 28. At the melting point, oxygen is incorporated as the interstitial and substitutional species. In addition, the presence of oxygen clusters proposed by Yatsurugi and by Shimanuki et al. (25), cannot be ruled out. Figure 27 indicates that the solubility of the substitutional species is likely to be lower than for interstitial oxygen at any given temperature, and that it is incorporated at a concentration much higher than the equilibrium value at any lower temperature. The oxygen solubility versus temperature curves of Logan and Peters (4), and Bean and Newman (5), may actually represent the substitutional oxygen solubility function.

At temperatures between about 1250°C and the melting point of silicon, equilibrium between interstitial and substitutional oxygen can be established, resulting in lower values of substitutional oxygen concentration. See, for example, the results of Shimura et al. (69). If as-grown samples are subject to extended heating in this range, the material properties are perhaps permanently altered with respect to oxygen precipitation and donor formation at lower temperatures.

Silicon which is cooled quickly from the melting point to room temperature and subsequently reheated will show precipitation effects and precipitate morphologies which are a function of the thermal history, indicated schematically in Figure 28. Reheating to 1050-1150°C allows surface denuding of oxygen to occur.

Low temperature cluster formation produces thermal donors and other complexes, such as the "new donor". These

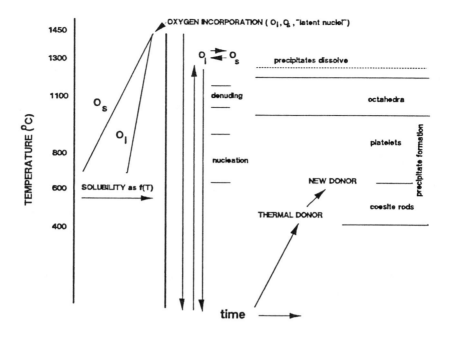

Figure 28. Schematic representation of the behavior of oxygen in silicon as a function of temperature.

grow into nuclei and precipitates at higher temperatures. Precipitate morphology appears to be controlled by the strain energy of the lattice or the effective "pressure" exerted by the surrounding lattice. Thus, extended heating at low temperatures, where lattice strain cannot be annealed out, results in coesite rods, the high pressure form of crystobalite. At temperatures above 800°C, an additional degree of freedom is provided for precipitate growth, and partial relief of strain produces SiO_x platelets. High temperature precipitation produces octahedra, with maximum three degrees of freedom due to minimal lattice strain. Octahedra thus exhibit the minimum surface to volume ratio, or lowest energy state possible for a second phase.

4.3.6 The A Center: The A center is a silicon-oxygen complex found in radiation-damaged silicon, Although it is not generally present in silicon used for devices, its structure is interesting in comparing the interstitial and substitutional oxygen species. The A center can be produced by irradiation of silicon which contains oxygen. Watkins, Corbett and coworkers, who determined the structure of this center, used 1.5 MeV electrons, whose effect is to create vacancies by knocking silicon atoms out of their lattice sites (81). The defects were studied by electron spin resonance and by infrared absorption. Results showed a decrease in intensity in the 9μm interstitial absorption and a new infrared band at 12μm. This latter band annealed out at temperatures of 300-400°C, and the 9μm band was restored to its original intensity when the 12μm band was completely annealed.

The structure of the A center consists of a vacancy-oxygen complex. The interstitial oxygen atom is connected to two near neighbor silicon atoms. When one of these atoms is knocked out of position by irradiation, the oxygen bonds to a next-near neighbor silicon. It is as if the oxygen has tried to move into the substitutional site, but cannot become fully substitutional because of its bond to one of the original silicon atoms. The Si-O bond length and the observed infrared absorption frequency are intermediate between those of interstitial and substitutional oxygen.

5.0 OXYGEN THERMAL DONOR

In 1954, Fuller et al. reported that donors were created in crucible grown silicon when samples were heated in the temperature range 300-500°C. They postulated that oxygen dissolved in the material was responsible for this effect. In the thirty years that have followed, the agreement that oxygen is indeed the cause of thermal donor formation in silicon is about the only thing that investigators have been able to establish beyond doubt. This is in spite of considerable effort to understand thermal donor structure, kinetics of formation and annihilation, diffusivity of oxygen at this temperature, and a host of other parameters. Indeed, the thermal donor represents perhaps nature's best kept secret in terms of the behavior of oxygen in silicon, or the behavior of any "impurity" in a semiconductor. This section is a review of some of what is known about the thermal donor.

5.1 Occurrence and Properties

Crucible grown silicon normally contains oxygen in amounts of $5-12 \times 10^{17}$ atom/cm^3. If the material is heated at about 450°C for times of a few minutes to several hours, donors are formed. The rate of formation can be as high as

10^{13} donors/cm³/s. The rate of formation is proportional to the fourth power of the oxygen concentration. Maximum donor levels of $2\text{-}5 \times 10^{16}$ added electrons/cm³ are achieved after about 50-70 hours of heating. Maximum donor concentration is proportional to the third power of the oxygen concentration.

Heating silicon samples with thermal donors at temperatures above 550°C and quenching causes donor annihilation, at a rate proportional to an exponential of temperature. This annihilation does not return the sample to the initial state, because subsequent donor formation is retarded and fewer donors are produced. When silicon ingots are pulled from the melt, the material cools through 450°C at a rate which is slow enough to induce donor formation. Manufacturing practice dictates an annihilation step, usually in wafer form. This step can influence the subsequent behavior of oxygen in silicon in thermal cycles used for device manufacturing. The kinetics of donor formation are rapid compared to the "normal" oxygen diffusivity expected at the formation temperature. The donor, which is stable below its formation temperature, is a double donor, with helium-like absorption spectra in the infrared at low temperatures. The energy levels lie at about 0.06 eV and 0.12 eV below the conduction band. The donor complex probably involves four oxygen atoms, and the simplest model consists of one substitutional oxygen atom surrounded by three next-near-neighbor interstitial oxygen atoms.

5.1.1 Kinetics of Formation:

Thermal donors are formed in the temperature range 300-500°C with a rate that is a strong function of temperature. Figure 29 shows data of Kaiser, Frish and Reiss for three temperatures (83). This is one of the first reports on the kinetics of formation of the donor. The initial rate of formation is proportional to the fourth power of the initial oxygen concentration, as measured by 9μm infrared absorption. Figure 30, also from Kaiser et al, shows this initial rate behavior. It is interesting that most subsequent analyses of the kinetics of donor formation depend on this early experimental data or that of Fuller and Logan (84). At temperatures between 430-500°C, extended heating produces a maximum donor concentration. In this temperature range, donor formation is a maximum at about 450°C. For 450°C, the time required for maximum donor formation is about 100 hours. Heating beyond this time causes a decrease in donor concentration. The maximum donor concentration is a function of the oxygen content, and is proportional to the third power of the oxygen concentration, as shown in Figure

31 (83). The time required to achieve the maximum donor concentration at 450°C increases with decreasing oxygen concentration.

More recent work has established that the oxygen donor is helium-like in character, that is, is a doubly ionized species rather than two singly ionized atoms. The energy levels associated with donor formation are measured to be approximately 0.06 and 0.12 eV below the conduction band; the levels decrease in energy or move toward the conduction band slightly with increasing time of heat treatment at 450°C (85),(86).

5.1.2 Kinetic Models for Donor Formation: The kinetics of donor formation were first addressed by Kaiser, Frisch and Reiss (KFR), (80). In order to account for the observed power law behavior of the initial formation rate and maximum concentration as a function of oxygen level, they postulated a sequence of reactions that involved the addition of oxygen atoms to form an SiO_4 species. The only mobile constituent on this model is interstitial oxygen. After the SiO_4 electrically active species is formed, it can be "annihilated" by the addition of further interstitial oxygens atoms to produce an

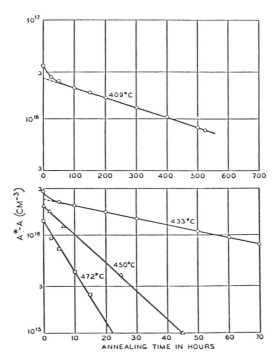

Figure 29. Oxygen thermal donor concentration as a function of annealing time and temperature (80).

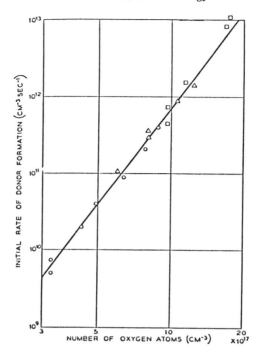

Figure 30. Oxygen thermal donor concentration as a function of initial interstitial oxygen content (80).

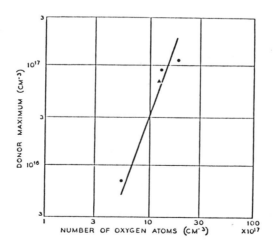

Figure 31. Maximum oxygen thermal donor concentration as a function of initial interstitial oxygen concentration (80).

electrically inactive complex of five or more oxygen atoms. The observed formation rate requires an oxygen diffusivity at least ten times the value extrapolated from high temperature measurements of the diffusion coefficient of interstitial oxygen.

The KFR model postulates complex formation by successive addition of interstitial oxygen atoms, forming two, three, and finally four atom clusters. It is the four atom species that is electrically active. They tacitly assume one added electron per SiO_4 species. The rate equations are

$$2O_i \underset{k_{-1}}{\overset{k_1}{\leftrightarrow}} A_2 \tag{33}$$

$$A_2 + O_i \underset{k_{-2}}{\overset{k_2}{\leftrightarrow}} A_3 \tag{34}$$

$$A_3 + O_i \underset{k_{-3}}{\overset{k_3}{\leftrightarrow}} TD \tag{35}$$

In these expressions, A_2 refers to the two oxygen atom complex, and so forth. TD is the four oxygen atom thermal donor species. Additional oxygen atoms can be added to the TD complex, forming "polymers" of five or more oxygen atoms. These larger clusters are electrically inactive. The rate equation for the formation of the five oxygen atom complex from the thermal donor is

$$TD + O_i \underset{k_b}{\overset{k_p}{\leftrightarrow}} P_5 \tag{36}$$

Donor formation predominates in the temperature range 300-500°C. The overall reaction of formation is

$$4O_i \underset{k_b}{\overset{k_f}{\leftrightarrow}} TD \tag{37}$$

This reaction leads to a formation rate which is proportional to the fourth power of the initial, interstitial oxygen concentration.

$$\frac{d[TD]}{dt} = (k_3 k_1 k_2) O_i^4 - (k_{-3} + k_p O_i) TD \tag{38}$$

KFR determined the maximum donor concentration to be

$$TD^* = \frac{k_f}{k_p}[O_i^0]^3 \qquad (39)$$

$$\ln[TD^* - TD(t)] = \ln TD^* - \kappa t \qquad (40)$$

where $\kappa = k_b + k_p[O_i]$

The value of κ was determined to be $\kappa = 1.8 \times 10^{-5} s^{-1}$. Equation 40 can be integrated to give

$$TD(t) = TD^0 e^{-\kappa t} \qquad (41)$$

This result is show graphically for temperatures of 433-472°C in Figure 29.

For temperatures of 550°C and higher, donors decay by rapid formation of P_5, the electrically inactive oxygen complex. For this temperature range, an additional term is needed to modify equation 41.

$$TD(t) = TD^*(1 - e^{-\kappa t}) + TD^0 e^{-\kappa t} \qquad (42)$$

For the high temperature range, $TD \ll TD^0$.

The activation energy for decay was determined to be 2.8eV. TD^0 is the original donor level prior to the decay or annihilation process. There's more on donor annihilation in the following section.

Several other investigators have attempted to modify, refine or extend the model of Kaiser et al. Oehrlein (87) used data obtained by Cazcarra and Zunio (88) to obtain a best fit by computer modeling to complexes with from three to five oxygen atoms. He concluded that the best fit was obtained with an SiO_3 structure for heating times of up to 100 hours. The double donor character was taken into account, and the oxygen content was measured by the calibration of Baker, termed the "Old ASTM" Calibration.

Helmreich and Sirtl proposed a single oxygen atom as responsible for thermal donor formation (89). In this model, vacancies diffuse to interstitial oxygen atom sites, and the oxygen atoms shift to substitutional positions. An additional vacancy is then trapped to form an electrically active oxygen-vacancy pair. Although this model is difficult to use for computations of formation kinetics, it does address the low diffusivity of interstitial oxygen and the observed suppression of donor formation by carbon, discussed later in this section.

Gosele and Tan, in an extensive review of oxygen diffusivity and thermal donor formation, suggested that two species of oxygen were involved (90). The interstitial species diffuses at the rate extrapolated from high temperatures, while a postulated molecular oxygen species exhibits a rate eight to ten orders of magnitude faster. The effective diffusivity of oxygen is a resultant of these two rates. The donor is suggested to result from an encounter of two molecular species, resulting in an O_4 complex.

Wada has also reviewed the literature on donor formation and has modified the Kaiser, Frisch, Reiss model to account for the double donor behavior (91). According to Wada's model, the formation rate and maximum concentration of donors depend on the intrinsic electron concentration at the formation temperature. The value of the diffusion coefficient of interstitial oxygen needed to fit the observed data is two orders of magnitude higher than the high temperature interstitial oxygen diffusivity. One interesting feature of this model, discussed below, is its explanation of the influence of electrical dopants on thermal donor formation. In addition, the theory addresses nucleation phenomena in the temperature range 650-800°C as extensions of donor formation mechanisms. These are also connected with donor annihilation, discussed below. A recent measurement of initial donor generation kinetics is given in (92).

Wada modified the KFR equations to include the ionization of electrons during the formation steps. for example,

$$A_3^{h+} + he^- + O_i \underset{k_{-3}}{\overset{k_3}{\leftrightarrow}} TD^{m+} + me^- \qquad (43)$$

and

$$TD^{m+} + me^- + O_i \underset{k_d}{\overset{k_p}{\leftrightarrow}} P_5 \qquad (44)$$

He then took m as a fitting parameter, and reviewed data of KFR and others for formation and decay of the donor. The best fit was obtained for m=2. This agrees with spectroscopic determination of the double donor character of the TD species.

Wada's rate equation for donor annihilation has the form

$$TD(t) = \frac{a}{b}[O_i]^3 n^{-3}\left(1 - \exp(-bD_i[O_i]t)\right) \qquad (45)$$

or

$$TD^* - TD = TD^* e^{(-\varkappa t)} \tag{46}$$

which is formally the same result as obtained by KFR. Wada took data from KFR and Logan and Peters, determining the value of the activation energy to be 2.54eV, or about the same result as KFR. Using the recently determined "high temperature" value of the diffusion coefficient of oxygen, the activation energy is right, but the value is two orders of magnitude too small to account for the annihilation rate. The "low temperature" diffusion coefficient has the right magnitude, but the activation energy is only 1.96eV. See below for a further discussion.

5.1.3 Kinetics of Annihilation: At temperatures above 550°C, donors are destroyed faster than they are created, at a rate that increases with temperature. Fuller and Logan showed that the annihilation process was irreversible; a second donor formation cycle after annihilation produced fewer donors than the first heating, and so on with subsequent cycles. These investigators found the process to be first order, and Kaiser et al. took the annihilation rate to be equal to the formation rate.

Kanamori recently investigated donor decay at 550°C, concluding with the following observations (93).

Using the rate equation of KFR, equation 40, samples annealed at 475°C for up to 144 hr did not exhibit the exponential decay reported by early investigators, except for the initial decay in some samples. In all cases, complete donor annihilation could not be achieved at 550°C. But interstitial oxygen is present at concentrations of 10^{18} or so, so the the KFR model was questioned by Kanamori as far as donor decay is concerned. He postulated that an unknown species is dominant for the decay process, according to the reaction

$$TD + X \underset{k'}{\overset{k}{\leftrightarrow}} P_5 \tag{47}$$

where P^5 indicates the neutral species according to the KFR notation. This results in second order decay kinetics, which fit the behavior of samples measured by Kanamori. He speculated that the silicon material available today is somehow different than that used by early investigators, perhaps in the concentration of the unknown species.

Rapid thermal annealing produces different donor decay characteristic than those observed by Kanamori, but his observations and the second order kinetics still apply. RTA at

550, 600, and 650°C results in decay curves shown in Figure 32. The initial decay regime shows that an incubation period occurs for some samples. This can be compared to Kanamori's observation that the donor annealing properties varied slightly from sample to sample in his investigation. Using the reaction of Kanamori, O'Mara et al. applied the following equation to the decay curves of Figure 34.

$$-\frac{d(TD)}{dt} = k[TD][X] \qquad (48)$$

The change in resistivity is fitted with a rate constant, k, whose activation energy is 1.7eV.

Annihilation is quite rapid, as seen for several temperatures in Figure 32. This figure shows the decay of electron concentration with time for samples previously heated at 450°C for 64 hours (94). For some samples a very brief initial period (< 1 second) is seen for which no decay is observed. At longer times, the decay follows second order kinetics as observed by Kanamori.

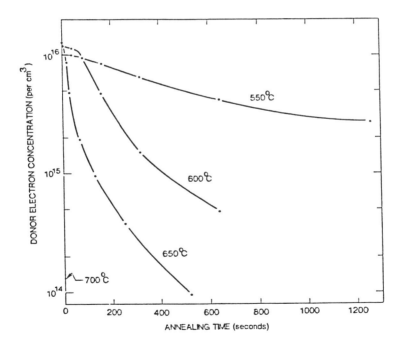

Figure 32. Thermal donor annihilation as a function of time and temperature (91).

Rapid annealing of donors is preferable to previous industrial practice, which consisted of the heating of silicon wafers at 600-700°C for one to two hours to destroy donors formed as the silicon ingot was cooled after crystal growth. Such extended heating serves to nucleate precipitates which grow in subsequent high temperature processing (95). Rapid annealing avoids unintentional nucleation. The fact that nucleation is coupled with donor annihilation, as well as the irreversible character of annihilation as observed by Fuller and Logan suggests that larger complexes of oxygen with silicon are indeed formed, rather than a dissolution and redispersion of oxygen atoms.

5.2 Influences on Donor Formation

5.2.1 Effect of Dopants: Thermal donor formation is strongly affected by the presence of electrical dopants at levels above 10^{17} carriers per cubic centimeter. The kinetic model of Wada explains this effect as being due to the difference in electron concentration in p and n-type material. This model predicts enhancement of thermal donor formation in heavily p-type silicon, and suppression in heavily n-type material. Available data support the enhancement effect, but the suppression of donor formation in n-type material is difficult to measure.

The influence of dopants on donor formation is connected with oxygen precipitation effects in these materials. Recent investigations of precipitation rates and degree of oxygen precipitation show an enhancement in heavily doped p-type silicon and suppression in n-type (96). The ease of thermal donor formation is intimately connected to precipitate nucleation, as previously mentioned, and results are consistent with the thermal donor as a precipitate precursor.

5.2.2 Effect of Carbon: If silicon contains a substantial amount of dissolved carbon, thermal donor formation is suppressed [89]. This effect is observable above 5×10^{16} atoms of carbon/cm^3. Some kind of complex formation between carbon and oxygen prevents the formation of electrically active donor complexes. This is an important clue to the structure of thermal donors, and can be incorporated into a model discussed below.

5.2.3 Oxygen Behavior at Donor Formation Temperature: As previously mentioned, the diffusion coefficient of oxygen obtained at high temperatures is too small to fit donor formation kinetics at 450°C. Attempts to measure the diffusion coefficient at this low temperature give results that depend on the prior thermal history of the sample. Figure 3 shows one such result, in which the "high temperature" diffusivity could be observed after high temperature (1350°C) heat treatment, followed by heating at 300-400°C (15). A diffusivity two orders of magnitude faster was observed in an "as received" crystal, which had been heat-treated at 900°C for two hours, presumably to annihilate donors formed during crystal growth. Wada's kinetic model as well as that of Kaiser et al. requires the faster diffusion coefficient to fit observed results. However, Wada argues that the activation energy of the diffusion coefficient is too low; instead of the observed value of 1.96 eV, his model requires the "high temperature" value of 2.54 eV.

5.3 Structural Models for Oxygen Thermal Donor

The structure of the thermal donor has been a subject of controversy since the initial observations of electrical activity in silicon. The original model for the thermal donor was derived by analogy with the thermal donor observed in germanium.

5.3.1 Thermal Donor in Germanium:
Oxygen can be incorporated in germanium crystals, from dissolution of the quartz crucible, if one is used, or by control of the oxygen partial pressure above the melt (75). Oxygen is more soluble in germanium than in silicon, and up to 10^{19} atom/cm^3 can be incorporated (74). The segregation coefficient is small, however; Darken has quoted a theoretically derived value of k=0.11 (75). Dissolved oxygen shows a very sharp infrared absorption peak at 855 cm^{-1} (12μm), with a weaker band at 1260 cm^{-1} (8μm). The 12μm band has a half-width of only six wavenumbers at 300 K, and sharpens at low temperatures to less than one wavenumber in width. No fine structure is observed, which indicates that the site symmetry of this band is unlike that of the 9μm oxygen in silicon band, and more like the 19μm oxygen in silicon absorption. In addition, the absorption is asymmetric, with a low wavenumber tail, also like the 19μm oxygen in silicon peak.

When germanium containing oxygen is heated in the range 300-500°C, a thermal donor is formed. At the same time, the 12μm absorption decreases in intensity. Calibration of the 12 μm line shows that one donor electron is produced for every four oxygen atoms lost from solution. In contrast to oxygen in silicon, the process appears to be reversible, and heating at 530°C nearly restores the initial intensity of the 12μm absorption. It is from these observations that Kaiser (73) suggested a GeO_4 complex; the question of double donor behavior was not addressed. The GeO_4 model was subsequently taken over to the silicon-oxygen thermal donor based on kinetics of formation.

5.3.2 Infrared Absorption: The double donor character of the oxygen complex in silicon has been established by a combination of Hall effect and infrared absorption measurements (85),(86). The exact energy values of the two donor levels, measured from the conduction band, depend on the time of heating at the donor formation temperature. Wruck and Gaworzewski reported the results shown in Table 8. The levels are seen at about 60 and 130 meV, and grow shallower with extended heating. Simultaneous observation of low temperature infrared absorption indicated the formation of multiple donor species.

Table 8
Electrically Determined Concentrations and Ionization Energies of Oyxgen Thermal Donors in Silicon

Annealing Time (h)	$n_{TD}(cm^{-3})$	E_1(meV)	E_2(meV)	E_2/E_1
0.75	10^{14}	60	130	2.2
7.5	10^{15}	55	120	2.2

Wruck and Garworzewski [84].

Observation of donor species by infrared depends on the electronic transitions between ground and excited states. These are seen only at low temperatures. The spectra could be interpreted by assuming a helium-like electronic structure for the donor with multiple ground states separated very slightly in energy. Up to nine distinct ground states have been observed. The distribution of thermal donors among the various ground states is a complicated function of the heating cycle. Absorptions are seen in two groups, from 380-520 cm^{-1} and from 710-1160 cm^{-1}. Most of the bands can be grouped into pairs of lines having energy separations of

$$E_{2p\pm} - E_{2p0}$$

and

$$4x\left(E_{2p\pm} - E_{2p0}\right) \tag{49}$$

corresponding to helium-like electronic transitions. The temperature dependence of the absorptions indicates the double donor character spectroscopically. The absorptions were compared to known double donors in silicon; sulfur, which is substitutional, and magnesium, which is interstitial. Wruck and Gaworzewski concluded that the oxygen donor is of the interstitial type.

The low temperature lines were studied recently by Claybourn and Newman to obtain an activation energy of formation for the nine species. A single value of the activation energy was obtained, of 1.7 ± 0.1eV (97). This is the same value that I and my coworkers obtained for the activation energy of annihilation (94), and verifies the assertion of Kaiser, Frisch and Reiss that formation and annihilation follow the same kinetics.

5.3.3 **Other Experimental Results:** Thermal donors have been studied by deep level transient capacitance spectroscopy (DLTS) (98), electron spin resonance (ESR) (99), and pressure dependent Hall effect (100). The ESR results of Wagner et al. for donors preferentially aligned by uniaxial stress show that the strain field produced by the donor core is compressive along its (001) axis, and tensile perpendicular to it. They conclude that the strain tensor components are consistent with substitutional oxygen at the core.

5.3.4 **Model for the Oxygen Thermal Donor:** Any model for the structure of the oxygen thermal donor must account for the following properties; its electrical activity, double donor character, formation rate more rapid than oxygen diffusion, four oxygen atoms as constituents, nine distinct donor species, effective mass center, substitutional oxygen as the core, and suppression of donor formation by carbon.

A model introduced by Keller fulfills all these requirements (100). It consists of a set of atomic structures including a substitutional oxygen atom, with three next-near-neighbor interstitial Si-O-Si groups. Figure 33 shows one possible way to arrange the four oxygen atoms. Although not mentioned in Keller's paper, a substitutional oxygen atom would have two extra valence electrons after satisfying the tetrahedral bond requirement; these would naturally constitute the pair of donor electrons and account for the observed electrical activity. The reason for the stability of the donor complex is explained in terms of strain compensation of the substitutional and interstitial oxygen species. The compressive strain of the Si-O substitutional bonds, which are shorter than Si-Si bonds, is compensated by the Si-O-Si bonds, which are longer than

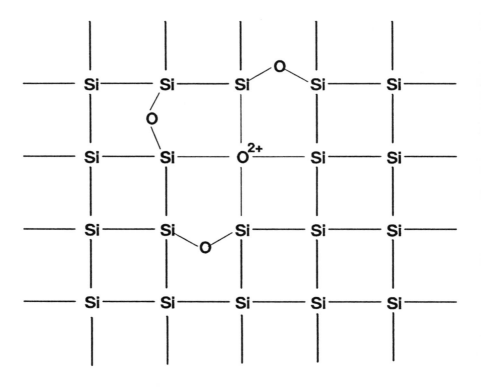

Figure 33. Structural model for the oxygen thermal donor (after 97).

Si-Si bonds. The structure is therefore stable, and the two electrons are derived from the ionization of the substitutional oxygen atom.

There are thirteen possible ways of arranging one substitutional oxygen and three neighboring interstitial oxygen atoms. Of these, four are ineffective in compensating the compressive stress of the central atom, leaving a total of nine species - the number seen in low temperature IR absorption. The successive increase and decrease of the different IR lines is explained in the following way. At temperatures between 300-500°C, oxygen atoms diffuse towards each other and form all kinds of the nine possible structures, since their energies are close together. Subsequent annealing allows rearrangement to form donors of successively lower energies, changing the relative strength of the IR lines.

The model of Keller can be extended to account for other observed properties of the donor. The fact that its formation rate is faster than accounted for by oxygen diffusivity is due to the rapid diffusion (or perhaps reorientation) of a substitutional species at low temperature. The fact that carbon suppresses donor formation is because similar complexes of a substitutional carbon and three interstitial oxygens would form, with similar strain compensation and stability, but *would not be electrically active* because carbon has only four valence electrons instead of six in the case of oxygen. Thus carbon competes with substitutional oxygen in the formation of these complexes. Since substitutional oxygen is presumably present at concentrations of $2-5 \times 10^{16}$ atom/cm^3 and quantitatively limits the amount of donors which can form, it is not surprising that the presence of carbon at similar concentrations will effectively suppress electrical activity.

When silicon in which donors have formed is heated to higher temperatures, oxygen continues to diffuse to the complex, and SiO$_x$ nuclei begin to form in the temperature range 600-900°C. Strain compensation assures the stability of the donor complex, but the clusters of larger size eventually form a second phase and become discrete precipitates. Thus the donor does not dissolve on heating above 550°C, but serves as a principle nucleation center for subsequent precipitation at high temperature. While Keller's is the most satisfactory donor model, several others have been proposed, as mentioned in the preceding portions of this section.

6.0 MECHANICAL STRENGTHENING AND WAFER WARPAGE

6.1 Dislocation Generation in Silicon

Silicon wafers are typically supplied free of dislocations, but subsequent handling can introduce them into the material. Isolated dislocations at low density have little effect on device properties, but under certain conditions, multiplication centers can develop which produce endless numbers of mobile dislocations. Sumino has found that the rate of dislocation movement is a sensitive function of temperature, and is relatively insensitive to the applied stress (101). In the temperature range 600-800°C, the rate of dislocation movement ranges from 10^{-6} to 10^{-3} cm/s! At the high end of this range, a dislocation can therefore move through a typical wafer thickness of 500μm in less than a minute. The propagation rate was found to be the same in float zone (oxygen-free) and crucible-grown wafers. Therefore, the presence of dissolved oxygen does not impede the movement of dislocations. Similarly, the yield strength of the two kinds of wafers was found to be the same prior to heat-treating steps which could introduce multiplication centers. Oxygen acts to lock these centers, preventing the generation of additional dislocations.

6.2 Slip and Bow

Slip is the term used to describe dislocation lines in silicon. These lines are typically revealed after a defect etch. Slip can result from dislocation generation, especially when wafers are subject to severe thermal gradients on cooling from high temperatures. Bow is a curvature of the wafer which can result if dislocations are generated during processing. This curvature can make subsequent handling and processing difficult.

6.3 Role of Oxygen

The primary reason that crucible grown silicon has become preferred for IC manufacturing, with its multiple high temperature steps, is the fact that oxygen inhibits the production of dislocations and minimizes slip and bow. Originally, it was thought that dissolved, interstitial oxygen atoms played this role. Recent work has indicated that some kind of oxygen clusters are required. Sumino suggested that an "oxygen atmosphere" forms around dislocation generation sites and prevents their operation. In crucible-grown silicon, although dislocations migrate as fast as in float zone material, there is a critical shear stress below which they are not generated. Furthermore, even if the critical stress is exceeded at high temperature, cooling allows the locking function to be established and prevent further dislocation generation. Chiou and coworkers have investigated the macroscopic aspects of dislocation and slip in crucible-grown material (102). They concluded that small grown-in crystobalite precipitates suppress the formation of slip. Dissolved nitrogen can also perform this function in low oxygen material.

6.4 Effects Due to Precipitation of Oxygen

Heating of silicon wafers can cause the precipitation of

oxygen from solid solution, and this usually leads to a reduction in the mechanical strength of the material, especially for large precipitates. Figure 34 shows the onset of plastic deformation in silicon as a function of the extent of precipitation. The precipitation was carried out at 1050°C and the tensile deformation performed at 900°C. In addition, precipitates were dissolved at temperatures ranging from 1240-1320°C. Results from the dissolution steps fall on the same line as for precipitation, i.e., the material strength is a unique and reproducible function of the extent of oxygen precipitation. Heating at 1050°C causes the formation of large precipitates. At lower temperatures, smaller precipitates form. This research work was performed by Kondo, who found that heating at 800°C could actually increase the material strength at low values of stress, although under higher stresses the 800°C pretreatment reduced the strength (103).

The bowing of wafers during heat treatment can be due to reduced mechanical strength caused by oxygen precipitation. Moerschel and coworkers found that if the initial bow of the wafer due to slicing was low, then the bow would not increase during heat treatment (104). For an intermediate value of oxygen precipitation, $1-2 \times 10^{17}$ atoms of oxygen/cm^3, the wafer was subject to substantial warpage during heat treatment. Values of precipitation above and below this produced wafers less susceptible to warpage.

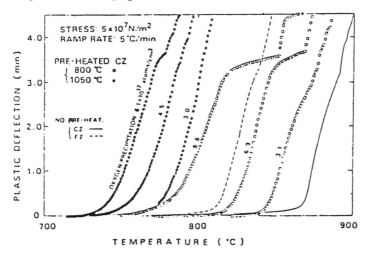

Figure 34. Yield strength of silicon as a function of dissolved and precipitated interstitial oxygen content (98). (This paper was originally presented at the Spring 1981 Meeting of the Electrochemical Society, Inc. held in Minneapolis, MN.)

7.0 DEVICE PROCESSING

7.1 Thermal Cycles and Process Simulation

Many factors influence the way that oxygen behaves in silicon during the manufacturing of IC's. Thermal processes vary from one device type to another, and a given device can be made using processes which differ from one wafer fabrication area to another. Furnace idle temperatures, ramp rates for heating and cooling, process temperatures and times and the gaseous ambient can affect denuding, nucleation and precipitation of oxygen. The thermal history of the wafer, its cooling experience in the ingot form as simulated in Figure 11, also influence the behavior of oxygen. Because of these complications, there have been recent efforts to develop physical and computer simulations which allow the prediction of effects due to oxygen. These efforts, sometimes termed "wafer engineering", are still in their infancy compared to process modeling for oxidation, diffusion and other steps in IC processing.

Physical simulation of CMOS and NMOS processes has recently been described by Chiou and Shive (105). They developed three and four step heat cycles for CMOS and NMOS respectively, which simulate the effects of twelve to fifteen thermal cycles actually used for device processing. Total time for the heat treatments was forty hours. Samples with a range of initial oxygen concentrations were tested for the extent of precipitation. Additional samples were subjected to shorter cycles consisting of the following sequences:

Test 1. 800°C/2hr + 1050°C/16hr

Test 2. 800°C/1hr + 1050°C/8hr

Again, the extent of oxygen precipitation was measured as a function of initial oxygen concentration. Results indicate the Test 1 is a good indicator of the behavior of oxygen in CMOS processes, while Test 2 is indicative of NMOS processes. Figure 35 shows a comparison of the results for the full CMOS simulation compared to the two-step Test 1 results. Oxygen concentration and precipitation values are derived from infrared absorption using the "Old ASTM" calibration. An S-shaped curve of precipitation versus initial concentration is seen. This is similar to results for the NMOS cycle, but the curve is shifted to the right in the latter case. That is, for a given oxygen concentration, less precipitation occurs for NMOS than for CMOS processing. Small amounts of carbon, in the 1-2 ppma range, have the effect of uniformly increasing the amount of precipitated oxygen to near the maximum observed level, regardless of the starting oxygen concentration (106). Additional tests are required to allow simulation of denuding and nucleation. Today, to obtain an appropriate denuded zone depth and extent of precipitation, a try and see approach is needed for each process line.

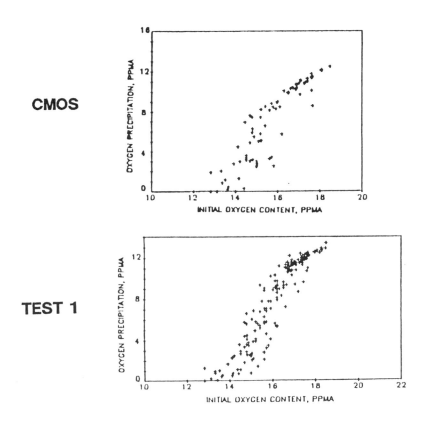

Figure 35. Reduction in concentration of interstitial oxygen in silicon as a function of initial oxygen content for thermal simulations of device processing (102). (This paper was originally presented at the Spring 1985 Meeting of the Electrochemical Society, Inc. held in Toronto, Canada.)

7.2 NMOS Circuits

High density dynamic random access memories (DRAMs) are built using n-channel MOS technology. These circuits operate as memory devices by storing charge near the silicon wafer surface in small capacitors. Defects and impurities can cause the loss of stored charge by leakage, and periodic refresh of the stored data is required. The rate at which refresh is required is a strong function of material properties. Huff and coworkers found that internal oxygen precipitation, together with a precipitation free surface, resulted in very long refresh times and higher yields [40]. In some cases, internal precipitation was found superior to epitaxial silicon for control of stray currents in DRAMs (45).

7.3 CMOS Circuits

Complementary MOS devices exhibit low power consumption and are used for static memories (SRAMs) and logic (microprocessors, gate arrays, etc.). Lee and Tobin have investigated the behavior of oxygen in silicon wafers for a CMOS process (107). Wafers with oxygen levels below 9×10^{17} atom/cm^3 levels showed little or no internal precipitation during the process. Above this concentration, precipitation amounting to about 40% of the initial oxygen occurred, most of which happened during the well drive step. Wafers with high oxygen concentration (>9×10^{17} atom/cm^3) showed warpage near the end of the process sequence. Warpage was eliminated by changing the push-pull rate of the step involved.

7.4 Bipolar Circuits

Bipolar devices require higher temperatures and longer process times than do MOS circuits. In addition, they are more susceptible to circuit failures due to defects and contamination. In order to minimize near surface defects, relatively high levels of initial oxygen in the wafer are required to provide precipitate growth throughout the circuit manufacturing process. In a recent study, Jastrzebski et al. showed that a gradual precipitation throughout the process was most beneficial (108). Optimum circuit yield was obtained when about 30% of the initial oxygen content precipitated during the process.

7.5 CCD Devices

Complex arrays of charge coupled devices (CCDs) are used as imagers and for specialized memories. The process sequence for such devices involves low temperatures (< 1000°C), and prior denuding and nucleation steps are added to ensure a defect free surface zone and the desired internal precipitation. The denuded zone ensures absence of white spots and internal precipitation can provide a sink for stray electrons by providing a low lifetime recombination zone. Researchers have discussed the techniques of wafer engineering for these devices (46),(108).

8.0 CARBON IN SILICON

Carbon is an impurity that can be incorporated in single crystals of silicon in a number of ways. In some cases, polysilicon can contain carbon due to traces of ethylchlorosilanes in the trichlorosilane starting material. During crystal growth, traces of oxygen or water vapor in contact with the hot graphite heater or susceptor will form carbon monoxide, which can dissolve in the molten silicon. Proper manufacturing and quality assurance methods can reduce or eliminate these sources of carbon.

8.1 Solubility

Carbon is slightly soluble in liquid and solid silicon. Equilibrium values of 40×10^{17} atom/cm^3 and 3.5×10^{17} atom/cm^3 respectively at the melting point have been reported (109),(110), although metastable solid solutions of up to 20×10^{17} atom/cm^3 have been measured (111). During normal growth of dislocation free silicon crystals, if the carbon level exceeds the solid solubility value, SiC particles will form and polycrystalline growth will result.

The solubility of carbon in silicon decreases below the melting point as a function of temperature as shown in Table 9. Whether or not precipitation occurs depends on factors such as the presence of oxygen. The carbon-silicon phase diagram is shown in Figure 36 (109).

8.2 Segregation Coefficient

Nozaki et al. determined the equilibrium segregation coefficient of carbon in silicon to be k=0.07±0.01 (106),(107).

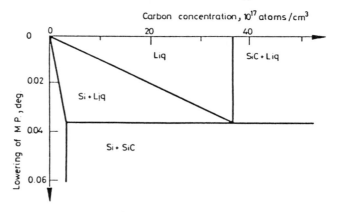

Figure 36. Phase diagram for the silicon-carbon system (106).

Table 9

Properties of Carbon in Silicon

Property	Value	Ref.
Solubility in Solid Si at M.P.	$C_s = 3.5 \times 10^{17}$ atom/cm^3	109
Solubility in Liquid Si at M.P.	$C_l = 40 \times 10^{17}$ atom/cm^3	110
Segregation Coefficient	$k_o = 0.07 \pm 0.01$ $k_o = 0.058 \pm 0.01$	109, 112
Solid Solubility as f(T)	$C_s = 4 \times 10^{24} \exp[-2.4 \text{eV}/kT]$ cm^{-3}	111
Diffusion Coefficient	$D = 1.9 \exp[-3.1 \text{eV}/kT]$ cm^{2}s^{-1}	113
Heat of Solution	$\Delta H_{sol'n} = 55\, kcal/mol$	109
Concentration by IR Absorption at 607 cm^{-1}	$C_s = \alpha \cdot 0.83 \times 10^{17}\, atom/cm^3$	117
	$C_s = \alpha \cdot 1.0 \times 10^{17}\, atom/cm^3$	118, 54

Kolbesen and Muhlbauer reexamined their data and determined a value of k=0.057; an independent experiment by the same investigators produced a value of k=0.058±0.005 (112). This equilibrium value of k=0.06-0.07 obtains for actual conditions of crystal growth for low growth velocities. At high growth rates, the effective segregation coefficient increases and can equal 1.0 for very fast growth rates such as occur in certain types of silicon sheet growth. In this way, concentrations of dissolved carbon higher than the solubility value are achieved. Carbon build-up in the melt for low growth rates can exceed the solid solubility level in poorly maintained crystal growth furnaces. This produces twinning and polycrystalline growth near the end of the ingot.

The low value of the segregation coefficient can lead to microscopic inhomogeneities in carbon inclusion with fluctuations in local growth rate. This is especially true for float zone crystal growth where radial variations in carbon can fluctuate ±70% locally. Small diameter crucible grown crystals are also subject to such fluctuations.

8.3 Diffusivity

Using ^{14}C tracer, Newman and Wakefield determined the diffusivity of carbon in silicon (113), with the value shown in Table 9. Carbon is a fast diffuser for a substitutional species. Gosele and Tan have suggested diffusion by a silicon-carbon complex, or by the kick-out mechanism, like gold in silicon (114).

8.4 State of Carbon in Silicon

Carbon occupies a substitutional site in the silicon lattice, and is therefore electrically inactive. Direct evidence for this is limited to observations of changes in lattice parameter and infrared absorption. Baker and coworkers measured the change in the lattice parameter of silicon as a function of carbon concentration [79]. Increasing the carbon concentration caused a decrease in the lattice parameter. Vegard's law says that such behavior means a substitutional impurity (115). Carbon levels above the solubility level caused the formation of a second phase, identified as SiC.

Dissolved carbon gives rise to an infrared absorption at 607 cm^{-1}, with intensity proportional to concentration. Care must be taken in measuring the absorption quantitatively, since it coincides with an intense lattice absorption. The latter is subtracted by double beam spectroscopy or by electronic methods using FTIR. The calibration of this line was originally performed by Newman and Willis in samples containing oxygen (116). Because of this, some C-O complexes formed which altered the apparent value of the dissolved carbon level. The calibration of Newman and Willis was recently reproduced by Regolini and coworkers using deuterium activation analysis for silicon samples also containing both carbon and oxygen (54). Endo et al. have produced the calibration given in Table 8 (117). This calibration is *lower by about 30%* than the calibration given by ASTM F123-81, which was derived from the data of Newman and Willis (118).

The infrared absorption of carbon at 607 cm^{-1} is a single line, which indicates that the center has the full symmetry of the lattice. Substitution of isotopes produces the frequency shifts shown in Figure 27. Comparison with boron absorptions led to the conclusion that carbon is substitutional, but the only compelling evidence for this assignment is the X-ray lattice constant measurements mentioned earlier; infrared analysis of boron and carbon absorptions could equally lead to the conclusion that both are due to interstitial impurities.

8.5 Complexes with Oxygen

Infrared absorption of silicon samples containing carbon and oxygen show bands due to complexes at 1103, 1052, and a small absorption at 1099 cm^{-1} (113). The interpretation of these bands is open to some question. As mentioned in the section on thermal donors, carbon-oxygen complexes which are evidently not infrared active form when silicon is heated in the 300-500°C range. These complexes are also electrically inactive and suppress the formation of thermal donors. On the other hand, the presence of carbon *enhances* the formation of new donors at 700-800°C, which are thought to consist of positively charged SiO_x precipitates (119),(120). Charged interfaces between larger silica precipitates, not necessarily containing carbon, and the surrounding lattice have been reported by Hwang et al. These phenomena suggest that carbon can serve as a nucleus for oxygen complexing and precipitation; this has recently been pointed out for "low oxygen" samples by Chiou (103). Significant precipitation in simulation heat-treatments occurred in the concentration range of 0.5-1.0 ppma of carbon (ASTM calibration).

8.6 Formation of Precipitates Due to Carbon

The presence of carbon in polysilicon used for crystal growth or the contamination of the melt by carbon from the graphite fixtures in the furnace can lead to levels of dissolved carbon high enough for precipitation in subsequent heating of the silicon. Figure 37 shows the results of an experiment to determine the effects of carbon incorporation (122). The upper curve corresponds to crystal growth from carbon-containing polysilicon. High concentrations of 1-5 ppma (ASTM) are seen by infrared absorption. The middle curve is from a crystal grown from carbon-free polysilicon at one atmosphere ambient and shows carbon incorporation from the furnace parts during crystal growth. The lower curve is from a crystal grown with carbon-free polysilicon under reduced pressure of 30 torr, used for most commercial crystal growth. Low carbon levels result from exhausting the furnace ambient at reduced pressure.

The late David Guidici and I took samples of silicon from these ingots, and heat treated them at 1050°C in a steam/O_2 environment for ten hours. Secco etching revealed a new type of defect in samples with initial carbon concentration of 2 ppma or greater. Figure 38 shows an optical micrograph of the tear-drop shaped etch pits, which are oriented 45° to the <110> directions in this (100) wafer. Transmission electron microscopy in these samples revealed precipitates different in morphology from those normally seen in crucible grown silicon of low carbon content. Figure 39 shows one such precipitate with the characteristic hexagonal platelet structure of beta silicon carbide. Bean and coworkers have noted similar defects in epitaxial silicon layers containing high levels of carbon (123). Using Wright etch, they showed that the teardrop shaped etch features lie at 45° angles to the <110> directions, identifying them as identical to those observed in bulk grown samples. It should be emphasized that ordinarily, both bulk and epitaxial silicon samples have low values of carbon content. For crucible grown ingots, reduced pressure growth, quartz shields around the hot graphite parts, careful meltdown procedures which avoid very high temperatures, regular leak-checking of water cooled chamber vessels, and routine analysis of finished ingots will minimize carbon inclusion. However, defect etching after oxidation remains a useful tool to identify potentially harmful defects due to this element.

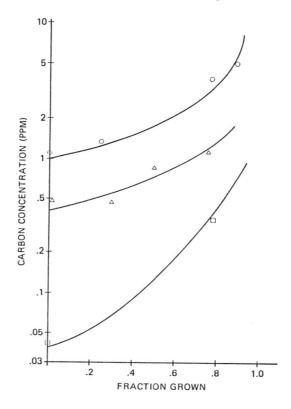

Figure 37. Axial variation of carbon content in a crucible grown silicon ingot as a function of starting material and growth conditions.

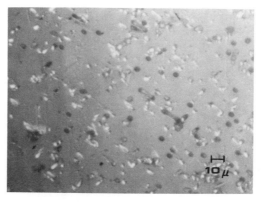

Figure 38. Optical micrograph of defects due to carbon in silicon wafers with high carbon content (117).

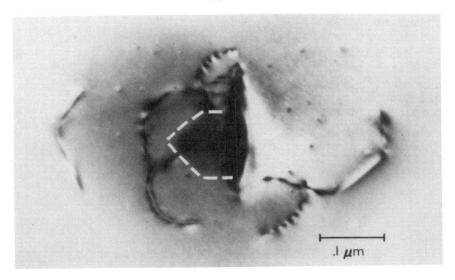

Figure 39. TEM photo of characteristic beta-SiC platelet in high-carbon silicon samples after heating (23).

A recent review of the properties of carbon by Kolbesen includes an analysis of the impact of carbon on device manufacturing (124). He observes that for discrete device manufacturing, where float zone silicon is used, carbon-induced defects are formed during device processing. These are termed "swirl" defects, and are harmful to yield and performance of discrete devices. On the other hand, for IC manufacturing using float zone substrates, carbon is a relatively benign impurity.

9.0 NITROGEN IN SILICON

Nitrogen can be incorporated into silicon crystals in small amounts during crystal growth. Float zone crystals can be nitrogen doped from the gas phase (18),(125). Crucible grown silicon containing nitrogen has been prepared by growth from a crucible of silicon nitride (126). The crucible was composed of hot pressed silicon nitride coated with α Si_3N_4, deposited at 100°C from SiH_4 and NH_4. Whiskers of silicon nitride grew from the crucible wall in contact with the molten silicon. Nitrogen can also be introduced into the silicon lattice by ion implantation (127). Stein has recently reviewed the properties of implanted nitrogen (128).

9.1 Solubility and Phase Diagram

The solubility of nitrogen in liquid silicon was measured by Kaiser and Thurmond (120), and Yatsurugi et al. (2),(106). The latter group also determined the solid solubility of nitrogen at the melting point of silicon. Solubility in the liquid is $C_l = 6 \times 10^{18}$ atom/cm^3, determined by rapid solidification of a liquid silicon sample saturated with nitrogen. Solid solubility was determined by very slow freezing of a molten zone in float zone growth. The value is $C_s = 4.5 \times 10^{15}$ atom/cm^3. These and other properties of nitrogen in silicon are summarized in Table 10.

Table 10

Properties of Nitrogen in Silicon

Property	Value	Ref.
Solubility in Solid Si at M.P.	$C_s = 4.5 \times 10^{15}$ atom/cm^3	2,106
Solubility in Liquid Si at M.P.	$C_l = 6 \times 10^{18}$ atom/cm^3	2,106
Segregation Coefficient	$k_0 = 7 \times 10^{-4}$	2,106
Diffusion Coefficient	$D = 0.87 \exp(-3.29/kT)$ cm^{2}s^{-1}	122
Concentration by IR Absorption at 963 cm^{-1}	$C_s = \alpha \cdot 1.82 \times 10^{17}$ atom/cm^3	125

Nitrogen concentrations at these low levels were determined by activation analysis, using high energy protons causing the reaction

$$^{14}N\ (p,\alpha)\ ^{11}C \tag{50}$$

Activation analysis had a detection limit of 1×10^{14} atom/cm^3 for nitrogen, according to Yatsurugi.

The solubilities of nitrogen in liquid and solid silicon indicated a segregation coefficient of $k = 7 \times 10^{-4}$. As mentioned previously, the small size of the nitrogen atom and the tendency to form stable Si_3N_4 compounds lead to a low segregation coefficient and low solubility.

Measured solubilities and segregation coefficient were used to construct a phase diagram. Lowering of the melting point was calculated from

$$\Delta T = \frac{x(1-k_0)RT_m^2}{L} \quad (51)$$

where

ΔT = melting point depression

x = mole fraction of nitrogen

k_0 = segregation coefficient

R = gas constant

T_m = melting point of silicon

 = 1683 K

L = molar heat of fusion of silicon

 = 12.1 kcal/mol

The phase diagram of the N-Si system at low nitrogen concentrations is shown in Figure 40.

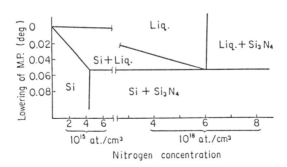

Figure 40. Phase diagram of the silicon-nitrogen system (106).

Diffusivity of nitrogen in silicon has been measured only in ion-implanted material. A value of the diffusion coefficient of

$D_O = 0.87\exp(-3.29eV/kT)cm^2s^{-1}$ has been reported (122).

9.2 Infrared Absorption

Nitrogen in silicon exhibits infrared absorption bands at 963 and 764 cm^{-1} (18). The strength of both bands is proportional to the concentration of nitrogen (129), and the peak at 963 cm^{-1} has been calibrated by activation analysis for the measurement of nitrogen concentration (130). The calibration coefficient is shown in Table 10. An isotope shift due to ^{15}N doping moves the 963 cm^{-1} absorption to 938 cm^{-1}. For silicon simultaneously doped with ^{14}N and ^{15}N, absorptions are observed at 963 cm^{-1} (^{14}N-^{14}N), 938 cm^{-1} (^{15}N-^{15}N) and 948 cm^{-1} (^{14}N-^{15}N). Low temperature absorption measurements showed no splitting of the 963 cm^{-1} N-N band due to interaction with the silicon lattice. Such interaction is observed for the 9μm oxygen in silicon absorption.

Interpretation of the infrared absorption leads to the conclusion that a nitrogen molecule is the species which is incorporated during crystal growth. However, the exact site symmetry of the impurity has not been unambiguously determined.

9.3 Mechanical Strengthening

Abe and coworkers have shown that the presence of nitrogen at parts per billion levels contributes added mechanical strength to oxygen-free float zone crystals (18),(131). Typical heating cycles used for integrated circuit manufacturing ordinarily cause massive dislocation generation, termed slip, in such material. Abe's studies included simulation of such heat cycles as well as experiments with dislocation movement after diamond indentation. Dislocation locking was found to be constant over a wide range of nitrogen concentration. When crucible grown crystals containing both oxygen and nitrogen were investigated, results indicated that nitrogen served as a nucleating center for oxygen precipitation.

REFERENCES

1. Fischler, S., Correlation between Maximum Solid Solubility and Distribution Coefficient for Impurities in Ge and Si, *J. Appl. Phys.* **33**, 1615 (1962).

2. Yatsurugi, Y., Akiyama, N., Endo, Y., and Nozaki, T., Concentration, Solubility, and Equilibrium Distribution Coefficient of Nitrogen and Oxygen in Semiconductor Silicon, *J. Electrochem. Soc.* **120**, 975-979 (1973).

3. Kaiser, W. and Keck, P.H., Oxygen Content of Silicon Single Crystals, *J. Appl. Phys.* **28**, 882-887 (1957).

4. Logan, R.A., and Peters, A.J., Diffusion of Oxygen in Silicon, *J. Appl. Phys.* **30**, 1627-1630 (1959).

5. Bean, A.R., and Newman, R.C., The Solubility of Carbon in Pulled Silicon Crystals, *J. Phys. Chem. Solids* **32**, 1211-1219 (1971).

6. Hrostowski, H.J., and Kaiser, R.H., The Solubility of Oxygen in Silicon, *J. Phys. Chem. Solids* **9**, 214-216 (1959).

7. Takano, Y., and Maki, M., Diffusion of Oxygen in Silicon, in, *Semiconductor Silicon 1973*, H.R. Huff and R.R. Burgess, ed., Electrochem. Soc., Pennington N.J., 1973, p. 469-481.

8. Mikkelsen, J.C., Jr., Excess Solubility of Oxygen in Silicon During Steam Oxidation, *Appl. Phys. Lett.* **41**, 871-873 (1982).

9. Craven, R.A., Oxygen precipitation in Czochralski Silicon, in *Semiconductor Silicon 1981*, H.R. Huff, R.J. Kriegler, and Y. Takeishi, ed., Electrochem. Soc., Pennington NJ, 1981, p. 254-271.

10. Itoh, Y., and Nozaki, T., Solubility and Diffusion Coefficient of Oxygen in Silicon, *Japan J. Appl. Phys.* **24**, 279-284 (1985).

11. Gass, J., Muller, H.H., Stussi, H. and Schweitzer, S., Oxygen Diffusion in Silicon and the Influence of Different Dopants, *J. Appl. Phys.* **51**, 2030 (1980).

12. Chikawa, J-I, Abe, T. and Harada, H., Oxygen Precipitation Enhanced with Vacancies in Silicon, in, *Semiconductor Silicon 1986*, H.R. Huff, T. Abe, and B. Kolbesen, ed., Electrochem. Soc., Pennington, NJ, 1986, p. 61-75.

13. Carlberg, T., Calculated Solubilities of Oxygen in Liquid and Solid Silicon, *J. Electrochem. Soc.* **133**, 1940-1942 (1986).

14. Mikkelsen, J.C., Jr., Diffusivity of Oxygen in Silicon During Steam Oxidation, *Appl. Phys. Lett.* **40**, 336-337 (1982).

15. Stavola, M., Patel, JR., Kimerling, L.C. and Freeland, P.E., Diffusivity of Oxygen in Silicon at the Donor Formation Temperature, *Appl. Phys. Lett.* **42**, 73-75 (1983).

16. Corbett, J.W., McDonald, R.S. and Watkins, G.D., The Configuration and Diffusion of Isolated Oxygen in Silicon and Germanium, *J. Phys. Chem. Solids* **25**, 873 (1964).

17. Gosele, U. and Tan, T.Y., Oxygen Diffusion and Thermal Donor Formation in Silicon, *Appl. Phys.* **A28**, 79-92 (1982).

18. Abe, T., Kikuchi, K., Shirai, S. and Muraoka, S., Impurities in Silicon Single Crystals, in, *Semiconductor Silicon 1981*, op. cit., p. 54-71.

19. Lin, W. and Stavola, M., Oxygen Segregation and Microscopic Inhomogeneity in Czochralski Silicon, *J. Electrochem. Soc.* **132**, 1412-1416 (1985).

20. Lin, W. and Hill, D.W., Oxygen Segregation in Czochralski Silicon Growth, *J. Appl. Phys.* **54**, 1082 (1983).

21. Jaccodine, R.J. and Pearce, C.W., The Segregation Coefficient of Oxygen in Silicon, in, *Defects in Silicon*, W.M. Bullis and L.C. Kimerling, ed., Electrochem. Soc., Pennington, NJ, 1983, p. 115-119.

22. Inoue, N., Wada, K., and Osaka, J., Oxygen Precipitation in Czochralski Silicon - Mechanism and Application, in *Semiconductor Silicon 1981*, op. cit., p. 282-293.

23. TEM photograph made by Joseph Peng.

24. Rogers, B., Fair, R.B., Dyson, W. and Rozgonyi, G.A., Computer Simulation of Oxygen Precipitation and Denuded Zone Formation, in, *VLSI Science and Technology/1984*, K.E. Bean and G.A. Rozgonyi, ed., Electrochem. Soc., Pennington, NJ, 1984, p.74-84.

25. Shimanuki, Y., Furuya, H., Suzuki, I. and Murai, K., Effects of Thermal History on Microdefect Formation in Czochralski Silicon Crystals, *Japan J. Appl. Phys.* **24**, 1594-1599 (1985).

26. Cavitt, J.H., Calculation of Oxygen Precipitate Density in Silicon and Its Applications to Nucleation Studies, in, *VLSI Science and Technology/1984*, op. cit., p. 66-73.

27. Lavine, J.P., Anagnostopoulos, C.N. and Rivaud, L., Oxygen Precipitation in *Silicon: Numerical Models*, Mat. Res. Soc. Symp. Proc. **59**, 301-307 (1986).

28. Hartzell, R.A., Schaake, H.F. and Massey, R.G., A Model that Describes the Role of Oxygen, Carbon, and Silicon Interstitials in Silicon Wafers during Device Processing, *Mat. Res. Soc. Symp. Proc.* **36**, 217 (1985).

29. Tan, T.Y. and Kung, C.Y., On Oxygen Precipitation Retardation/Recovery Phenomena, in, *Semiconductor Silicon 1986*, op. cit., p. 864-873.

30. Tsuya, H., Behavior of Oxygen and Dopants in Heavily Doped Silicon Crystals, in, *Semiconductor Silicon 1986*, op. cit., p. 849-863.

31. Walitzki, H., Rath, H., Reffle, J., Pahlke, S. and Blatte, M., Control of Oxygen and Precipitation Behavior of Heavily Doped Silicon Substrate Materials, in, *Semiconductor Silicon 1986*, op. cit., p. 86-99.

32. Wada, K. and Inoue, N., Oxide Precipitate Nucleation in Czochralski Silicon - An Insight From Thermal Donor Formation Kinetics, in *Semiconductor Silicon 1986*, op. cit., p. 778-789.

33. Daido, K., Snoyama, and Inoue, N., Czochralski Silicon Growth and Characterization, *Rev. Elec. Comm. Labs* (Japan) **27**, 33-40 (1979).

34. Patel, J.R., Jackson, K.A. and Reiss, H., Oxygen Precipitation and Stacking-Fault Formation in Dislocation-Free Silicon, *J. Appl. Phys.* **48**, 5279 (1977).

35. Wada, K., Inoue, N., and Kohra, K., Diffusion Limited Growth of Oxide Precipitates in Czochralski Silicon, *J. Cryst. Growth* **49**, 749 (1980).

36. Shiraki, H., Yoshinaka, A., and Sugita, Y., Formation of a Stacking Fault-Free Region in Thermally Oxidized Silicon, *Japan J. Appl. Phys.* **17**, 767-771 (1978).

37. Shiraki, Y., Silicon Device Consideration on Grown-In and Process Induced Defects and Fault Annihilation, in, *Semiconductor Silicon 1977*, H.R. Huff and E. Sirtl, ed., Electrochem. Soc., Pennington, NJ, 1977, p. 546-558.

38. Sugita, Y., Kawata, H., Nakamichi, S., Okabe, T., Watanabe, T., Yoshikawa, S., Itoh, Y. and Nozaki, T., Measurement of the Out-Diffusion Profile of Oxygen in Silicon, *Japan J. Appl. Phys.* **24**, 1302-1306 (1985).

39. Andrews, J., Oxygen Out-Diffusion Model for Denuded Zone Formation in Czochralski-Grown Silicon with High Interstitial Oxygen Content, in, *Defects in Silicon*, op. cit., p. 133-141.

40. Huber, D. and Reffle, J., Precipitation Process Design for Denuded Zone Formation in Cz-Silicon Wafers, *Sol. State Technol.*, August, 1983, p. 137-143.

41. Huff, H.R., Schaake, H.F., Robinson, J.T., Baber, S.C. and Wong, D., Some Observations on Oxygen Precipitation/Gettering in Device Processed Czochralski Silicon, *J. Electrochem. Soc.*, **130**, 1551-1555 (1983).

42. Kishino, S., Aoshima, T., Yoshinaka, A., Shimizu, H., and Ono, M., A Defect Control Technique for Intrinsic Gettering in Silicon Device Processing, *Japan J. Appl. Phys.* **23**, L9 (1984).

43. Hirao, T. and Maegawa, S., A Study of HCl Intrinsic Gettering for Application to Bipolar Devices and MOS LSI, in, *Semiconductor Silicon 1986*, op.cit., p. 927-938.

44. Borland, J.O., Kuo, M., Shibley, J., Roberts, B., Schindler, R. and Dalrymple, T., An Intrinsic Gettering Process to Improve Minority Carrier Lifetimes in MOS and Bipolar Silicon Epitaxial Technology, in, *Semiconductor Processing*, D. C. Gupta, ed., Amer. Soc. Test. Matl., Philadelphia, PA, 1984, p. 49-62.

45. Harada, H., Itoh, T., Ozawa, N., and Abe, T., Incorporation of Oxygen Impurity into Silicon Crystals During Czochralski Growth, in, *VLSI Science and Technology/ 1985*, Electrochem. Soc., Pennington NJ 1985, p. 526-535.

46. Otsuka, H., Watanabe, K., Nishimura, H., Iwai, H. and Nihira, H., The Effect of Substrate Materials on Holding Time Degradation in MOS Dynamic RAM, *IEEE Elec. Dev. Lett.* **EDL-3**, 182-184 (1982).

47. Anagnostopoulos, C.N., Nelson, E.T., Lavine, J.P., Wong, KY., and Nicholas, D.N., Latch-Up and Image Crosstalk Suppression by Internal Gettering, *IEEE Trans. Elec. Dev.* **ED-31**, 225-231 (1984).

48. Graff, K., Grallath, E., Ades, S., Goldbach, G. and Tolg, G., Determination of Oxygen in Silicon at Parts-Per Billion Level Through Calibration of IR-Absorption at 77K, *Sol. State Electron.* **16**, 887-893 (1973).

49. Baker, J.A., Oxygen and Carbon Content of Czochralski Silicon Crystals, *Sol. State Electron.* **13**, 1431 (1970).

50. Kim, C.K., Determination of Oxygen in Gallium Phosphide and Silicon by Helium-3 Activation, *Anal. Chim. Acta* **54**, 407-414 (1971).

51. Nozaki, T., Yatsurugi, Y., and Akiyama, N., Charged Particle Activation Analysis for Carbon, Nitrogen and Oxygen in Semiconductor Silicon, *J. Radioanal. Chem* **4**, 87-98 (1970)

52. Iizuka, T., Takasu, S., Tajima, M., Arai, T., Nozaki, M., Inoue, N., and Watanabe, M., Conversion Coefficient for IR Measurement of Oxygen in Si, in, *Defects in Silicon*, op. cit., p. 265-274.

53. Rath, H.J., Stallhofer, P., Huber, D., and Schmitt, B.F., Determination of Oxygen in Silicon by Photon Activation Analysis for Calibration of Infrared Absorption, *J. Electrochem. Soc.* **131**, 1920-1923 (1984).

54. Regolini, J.L., Stoquert, J.P., Ganter, C. and Siffert, P., Determination of the Conversion Factor for Infrared Measurements of Carbon in Silicon, *J. Electrochem. Soc.* **133**, 2165-2171 (1986).

55. See J.I. Pankov, *Optical Processes in Semiconductors*, Dover Publ., NY, NY 1971, p. 75.

56. ASTM Method F121-70T through F121-79

57. ASTM Method F121-80

58. Bullis, W.M., Watanabe, M., Bagdadi, A., Yue-zhen, Li, Scace, R.I., Series, R.W. and Stallhofer, P., Calibration of Infrared Absorption Measurements of Interstitial Oxygen Concentration in Silicon, in, *Semiconductor Silicon 1986*, op. cit., p. 166-180.

59. He Huan-nan, Li Yue-zhan, Xhao Guan-di, Yan Rong-hua, Lu Quing-nan, and Qi Ming-wei, Talanta, **30**, 761 (1983). Also Li Yue-zhen and Wang Qi-min in *Twelfth Annual Review of Progress in Quantitative Nondestructive Evaluation*, **5**, 957 (1986).

60. Pajot, B., Characterization of Oxygen in Silicon by Infrared Absorption, *Analusis* **5**, 32 (1977).

61. Tempelhoff, K. and Spiegelberg, F., Precipitation of Oxygen in Dislocation-Free Silicon, in, *Semiconductor Silicon 1977*, op. cit., p. 585-594.

62. Freeland, P.E., Oxygen Precipitation in Silicon at 650°C, *J. Electrochem. Soc.* **127**, 754-756 (1980).

63. Haas, W.E. and Schnoller, M.S., Silicon Doping by Nuclear Transmutation, *J. Elec. Mat.* **5**, 57 (1976).

64. Chen, C.S. and Corelli, J.C., Infrared Spectroscopy of Divacancy-Associated Radiation-Induced Absorption Bands in Silicon, *Phys. Rev.* **B5**, 1505 (1972).

65. Stallhofer, P. and Huber, D., Oxygen and Carbon Measurements on Silicon Slices by the IR Method, *Sol. State Technol.*, August (1983) p. 233-237.

66. Herzberg, G., *Infrared and Raman Spectra of Polyatomic Molecules*, Van Nostrand Reinhold, NY, NY 1945.

67. Kaiser, W., Keck, P.H., and Lange, C.F., Infrared Absorption and Oxygen Content in Silicon and Germanium, *Phys. Rev.* **101**, 1264-1268 (1956).

68. Bond, W.L. and Kaiser, W., Intersitital Versus Substitutional Oxygen in Silicon, *J. Phys. Chem. Solids* **16**, 44-45 (1960).

69. Bosomworth, D.R., Hayes, W., Spray, A.R.L. and Watkins, G.D., Absorption of Oxygen in Silicon in the Near and Far Infrared, *Proc. Roy. Soc.* **A317**, 133-152 (1970).

70. Thorson, W.R. and Nakagawa, I., Dynamics of the Quasi-Linear Molecule, *J. Chem. Phys.* **33**, 994-1004 (1960).

71. Hrowtowski, H.J. and Kaiser, R.H., Infrared Absorption of Oxygen in Silicon, *Phys. Rev.* **107**, 966-972 (1957).

72. Shimura, F., Ohnishi, Y. and Tsuya, H., Heterogeneous Distribution of Interstitial Oxygen in Annealed Czochralski-Grown Silicon, *Appl. Phys. Lett.* **38**, 867-870 (1981).

73. O'Mara, W.C., Oxygen in Silicon and Germanium, *Electrochem. Soc. Extended Abstracts* **80-1**, 475-478 (1980)

74. Kaiser, W., Electrical and Optical Investigation of the Donor Formation in Oxygen-Doped Germanium, *J. Phys. Chem. Solids* **23**, 255-260 (1960).

75. Edwards, W.D., The Distribution Coefficient of Oxygen in Germanium, *J. Electrochem. Soc.* **115**, 753 (1968). See also L.S. Darken, Oxygen Equilibrium Between Germanium and Gas Over Melt During Growth, *J. Electrochem. Soc.* **126**, 373 (1965).

76. Newman, R.C. and Willis, J.B., Vibrational Absorption of Carbon in Silicon, *J. Phys. Chem. Solids* **26**, 373-379 (1965).

77. Smith, S.D. and Angress, J.F., Vibrational Absorption of Substitutional Boron and Phosphorus in Silicon, *Phys. Lett.* **6**, 131-132 (1963).

78. O'Mara, W.C., Oxygen in Silicon, in, *Defects in Silicon*, op. cit., p. 120-129.

79. Baker, J.A., Tucker, T.N., Moyer, N.E. and Buschart, R.C., Effect of Carbon on the Lattice Parameter of Silicon, *J. Appl. Phys.* **39**, 4365-4368 (1968).

80. Stavola, M., Infrared Spectrum of Interstitial Oxygen in Silicon, *Appl. Phys. Lett.* **44**, 514-516 (1984).

81. Watkins, G.D., and Corbett, J.W., Electron Spin Resonance of the Si-A Center, *Phys. Rev.* **121**, 1001-1014 (1961), and Corbett, J.W., G. D. Watkins, R.M. Chrenko, and R.S., McDonald, Infrared Absorption of the Si-A Center, *Phys. Rev.* **121**, 1015-1022 (1961).

82. Fuller, C.S., Ditzenberger, N.B., Hannay, N.B. and Buehler, E., Resistivity Changes in Silicon Induced by Heat Treatment, *Phys. Rev.* **96**, 833 (1954).

83. Kaiser, W., Frisch, H.L. and Reiss, H., Mechanism of the Formation of Donor States in Heat-Treated Silicon, *Phys. Rev.* **112**, 1546-1554 (1958).

84. Fuller, C.S. and Logan, R.A., Effect of Heat Treatment Upon the Electrical Properties of Silicon Crystals, *J. Appl. Phys.* **28**, 1427-1436 (1957).

85. Wruck, D. and Gaworzewski, P., Electrical and Infrared Spectroscopic Investigations of Oxygen-Related Donors in Silicon, *Phys. Stat. Solidi* **A56**, 557-564 (1979).

86. Shaake, H.F., Baber, S.C. and Pinizzotto, R.F., The Nucleation and Growth of Oxide Precipitates in Silicon, in, *Semiconductor Silicon 1981*, op. cit., p. 273-281.

87. Oehrlein, G.S., Silicon-Oxygen Complexes Containing Three Oxygen Atoms as the Dominant Thermal Donor Species in Heat-Treated Oxygen-Containing Silicon, *J. Appl. Phys.* **54**, 5453-5455 (1983).

88. Cazcarra, V. and Zunio, P., Influence of Oxygen on Silicon Resistivity, *J. Appl. Phys.* **51**, 4206-4211 (1980).

89. Helmrich, D. and Sirtl, E., Oxygen in Silicon: A Modern View, in, *Semiconductor Silicon 1977*, op. cit., p. 626-632.

90. Gosele, U. and Tan, T.Y., Oxygen Diffusion and Thermal Donor Formation in Silicon, *Appl. Phys.* **A28**, 79-92 (1982).

91. Wada, K., Unified Model for Formation Kinetics of Oxygen Thermal Donors, *Phys. Rev.* **B30**, 5884-5895 (1984).

92. Kamiura, A., Hashimoto, F., and Endo, K., Initial Generation Kinetics of Oxygen-Related Thermal Donors at 43°C in Silicon, *J. Appl. Phys.* **61**, 2478-2485 (1987).

93. Kanamori, A., Annealing Behavior of the Oxygen Donor in Silicon, *Appl. Phys. Lett.* **34**, 287-289 (1979).

94. O'Mara, W.C., Parker, J.E., Butler, P. and Gat, A., Investigation of Short Time Donor Annihilation in Silicon, *Appl. Phys. Lett.* **46**, 299-301 (1985).

95. O'Mara, W.C., Parker, J.E., Butler, P. and Gat, A., Rapid Thermal Annealing for Oxygen Donor Annihilation, in, *VLSI Science and Technology/1985*, W.M. Bullis and S. Broydo, ed., Electrochem. Soc., Pennington NJ, 1985, p. 456-464.

96. Pearce, C.W. and Rozgonyi, G., Electrochem. Soc. Extended Abstracts **82-2**, 228 (1982), Intrinsic Gettering in Heavily Doped Silicon Substrates for Epitaxial Devices, in, *VLSI Science and Technology/1982*, C.J.

Dell'Oca and W.M. Bullis ed., Electrochem. Soc., Pennington NJ, 1982, p. 53.

97. Claybourn, M. and Newman, R.C., Activation Energy for Thermal Donor Formation in Silicon, *Appl. Phys. Lett.* **51**, 2197-2199 (1987).

98. Kimerling, L.C. and Benton, J.L., Oxygen-Related Donor States in Silicon, *Appl. Phys. Lett.* **39**, 410-412 (1981).

99. Wagner, P., Gottschalk, H., Trombetta, J. and Watkins, G.D., Alignment of Thermal Donors in Si by Uniaxial Stress, *J. Appl. Phys.* **61**, 346-353 (1987).

100. Keller, W.W., Pressure Dependence of Oxygen-Related Defect levels in Silicon, *J. Appl. Phys.* **55**, 3471-3477 (1984).

101. Sumino, K., Dislocation Behavior and Mechanical Strengths of Float-Zone Silicon Crystals and Czochralski Silicon Crystal, in, *Semiconductor Silicon 1981*, op. cit., p. 208-219.

102. Chiou, H-D., Moody, J., Sandfort, R. and Shimura, F., Effects of Oxygen and Nitrogen on Slip in Cz Silicon Wafers, in, *VLSI Science and Technology/1984*, op. cit., p. 59-65.

103. Kondo, Y., Plastic Deformation and Preheat Treatment Effects in Cz and Fz Crystals, in, *Semiconductor Silicon 1981*, op. cit., p. 220-231.

104. Moerschel, K.G., Pearce, C.W., and Reusser, R.E., A Study of the Effects of Oxygen Content, Initial Bow, and Furnace Processing on Warpage of Three-Inch Diameter Wafers, in *Semiconductor Silicon* 1977, op. cit., p. 170-181.

105. Chiou, H-D and Shive, L.W., Test Methods for Oxygen Precipitation in Silicon, in, *VLSI Science and Technology/1985*, op. cit., p. 429-435.

106. Chiou, H-D, Oxygen Precipitation Behavior and Control in Silicon Crystals, *Sol. State Technol.*, March 1987, p. 77-81.

107. Lee, C.O. and Tobin, P.J., The Effect of CMOS Processing on Oxygen Precipitation, Wafer Warpage and Flatness, *J. Electrochem. Soc.* **133**, 2147-2152 (1986).

108. Jastrzebski, R., Soydan, R., McGinn, J., Kleppinger, R., Blumenfeld, M., Gillespie, G., Armour, N., Goldsmith, B., Henry, W. and Vecrumba, S., A Comparison of Internal Gettering during Bipolar, CMOS, and CCD Processes, *J. Electrochem. Soc.* **134**, 1018-1027 (1987).

109. Nozaki, T., Yatsurugi, Y., Akiyama, N., Endo, Y., and Makida, Y., Behavior of Light Impurity Elements in the Production of Semiconductor Silicon, *J. Radioanal. Chem.* **19**, 109-128 (1974).

110. Nozaki, T., Yatsurugi, Y., and Akiyama, N., Concentration and Behavior of Carbon in Semiconductor Silicon, *J. Electrochem. Soc.* **117**, 1566-1568 (1970).

111. Bean, A.R. and Newman, R.C., The Solubility of Carbon in Pulled Silicon Crystals, *J. Phys. Chem. Solids* **32**, 1211-1219 (1971).

112. Kolbesen, B.O. and Muhlbauer, A., Carbon in Silicon: Properties and Impact on Devices, *Solid State Electron.* **25**, 759 (1982).

113. Newman, R.C. and Wakefield, J., Solubility of Carbon in Silicon and Germanium, *J. Chem. Phys.* **30**, 1551 (1959).

114. Gosele, U. and Tan, T.Y., The Role of Vacancies and Self-Interstitials in Diffusion and Agglomeration Phenomenal in Silicon, in, *Aggregation Phenomena of Point Defects in Silicon*, E. Sirtl and J. Goorissen ed., Electrochem. Soc., Pennington NJ, 1983, p. 17-36.

115. See Pearson, G.L. and Bardeen, J., Electrical Properties of Silicon and Silicon Alloys Containing Boron and Phosphorus, *Phys. Rev.* **75**, 865-883 (1949).

116. Newman, R.C. and Willis, J.B., Vibrational Absorption of Carbon in Silicon, *J. Phys. Chem. Solids* **26**, 373-379- (1965).

117. Endo, Y., Yatsurugi, Y., Akiyama, N., and Nozaki, T., Infrared Spectrophotometry for Carbon in Silicon as Calibrated by Charged Particle Activation, *Anal. Chem.* **44**, 2258-2262 (1972).

118. Amer. Soc. Test. Mat. Method F123-81, Philadelphia PA, 1981.

119. Hoelzlein, K., Pensl, G. and Schulz, M., Trap Spectrum of the New Oxygen Donor in Silicon, *Appl. Phys.* **A34**, 155 (1984).

120. Lei, Z., Jinxing, S. and Seming, S., Investigation of New Donor in P-Cz Silicon, in, *Semiconductor Silicon 1986*, op. cit., p. 813-825.

121. Hwang, J.M., Schroder, D.K. and Goodman, A.M., Recombination Lifetime in Oxygen-Precipitated Silicon, *IEEE Elec. Dev. Lett.* **EDL-7**, 172-174 (1986).

122. O'Mara, W.C. and Guidici, D., Swirl Defects in Carbon-Rich Silicon, *Electrochem. Soc. Extended Abstracts* **79-1**, 1979.

123. Bean, K.E., private communication.

124. Kolbesen, B.O., Carbon in Silicon, in, *Aggregation Phenomena of Point Defects in Silicon*, op. cit., p. 155-175.

125. Kaiser, W. and Thurmond, C.D., Nitrogen in Silicon, *J. Appl. Phys.* **30**, 427 (1959).

126. Watanabe, M., Usami, T., Muraoka, H., Matsuo, S., Imanishi, Y. and Nagashima, H., Oxygen-Free Silicon Single Crystals Grown from Silicon Nitride Crucible, in, *Semiconductor Silicon 1981*, op. cit., p. 126-137.

127. Pavlov, P.V., Zorin, E.I., Tetelbaum, D.I. and Khokhlov, A.F., Nitrogen as Dopant in Silicon and Germanium, *Phys. Stat. Solidi* **A35**, 11 (1976).

128. Stein, H., Nitrogen in Crystalline Si, in, Oxygen, Carbon, Hydrogen, and Nitrogen in Crystalline Silicon, J.C. Mikkelsen Jr., S.J. Pearton, J. W. Corbett, and

S.J. Pennycook, ed., *Mat. Res. Soc. Symp. Proc.* **59**, 523-535 (1985), Materials Research Society, Pittsburgh PA, 1985.

129. Abe, T., Harada, H., Ozawa, N., and Adoni, K., Deep Level Generation-Annihilation in Nitrogen Doped Fz Crystals, in, *Oxygen, Carbon, Hydrogen, and Nitrogen in Crystalline Silicon*, op. cit., p. 537-544.

130. Itoh, Y., Nozaki, T., Masui, T. and Abe, T., Calibration Curve for Infrared Spectrophotometry of Nitrogen in Silicon, *Appl. Phys. Lett.* **47**, 488-489 (1985).

131. Abe, T., Masui, T., Harada, H., and Chikawa, J., The Characteristics of Nitrogen in Silicon, in, *VLSI Science and Technology/1985*, op. cit., p. 543-551.

8

Carrier Lifetimes in Silicon

Dieter K. Schroder

1.0 INTRODUCTION

This chapter provides an overview of lifetimes in silicon. We have attempted to give a fairly comprehensive discussion of lifetimes and their effects on semiconductor device behavior. For completeness we have also included some of the more popular lifetime measurement techniques that are used today.

We start the chapter by briefly describing the various types of *defects* that can exist in semiconductor devices because it is the defects that largely determine the lifetime. Many of these defects can be found in most silicon devices. Next we introduce lifetimes and divide them into two broad categories: *recombination* and *generation lifetimes*, providing a discussion of both of them. The recombination lifetime is further subdivided into *multiphonon* or *Shockley-Read-Hall*, *radiative* and *Auger* recombination lifetimes. Multiphonon recombination dominates at low injection levels and low doping concentrations. Auger recombination is important for injection levels and for doping concentrations above about 10^{17}-10^{18} cm^{-3}. Auger recombination is the ultimate lifetime limiting mechanism in silicon and some other semiconductors. Radiative lifetime is not important in silicon.

Only the multiphonon generation lifetime is important for silicon devices. We discuss the concept of the generation lifetime and point out where in the device it is active. Then we compare the generation and the recombination lifetimes and show that typically the generation lifetime is much larger than the recombination lifetime.

Then we describe how the recombination and generation lifetimes influence the forward and reverse-biased junction currents. Both bulk and surface recombination and generation are considered. In addition we discuss the influence of denuded zones and precipitated bulk regions in intrinsically gettered materials on leakage current. Finally we compare the leakage current behavior of the two types of substrates most commonly used for MOS circuits: intrinsically gettered substrates and p-epi on p^+ substrates.

Having discussed the lifetimes and their influence on device leakage current we then turn to measurement techniques. There we describe the more commonly used methods. There are many other methods, but most of them are rarely used. For recombination lifetime measurements we describe: photoconductive decay, open-circuit voltage decay, diode reverse-recovery, surface photovoltage, and the pulsed MOS capacitor. The pulsed MOS capacitor and gate-controlled diode methods are two methods for generation lifetime measurements. Finally we end with a few concluding remarks.

2.0 DEFECTS

In a perfect crystal, shown schematically in Figure 1a, all host atoms are located on their proper lattice sites and the crystal contains neither impurities nor structural imperfections. The band diagram for such a crystal consists of the conduction and valence band with no energy levels within the band gap. A real crystal is not like that. It contains structural imperfections as well as foreign impurities. The foreign impurities consist of intentionally introduced dopant atoms and unintentionally introduced impurity atoms as well as structural imperfections that are generated during crystal growth and during device processing.

Some structural defects are unavoidable while others are process-dependent. In principle it is possible to eliminate all foreign impurities and most structural defects. Thermodynamic considerations, however, demand a certain concentration of self-interstitials and vacancies as shown in Figure 1b. In other words the perfect crystal of Figure 1a is thermodynamically impossible to achieve. Self-interstitials are atoms of the host lattice - silicon atoms in a silicon crystal - displaced from their normal substitutional sites to interstitial sites.

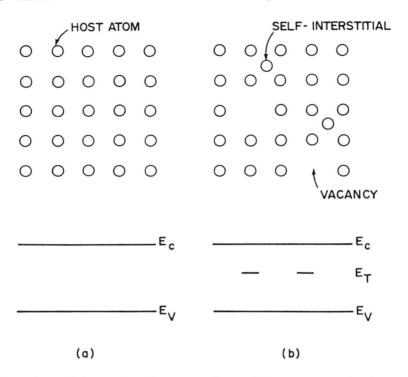

Figure 1: Schematic diagram of a silicon crystal lattice and the corresponding energy band diagrams (a) without, and (b) with vacancies and self-interstitials. The energy level E_T in (b) is only a schematic representation. Actual energy levels are shown in Figure 3.

Thermodynamics tells us that the *free energy* of a crystal depends on its internal energy, E, entropy, S, temperature, T, pressure, p, and volume, V. A reduction in self-interstitials and vacancies reduces the crystal's internal energy because the crystal takes on a more ordered structure. However, the entropy of a more ordered crystal is higher than that of a disordered crystal. The free energy, G, sometimes called the Gibbs free energy, is given by G=E-TS+pV. Aside from the pV product, the free energy is proportional to the difference between the internal energy, E, and TS - a term proportional to the entropy. We do not consider the pV term in this discussion. The equilibrium state of the crystal corresponds to a minimum in the free energy which requires a certain amount of disorder. Recall that entropy is a measure of disorder. In other words, the most favorable free energy state of the crystal is a compromise between crystal perfection and disorder. A single crystal is largely a perfect crystal, but contains a certain concentration of self-interstitials and vacancies to satisfy the "minimum free energy" criterion.

Besides vacancies and self-interstitials, there are other defects. We distinguish between point, line, area and volume defects schematically illustrated in Figure 2.

- o Point Defects
    - o Self-interstitials
    - o Vacancies
    - o Foreign substitutionals and interstitials
- o Line Defects
    - o Dislocations
- o Area Defects
    - o Stacking faults
    - o Grain Boundaries
    - o Twin planes
- o Volume Defects
    - o Precipitates
    - o Impurity clusters

Grain boundaries and twin planes are not found in single crystal Czochralski or float-zone grown Si. All other defects can be and usually are encountered in a processed device.

The consequences of defects from lifetime considerations are the introduction of energy levels into the semiconductor band gap as indicated in Figure 1b. Each energy level represents some type of imperfection in the semiconductor characterized by its energy, E_T, its concentration, N_T, and its electron and hole capture cross-sections, σ_n and σ_p. Such imperfections are more commonly known as *generation-recombination centers* (G-R centers). As their name implies, they act as recombination and generation sites. For most semiconductor devices it is desirable to keep the G-R center concentration as low as possible. Their concentration is occasionally deliberately increased to tailor certain device or

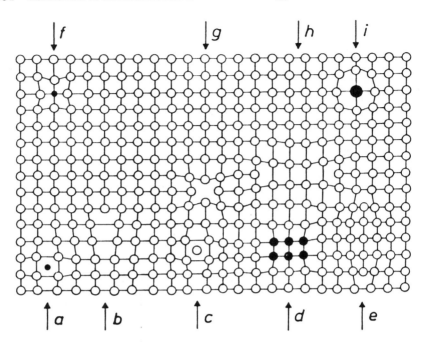

Figure 2: A pictorial representation of the various types of point, line, area and volume defects in a semiconductor. (a) foreign interstitial, (b) dislocation, (c) self-interstitial, (d) precipitate, (e) extrinsic stacking fault and partial dislocation, (f) foreign substitutional, (g) vacancy, (h) intrinsic stacking fault surrounded by a partial dislocation, (i) foreign substitutional. Reprinted after Reference (1) by permission of Springer Verlag.

circuit operating characteristics. For example, the switching time of diodes and bipolar junction transistors is reduced when the G-R center concentration is increased.

We show in Figure 3a the energy levels of some metallic impurities (2) in silicon and in Figure 3b the energy levels of some structural imperfections (3). These levels appear as distinct and precise in these figures. In reality there is a fair amount of scatter in the energies of most of these impurities because the measurements are not trivial and different experimenters frequently find different levels for identical impurities. Furthermore the results are often influenced by the sample preparation techniques. Nevertheless, the data should provide some guidance for those readers interested in energy levels.

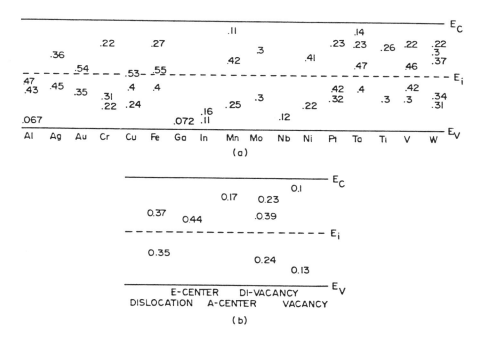

Figure 3: (a) Metallic impurity energy levels and (b) structural impurity energy levels in silicon. The numbers represent the energy E_T. If the energy level is above E_i, the number is $E_c - E_T$, if below E_i it is $E_T - E_v$.

3.0 RECOMBINATION LIFETIME

We classify the carrier lifetimes into two broad categories: *recombination lifetimes* and *generation lifetimes*. The recombination lifetime, τ_r, is generally used when there are excess carriers in the semiconductor and recombination dominates. The generation lifetime, τ_g, obtains when thermal generation dominates. Occasionally we will subdivide the lifetimes further. For example, it is useful to distinguish between low-level and high-level injection lifetimes. But such finer subdivision is only used if it clarifies the concepts.

Excess carriers are those electron and hole concentrations in excess of their thermal equilibrium concentrations. Excess carriers are introduced into a semiconductor by one of several means. Most commonly, minority carriers are injected from a forward-biased junction. They can also be generated by light absorbed in the device as in a solar cell, for example. These excess carriers are in a non-equilibrium state and will therefore recombine in an attempt to return the semiconductor to equilibrium.

The *recombination lifetime* is the average time an excess electron-hole pair (ehp) exists. The recombination lifetime is frequently referred to as the minority carrier lifetime when minority carriers dominate the recombination process. This is generally true under low-level injection conditions when the minority carrier concentration is small compared to the equilibrium majority carrier concentration. For high level injection, it is no longer the minority carrier lifetime, but the combined minority-majority carrier lifetime that dominates. We will use the general term *recombination lifetime* to cover all of these possibilities.

The creation of an excess ehp requires an energy equal to the semiconductor band gap. When excess electron-hole pairs recombine they release this energy by one of several distinct physical mechanisms. Three basic mechanisms are shown in Figure 4. In Figure 4a the energy is given off as phonons or lattice vibrations. The phonons are indicated by the small arrows. This recombination mechanism is known as *multiphonon recombination*. The recombination event can proceed by a conduction band electron falling directly into a hole in the valence band or it can proceed through intermediate generation-recombination centers. Both mechanisms are shown on Figure 4a.

A conduction band-to-valence band multiphonon recombination event requires the simultaneous creation of more phonons than a conduction band-to-G-R center followed by a G-R center-to-valence band event. The total number of phonons is approximately the same in both cases. But for multiphonon recombination via G-R centers there are two distinct steps. First an electron is captured by the G-R center. Then a hole is captured by the G-R center. Multiphonon recombination via G-R centers is therefore much more likely than band-to-band multiphonon recombination for indirect band gap semiconductors like Si. Multiphonon recombination is also known as Shockley-Read-Hall (SRH) recombination.

The SRH recombination lifetime is given by (4)

$$\tau_{SRH} = [\tau_{po}(n_o+n_1+\Delta n) + \tau_{no}(p_o+p_1\Delta p)]/[p_o+n_o+\Delta n] \qquad (1)$$

where $\tau_{po}=1/(\sigma_p v_p N_T)$, $\tau_{no}=1/(\sigma_n v_n N_T)$, σ_p and σ_p are the hole and electron capture cross-sections, respectively, v_p,

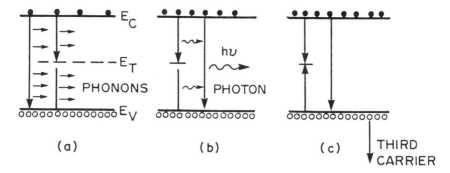

Figure 4: Excess electron-hole recombination mechanisms. (a) multiphonon recombination, (b) radiative recombination, and (c) Auger recombination.

v_n=thermal velocity of the holes and electrons, respectively, and N_T=G-R center concentration. The capture cross-sections vary over a wide range. For coulombic attractive G-R centers they may be as high as 10^{-12} cm² while for coulombic repulsive G-R centers they are as low as 10^{-18} cm². For neutral centers they are around 10^{-15} cm². The equilibrium hole and electron concentrations are p_o and n_o, respectively, and the total concentrations are $p=p_o+\Delta p$ and $n=n_o+\Delta n$. The excess concentrations Δp and Δn are equal to each other in the absence of trapping. Our usual assumption will be $\Delta n=\Delta p$. The effect of trapping will be discussed later. The quantities n_1 and p_1 are defined by (4)

$$n_1 = n_i \exp[(E_T-E_i)/kT] \tag{2a}$$

$$p_1 = n_i \exp[-(E_T-E_i)/kT] \tag{2b}$$

where E_i=intrinsic Fermi energy.

The more effective recombination centers are those whose energy levels are near E_i. Equations (2) show n_1 and p_1 to be on the order of n_i for E_T near E_i. The main point for our purpose here is that n_1 and p_1 are generally negligibly small compared to the majority carrier concentration even if they are not equal to n_i. The assumption n_1, $p_1 \ll$ majority carrier concentration is poor, however, for those G-R centers with E_T significantly above or below E_i and for high resistivity substrates with low majority carrier concentrations.

We will use moderately doped extrinsic p-type substrates

in most of our discussions. There we always find $n_o \ll p_o$ and generally $n_1 \ll p_o$ and $p_1 \ll p_o$. This allows Equation (1) to be written as

$$\tau_{SRH} \approx \tau_{po}/(1+p_o/\Delta_n) + \tau_{no} \tag{3}$$

When the ehp energy is released through the emission of photons, as in Figure 4b, the recombination mechanism is known as *radiative recombination.* Radiative band-to-band recombination is dominant in direct band gap semiconductors like GaAs and forms the basis for light emitting diodes and semiconductor lasers. Radiative recombination through G-R centers is dominant in phosphors where the photon energy can be varied by choosing G-R centers with appropriate energy levels within the band gap. This radiative recombination mechanism is utilized in the three-color phosphors in television display tubes, for example.

The radiative lifetime for band-to-band recombination is given by (5)

$$\begin{aligned}\tau_{rad} &= \Delta n/[B(pn = p_o n_o)] \\ &= 1/[B(p_o + n_o + \Delta n)]\end{aligned} \tag{4}$$

where B is the radiative recombination coefficient. For Si, $B \approx 10^{-15}$ cm^3/s and the lifetime due to radiative recombination is very high and not important compared to τ_{SRH} (6). We should point out that radiative recombination is not zero. The radiative recombination light emanating from a silicon device is used in a number of measurements including lifetime measurements where the temporal or spatial behavior is measured.

For extrinsic p-type substrates Equation (4) becomes

$$\tau_{rad} \approx 1/[B(p_o + \Delta n)] \tag{5}$$

The band-to-band radiative lifetime is inversely proportional to the carrier concentration. This lifetime dependence on carrier density comes about for the following reason. For minority electrons in the conduction band to recombine directly with holes in the valence band there must be majority holes available. The more majority holes there are the lower is the lifetime as shown by Equation (5).

The third combination mechanism is shown in Figure 4c. The ehp energy is given to either electrons in the conduction band or holes in the valence band for both band-to-band and band-to-impurity recombination events. This recombination mechanism is known as *Auger recombination*. It is dominant in narrow band gap semiconductors, and at high injection levels or high doping concentrations in semiconductors like Si and GaAs. For band-to-band Auger recombination, the lifetime is (5)

$$\tau_{Auger} = \Delta n/[C_p(p^2n - p_o^2 n_o) + C_n(pn^2 - p_o n_o^2)] \quad (6)$$

where C_p and C_n are the Auger recombination coefficients in p-type and n-type substrates, respectively. For extrinsic p-type material, Equation (6) becomes

$$\tau_{Auger} \approx 1/[C_p(p_o^2 + 2p_o\Delta n + \Delta n^2) + C_n(n_o^2 + 2n_o\Delta n + \Delta n^2)] \quad (7)$$

Note the inverse dependence on the square of the carrier concentration. This inverse square-law dependence comes about for the following reason. Electrons recombining directly with holes require not only that there be holes in the valence band with which to recombine. They require additionally that there be holes to acquire their excess energy in the Auger recombination process.

The Auger recombination coefficients C_n and C_p have been measured by a variety of experimental techniques. Beck and Conradt (7) used weak pulsed optical excitation on highly-doped silicon and measured the time-dependent band-to-band recombination radiation. They found $C_n = 1.7 \times 10^{-31}$ and $C_p = 1.2 \times 10^{-31}$ cm^6/s. For n-silicon they found essentially identical lifetimes whether the material was doped with P, As, or Sb. For p-silicon both B-doped and Ga-doped silicon gave comparable lifetimes. However, Beck and Conradt (7) had difficulty measuring lifetimes below 10^{-7}s. Dziewior and Schmid (8) used a similar, but more sophisticated experimental technique and obtained $C_n = 2.8 \times 10^{-31}$ and $C_p = 0.99 \times 10^{-31}$ cm^6/s. Both of these papers report on measurements made on highly doped or highly excited samples.

More recent measurements (9) of the Auger recombination coefficient in n-type Si for doping concentrations of 10^{18} cm^{-3} or less, i.e. for the lower doping range of Auger-dominated recombination, give $C_n = 1.5 \times 10^{-30}$ cm^6/s. Other recent measurements (10) on lowly-doped samples where modest optical excitation produced high level injection conditions gave $C_n + C_p = 2 \times 10^{-30}$ cm^6/s. This value agrees quite well with that in Reference 9 and is significantly higher than the earlier values. A discrepancy between the experimental data and the extrapolated lifetimes using the value 2.8×10^{-31} cm^6/s can be seen in the data of Reference 8 for doping concentrations in the range of 5×10^{17} to 5×10^{18} cm^{-3}. That discrepancy is consistent with a higher value of C_n but was not alluded to in that paper. The lifetimes in Reference 8a were determined by exciting the Si with highly penetrating, pulsed 1.06 μm wavelength laser light to generate excess carriers. The time dependence of the excess carriers was monitored by the transmission of a 2 μm wavelength probe beam. The transmissivity of the sample at the probe beam wavelength depends on the free carrier concentration through free carrier absorption.

The implication of the higher Auger recombination coefficient is a lower lifetime than generally predicted for doping or excess carrier concentrations less than approximately 10^{18} cm^{-3}. This lower lifetime is beginning to be experienced in high efficiency solar cells in which lifetime is very important. It appears that the Auger coefficients are carrier concentration-dependent and not constant as generally assumed.

Band-to-band Auger recombination has been questioned. Hu and Oldham (11) proposed recombination through shallow level G-R centers as the dominant recombination mechanism in heavily-doped silicon. Possin et al. (12) and Weaver and Nasby (13) also questioned band-to-band Auger recombination and proposed G-R center-assisted Auger recombination. This argument for G-R center-assisted Auger recombination has been refuted by several well argued papers (14),(15),(16). Band-to-band Auger recombination is now generally accepted to be the limiting recombination mechanism for heavily-doped and highly excited silicon. The case for band-to-band Auger recombination rests partially on the fact that the Auger coefficients measured for low optical excitation of highly-doped silicon substrates agree very closely with those coefficients measured by intense optical excitation of lightly-doped silicon (17),(18). The Auger coefficient is C_n+C_p in the latter case since equal electron and hole concentrations are optically generated. Numerical values of $C_n+C_p = 2-3.9 \times 10^{-31}$ cm^6/s agree reasonably well with the sum of the individual coefficients measured on highly-doped silicon. The Auger coefficients follow a $T^{0.6}$ temperature dependence (18).

Equations (3),(5) and (7) simplify further for both low-level and high-level injection. Low-level injection (ll) conditions hold when the excess minority carrier concentration is small compared to the equilibrium majority carrier concentration, i.e. $\Delta n \ll p_o$. High-level injection (hl) hold when $\Delta n \gg p_o$. The injection level is important during lifetime measurements as shown by the ll and hl injection lifetimes.

For low-level electron injection into p-type substrates we find

$$\tau_{SRH}(ll) \approx \tau_{no} \tag{8}$$

$$\tau_{rad}(ll) \approx 1/[Bp_o] \tag{9}$$

$$\tau_{Auger}(ll) \approx 1/[C_p p_o^2] \tag{10}$$

and for high-level injection

$$\tau_{SRH}(hl) \approx \tau_{no} + \tau_{no} \tag{11}$$

$$\tau_{rad}(hl) \approx 1/[B\Delta n] \tag{12}$$

$$\tau_{Auger}(hl) \approx 1/[(C_p + C_n)\Delta n^2] \tag{13}$$

The SRH recombination lifetime dependence on injection level is shown in Figure 5 (19). Notice the increase in τ_{SRH} as the injection level is varied from $\Delta n/n_o=0.01$ to 100. The lifetime increases from τ_{po} at low injection level to ($\tau_{po} + \tau_{no}$) at high injection level for all three samples. The data in Figure 5 are for n-substrates. The ratio $(\tau_{po}+\tau_{no})/(\tau_{po})$ varies with the sample processing but it is always larger than unity as expected. We also show the capture cross-section ratio σ_p/σ_n. It lies between 0.6 and 15.7.

The recombination lifetime for either injection level is

$$\tau_r = 1/[1/\tau_{SRH} + 1/\tau_{rad} + 1/\tau_{Auger}] \tag{14}$$

Equation (14) is an expression for parallel processes. The lowest of the three lifetimes dominates the recombination lifetime in such an expression.

Equations (8-10) and (14) are plotted in Figure 6 as a function of the majority carrier hole concentration. The SRH lifetime is determined by the G-R center concentration, N_T, and is independent of the carrier concentration. N_T is determined by the device process cycle and by the process cleanliness. Typical τ_{SRH} values lie in the range of 1 to 10 μs for processes where high lifetimes are not important and 10 to 100 μs when high lifetimes are crucial, as in high efficiency solar cells, for example.

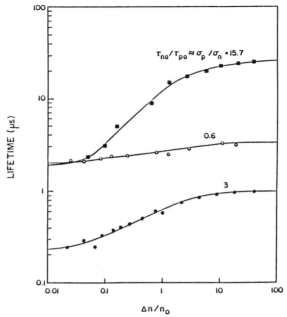

Figure 5: The recombination lifetime as a function of injection level for n-substrates. The lifetimes here are SRH lifetimes and the data refer to three different diodes.

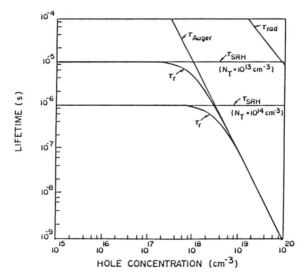

Figure 6: Theoretical multiphonon, radiative and Auger lifetimes as a function of the majority carrier concentration in p-Si. For τ_{SRH}: $\sigma_n = 10^{-15} cm^2$; for τ_{rad} : $B = 10^{-15} cm^3/s$, and for τ_{Auger} : $C_p = 10^{-31} cm^6/s$ were used.

The radiative lifetime is very high in Si and does not play any significant role in the overall recombination lifetime. The Auger lifetime dominates at high concentrations and theory predicts it to vary inversely with the square of the majority carrier concentration. Figure 6 clearly shows the recombination lifetime to be determined by τ_{SRH} for majority carrier concentrations below approximately 10^{17}-10^{18} cm^{-3} and by τ_{Auger} for higher concentrations. Neither radiative nor Auger lifetimes depend on G-R centers and are therefore not dependent on the fabrication process. They are a property of the silicon and its doping concentration or injection level only.

For low doping concentrations $p_o = N_A$ but $p_o < N_A$ at those doping levels where Auger recombination dominates. This comes from the fact that not all acceptor atoms are ionized at high doping concentrations. One find the recombination lifetime plotted against both carrier concentrations and doping concentrations in the literature. The inverse square law dependence of τ_{Auger} applies only to the carrier concentration, not the doping concentration. That may explain why the inverse square law dependence is not always observed when τ_r is plotted against N_A.

Experimental lifetime data for both n- and p-type Si are shown in Figure 7 (20)-(23). The lifetimes follow the predicted lifetimes of Figure 6 quite well. Although there is a fair amount of scatter in the experimental data, there is clearly a trend of fairly constant lifetime at low carrier concentrations and a lifetime decrease at the higher carrier concentrations. The low-concentration "constant lifetime" is the SRH lifetime determined by G-R processes. As such it depends very sensitively on the device process cycle and cleanliness. The scatter in the data is a reflection of that. At the higher concentrations — typically above about 10^{17}cm^{-3} — the lifetime decreases very closely to the predicted $1/n^2$ or $1/p^2$ dependence.

A number of empirical lifetime expressions have been proposed. Although they are chiefly based on experimental data they are useful predictors of lifetime. Passari and Susi give (22)

$$\tau = 1/[3.356 \times 10^4 + 1.072 \times 10^{-26} n^{1.76}] \tag{15}$$

for n-type silicon, and

$$\tau = 1/[2.841 \times 10^4 + 1.716 \times 10^{-26} p^{1.67}] \tag{16a}$$

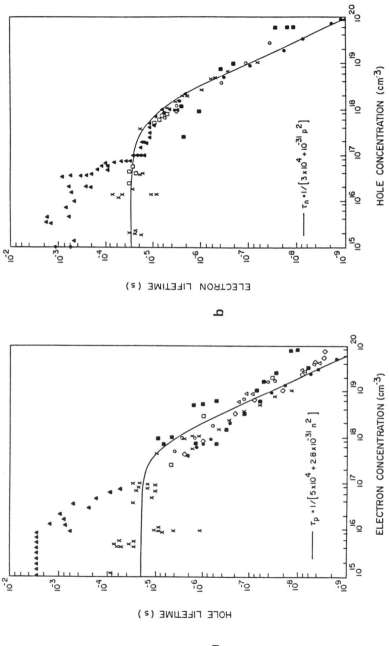

Figure 7: Experimental lifetime values as a function of the carrier concentration for silicon. Data from Reference (8)-(10), (21)-(23). (a) Hole lifetimes in n-Si, (b) electron lifetimes in p-Si.

$$\tau = 1/[3.333 \times 10^3 + 1.716 \times 10^{-26} \, p^{1.67}] \tag{16b}$$

for Czochralski-grown and float-zone-grown p-type silicon, respectively.

Others use a modified Equation 14 by only considering the SRH and Auger lifetimes (24)

$$\tau_r = 1/[1/\tau_{SRH} + 1/\tau_{Auger}] \tag{17}$$

The Auger lifetime is given by Equation (6) and the SRH lifetime is expressed as

$$\tau_{SRH} = \tau_1/[b + (N_D/N_{ref})^a] \tag{18}$$

where a, b, τ_1 and N_{ref} are fitting parameters determined from electrical measurements. For example a=0.3, b=0, $\tau_1 \approx 3 \times 10^{-7}$ s and $N_{ref} = 2 \times 10^{14}$ cm^{-3} in Reference 25 and a=1, b=1, $\tau_1 \approx 1.34 \times 10^{-7}$ s and $N_{ref} = 4.16 \times 10^{17}$ cm^{-3} in Reference 26. Another paper gives the lifetime that can be expected in silicon as (27)

$$\tau_n = 21.2 \times 10^{-6}/(1 + N_A/7.1 \times 10^{15}) \tag{19a}$$

$$\tau_p = 15.5 \times 10^{-6}/(1 + N_D/7.1 \times 10^{15}) \tag{19b}$$

where N_A and N_D are the acceptor and donor doping concentrations in p- and n-type Si, respectively.

We find a reasonable fit to the experimental data of the hole lifetime in n-silicon in Figure 7a using

$$\tau_p = 1/[5 \times 10^4 + 2.8 \times 10^{-31} n^2] \tag{20a}$$

For the electron lifetime in p-silicon of Figure 7(b) we find

$$\tau_n = 1/[3 \times 10^4 + 10^{-31} p^2] \tag{20b}$$

These equations are reasonably representative of today's processed silicon and are shown by the solid curves on the figures. They were fitted by the Auger coefficients applicable at high doping concentrations and therefore overestimate the lifetimes in the intermediate doping concentration ranges. Lifetime expressions in the lowly-doped regions are more difficult to generate because the lifetimes there are very process-dependent and will therefore vary a great deal from process to process. We have chosen equations that describe lifetimes generally encountered after conventional silicon processing. There are, of course, high lifetime fabrication processes that yield higher lifetimes at low doping concentrations. The highest lifetimes we are aware of for as-grown silicon are shown on Figure 7. They require a different lifetime expression.

The fundamental upper lifetime limits have been proposed as 300-500μs (28). However, Figure 7 clearly shows much higher values. These high lifetimes were measured on unprocessed, float-zone grown silicon (23). We see from Figure 7 the electron lifetimes in p-Si to be lower than the hole lifetimes in n-Si. One might speculate that these "ultimate" lifetimes may be limited by some intrinsic defect in silicon, not through contamination by foreign impurities. Such an intrinsic defect has been postulated (28)-(29). However, we do not understand this defect. The defect postulated in Reference 28 predicts the electron lifetime to exceed the hole lifetime. According to Figure 7 this is not the case.

4.0 GENERATION LIFETIME

Each of the recombination processes indicated in Figure 4 has a generation counterpart. We show in Figure 8 the multiphonon, the band-to-band radiative, and the band-to-band Auger generation. The inverse radiative band-to-band recombination is ehp generation by absorption of photons shown in Figure 8b. This is of course utilized in photodiodes and photoconductors where the electron-hole pairs generated by incident photons constitute the signal. For a semiconductor device in the dark there is no optical eph generation. The term "dark" means that there should be no photons of energies larger than the band gap incident on the semiconductor. For semiconductors like Si with a relatively wide band gap it is easy to keep the device in the dark by containing it in a light-tight enclosure.

For narrow band gap semiconductors life HgCdTe, for example, it is more difficult to keep the device in the dark, because the device is sensitive to infrared irradiation. For example, a semiconductor with a 0.1 eV band gap is sensitive to photons with energies as low as 0.1 eV. Any material at room temperature radiates photons of such energies. This type of radiation is more commonly known as black-body radiation. In other words, optical ehp generation by photons emanating from the device package or the room-temperature

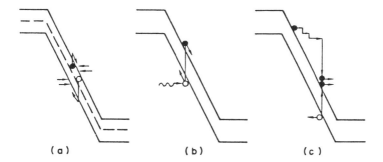

Figure 8: Three generation mechanisms. (a) Multiphonon generation via G-R centers, (b) optical generation, and (c) impact ionization.

light-tight box, for example, is no longer negligible for narrow band gap semiconductors. In order for these semiconductors to be in the "dark", the enclosure must not only be light-tight to visible radiation but it must also be cooled. A cooled body radiates much lower black-body radiation. Fortunately, room temperature black body radiation contributes an insignificant photon density at those energies where Si is sensitive. Hence we can neglect optical generation for Si devices kept in the dark.

The inverse of Auger recombination is impact ionization generation shown in Figure 8c. As long as the bias voltage keeps the reverse-biased device biased below impact ionization breakdown, we find impact ionization generation to contribute an insignificant ehp generation component. We should point out, however, that even at those voltages where avalanche multiplication is generally considered to be small, it may not be negligible. For example, it is known that weak avalanche multiplication can degrade the drain current-drain voltage characteristics of short-channel MOSFETs (30). The I_D-V_D curves exhibit a snapback breakdown.

Multiphonon recombination has multiphonon generation as its inverse process, as shown in Figure 8a. Multiphonon generation is a thermally activated process. Electrons are thermally excited from G-R centers into the conduction band and holes are similarly thermally excited from G-R centers into the valance band. Those readers who wonder where the electrons and holes come from to be continually excited may want to view this generation process slightly differently. Hole emission from G-R centers into the valence band may be thought of as electron emission from the valence band onto the G-R centers. From this perspective one can look on the generation process as one in which electrons are first thermally excited from the valence band onto G-R centers and subsequently from G-R centers into the conduction band.

The key feature of SRH generation is *thermal excitation.* During the recombination process there is no thermal excitation. Carriers recombine through intermediate G-R centers. Electrons drop and holes rise in energy. This is a desired state and recombination proceeds relatively rapidly. This is not true for generation. A thermally activated process is very dependent on the temperature and on the activation energy. We can see this by examining the SRH generation rate. SRH statistics give the thermal ehp generation rate as (4)

$$G = (n_i^2 - pn)/[\tau_{po}(n + n_1) + \tau_{no}(p + p_1)] \tag{21}$$

where p, n, n_1 and p_1 were defined earlier.

G is zero in equilibrium when $pn = n_i^2$. For $pn < n_i^2$ generation proceeds. The pn product need not be zero but it must be less than n_i^2 during generation. The further the device deviates from thermal equilibrium the smaller is the pn product; the smaller the pn product, the higher the generation rate. In other words, a device in non-equilibrium attempts to establish equilibrium by generating electron-hole pairs.

For reverse-biased devices $pn < n_i^2$. For small pn products we can approximate $pn \approx 0$ and write the generation rate as

$$G \approx n_i^2/(\tau_{po} n_1 + \tau_{no} p_1) \tag{22}$$

which can be written as (31)

$$G = n_i/\tau_g \tag{23}$$

where τ_g is the *generation lifetime* which, according to Equations (2) and (22) is defined as

$$\tau_g = \tau_{po} \exp[(E_T - E_i)/kT] + \tau_{no} \exp[-(E_T - E_i)/kT] \tag{24}$$

What is τ_g? Just as τ_r represents the average time for ehp to recombine, so τ_g represents the average time for ehp to be generated. It is intuitively obvious that the mechanisms responsible for recombination and generation are quite different. We should, therefore, expect the recombination and generation lifetimes to be quite different as well. If in fact they are different what do they mean, how are they measured and how do they apply to device operation? We will address all of these issues in this chapter.

First we discuss typical value for τ_r and τ_g. The ratio of τ_g/τ_r for low-level injection can be written as

$$\tau_g/\tau_r = 2(\tau_{po}/\tau_{no})^{\frac{1}{2}}\cosh[(E_T-E_i)/kT + 0.5\ln(\tau_{po}/\tau_{no})] \tag{25}$$

This ratio can only be calculated if the τ_{po}/τ_{no} ratio is known. From $\tau_{po} = 1/\sigma_p v_p N_T$ and $\tau_{no} = 1/\sigma_n v_n N_T$ we find $\tau_{po}/\tau_{no} = \sigma_n \sigma_p$ assuming the electron and hole thermal velocities are equal. The σ_n/σ_p ratio is known for some G-R centers. It lies typically between 0.01 and 100. Since the ratio appears either in the square root or in the natural logarithm in Equation (25) its effect is small. To first order we assume σ_n/σ_p to be unity. This allows Equation (25) to be simplified to

$$\tau_g/\tau_r \approx 2\cosh[(E_T-E_i)/kT] \tag{26}$$

In other words, for G-R centers with $E_T \neq E_i$, the generation lifetime can be much larger than the recombination lifetime. This has been experimentally confirmed (31). Ratios of $\tau_g/\tau_r \approx$ 50-500 are frequently observed. Even if $E_T = E_i$ we find the two lifetimes still not equal to one another as shown by Equation (25). For that special case we find $\tau_g = \tau_{po} + \tau_{no}$. What is implied by unequal generation and recombination lifetimes? The lifetime measurement techniques and the resulting lifetimes measured with them must be clearly understood. For example, the *recombination lifetime* is measured by such methods as photoconductive decay, open-circuit voltage decay, diode reverse-recovery, forward-biased diode current-voltage, electron-bean induced current and others. Each lifetime measurement technique gives somewhat different numerical values, but the variation is not severe provided the same entity is measured.

The *generation lifetime* is measured with the pulsed MOS capacitor or the gate-controlled diode technique. It is important to understand that the lifetime measured by this technique can, and generally does, give very different values from the recombination lifetime measured by one of the techniques indicated above.

Experimental values of $\tau_g/\tau_r \approx$ 50-500 are typically measured in uniformly doped material with uniform G-R center densities. The ratio is essentially determined by the exponential dependence of the generation lifetime on energy and temperature. The τ_g/τ_r lifetime ratio gets even higher when the G-R center distribution in the material is non-uniform. For example, most of today's ICs are fabricated on Czochralski-grown Si which contains high densities of oxygen. During certain high temperature anneals, it is possible for the oxygen to out-diffuse from the upper 10-20μm of the wafer. This leaves a low-oxygen region over that depth. Subsequent heat treatments cause some of the remaining oxygen in the wafer interior to precipitate (32). The result is a region of high perfection near the surface called the *denuded zone* which is typically 10-20μm thick, and a *heavily precipitated bulk*.

The precipitation effect is demonstrated in Figure 9 (33). We used two adjacent wafers cut from a specially grown ingot. All the wafers from this ingot were numbered and the location of each wafer in that ingot were known. The generation and recombination lifetimes of one wafer were measured across a three-inch diameter wafer after a thermal oxide was grown to allow pulsed MOS capacitor measurements to be made. The generation lifetime was significantly higher than the recombination lifetime as seen by the solid circles in Figure 9. The two lifetimes were then measured on the second wafer which had undergone a denuding and precipitation anneal. The lifetimes of this wafer are also shown on Figure 9 by the open circles. The generation lifetime has increased further due to gettering of mobile lifetime-killing impurities from the denuded zone into the precipitated bulk. The recombination lifetime has decreased significantly due to precipitate formation.

This example shows the regions with very different lifetimes. In the denuded zone there are very few G-R centers while the wafer interior is loaded with G-R centers- the oxygen precipitates. Lifetime interpretation becomes even more difficult for such wafers. For example, the usual pulsed MOS capacitor method determines the generation lifetime in a reverse-biased scr. The scr width is typically a few microns and this method samples only a few microns of the wafer from its surface. The measured τ_g value will be very high. Recombination lifetime measurements probe the denuded zone **and** the precipitated interior and the lifetime will be very low. τ_g/τ_r ratios as high as 1500 have been reported (34).

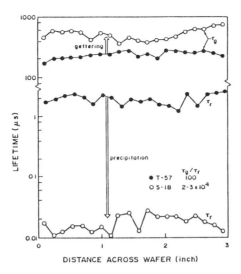

Figure 9: Generation and recombination lifetimes as a function of distance across three-inch diameter wafers. Annealing cycles: T-57:1000°C/2 hrs in O_2; S-18:1100°C/2 hrs in O_2 + 750°C/20 hrs in N_2 + 1000°C 20 hrs in N_2.

Comparing lifetimes measured by different techniques on different substrates is fraught with difficulties. It should be appreciated by the experimenter what is measured, and the results should be represented unambiguously. Lifetimes should not be compared without a thorough description of the measurement technique and the type of substrate measured. Unfortunately this is not always followed leading to confusion in lifetime data comparison and interpretation.

5.0 THE ROLE OF LIFETIME ON DEVICE CURRENTS

5.1 Forward-Biased Diodes

We will use the n^+p junction in Figure 10 to illustrate the role of the lifetime on the junction current-voltage characteristics. We will call the n^+ region the n-emitter and the p-substrate the p-base for simplicity. Electrons are injected from the n-emitter into the space-charge region (scr) and into the quasi-neutral p-base for a forward-biased n+p junction. Similarly, holes are injected from the p-base into the scr and into the n-emitter.

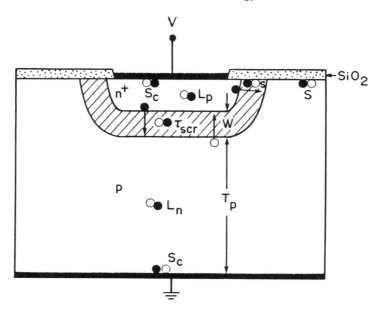

Figure 10: An n+p junction showing the recombination mechanisms and the geometry used in the discussion.

Three current components are of interest: (i) minority hole recombination in the n-emitter (indicated by L_p on Figure 10) and at the n^+/metal contact (indicated by s_c), (ii) minority electron recombination in the p-base (indicated by L_n) and at the substrate/metal contact (indicated by s_c), and (iii) electron-hole recombination in the space-charge region (indicated by τ_{scr}). For each component we can further distinguish between recombination in the semiconductor and at the semiconductor surface. Surface recombination is indicated by the surface recombination velocity, s, on Figure 10.

If we assume the quasi-neutral region currents to be diffusion-limited they are given by

$$I_{diff} = qAn_i^2[(D_p/L_p^*N_D) + (D_n/L_n^*N_A)][\exp(qV/kT)-1] \qquad (27)$$

where $q=1.6 \times 10^{-19}$ eV, A=device area, D_p and D_n are the hole and electron diffusion constants, L_p^* and L_n^* are the effective hole and electron minority carrier diffusion lengths, and N_D and N_A are the doping concentrations in the n-emitter and the p-base, respectively. The *effective* minority carrier diffusion lengths account for both bulk recombination and interface recombination at the semiconductor-metal contacts. We will demonstrate this by considering the effective electron minority carrier diffusion length in the p-base for the geometry of Figure 10. L_n^* is given by

$$L_n^* = L_n[1 + (s_c L_n/D_n)\tanh(\alpha)]/[(s_c L_n/D_n) + \tanh(\alpha)] \qquad (28)$$

with $L_n = \sqrt{(D_n \tau_n)}$, $\alpha = T_p/L_n$, T_p = quasi-neutral p-base thickness and s_c = surface recombination velocity at the semiconductor-metal interface. The recombination lifetime in the p-base is designated as τ_n to distinguish it from τ_p, the recombination lifetime in the n-emitter.

Equation (28) is a rather unwieldy equation and several simplifications are possible. For very thick substrates or very short diffusion lengths where $L_n \ll T_p$ we have the usual *long-base* solution

$$L_n^* = L_n \qquad (29)$$

For long diffusion length devices in which $L_n \gg T_p$ we have the *short-base* solution and Equation (28) simplifies to

$$L_n^* = T_p(s_c + D_n/T_p)/(s_c + T_p/\tau_n) \qquad (30)$$

This approximation can be further simplified by considering two special cases of Equation (30). For low surface recombination where $s_c \ll D_n/T_p$ and $s_c \ll T_p/\tau_n$ we find

$$L_n^* \approx (L_n/T_p)^2 T_p \qquad (31)$$

For this case L_n^* is much longer than the actual diffusion length L_n. Such low surface recombination velocities are approximated by low-high p-p$^+$ junctions. Values of $s_c \approx 80$ cm/s, which satisfy the inequality leading to Equation (31), have been reported (35). It should be noted that s_c in this case is not the recombination velocity at the semiconductor-metal interface but at the p-p$^+$ interface. The minority carriers experience this interface, not the semiconductor-metal interface. For high surface recombination where $s_c \gg D_n/T_p$ and $s_c \gg T_p/\tau_n$ we find the effective diffusion length to be equal to the substrate thickness

$$L_n^* \approx T_p \qquad (32)$$

This condition is approximated by ohmic contacts with $s_c \approx 10^6 - 10^7$ cm/s.

The scr region current component is approximately

$$I_{scr} = [qAn_iW/\tau_{scr}][\exp(qV/nkT)-1] \tag{33}$$

where τ_{scr} is an effective scr recombination lifetime and W is the scr width. The ideality factor n is often quoted as n=2 for scr recombination. It is more likely to be in the range of n=1.5-1.8 for many Si diodes.

In contrast to Equation(28), Equation(33) has no simplifications because the scr width W is very small. A more accurate analysis gives a somewhat more complicated expression for the scr current (36). We will limit ourselves to the simpler and more commonly used expression in Equation (33) for simplicity.

5.2 Reverse-Biased Diodes

Equations (27) and (33) are the forward-biased junction currents. For reversed-biased junctions with V being negative we find the two current components to be

$$I_{diff} = -qAn_i^2[(D_p/L_p^*N_D) + (D_n/L_n^*N_A)] \tag{34}$$

$$I_{scr} = -qAn_iW/\tau_g \tag{35}$$

where W is now the width of the reverse-biased space-charge region. The reverse-biased junction *diffusion current*, also known as the *saturation current*, can be used to determine the minority carrier diffusion length and therefore the *recombination lifetime*. The reverse-biased junction *space-charge region current* can be used to determine the *generation lifetime*.

How is it possible to determine the recombination lifetime as well as the generation lifetimes from reverse-biased junction currents when obviously there is no recombination in reverse-biased junctions? To answer that question we must look at the generation rate expression given in Equation (21) and repeated here

$$G = (n_i^2-pn)/[\tau_{po}(n + n_1) + \tau_{no}(p + p_1)] \tag{36}$$

The electron and hole concentrations are very low in the reverse-biased scr. For $p \approx n \approx 0$, Equation (36) becomes

$$G = n_i/\tau_g \tag{37}$$

and the generation rate is clearly determined by the generation lifetime. What we used in going from Equation (36) to Equation (37) is the assumption that both *majority-carrier* and *minority-carrier concentrations* are negligibly small in the reverse-biased scr. This is a good assumption over most of the reverse-biased space-charge region.

The carrier concentrations vary from approximately zero over most of the reverse-biased scr to their equilibrium concentrations in the quasi-neutral regions far from the scr. In the quasi-neutral regions - the regions where the diffusion or saturation currents originate - we have the following situation. The *minority carrier concentration* varies from near zero at the edge of the scr to its equilibrium value within a few minority carrier diffusion lengths from the scr edge. Charge neutrality considerations require the *majority carrier concentration* to follow the spatial dependence of the minority carrier concentration fairly closely. This implies that the majority carrier concentration varies by about the same amount as the minority carrier concentration. But the minority carrier concentration is much lower than the majority carrier concentration. In the p-base of the n^+p junction we have $n \approx n_o - \Delta n$ and $p \approx p_o - \Delta p$, where Δn varies from approximately n_o at the scr edge to zero at several diffusion lengths from the scr edge. For $\Delta n \approx \Delta p$ we find $n \approx 0$ and $p \approx p_o$ at the scr edge because $\Delta n \approx n_o \ll p_o$. With these assumptions we can write Equation (36) as

$$G \approx p_o \Delta n / [\tau_p(n_o - \Delta n + n_1) + \tau_n(p_o + p_1)] \quad (38)$$

For $p_o \gg p_1, n_1$ the generation rate can be approximated by

$$G \approx \Delta n / \tau_n \approx n_o / \tau_n \quad (39)$$

Comparing Equation (37) with Equation (39) reveals the interesting observation that scr generation is described by the generation lifetime while quasi-neutral region generation is described by the recombination lifetime or the minority carrier diffusion length. This is why reverse-biased junction leakage current measurements can be used to determine both τ_g and τ_r provided the diffusion current and the scr current can be measured separately.

The total n^+p junction leakage current is the sum of both scr and quasi-neutral region generation currents, i.e.

$$I = -\{qAn_iW/\tau_g + qAn_i^2[(D_p/L_p^*N_D) + (D_n/L_n^*N_A)]\} \quad (40)$$

Most leakage current measurements are made at room temperature where the scr generation dominates. It is usually assumed that the diffusion current is negligibly small and $I \approx qAn_iW/\tau_g$. This may not be true if τ_g is very high and L_n^* and/or L_p^* is very small. If those conditions are indeed true, then we should write Equation (40) as

$$I = -qAn_iW/\tau_{g,eff} \tag{41}$$

where the effective generation lifetime is

$$\tau_{g,eff} = \tau_g/[1 + (n_i\tau_g/W)[(D_p/L_p^*N_D) + (D_n/L_n^*N_A)]] \tag{42}$$

The effective generation lifetime $\tau_{g,eff}$ is reduced from its true value τ_g by the quasi-neutral diffusion current.

A short diffusion length can have a significant influence on the generation lifetime. The MOS capacitor is a common test structure used to measure the generation lifetime. For such a device there is no n-emitter, only the p-base. Equation (42) normalized by τ_g becomes

$$\tau_{g,eff}/\tau_g = 1/[1 + (n_i\tau_g/W)(D_n/L_n^*N_A)] \tag{43}$$

The normalized effective generation lifetime calculated from Equation (43) is plotted in Figure 11 as a function of the minority carrier diffusion length L_n^*. Note the significant influence the L_n^* has on $\tau_{g,eff}$ at room temperature as shown in Figure 11a. This effect is even more pronounced at higher temperatures as shown in Figure 11b. This effect is further enhanced when the generation-recombination properties of the wafer are not uniform, as is always the case for wafers with high-perfection denuded zones and heavily precipitated bulk.

We should discuss one more consideration before leaving this section. The diffusion current in Equations (27) or (34) consists of the n-emitter and the p-base components. The doping concentration in the p-base is generally constant and hence the diffusion coefficient D_n is constant, too. We exclude the dependence or D_n on carrier-carrier scattering at very high injection levels. Measurements of the diffusion length in the p-base are therefore unambiguous. This is not so for the diffusion length in the n-emitter where the doping concentration of a diffused or ion-implanted layer is very non-uniform. The doping concentration profile can of course be determined from spreading resistance or secondary ion

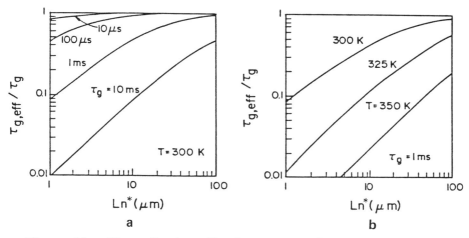

Figure 11: Normalized effective generation lifetime as a function of the effective minority carrier diffusion length and the recombination lifetime, where $L_n^*=(D_n\tau_r)^{\frac{1}{2}}$ was used to convert from L_n^* to τ_r. (a) T=300K, (b) temperatures as indicated. $N_A=10^{15}$cm^{-3}. $D_n=30$cm^2/s, W=3.5μ.

mass spectroscopy (SIMS) measurements. But the spatial variation of the diffusion coefficient D_p cannot be incorporated into the diffusion length measurement analysis and hence τ_r cannot be determined from DC measurements. Only the ratio N_D/D_p and L_p^* can be found (37). To find all three parameters it is necessary to use an additional AC measurement.

Frequently it is not necessary to know L_p^* explicitly. It may be sufficient to know the emitter diffusion contribution. In that case it is useful to rewrite the n-emitter diffusion current component as

$$I_{diff} = qAn_i^2 h \qquad (44)$$

where the recombination parameter $h=D_p/L_p^*N_D$ incorporates the spatial variation of all three n-emitter parameters D_p, L_p^*, and N_D. The parameter h has been found to have values that lie between 1×10^{-14} to 6×10^{-14} cm^4/s for highly doped emitters regardless whether the emitters are n$^+$ or p$^+$ used for high-power diodes and thyristors or low-power transistors (21). This relative constancy over a wide range of different emitter structures with widely varying thicknesses and doping concentrations appears to be associated with Auger recombination.

5.3 Non-Uniform Substrates

Let us see what lifetime is measured for the non-uniform substrate illustrated in Figure 12. We consider the case in which the substrate consists of a denuded zone of thickness d_z with diffusion length L_{n1} and a precipitated bulk with diffusion length L_{n2}. When the lifetime is measured with the pulsed MOS capacitor, the device is pulsed into deep-depletion. Through thermal ehp generation both in the scr and in the quasi-neutral bulk, the device relaxes back to its equilibrium state. During the measurement there is a scr of width W which we assume to be entirely contained within the denuded zone as shown in Figure 12.

The undepleted denuded zone thickness is d_z-W and the thickness of the precipitated bulk is T_p-d_z. The effective minority carrier diffusion length in the region of the wafer beyond the depleted scr now becomes (38)

$$L_{n,eff} = L_{n1}[1 + \Upsilon\tanh(\beta_1)]/[\Upsilon + \tanh(\beta_1)] \tag{45}$$

where

$$\Upsilon = (L_{n1}/L_{n2})[\tanh(\beta_2) + (s_c L_{n2}/D_n)]/[1 + (s_c L_{n2}/D_n)\tanh(\beta_2)] \tag{46}$$

and $\beta_1=(d_z-W)/L_{n1}$ and $\beta_2 = (T_p-d_z)/L_{n2}$.

$L_{n,eff}$ accounts for the two different diffusion lengths in a wafer containing non-uniform G-R center concentrations. Equation(45) reduces to the simple form of Equation(28) for the special case of $L_{n1}=L_{n2}=L_n$ as expected. The diffusion length of precipitated wafers is usually very low and $\beta_2 \gg 1$. This allows Υ to be written as $\Upsilon \approx L_{n1}/L_{n2}$. Furthermore, in the denuded zone we have $\beta_1 \ll 1$. This allowing Equation(45) to be simplified to

$$L_{n,eff} \approx [(d_z-W) + L_{n2}]/(1 + L_{n2}/L_{n1})$$
$$\approx (d_z-W) + L_{n2} \tag{47}$$

because the high perfection denuded zone assures that $L_{n1} \gg L_{n2}$.

Equation (47) states simply that for $d_z \leq 0.2 L_{n1}$ and $(T_p-d_z) \geq 2L_{n2}$, the effective diffusion length is the sum of the

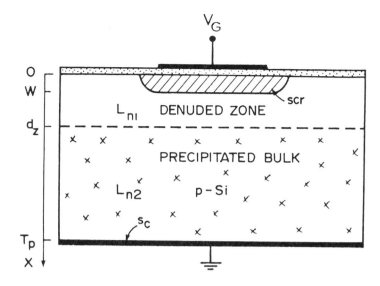

Figure 12: A pulsed MOS capacitor on a wafer consisting of a precipitated bulk and a surface denuded zone. The scr region, shown by the shaded region, is contained with the denuded zone.

undepleted denuded zone and the diffusion length of the precipitated bulk. The concepts embodied in Equation (45) apply, of course, to any wafer with unequal diffusion lengths. It need not be a denuded zone/precipitated wafer. Considerations similar to those expressed by Equation (45) have been used in the interpretation of junction leakage current measurements on oxygen-precipitated wafers (39). Similarly, the opposite effect of the high lifetime or diffusion length of the denuded zone enhancing the diffusion length of the precipitated bulk has also been addressed (40).

A plot of the normalized diffusion length, $L_{n,eff}/L_{n1}$, given by Equation (45) is shown in Figure 13. It is obvious from these curves that for low L_{n2} and denuded zones not entirely depleted during the lifetime measurement the effective diffusion length is considerably reduced from the value in the denuded zone. Surface recombination at the back surface can further degrade the diffusion length, but only if $L_{n2} > (T_p - d_z)$. This is very unlikely for short L_{n2} values.

Present high density MOS circuits are typically made on one of two substrates: (i) a p-substrates of uniform doping with oxygen precipitation anneals to form a denuded zone

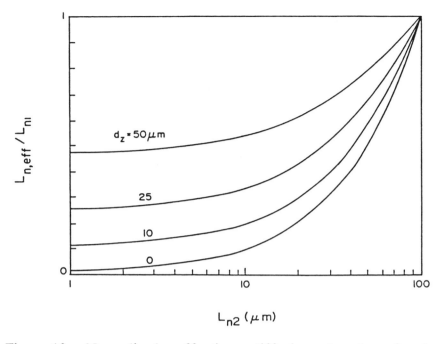

Figure 13: Normalized effective diffusion length of the device of Figure 12 as a function of L_{n1}, L_{n2} and d_z. It is a plot of Equation (45) with $L_{n1}=100\mu m$, $s_c=10^5 cm/s$, $T=350\mu m$ and $D_n=30 cm^2/s$. Reprinted after Reference (38) by permission of Pergamon Press, Ltd.

and a precipitated bulk, sometimes referred to as intrinsically gettered substrates, and (ii) a p-epitaxial layers on p$^+$ substrates. The choice of substrate depends on many factors including wafer cost, alpha particle immunity, latch-up immunity in CMOS circuits, yield, leakage currents, etc. We will compare the two substrates from leakage current considerations. The room temperature scr leakage current is very similar for both substrates since the quality of the denuded zone is similar to that of the epi layer.

The high temperature leakage current, however, differs. We saw for the intrinsically gettered substrates from Equation (47) that $L_{n,eff} \approx (d_z-W)+L_{n2} \approx L_{n2}$ for $(d_z-W)<L_{n2}$. The diffusion current is therefore

$$I_{diff}(\text{intrins.}) \approx qAn_i^2 D_n/L_{n2}N_A \qquad (48)$$

where L_{n2} is the diffusion length in the precipitated bulk. For the p-epi/p+ substrate we have for the p+ substrate

$$I_{diff}(p/p+) \approx qAn_i^2 D_n^+/L_n^+ N_A^+ \tag{49}$$

In Equation (48) we have neglected the diffusion current in the denuded zone and in Equation (49) we have neglected the diffusion current in the epi layer because both are small. The diffusion lengths L_{n2} and L_n^+ are much smaller than the substrate thickness allowing us to disregard the back surface recombination velocity s_c.

The ratio $I_{diff}(p/p^+)/I_{diff}(intrins.)$ is

$$I_{diff}(p/p^+)/I_{diff}(intrins.) \approx D_n^+ L_{n2} N_A / D_n L_n^+ N_A^+ \tag{50}$$

where we have neglected band gap narrowing in the p+ substrate which influences n_i. We know the diffusion coefficient in highly-doped substrates to be smaller than that in lowly-doped substrates, $(D_n^+/D_n)<1$, and we also have $(N_A/N_A^+)\ll 1$. These inequalities give

$$I_{diff}(p/p^+)/I_{diff}(intrins.) \ll 1 \tag{51}$$

The ratio L_{n2}/L_n^+ is likely to be on the order of unity since the diffusion length is small for both intrinsically gettered as well as p+ substrates.

Equation (51) shows devices on p/p+ substrates to have lower diffusion currents than those made on intrinsically gettered substrates. This has been experimentally confirmed for those device operating temperatures at which diffusion currents dominate (41)-(42). Those temperatures are typically 50-70°C and above. The current ratio is largely determined by the doping ratio N_A/N_A^+ and N_A^+ should be 10^{19} cm^{-3} or above (41).

6.0 LIFETIME MEASUREMENT TECHNIQUES

A semiconductor junction device in the non-equilibrium state is either forward or reverse-biased. During forward-bias of minority-carrier devices there are excess carriers in the device. The source of the excess carriers is either a forward-biased junction or light incident on, and absorbed by, the device. A semiconductor without a pn junction but with ohmic contacts or one without any contacts can only be excited by light or some other form of radiation such as X-rays, electron beams etc.

One technique to determine the recombination lifetime is to measure a property of the device or the semiconductor material that depends on excess carriers. Once excess carriers have been created they are detected in one or two basic methods. In the first method the excitation source is abruptly terminated and the temporal behavior of the excess carriers is monitored. This is done in the photoconductivity decay technique. In the second method, the steady-state excess carrier density is monitored. The surface-photovoltage method, for example, relies on the establishment of a steady-state forward-biased junction by light-generated excess carriers. One can also monitor the current as in the reverse-recovery method or the voltage as in the open-circuit voltage decay technique, to mention just a few. So we see that both electrical as well as optical excitation are useful for the measurement of the recombination lifetime.

A reverse-bias can only be established in a device containing a junction. The junction may be a pn junction, a Schottky-barrier junction or a gate-induced junction. To measure the generation lifetime it is necessary to establish a reverse-bias condition and detect the thermal electron-hole pair generation. This can be done by measuring the junction current, as in the gate-controlled diode method, or by monitoring the discharge of a capacitor by the junction current, as in the pulsed MOS capacitor technique. We will describe a few of the more commonly employed lifetime characterization techniques.

6.1 Recombination Lifetime

6.1.1 Photoconductive Decay: The photoconductive decay (PCD) lifetime measuring technique was one of the first lifetime characterization methods to be used (43). As the name implies it uses optical excitation of e-h pairs. The carrier decay is monitored as a function of time following the termination of the optical pulse. Traditionally the sample is provided with contacts and the current is measured as a function of time. More recently, non-contacting techniques have been developed that make the method attractive because it is fast and non-destructive.

PCD relies on a measurement of the sample's conductivity, σ, given by

$$\sigma = q(\mu_n n + \mu_p p) \tag{52}$$

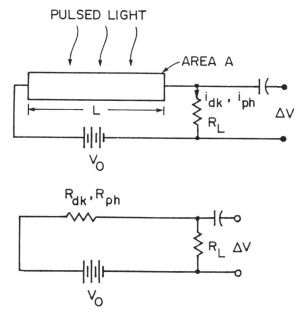

Figure 14: Schematic photoconductive decay measurement circuit.

The carrier concentrations are given by $n=n_o+\Delta n$ and $p=p_o+\Delta p$. The conductivity increases in the presence of light because both n and p increase. Let us consider the circuit shown in Figure 14. For the derivation of the relevant equations we follow Ryvkin (44). The sample has a dark resistance R_{dk} and a photoresistance R_{ph} with $R_{dk}>R_{ph}$. When light strikes the sample its resistance changes by the amount $\Delta R=R_{dk}-R_{ph}$. The applied voltage is V_o and the change in resistance gives a voltage change of

$$\Delta V = (i_{ph} - i_{dk})R_L \tag{53}$$

where i_{ph} and i_{dk} are the photo and the dark current, respectively, given by

$$i_{ph} = V_o/(R_L+R_{dk}-\Delta R) \tag{54a}$$

$$i_{dk} = V_o/(R_L+R_{dk}) \tag{54b}$$

Substituting Equation (54) into (53) give the voltage change across the load resistance, R_L, as

$$\Delta V = -R_L \Delta R V_o / [(R_L + R_{dk})(R_L + R_{dk} - \Delta R)] \tag{55}$$

In PCD measurements we are interested in the excess carrier decay with time. The excess carrier density is related to the sample conductance through

$$\Delta G = \Delta \sigma A / L = (qA/L)(\mu_n + \mu_p) \Delta n \tag{56}$$

where σ is defined in Equation (52), A=sample area and L=sample length. We also assume $\Delta n = \Delta p$. This latter assumption is not true if the sample exhibits trapping. With ΔG given by

$$\Delta G = G_{ph} - G_{dk} = (R_{dk} - \Delta R)^{-1} - R_{dk}^{-1} \tag{57}$$

we arrive at

$$\Delta V = R_L R_{dk}^2 V_o \Delta G / [(R_L + R_{dk})(R_L + R_{dk} + R_L R_{dk} \Delta G) \tag{58}$$

Equation (58) shows that there is no simple relationship between the time-dependence of the measured voltage and the time-dependence of the excess carrier concentration.

In practice there are two main PCD characterization techniques: (i) the constant voltage method, and (ii) the constant current method. In the constant voltage method the load resistance R_L is chosen to be small compared to the sample resistance and Equation (58) becomes

$$\Delta V \approx \Delta G R_L V_o$$

$$\approx (qA/L)(\mu_n + \mu_p) R_L V_o \Delta n \tag{59}$$

valid for $\Delta G R_L \ll 1$. In the constant current method, the load resistance is large compared to the sample resistance and Equation (58) becomes

$$\Delta V \approx \Delta G (R_{dk}^2 / R_L) V_o$$

$$\approx (qA/L)(\mu_n + \mu_p)(R_{dk}^2 / R_L) V_o \Delta n \tag{60}$$

Equation (60) is valid for $\Delta G R_{dk} \ll 1$.

It should be noted that the simple relationship $\Delta V \sim \Delta n$ is valid only for low-level excitation. If bulk recombination were the dominant recombination mechanism we would expect

$$\Delta n(t) = \Delta n(0) \exp(-t/\tau_r) \tag{61}$$

for simple recombination. The actual decay is more complicated because surface recombination, trapping and higher order decay modes play a significant role, especially during the initial part of the decay (45).

Surface recombination can be separated from bulk recombination by measuring the PCD on identically prepared samples of varying thicknesses (46). The underlying assumption is that for identically prepared samples both bulk and surface lifetimes do not change. The varying thicknesses allow separation of the two. Another method to separate surface bulk from recombination has been proposed (46). The slope of the photoconductive decay signal at short and long times is used in this technique to separate the surface from the bulk lifetime.

The carrier decay can be monitored in two basic ways. In one, described above, the current through, or the voltage across, the sample is monitored. The sample requires two ohmic contacts. The contacts are usually evaporated or plated metal contacts. For lifetime measurements on grown ingot the contacts are often in the form of clamps. The contacts need not be highly ohmic, but they should not inject minority carriers to any significant extent nor should they lead to sample heating. A recent circuit implementation that is used for silicon ingots, but can also be used for wafers, is shown in Figure 15 (47).

It is important that the illumination be restricted to the semiconductor itself to avoid contact effects. Furthermore, the electric field in the sample should be restricted to a value $E \leq 0.3\sqrt{(\mu\tau_r)}$, where μ is the minority carrier mobility and τ_r is the minority carrier lifetime (48).

The carrier decay can also be monitored directly without the need for contacts. Contactless measurements are very desirable because they are fast and non-destructive. For example, a wafer can be pulled from the process cycle, measured and then re-inserted into the process. One method of PCD contactless measurements is the microwave reflection technique of Figure 16 (49)-(51). Excess carriers are created by light pulses as in the conventional PCD method. The time-dependent photoconductivity is monitored by detecting the time-dependent microwave reflection from the sample's surface. The microwave

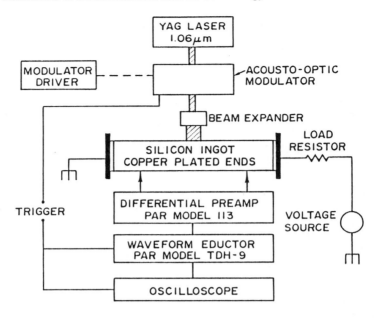

Figure 15: An experimental measurement circuit with ohmic contacts. After Reference (47).

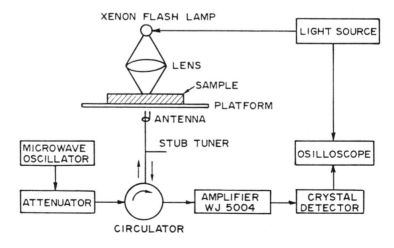

Figure 16: An experimental microwave reflectance contactless PCD circuit. After Reference (51).

reflectivity R is a function of the conductivity σ given by

$$R = 1 - \Upsilon/\sqrt{\sigma} \tag{62}$$

where Υ is a constant whose magnitude depends on the microwave frequency and the waveguide impedance. The microwaves are not reflected from the surface itself. They penetrate a skin depth into the sample. For silicon irradiated with microwaves of 10GHz frequencies, the skin depth is around 350μm for 0.5 Ω-cm and 2200μm for 100 Ω-cm resistivity. Consequently, a significant fraction of the wafer thickness is sampled by the microwaves and the reflected microwave signal is characteristic of the bulk carrier concentration. If a resonant cavity is used in the experimental set-up then it is important to ascertain that the microwave signal decay is indeed the photoconductive decay and not the decay of the measurement circuit. It has been observed that the system response is very fast when the cavity is off-resonance, while an on-resonance cavity results in a large increased system fall time (52).

A comparison of lifetimes measured with the conventional photoconductive decay and the microwave PCD methods has shown the conventional PCD derived lifetimes to be always higher then those obtained from microwave PCD (23). This discrepancy was found to increase for higher lifetimes.

The microwave reflection PCD measurement technique is available as a commercial instrument and lifetimes to a lower limit of about 1μs can be measured. As in all PCD methods, the shortest lifetime that can be measured is determined by either the fall time of the light pulse or the response time of the detection circuit. The longer of these two times determines the lower lifetime measurement limit.

A semiconductor can contain *traps* in addition to G-R centers. Traps capture excess carriers and then emit them back to the band from which they were captured. For example, an excess electron temporarily captured on a trap is removed from the conduction band during the time it is trapped and is therefore not available for conduction or recombination. Once it is emitted back to the conduction band it becomes available for conduction or recombination. The electron lifetime is effectively extended by the trapping time.

Trapping can considerably falsify recombination lifetimes measured by transient methods such as PCD. Artificially long lifetimes are sometimes obtained. Trapping effects can be considerably reduced by steady-state illumination of the sample during the PCD measurement. The steady-state light keeps the traps occupied during the transient decay measurement thus preventing excess carrier trapping during the PCD measurement.

6.1.2 Open Circuit Voltage Decay: The open-circuit voltage decay (OCVD) lifetime characterization technique was one of the earliest lifetime measurement methods (53). It is simple and is among the preferred lifetime characterization techniques because it is easy to implement. It requires only a simple measurement circuit. The interpretation of the experimental data is fairly straightforward and a commercial instrument is available.

A steady-state excess carrier concentration is established in the device during the initial phase of measurement by passing a current through a pn junction as shown in Figure 17a. At t=0 the switch S is opened and the pn junction becomes open-circuited. Excess carriers in the device at this time can only recombine, they cannot flow out of the device. Carrier recombination is detected by monitoring the open-circuit voltage. The current pulse when switch S is closed can be replaced by pulsing the junction with a light pulse having a short turn-off time.

The open-circuit voltage decay V-t curve of Figure 18 shows several distinct regions. Upon opening the switch at t=0, there is a sudden voltage drop. This voltage drop corresponds to the "IR" drop across the diode. When the current flows it develops a voltage across the diode's resistance. When the current becomes zero, this voltage drop becomes zero. Next a non-linear region follows which is largely due to higher order decay modes and recombination in the heavily doped n-emitter. Then comes a fairly linear region which is used for lifetime extraction. Finally the voltage decays rapidly to zero in a non-linear fashion.

A voltage exists across the diode when the switch is opened because there are excess carriers within the device. This open-circuit voltage is the sum of the junction voltage, V_j and the Dember voltage V_D. The junction voltage depends on the excess minority carrier concentrations in the quasi-neutral regions at the space-charge region edges. For simplicity we will consider only the electron concentration in the p-base as shown in Figure 17b. This is a reasonable assumption since the hole concentration in the n-emitter decays rapidly during the early phase of the voltage decay due to the low hole lifetime in the heavily-doped n-emitter.

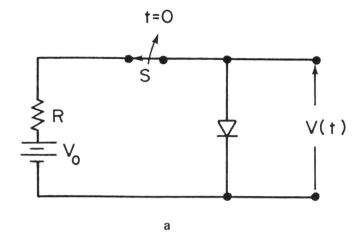

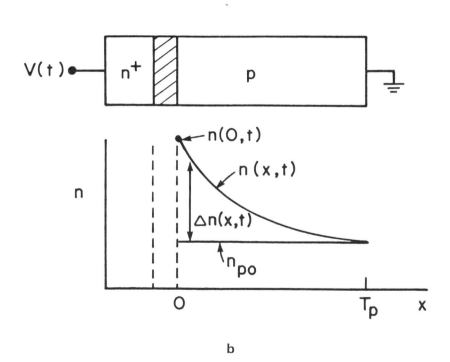

Figure 17: The open-circuit voltage decay measurement technique. (a) the circuit, and (b) the n⁺p junction and the minority electron distribution in the p-base.

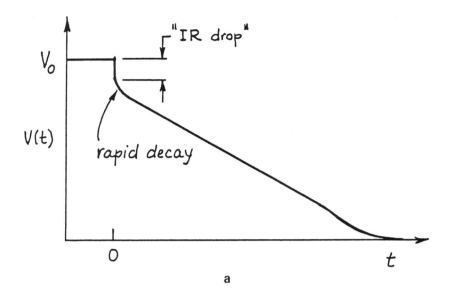

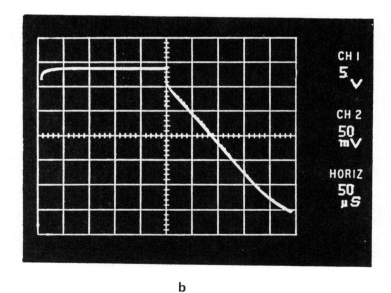

Figure 18: The voltage-time response of the open-circuit voltage decay method. (a) Schematic response, (b) experimental response. $\tau_r = 21 \mu s$. Courtesy of R.C. Dondero, Arizona State University.

The excess electron concentration at the space-charge region edge is given by

$$\Delta n(x=0,t) = n_{po}[\exp(qV_j/kT)-1] \qquad (63)$$

where n_{po} is the equilibrium electron concentration in the p-base.

The diode voltage is

$$V = V_j + V_D \qquad (64)$$

where the Dember voltage depends on unequal electron and hole mobilities. V_D is given by (54)

$$V_D = (kT/q)[(b-1)/(b+1)]\ln\{[n_{po}+bp_{po}+ (b+1)\Delta n(0,t)]/(n_{po}+bp_{po})\} \qquad (65)$$

Here b is the ratio of electron to hole mobility, $b=\mu_n/\mu_p$, and p_{po}=majority carrier concentration in the p-base. Equation (65) shows that the Dember voltage contributes only under high-level injection conditions, i.e. $\Delta n(x=0,t) > p_{po}$. Such high injection levels are usually not used for OCVD measurements and, consequently, V_D is generally neglected. The junction voltage for $V_j(t) \gg kT/q$ can be written as (55)

$$V_j(t) = V_o - (kT/q)\ln[\mathrm{erfc}(t/\tau_r)^{\frac{1}{2}}] \qquad (66)$$

A first-order expansion of the erfc-function for $x \geq 2$ is $\mathrm{erfc}(x) \approx \exp(-x^2)/(x\sqrt{\pi})$. This gives $\ln[\mathrm{erfc}(t/\tau_r)^{\frac{1}{2}}] \approx t/\tau_r + \ln\sqrt{(\pi t/\tau_r)} \approx t/\tau_t$ for $t \geq 4\tau_r$ and Equation (66) becomes

$$V_j(t) \approx V_o - (kT/q)(t/\tau_r) \qquad (67)$$

The recombination lifetime is obtained from the expression

$$\tau_r \approx -(kT/q)/(dV_j(t)/dt) \qquad (68)$$

i.e. τ_r is inversely proportional to the slope of the OCVD curve.

The conventional OCVD method utilizes a measurement of the slope of the V_j vs. t decay curve for $t \geq 4\tau_r$. An experimental voltage-time curve is shown in Figure 18b. Note on this curve the initial rapid voltage drop when the switch is opened, then a nonlinear portion during the early decay, following by a reasonably straight line. Eventually the decay becomes nonlinear again during the final decay stage.

The simple voltage decay approximated by the base recombination lifetime alone, as shown by Equation (67) or (68), is an approximation and applies only for bases longer than the minority carrier diffusion length. Surface recombination at the back contact plays an important role for $T_p \leq L_n$. The decay time constant is no longer simply τ for that case. It becomes an effective lifetime τ_{eff} given by (56)

$$\tau_{eff} \approx \tau_r \qquad (69)$$

for a very low surface recombination velocity at the back surface, i.e. $s_c \to 0$. For high s_c, i.e. $s_c \to \infty$

$$\tau_{eff} \approx 4T_p^2/\pi^2 D_n \qquad (70)$$

Yet another complication is recombination in the highly doped n-emitter which we have so far neglected. The n-emitter lifetime is usually much lower than the base lifetime. The excess carriers in the emitter recombine much more rapidly than those in the base and some of the excess carriers are injected into the emitter. The junction voltage decay is reduced. This effect becomes negligible for $t \geq 2.5\tau_{r,base}$ and the $V_j(t)$ - t decay becomes linear with slope $kT/q\tau_{r,base}$ regardless of emitter recombination or emitter band gap narrowing (57).

For high injection levels in p^+in^+ power iodes, the recombination in both p^+ and n^+ end regions is important. For those devices with "i" base width less than the ambiopolar diffusion length, the base lifetime can only be determined if the end region recombination parameters are known (58).

The simplified lifetime analysis for high injection levels, neglecting end-region recombination, is given by (59)

$$\tau_r = -(2kT/q)/(dV_j(t)/dt) \qquad (71)$$

Additional complications are encountered for diodes in which the diode capacitance is appreciable (60) and when there is a shunt resistance across the diode as is usually encountered in solar cells. The voltage decay can then given unreliable lifetime measurements. Green (61) proposes to use an external resistor and capacitor to linearize the measurement. Moore (62) has proposed the use of a bias light on the device upon which is superimposed a small, repetitively pulsed forward bias voltage. One circuit implementation is shown in Figure 19.

The voltage of the device is changed only slightly during this measurement and the voltage decay is exponential and given by (62)

$$V_j(t) = V_{dc} + (kT/q)(\Delta n/n_{bias})\exp(-t/\tau_r) \tag{72}$$

where n_{bias} is the bias light-generated minority carrier concentration. The linear and semi-logarithmic voltage-time curves of the diode of Figure 18b with bias light are shown in Figure 20. The bias light for this case is equivalent to concentrated solar irradiation of six suns. Note the lifetime increase from $\tau_r = 21 \mu s$ in the dark to $31 \mu s$ with bias light. This bias light has caused a higher injection level and as shown by Equations (8) and (11) there is a recombination lifetime increase with increased injection level.

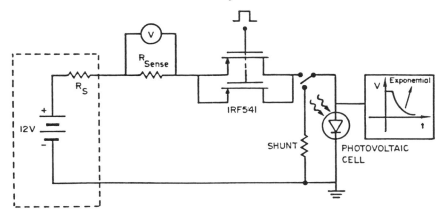

Figure 19: An open circuit voltage decay circuit using a bias light and a pulsed forward bias voltage. Courtesy of R.C. Dondero, Arizona State University.

The OCVD method is considered one of the most reliable recombination lifetime characterization techniques. The voltage decay of the dark voltage decay curve generally has a linear region from which the lifetime is calculated. However, care must be taken to keep the device in low-level injection and to measure the slope after the initial rapid decay has

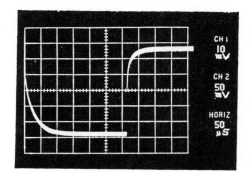

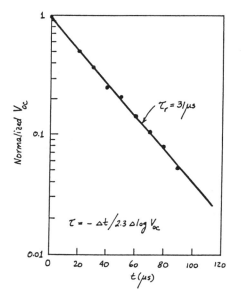

Figure 20: The voltage-time response of the diode of Figure 18 under a bias light of concentrated sunlight corresponding to six sun concentration. Courtesy of R.C. Dondero, Arizona State University.

died out. Then a reliable lifetime value is obtained. The voltage decay is very much influenced by the surface recombination velocity at the back surface for base widths much less than the minority carrier diffusion length. Reliable lifetime measurements are then difficult. Separation of the recombination lifetime and surface recombination velocity is possible by utilizing both the open-circuit voltage and short-circuit current decay methods (63)-(64).

6.1.3 Diode Reverse Recovery: The diode reverse-recovery (RR) lifetime characterization method was one of the earliest lifetime measurement techniques (65)-(67). The method is illustrated in Figure 21. A forward current, I_f, is forced through the diode. Excess carriers are injected into both the n-emitter and the p-base. The diode resistance is consequently very low. At t=0 the current is either abruptly switched from forward to reverse conduction, as shown in Figure 21b, or it is gradually switched from forward to reverse conduction as in Figure 21c. The latter is more typical of power devices in which large currents cannot be switched very abruptly.

We will use the abrupt current transition of Figure 21b in our discussion. The diode is initially forward-biased to voltage V_d by the current I_f. When the current id abruptly switched to the reverse direction, the diode will remain forward-biased for a short time in spite of the fact that the current direction has changed because the diode is still flooded with excess carriers. The reverse current, I_r, is initially given by $I_r \approx (V_r - V_d)/R$. We neglect the diode resistance during this portion of the diode response because the resistance of the forward-biased diode, r_d, is small. The resistive drop across the diode is $V_{dr}(\text{forward}) - I_f r_d$ during the forward current phase and $V_{dr}(\text{reverse}) = -I_r r_d$ during the reverse current phase. The voltage change at t=0 is $\Delta V_d = V_{dr}(\text{forward}) - V_{dr}(\text{reverse}) = (I_f + I_r) r_d$ and is shown in the lower portion of Figure 21b.

The current remains constant at I_r for some time after it is switched form I_f to I_r at t=0. The voltage remains positive and the diode stays forward-biased because the device contains excess carriers. The forward-bias voltage decreases with time as the excess carrier concentration decreases by being swept out of the device by the reverse current and by recombination within the device.

Equation (63) shows a pn junction to remain forward-biased as long as there are excess carriers in the quasi-neutral region at the scr edge. The diode voltage becomes zero when the excess carrier density at the scr edge vanishes at $t=t_s$, the diode storage time. The diode becomes reverse-biased for $t>t_s$ and the carrier density at the scr edge drops below its equilibrium value. However, there are still excess carriers deeper within the quasi-neutral region causing the current to only gradually approaches its leakage current, I_{rev}, as those excess carriers recombine. The voltage approaches and finally reaches the bias voltage, V_r.

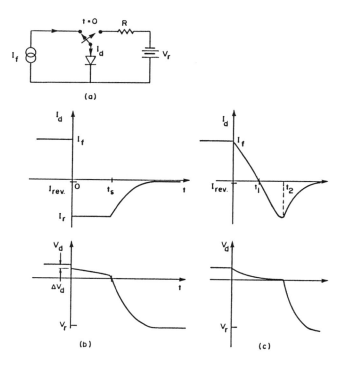

Figure 21: The diode reverse-recovery lifetime measurement method. (a) the measurement circuit, (b) the current-time and voltage-time behavior for abruptly switched current, and (c) for gradually switched current.

The reader may wonder why the current through the device can change so rapidly while the voltage changes relatively slowly. We know from the semiconductor current equations that for a device whose current is largely diffusion-limited, the current is proportional to the gradient of the minority carrier concentration, i.e. $I \sim dn/dx$ in the quasi-neutral base region at the scr edge. The carrier distribution during the forward current regime is that of Figure 17b for t<0. A change of the slope of the curve at x=0 from a negative to a positive slope at x=0 is sufficient to change the current from I_f to I_r. The voltage, however, is proportional to the logarithm of the excess carrier concentration according to Equation (63). It takes much longer for the carrier concentration than for the slope to change. Hence, the voltage changes much more slowly than the current as shown in Figure 21b.

The current-time curve is conveniently divided into two phases: (i) the constant current phase for $0 \leq t \leq t_s$, and (ii) the current decay phase for $t > t_s$. The constant current phase is used for the lifetime analysis. A charge storage analysis gives the recombination lifetime as (68).

$$\tau_r = t_s/[\ln(1 + I_f/I_r) - \ln(1 + Q_s/I_r\tau_r)] \qquad (73)$$

where Q_s is the charge in the diode at $t=t_s$. The term $Q_s/I_r\tau_r$ can be considered a constant in many cases. If that is true, then a plot of t_s vs. $(1+I_f/I_r)$ yields a straight line whose slope is the recombination lifetime. An example of such a plot is shown in Figure 22. The method has been shown to be valid even where there is a drift field in the base (70).

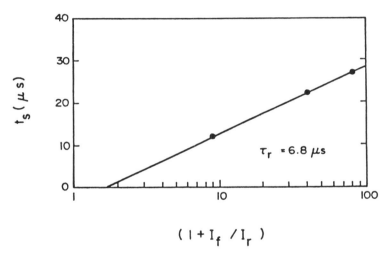

Figure 22: Diode reverse-recovery storage time as a function of $(1 + I_f/I_r)$. The recombination lifetime is calculated from the slope. Courtesy of C.J. Varker and J.D. Whitfield, Motorola, Reference (69).

If the term $Q_s/I_r\tau_r$ is not constant, then a plot of T_s vs. $(1+I_f/I_r)$ is no longer a straight line and the recombination lifetime cannot be easily obtained from such a plot. It is obvious that if I_r is changed in the term $(1+I_f/I_r)$ then clearly $Q_s/I_r\tau_r$ cannot be constant. Various approximations to this term have been derived (71) and for $I_r \ll I_f$, $Q_s/I_r\tau_r$ is approximately constant. It has also been shown (72) that the effect of end region recombination in the heavily doped emitter can be virtually eliminated by keeping $I_r \ll I_f$. Other potential sources of error are the scr capacitance discharge and scr recombination as well as the diode series resistance. Generally these are negligible.

For the gradual current change method of Figure 21c the extraction of the lifetime is slightly more complicated. We must solve the equation (73)

$$1 - \exp(-t_2/\tau_r) = (t_2-t_1)/\tau_r \qquad (74)$$

The RR method is used for lifetime measurements in spite of its shortcomings. By using coaxially mounted diodes in a matched circuit and measuring the current transient with a sampling oscilloscope, lifetimes as low as 1 ns have been measured (71). However, the method has diminished in popularity during the last few years and has been largely replaced by the open-circuit voltage decay method. The OCVD method gives lifetime values that agree more closely with lifetime values determined by other techniques (74). Another reason for the decline of the RR lifetime measurement technique is the non-availability of commercial test equipment.

In contrast to OCVD, where the diode is forward-biased during the entire voltage decay, in the RR method the diode is initially forward-biased and then is driven into reverse-bias. The range of carrier distribution is much wider than during the OCVD and the resulting lifetime is influenced more by the various recombination mechanisms that are active in the diode during its entire transient. It has been suggested (75) that in certain cases it is not possible to deduce the base recombination lifetime.

6.1.4 Surface Photovoltage:

The surface photovoltage lifetime method (SPV) is a *steady-state* technique in which optical excitation is used to determine the minority carrier diffusion length (76). The lifetime must be calculated from the diffusion length through the relation $L_n = \sqrt{(D_n \tau_r)}$. SPV is an attractive technique because it is non-destructive. Sample preparation is simple because contacts, junctions and high-temperature processing are not required. Being a steady-state method it is relatively immune to trapping effects that can influence the lifetime values determined by transient measurements.

The sample whose diffusion length is to be measured is usually in the form of a wafer or a portion of a wafer. It is

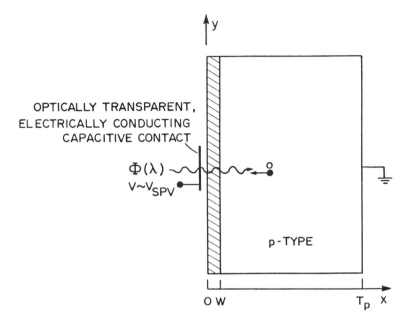

Figure 23: Schematic surface photovoltage measurement set-up. The optically transparent, electrically conducting capacitive contact allows light into the wafer and is used to measure the surface photovoltage.

schematically shown in Figure 23. One surface of the wafer is treated to induce a surface space-charge region of width W. The scr exists in the absence of any contacts because it is the result of surface charges, not the result of a bias voltage. The surface is uniformly illuminated by a chopped monochromatic light of energy larger than the band gap of the semiconductor. The back surface is kept dark. Electron-hole pairs are generated by the absorbed light. Those minority carriers that diffuse to the illuminated surface are separated from the majority carriers by the electric field of the surface scr. The minority carriers establish a surface photovoltage with respect to the grounded back surface. This surface photovoltage, V_{SPV}, is proportional to the excess carrier concentration, $\Delta n(W)$, at the edge of the scr

$$V_{SPV} = C_1 \Delta n(W) \tag{75}$$

where C_1 is a constant. The precise relationship between

V_{spv} and $\Delta n(W)$ need not be known. It need only be a monotonic function.

The optically generated excess carrier concentration distribution through the wafer in the x-direction is a rather complex expression. The distribution in the y-direction is assumed to be uniform. Several conditions must be met in order to make the SPV analysis tractable. First, the undepleted wafer must be much thicker than the minority carrier diffusion length. Generally the requirement $(T_p - W) \geq 4 L_n$ is sufficient. It is furthermore desirable that the scr width be small compared to L_n, i.e. $W \ll L_n$. Additional restrictions are placed on the optical absorption coefficient α. The conditions $\alpha W \ll 1$ and $\alpha T_p \gg 1$ should be met. Lastly the condition of low-level injection should be met.

With these assumptions it has been shown (77)-(78) that

$$\Delta n(W) \approx [(1-R)\Phi(\lambda)/(S_r + D_n/L_n)][\alpha L_n/(1 + \alpha L_n)] \qquad (76)$$

where $\Phi(\lambda)$=photon flux density incident on the semiconductor, R=semiconductor reflectivity, s_r=surface recombination velocity at the light-incident surface. The photon flux density is a function of the wavelength λ. From Equations (75) and (76) we find the surface photovoltage as

$$V_{SPV} \approx C_1[(1-R)\Phi(\lambda)/(S_r + D_n/L_n)][\alpha L_n/(1 + \alpha L_n)] \qquad (77)$$

To determine L_n the surface photovoltage is measured as a function of α. V_{SPV} is kept constant during the measurement by adjusting $\Phi(\lambda)$. Holding V_{SPV} constant keeps the surface recombination velocity, s_r, approximately constant during the measurement since the surface potential is approximately constant. If we further assume D_n, L_n and R to be independent of wavelength then we can rewrite Equation (77) as

$$\Phi(\lambda) = C_2(L_n + 1/\alpha) \qquad (78)$$

where C_2 is another constant.

The photon flux density or the normalized photon flux density is plotted against $1/\alpha$ for each constant-magnitude V_{SPV} during the measurement. α is varied by changing the wavelength of the incident light. Wavelengths between 0.8 and 1 μm are generally used for diffusion length measurements on silicon. The result is a linear plot whose extrapolated intercept on the negative $1/\alpha$ axis is L_n as shown by several examples in Figure 24. In order to measure V_{SPV} without

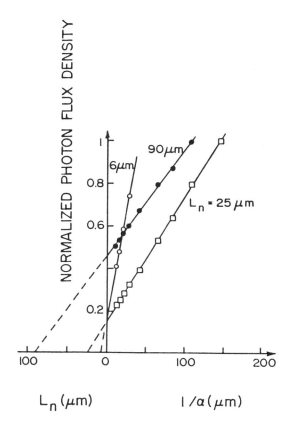

Figure 24: Typical surface photovoltage plots of the normalized photon flux density vs. the inverse absorption coefficient. The diffusion lengths are the intercepts on the "L_n" portion of the horizontal axis. Courtesy of A.M. Goodman, RCA Laboratories.

any permanent contacts the voltage is measured by a capacitively-coupled probe as shown in Figure 23.

Recent measurements (23) have compared lifetimes measured by the conventional photoconductive decay, the microwave PCD and the surface photovoltage techniques. The conventional PCD lifetimes were always higher than the microwave PCD lifetimes as mentioned earlier. Furthermore, the microwave PCD lifetimes were found to be lower than the SPV lifetimes for lifetimes higher than 10-20 μs. For lower lifetimes the microwave PCD and SPV values agreed.

Although the SPV method has been in use since 1961 it

has not found wide acceptance. One reason for this is the tedious and slow nature of the measurement technique. The wavelength for the appropriate α and the light intensity for the constant V_{SPV} had to be adjusted manually in earlier systems. These adjustments are made by a computer in more recent systems. Another reason for its limited use is that, as of 1989, there was no commercial instrument available.

The tediousness and slowness of the technique has been overcome by Goodman (79) with the computer-automated instrument, shown in Figure 25. Its basic components are a light source, a monochromator, a chopper to interrupt the light at chopping frequencies of typically 100-600Hz, the sample, a capacitive pick-up probe, and a lock-in amplifier for synchronous detection of the SPV signal which is typically only a few mV in magnitude. The monochromator can be replaced by a light source and narrow-band interference filters for a simplified system.

A crucial component of the SPV method is the surface treatment to create the surface space-charge region. Sample preparation is easy once a reproducible method is found to do this. The ASTM method (80) recommends boiling n-Si in water for 1h. For p-Si a 1 min etch in 20 ml concentrated HF + 80 ml water is recommended. Goodman (79) notes that this method works best when care is taken in earlier preparation steps not to produce a stain film on the wafer by withdrawing the sample from an HF-containing etch directly into air. A stained surface is likely to produce a low or unstable V_{SPV}. The stain can be avoided by quenching the HF-containing etch thoroughly with water before withdrawing the sample into the air. Another surface treatment for Si wafers is a standard Si clean/etch (81) removing any residual SiO_2 in buffered HF and treating n-Si wafers in an aqueous solution of $KMnO_4$. For p-Si the $KMnO_4$ step is omitted.

It is not the absorption coefficient but the wavelength that is experimentally varied during the measurement. The SPV plot, however, utilizes the absorption coefficient as shown in Figure 24. An accurate wavelength-absorption coefficient relation is therefore very important. Any error in that conversion relationship results in an error in the measured diffusion length. A good fit to the published data for silicon is given by (82).

$$\alpha(\lambda) = (84.732/\lambda - 76.417)^2 \tag{79}$$

where λ is in μm and α in cm^{-1}. This relationship is valid

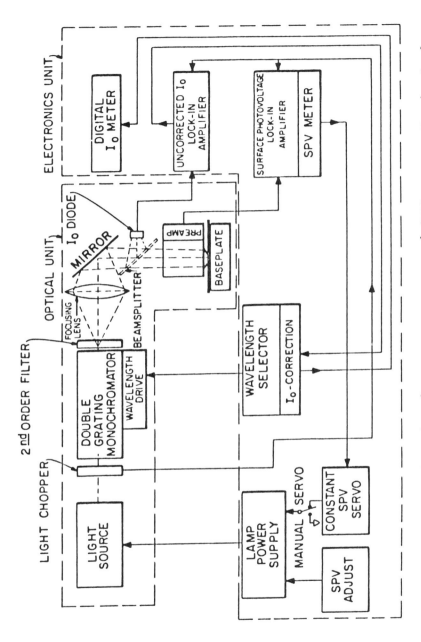

Figure 25: Measurement schematic of a computer-automated SPV system. Reprinted after Reference (79) by permission of IEEE (Copyright 1980 IEEE).

for the 0.7 to 1.1μm wavelength range that is normally used for SPV measurements of Si wafers.

The major strength of the surface photovoltage method lies in its ability to measure the diffusion length non-destructively in a steady-state mode. The measured lifetime is to first order independent of surface recombination and trapping effects. It has been used for Si (83), GaAs (77) and InP (84) wafers. It has also been used for amorphous silicon where diffusion lengths as low as 200Å have been determined (85).

Even though the minority carrier diffusion length is not of great importance to the MOS IC industry, SPV could and is being used as an in-line process monitor. Wafers are pulled from wafer lots at various stages during processing and the diffusion length is measured to check the status of furnaces and other process equipment.

We should reiterate that the diffusion length is important in MOS dynamic memories as we pointed out earlier in Section 5.2. It determines the quasi-neutral bulk diffusion current which contributes to the memory cell discharge. The diffusion current may only contribute a small portion to the total current at room temperature, but it becomes dominant at temperatures above 40-60°C. A knowledge of the diffusion length is therefore very useful.

The method's weakness presently lies in the absence of a commercial instrument. A number of systems exist in various laboratories, the most recent ones being computer-automated. A commercial instrument would no doubt extend the application of SPV significantly. An interesting adaptation of SPV is the mapping of the semiconductor photovoltaic response using penetrating sub-band gap radiation (86). A spatial response map is obtained by moving a water electrode probe across the wafer. The long wavelength radiation assures the response to be dominated by the minority carrier diffusion length and the map, therefore, represents a display of the G-R center distribution.

6.1.5 Pulsed MOS Capacitor: The pulsed MOS capacitor (MOS-C) is routinely used for generation lifetime measurements. We discuss this in Section 6.2.1. The pulsed MOS capacitor is also occasionally used for recombination lifetime measurements. This is the topic of the present section. We describe the method here not because it is very popular, but because the MOS capacitor is a very useful and common test structure and lends itself readily to τ_r measurements.

There are two very different methods to measure the recombination lifetime with an MOS capacitor. The first method is discussed with reference to Figure 26. The second method is discussed later in this chapter. We will consider a p-type substrate in our discussion. The MOS-C is biased with gate voltage V_{G1} into strong inversion at point A in Figure 26a and d. The inversion charge density for this bias condition is given by $Q_{n1} \approx (V_{G1}-V_T)C_{ox}$, where V_T is the threshold voltage and C_{ox} is the oxide capacitance/unit area. Next a voltage pulse of amplitude $-\Delta V_G$ and width t_p is superimposed on the V_{G1} gate voltage. During the pulse, the gate voltage is reduced from V_{G1} to $V_{G2}=V_{G1}-\Delta V_G$ shown in Figure 26b and by point B in Figure 26d. The inversion charge

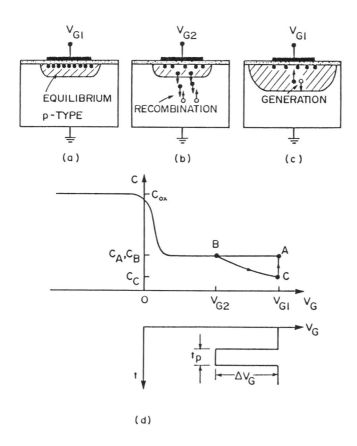

Figure 26: Pulsed MOS capacitor recombination lifetime measurement technique. The device behavior at various voltages and times is shown in (a), (b) and (c) and the $C-V_G$ and V_G-t curves are shown in (d).

during the pulse period is reduced from Q_{n1} to $Q_{n2} \approx (V_{G2} - V_T)C_{ox} < Q_{n1}$. The difference $\Delta Q_n = Q_{n1} - Q_{n2}$ not required during the time t_p is injected into the substrate as shown in Figure 26b. What happens to ΔQ_n? Minority electrons in the inversion layer on a p-type substrate do not recombine with majority holes because they are separated from the holes by the electric field in the space-charge region. However, when the electrons are injected into the p-substrate they find themselves surrounded by holes and are able to recombine.

Let us now consider two extremes. (i) the pulse period t_p is very long, and (ii) the pulse period is very short. For the long pulse period case where $t_p > \tau_r$ the minority electron charge ΔQ_n has had sufficient time to recombine with majority holes in the p-base. The gate voltage returns to V_{G1} upon pulse termination. The available inversion charge is only Q_{n2} and the capacitor is driven into partial depletion shown in Figure 26c and in Figure 26d by point C. Thermal electron-hole pair generation returns the device to equilibrium at position A as indicated in Figure 26d.

For the second case we let $t_p \ll \tau_r$. The device goes through the same stages as before with one major exception: the minority electrons injected into the p-substrate during the pulse period have insufficient time to recombine because the pulse period is much shorter than the recombination lifetime. Consequently the capacitance sequence in Figure 26d is $C_A \rightarrow C_B \rightarrow C_C$ as before, but $C_C = C_A$. In other words, the capacitances just before and just after the pulse are identical.

By changing the pulse width it is then possible to vary capacitance C_C from values smaller than C_A to values equal to C_A. In other words, the capacitance C_C is a measure of how many of the inversion electrons injected into the p-base and have recombined during the pulse period.

If we assume the recombination of the injected electrons to follow the simple exponential decay

$$\Delta Q_n(t) = \Delta Q_n(0) \exp(-t/\tau_r) \tag{80}$$

then it can be shown (87) that

$$\Delta Q_n(t_p) = \Delta Q_n(0) \exp(-t_p/\tau_r) = K(C_A^{-2} - C_C^{-2}) \tag{81}$$

where C_A is the capacitance for $t_p \rightarrow \infty$, i.e. long pulses with $t_p > \tau_r$, and K is a constant.

To determine the recombination lifetime the pulse width

t_p is varied and capacitance C_C is measured for each pulse width. C_C will eventually approach C_A as the pulse becomes longer than the recombination lifetime. The logarithm of $(C_A^{-2} - C_C^{-2})$ is plotted against t_p and the recombination lifetime is obtained from the slope $\Delta \log(C_A^{-2} - C_C^{-2})/\Delta t_p$ according to

$$\tau_r = -\Delta t_p / [2.3 \Delta \log(C_A^{-2} - C_C^{-2})] \tag{82}$$

where the 2.3 accounts for the "base 10 log" to the "base e ln" conversion. Such a plot is shown in Figure 27.

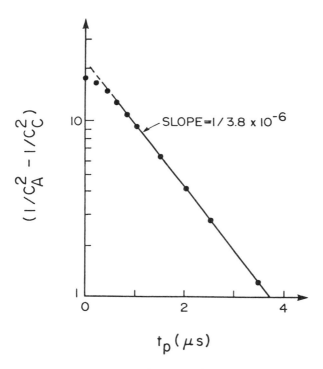

Figure 27: A plot of $(C_A^{-2} - C_C^{-2})$ vs. pulse width t_p for an MOS capacitor. The recombination lifetime is $1.6\mu s$.

A more detailed theory (88) shows the simple exponential time decay of the minority carriers in Equation (80) to be too simplistic because minority carriers recombine not only in the quasi-neutral p-substrate but also in the scr and at the surface.

This pulsed MOS-C recombination lifetime measurement

method has not found wide acceptance despite of the use of the simple and ubiquitous MOS capacitor test structure, because most capacitance meters and bridges do not easily pass the required narrow pulses through the instrument. It is generally easier to couple the MOS capacitor to the capacitance meter through a pulse transformer at its input terminals and apply the pulse there instead of passing it through the capacitance meter. A nice experimental arrangement is given in Reference (89). For recombination lifetimes on the order of microseconds, the pulse widths are sub-microsecond.

A variation of the pulsed MOS capacitor recombination lifetime measurement method has been proposed for MOSFETs (90). It is based on the charge pumping effect. When a MOSFET is pulsed from inversion into accumulation, most of the inversion charge leaves the device through the source and the drain. However, a small fraction is unable to reach either source or drain and recombines with majority carriers. This fraction is measured as a substrate current.

The substrate current is proportional to the pulse frequency for low frequency pulse trains. As the pulse repetition frequency increases to the point where the time between successive pulses becomes on the order of τ_r, then the substrate current-pulse frequency relationship becomes nonlinear and τ_r can be extracted from the measured current. Recombination lifetimes from 100ns to 100ms can be determined by this technique.

The second pulsed MOS capacitor method for determining the recombination lifetime is based on an entirely different principle. It relies indirectly on a current measurement even though it is the MOS capacitance that is actually measured. An MOS capacitor in deep-depletion is subject to a current flow given by

$$I = CdV/dt \qquad (83)$$

The current restores equilibrium in the device. The time to discharge the capacitor is the storage time, t_F. It is from Equation (83)

$$t_F = (C/I)\Delta V \qquad (84)$$

where ΔV is the voltage change across the capacitor. The main point for our discussion here is the inverse relationship between t_F and I, i.e. $t_F \sim 1/I$.

The current in the reverse-biased MOS capacitor is given by

$$I = I_{diff} = qAn_i^2 D_n/L_n^* N_A \qquad (85)$$

when the quasi-neutral region diffusion current dominates and by

$$I = I_{scr} = qAn_i W/\tau_g \qquad (86)$$

when scr current dominates as discussed in Section 5.2. We consider only the magnitude of the currents here and have not considered the negative sign of Equations (34)-(35). From Equations (84)-(86) we find

$$t_F \sim L_n^*/n_i^2 \qquad (87)$$

for I_{diff} and

$$t_F \sim \tau_g/n_i \qquad (88)$$

for I_{scr}.

Equation (88) dominates at room temperature and pulsed MOS capacitor measurements yield τ_g. This will be discussed in Section 6.2.1. Equation (87) dominates at higher temperatures of typically 70°C and above because n_i^2 increases much more rapidly with temperature than does n_i.

In the pulsed MOS capacitor (MOS-C) technique, a voltage pulse or voltage step is applied to the MOS capacitor to drive it into deep-depletion. We illustrate the technique in Figure 28 where the MOS-C is initially biased at point A in accumulation and is driven to gate voltage V_{G1} at point B by voltage step. Thermal generation returns the MOS-C to equilibrium, shown by the path B to C. The recovery or storage time, t_F, is determined by the thermal electron-hole pair generation properties of the bulk semiconductor and the oxide-semiconductor interface.

When the depleting voltage step or pulse is applied, majority carriers are repelled in the semiconductor over the depth of the depleted scr. The capacitance decreases very quickly from C_{ox} to C_i. It is the measurement instrument-typically a capacitance meter - that is the time-limiting element during this transition. The subsequent capacitance decay from C_i to C_F takes seconds or minutes for typical Si MOS-Cs.

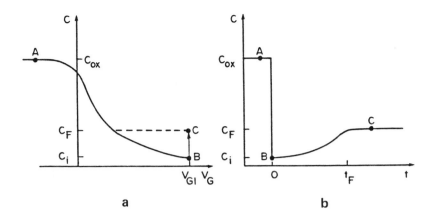

Figure 28: The C-V_G and C-t behavior of an MOS capacitor pulsed into deep-depletion.

Once the device is in deep-deletion, it reverts to the equilibrium state through thermal generation of electron-hole pairs. It is important that there be no other sources of minority carriers during the measurement. There should be no pn junctions nearby capable of injecting minority carriers into the MOS-C. For example, if the measurement is made on a MOSFET with source and drain connected to the substrate, then both the source and the drain inject minority carriers and the C-t decay time is very short. The time t_F becomes a measure of the injection efficiency of the source and drain, but not of the generation parameters of the device. By similar arguments, it should be evident that the measurement must be performed in the dark. Otherwise, photon-generated carriers will contribute to the capacitor discharge and again an erroneous reading is obtained.

The thermal generation components that participate in the capacitance discharge are shown in Figure 29. (1) is bulk scr generation characterized by the generation lifetime τ_g; (2) is lateral surface scr generation characterized by the surface generation velocity s_o; (3) is the scr surface generation under the gate characterized by s; (4) is the quasi-neutral bulk generation component characterized by the majority carrier diffusion length L_n, and (5) is the surface generation component at the back of the substrate described by the contact surface generation velocity s_c.

The semiconductor-oxide interface ehp generation is

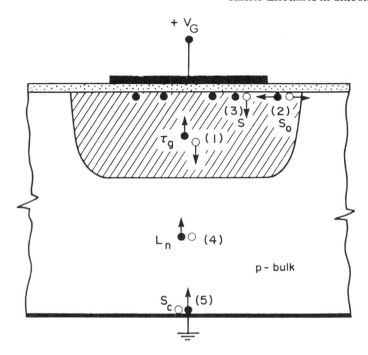

Figure 29: The thermal generation components in a deep-depletion MOS capacitor.

divided into two components (2) and (3) in Figure 29. The lateral portion of the surface scr beyond the inversion layer is always depleted of minority carriers. In contrast, the area under the gate is depleted only at the beginning of the capacitance-time transient. The inversion layer forms beyond that initial time. Carrier generation is maximum for a depleted surface and diminishes as an inversion layer forms. Hence surface generation is identical for the two areas at t=0. However, when the inversion layer forms in the gate area for t>0 surface generation decreases under the gate and then we find $s<s_o$.

The considerations for surface generation hold when the MOS capacitor is pulsed from accumulation into deep-depletion. The situation is slightly different when the device is pulsed from inversion into deep-depletion. The area under the gate is already inverted at t=0 when the device is initially biased in inversion and then pulsed into deep-depletion. The two generation components are never equal in that case even at t=0. Pulsing from inversion instead of from accumulation is commonly used when surface generation is to be reduced as much as possible. Generation components (4) and (5) are the quasi-neutral bulk generation components.

The MOS-C capacitance is related to the gate voltage V_G and inversion charge Q_n by

$$C = C_{ox}/[1 + 2(V_G + Q_n/C_{ox})/V_o]^{\frac{1}{2}} \tag{89}$$

where $V_o = qK_s\xi_o N_A/C_{ox}^2$ and we assume zero flatband voltage. Q_n in Equation (89) is negative because it stands for the electron inversion charge. Solving Equation (89) for V_G and differentiating with respect to time gives

$$dV_G/dt = -(1/C_{ox})dQ_n/dt - (qK_s\xi_o N_A/C^3)dC/dt \tag{90}$$

Equation (90) is an important equation relating the gate voltage rate of change with time to inversion charge and capacitance rate of change with time. Two measurement techniques are used. The gate voltage is pulsed in one method ($dV_G/dt=0$) and in the second method the gate voltage is ramped ($dV_G/dt=$constant). for the pulsed capacitor case $dV_G/dt=0$, and dQ_n/dt from Equation (90) becomes

$$dQ_n/dt = -(qK_s\xi_o C_{ox} N_A/C^3)dC/dt \tag{91}$$

dQ_n/dt stands for the thermal generation components in Figure 29.

The rate of inversion layer formation due to the diffusion current is given by

$$dQ_n/dt = qn_i^2 D_n/L_n^* N_A \tag{92}$$

Substituting Equation (92) into (91) and solving for C using the initial condition that $C=C_i$ at $t=0$ leads to the capacitance-time response function

$$C(t) = C_i(1-t/t_1)^{\frac{1}{2}} \tag{93}$$

where C_i is the capacitance immediately after the depletion pulse has been applied. A plot of $[1-(C_i/C)^2]$ vs. t has the slope $1/t_1$ where t_1 is given by

$$t_1 = K_s\xi_o A C_{ox} N_A^2 L_n^*/2D_n C_i^2 n_i^2 \tag{94}$$

L_n^* is calculated from this slope with the aid of Equation (94).

A typical 70°C C-t curve is shown in Figure 30a and the corresponding $[1-(C_i/C)^2]$ vs. t plot is shown in Figure 30b. We show in Figure 30(a) the room temperature C-t curve as well. Note the very significant decrease in the time for the capacitance to reach its steady-state value. It decreases from about 1000s to 4.5s. This is the effect predicted by Equations (87) and (88).

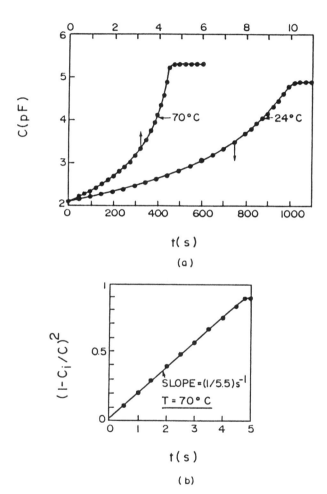

Figure 30: (a) The 24°C and 70°C MOS capacitance C-t response following a depleting gate voltage step. (b) The $[1-(C_i/C)^2]$ vs. t plot for the 70°C curve of (a). The slope gives $t_1=5.5$s leading to $L_n^*=75\mu$m. Reprinted after Reference (91) by permission of IEEE (Copyright 1984 IEEE).

This pulsed MOS-C technique measures the diffusion length. It is easily implemented because it utilizes the same measuring set-up and test structure as that for the conventional pulsed MOS-C generation lifetime method. The only additional requirement is substrate heating. The measurement is easy and the diffusion length extraction is simple. What is a suitable temperature for diffusion current to be dominant over scr generation current? We have found temperatures around 70°C to be suitable.

We have discussed only a few recombination lifetime and diffusion length measurement techniques. Many more have been tried. A nice summary of these can be found in references (92) and (93).

6.2 Generation Lifetime

6.2.1 Pulsed MOS Capacitor:
The pulsed MOS capacitor method can also be used to measure the generation lifetime. To determine τ_g it is necessary for scr generation to be dominant. For silicon devices this is generally true when the pulsed MOS-C measurement is done at room temperature. For scr bulk generation we use the bulk generation current from Equation (86) but modify it to give (94)

$$dQ_n/dt = -qn_i(W-W_F)/\tau_g \tag{95}$$

the term $(W-W_F)$ ensures that $dQ_n/dt \to 0$ when $t \to t_F$ because $W \to W_F$.

The scr width is related to the measured capacitance by

$$W = K_s \xi_o (C_{ox}-C)/C_{ox}C \tag{96}$$

The expression relating the generation rate and the capacitance transient is obtained by combining Equations (91), (95) and (96) into

$$(C_{ox}N_A/C^3)dC/dt = n_i(C_F/C-1)/C_F\tau_g \tag{97}$$

This is the chief equation that relates the capacitance to the bulk scr generation rate. To extract the generation lifetime we use the identity $2(1/C^3)dC/dt = -[d(1/C)^2/dt]$ and write Equation (97) as

$$-d(C_{ox}/C)^2/dt = (2n_iC_{ox}/N_AC_F\tau_g)(C_F/C-1) \tag{98}$$

This equation is the basis of the well-known *Zerbst Plot*, shown in Figure 31b, named after M. Zerbst, who first proposed it (94). It is a plot of $-d(C_{ox}/C)^2/dt$ vs. $(C_F/C-1)$. We also show in Figure 31a the original C-t plot from which the Zerbst Plot in Figure 31b is derived.

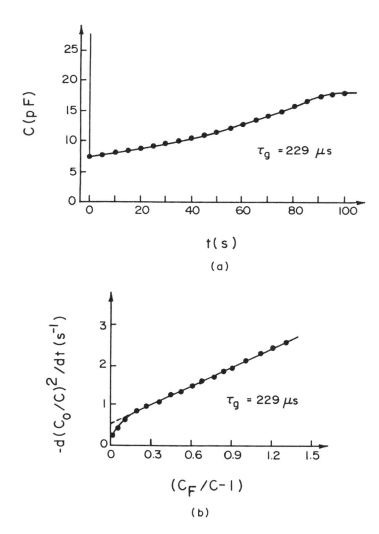

Figure 31: A C-t response of an MOS capacitor in (a) and its Zerbst plot in (b). An experimental C-t curve shown by the data points. The solid line is calculated from Equation (2.14) with $N_A = 4.8 \times 10^{14}$ cm^{-3}, $W_{ox} = 1020$ Å, $A = 3.42 \times 10^{-3}$ cm^2, T=300K. Reprinted after Reference (97) with permission.

It is instructive to examine the two axes of the Zerbst Plot in more detail for a better insight into the physical meaning of such a plot. If we use the identity that leads to Equation (98), we find from Equation (91)

$$-d(C_o/C)^2/dt \sim dQ_n/dt \qquad (99)$$

The Zerbst plot vertical axis is proportional to the total ehp carrier generation rate or the generation current.
From Equation (96) we find

$$(C_F/C - 1) \sim (W - W_F) \qquad (100)$$

i.e. the horizontal axis is proportional to the scr generation width. So we find the rather complicated *Zerbst Plot* to be nothing more than a plot of generation current vs. scr generation width. The current can, of course, be measured directly. The generation width cannot be measured directly and is most easily extracted from capacitance data. Recent implementations of this generation lifetime measuring technique have utilized computers to extract the relevant data from the experimental C-t curve directly.

The slope of a Zerbst Plot is given by $2n_i C_{ox}/N_A C_F \tau_g$ and is used to determine the generation lifetime. We like to point out that the intercept should not be interpreted as the surface generation velocity as is sometimes done. The intercept is a complicated function of surface generation, quasi-neutral bulk generation and it also depends on the assumptions used in the Zerbst analysis (95). To extract a surface generation velocity from the intercept is very misleading. The numerical value obtained from the intercept does not mean much.

There is a variation on the "Zerbst" technique in which the current and capacitance are measured directly and plotted against one another (96). No differentiation of the experimental data is required, but one must measure both current and capacitance. It is one of many pulsed MOS capacitance measurement variations that have been developed over the years. For a detailed review of the many lifetime extracted methods based in one form or another on the deep-depletion MOS-C, the reader is referred to Reference (97). One of the methods is based on driving the MOS-C into deep-depletion with a voltage ramp (98). Equation (90) must then be solved with dV_G/dt=constant.

The C-t response of the MOS-C is determined not only

by τ_g as we have discussed so far, but also by surface generation. The surface directly under the gate inverts early on during the C-t response. Hence, it is a good approximation to neglect surface generation for the surface area under the gate. However, the lateral scr remains depleted during the entire response. It has been shown that the measured generation lifetime is an effective lifetime given by (99)

$$\tau_g'=\tau_g/(1+2s_o\tau_g/r) \qquad (101)$$

where s_o=surface generation velocity in the lateral scr and r=radius of the gate whose area is assumed to be circular. We see from Equation (101) that the second term is only negligible if $s_o\tau_g \ll r/2$. For gates of radius r=250μm and $s_o \approx 2$cm/s we find $\tau_g \ll 6$ms. The lateral surface generation component can be neglected for generation lifetimes of 100μs or so. However, for generation lifetimes in the ms range it must be considered. Generation lifetimes lie in the ms range for well-gettered silicon material.

The measured C-t transient times are usually quite long with time of tens of seconds to minutes being common. Why are these times so long? The response time t_F is on the order of $(N_A/n_i)\tau_g$. Values of τ_g range over many orders of magnitude, of course, but representative values for high quality silicon devices lie in the range of 10^{-4} to 10^{-3} s leading to actual C-t transient times of 10 to 1000s for substrate doping concentrations of $N_A=10^{15}$cm^{-3} and $n_i \approx 10^{10}$ cm^{-3}. These long times point out the great virtue of the pulsed MOS capacitor lifetime measurement technique. To measure lifetimes in the microsecond range it is only necessary to measure capacitance transient or recovery times on the order of seconds. The circuit implementation to do this is very simple which accounts for the method's popularity.

The MOS pulsed C-t technique has found wide acceptance because it is easily implemented with commercially available capacitance meters, a dc power supply, and a microswitch or very low frequency function generator and an X-Y recorder. No fast response circuitry is required. For short C-t response times, the output is displayed on an oscilloscope instead of an X-Y recorder. A measurement schematic is shown in Figure 32. Recent systems are computer-automated in which data acquisition and calculations are all done by a computer.

6.2.2 Gate-Controlled Diode: The MOS-C is in deep-depletion or non-equilibrium only during its transient state. The generation lifetime determination by the pulsed MOS-C is

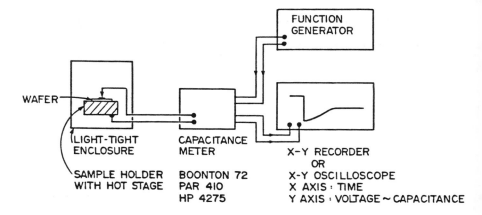

Figure 32: A pulsed MOS capacitor measurement schematic.

thus a transient measurement. The measurement is usually treated as a steady-state measurement because the transient is slowly varying with time. The gate-controlled diode is in non-equilibrium at steady-state voltages and the generation lifetime can be measured under truly steady-state conditions. The gate-controlled diode can be viewed as a diode with an adjacent gate or alternatively as an MOS capacitor with an adjacent diode. It is a very useful test device to extract several generation parameters (100) and is for this reason found on a number of test patterns. We will first treat the gate-controlled diode as an MOS-C, that is in deep-depletion for dc or steady-state gate voltages. This state is impossible for a conventional MOS-Cs because the oxide allows no dc current flow.

The three-terminal gate-controlled diode structure is shown in Figure 33. It consists of a p substrate, an n^+-emitter, a circular gate surrounding the n^+-emitter and a circular guard ring surrounding the gate. The gate can also be located in the center surrounded by the n^+-emitter. The n^+p junction and the gate constitute the gate-controlled diode. The gate should overlap the n^+-emitter slightly to prevent potential barriers from developing in the gap between n^+-emitter and gate. The guard ring is preferred, but it is not always used. It should be close to the gate and should be biased to keep the semiconductor surface in accumulation in order to isolate the gate-controlled diode from the rest of the wafer. For example, moderately doped and oxidized p-

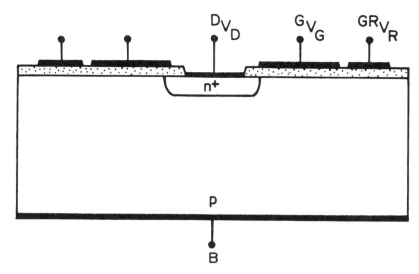

Figure 33: The gate-controlled diode. D is the n^+p junction diode, G is the gate, GR is the guard ring, and B is the substrate.

type silicon substrates are easily inverted by the positive fixed oxide charge that is always present in thermally grown SiO_2. Such inversion layers are capable of coupling adjacent devices to one another. The guard ring prevents such coupling. Adjacent devices can also be decoupled by doping the semiconductor between the devices more heavily.

The current-voltage characteristic of the gate-controlled diode is very well suited for measuring the generation lifetime and the surface generation velocity. For this purpose it is better to treat the device as a diode surrounded by a gate. The diode current is altered from that of a simple diode by the adjacent gate. The function of the diode is to collect the current that is generated in the semiconductor under the MOS capacitor. The diode bias voltage is kept constant during the measurement and the gate voltage is varied.

Let the diode be reverse-biased to V_{D1}. The surface under the gate is accumulated for negative gate voltage $-V_{G1}$ and the junction scr is slightly pulled in at the surface, as shown in Figure 34a. The measured current is the scr generated bulk current, I_J, shown by the electron-hole pairs in Figure 34a and by point A in Figure 34d. For more negative gate voltages one frequently observes a junction current increases that is attributed to junction breakdown

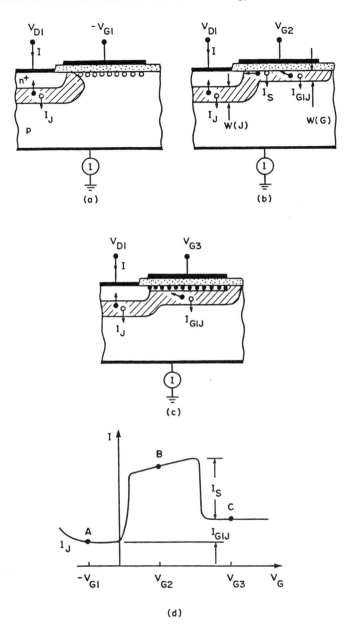

Figure 34: The gate-controlled diode for (a) accumulation, (b) depletion, and (c) inversion under the gate; (d) shows the current with points A, B, and C corresponding to (a), (b), and (c).

current between the n^+-emitter and the heavily accumulated p^+ surface. At the surface the junction is an n^+p^+ junction and such a junction has a low breakdown voltage.

For zero gate voltage, the semiconductor is at flatband assuming zero flatband voltage and the accumulation-induced scr width narrowing at the surface has disappeared. The scr width now has the same width at the surface as in the bulk and the current is slightly increased as a result of the slightly larger generation volume. For large gate voltage the surface under the gate begins to deplete and the current increases rapidly. This quite abrupt increase is due to surface generation current, I_s, and the gate-induced scr bulk current I_{G1J} in Figure 34b. Further gate voltage increases lead to only modest current increases because the surface is already depleted. The surface current remains constant. The bulk current component I_{G1J} continues to increase because the scr under the gate increases with gate voltage. Gate voltage V_{G2} is characteristic of this part of the current-voltage curve. The generation regions are now shown in Figure 34b to consist of three components: (i) the original diode scr of width $W(J)$, (ii) the gate-induced scr of width $W(G)$ and (iii) the depleted surface under the gate. The total current is $I = I_J + I_s + I_{G1J}$.

The surface inverts for more positive gate voltages and the gate scr width pins at its final value. Surface generation drops precipitously and generation component (iii) effectively disappears, as shown in Figure 34c. Further gate voltage increases beyond the inversion voltage, as for example V_{G3} in (c) give no further current change. To first order, we assume that surface generation has become zero. The current is now due to components (i) and (ii) and shown by C in Figure 34d and in cross-section (c). The surface generation current does, of course, not become exactly zero but it becomes very small and can be neglected.

For a quantitative description, we make several assumptions. The first is a sufficiently high reverse bias for the mobile carriers to be negligibly small in the scr and at the depleted surface. Then the bulk generation rate is given by Equation (37) as

$$G = n_i / \tau_g \qquad (102)$$

It can be shown that similar arguments that lead to Equation (102) give the surface generation rate of a depleted surface in terms of the surface generation velocity as (101)

$$G_s = n_i s_o \tag{103}$$

The bulk scr generation current is $I_b = qG(\text{volume})$ and the surface component is $I_s = qG_s(\text{area})$, where "volume" and "area" are the thermal generation volume and area, respectively.

The three currents become

$$I_j = qn_i[W(J)-W_F(J)]A_J/\tau_g(J) \tag{104}$$

$$I_{Glj} = qn_i[W(G)-W_F(G)]A_G/\tau_g(G) \tag{105}$$

$$I_S = qn_i s_o A_G \tag{106}$$

where "J" and "G" stand for the n⁺p junction and the gate, respectively.

By including the "W_F" term, the two scr currents approach zero as equilibrium is approached. For many gate-controlled diode measurements, the "W_F" term is neglected in the interpretation of the data and the bulk currents are then given by

$$I_J = qn_1 W(J)A_J/\tau_g(J) \text{ and } I_{Glj} = qn_i W(G)A_G/\tau_g(G) \tag{107}$$

To use the gate-controlled diode as a test structure, one must be able to measure or calculate the various scr widths in order to extract the generation parameters. They can be experimentally determined from capacitance measurements, but it is usually more convenient to calculate them. From pn junction theory we find for an n⁺p junction, in which the n⁺ emitter is much more heavily doped than the p substrate,

$$W(J) = [2K_S \xi_O(V_D + V_{bi})/qN_A]^{\frac{1}{2}} \tag{108}$$

with V_{bi} being the built-in or diffusion potential. for $V_D = 0$, W becomes $W_F(J)$. The gate-induced scr width is given by (100)

$$W(G) = [2K_S \xi_O(V_D + 2\phi_F)/qN_A]^{\frac{1}{2}} \tag{109}$$

for an inverted surface. This condition applies to point C in Figure 34d. $W_F(G)$ is the scr width for zero diode voltage

$$W_F(G) = [2K_S \xi_O 2\phi_F/qN_A]^{\frac{1}{2}} \tag{110}$$

Note the close relationship between Equations (108) and (109). The only variation between them is the built-in voltage V_{bi} in the n^+p junction and the Fermi potential $2\phi_F=2(kT/q)\ln(N_A/n_i)$ in the gate-induced junction.

The generation parameters are extracted from the current-voltage curve using the equations

$$\tau_g(J) = qn_i[W(J)-W_F(J)]A_J/I_J \tag{111}$$

$$\tau_g(G) = qn_i[W(G)-W_F(G)]A_G/I_G \tag{112}$$

$$S_o = I_s/qn_iA_G \tag{113}$$

The generation lifetimes under the n^+-emitter and under the gate are determined separately, because they may be different. For example, it is possible that for an n^+p junction, the formation of the n^+ layer causes gettering of metallic impurities from the immediate surroundings. We would then expect $\tau_g(J)>\tau_g(G)$. On the other hand it is possible that the junction formation introduces some damage into the junction region and its immediate neighborhood. This might be the case for ion-implanted junctions, for example. In that case we would expect $\tau_g(J)<\tau_g(G)$. It is for this reason that the generation lifetimes in the diode and in the gate regions should be evaluated separately. The gate-controlled diode allows this to be done.

The surface generation velocity calculated from Equation (113) lies typically in the 1-5 cm/s range for well-annealed Si devices. Equation (113) assumes the surface to be entirely devoid of majority and minority carriers. That is only a simplification because there is a lateral current flow shown by the electrons flowing into the n^+-emitter in Figure 34b and 34c. This corresponds to a weakly inverted surface. A more detailed analysis (102) has shown that the surface generation velocity of Equation (113) is an underestimate. However, most of the published data have used Equation (113) in their analysis without this correction.

Equations (111)-(113) show that all three generation parameters can be determined with the gate-controlled diode. If only the generation lifetime is desired, it is more easily measured with the pulsed MOS capacitor. However, that measurement does not allow for a simple, unambiguous separation of bulk generation lifetime and surface generation velocity. The gate-controlled diode allows this separation and whenever these parameters are required it is best to use the gate-controlled diode as the test structure.

7.0 SUMMARY

The lifetime is an important parameter in silicon devices. For some devices the lifetime has a direct influence on device operation. For example, the switching time in diodes and thyristors is very much influenced by the recombination lifetime as is the efficiency of solar cells. Lifetime has a second order influence on other devices. The gain of narrow base width bipolar transistors is only marginally dependent on the minority carrier lifetime. For MOSFETs it is often said that lifetime plays no role because the MOSFET is a unipolar device. That is only partially true. Certainly the drain junction leakage current is dependent on the generation lifetime and this leakage current determines the static power dissipation of CMOS circuits. The refresh time of dynamic random access memories and the dark current of charge-coupled devices is critically dependent on both generation and recombination lifetimes.

The lifetime of a silicon device is also a good indicator of process cleanliness. Routine lifetime or diffusion length measurements are used by some manufacturers to monitor the status of their process lines for this reason. Monitoring the lifetime is a very sensitive process indicator, because slight contamination problems show up immediately. Each process step can be monitored by pulling test wafers after each high temperature operation.

We have attempted to give a comprehensive overview of lifetime in silicon. The relationship of the lifetime to foreign impurities and structural defects was outlined during the early part of the chapter. There is, unfortunately, no good way to predict the lifetime following some fabrication process. The energy levels of a number of metallic impurities are reasonably well known but there is some discrepancy in the published data. The capture cross-sections are less well known and only a few impurities have been thoroughly investigated. Much less is known about the energy levels and capture cross-sections of structural defects. Structural defects do not always behave in the same manner in different devices. For example, the recombination properties of dislocations are very much dependent on whether the dislocations are decorated with metallic impurities or whether they are clean. Decorated dislocations are generally bad for devices while undecorated dislocations may have little or no electrical activity. Relatively little is known about the recombination/generation properties of self-interstitials and vacancies.

The concept of recombination and generation lifetime is frequently confused and we have attempted to clarify it. We have, furthermore, indicated the role of the two lifetimes in device operation. The device designer should be aware of these concepts, especially how they apply when the wafer consists of regions of varying defect concentrations as in intrinsically gettered silicon, for example. The high purity denuded zone is necessary for good device performance. However, the heavily precipitated bulk may influence device performance adversely. This is especially so at higher device operating temperatures.

Finally, we mention the more popular lifetime measurement techniques. Many more methods have been tried, but the ones discussed here are the more popular ones. commercial measurement equipment exists for some of them and those methods for which commercial equipment is available tend to be used more frequently. Nevertheless, there are many "home-built" lifetime measurement set-ups that are very effective.

Acknowledgment

The critical reading of the manuscript by J.S. Kang is appreciated.

8.0 REFERENCES

1. Zulehner, W. and Huber, D., Czochralski-Grown Silicon, in: *Crystals: Silicon, Chemical Etching*, (Grabmaier, J., ed.), pp. 1-43, Springer Verlag, Berlin (1982)

2. Chen, J.W., Milnes, A.G., Energy Levels in Silicon, in *Annual Rev. Mater. Sci.* 10:157-228 (1980); Weber, E.R., Transition Metals in Silicon. *Apply. Phys.* A30:1-22 (1983); Rohatgi, A., Davis, J.R., Hopkins, R.H. and McMullin, P.G., A study of grown-in impurities in silicon by deep-level transient spectroscopy. *Solid-State Electron.* 26:1039-1051 (1983)

3. Ourmazd, A., The electrical properties of dislocations in semiconductors. *Contemp. Phys.*, 25:251-268 (1984); Chen, J.W., and Milnes, A.G., op.cit.

4. Shockley, W. and Read, W.T., Statistics of the recombination of holes and electrons. *Phys. Rev.*, 87:835-842 (1952); Hall, R.N., Electron-hole recombination in germanium. *Phys. Rev.*, 87:387 (1952)

5. Hall, R.N., Recombination processes in semiconductors. *Proc. IEE.*, 106B:923-931 (1960)

6. Varshni, Y.P., Band-to-band radiative recombination in groups IV, VI and III-V semiconductors (I) and (II). *Phys. Stat. Sol.* 19:459-514 (1967) and 20:9-36 (1967)

7. Beck, J.D., and Conradt, R., Auger-recombination in Si (in German). *Solid-State Communic.* 13:93-95 (1973).

8. Dziewior, J., and Schmid, W., Auger coefficients for highly doped and highly excited silicon. *Applied Phys. Lett.* 31:346-348 (1977)

9. Grekhov, I.V. and Delimova, L.A., Auger recombination in Silicon. *Sov. Phys. Semicond.* 14:529-532 (1980); Delimova, L.A., Auger recombination in silicon at low temperatures. *Sov. Phys. Semicond.* 15:778-780 (1981)

10. Yablonovitch, E. and Gmitter, T., Auger recombination coefficients in silicon at low carrier densities. *Appl. Phys. Lett.* 49:587-589 (1986)

11. Hu, C.M. and Oldham, W.G., Carrier recombination through donors/acceptors in heavily doped silicon. *Appl. Phys. Lett.* 35:636-639 (1979)

12 Possin, G.E., Adler, M.S. and Baliga, B.J., Measurement of heavy doping parameters in silicon by electron-beam-induced current. *IEEE Trans. Electr. Dev.* ED-27:983-990 (1980)

13. Weaver, H.T. and Nasby, R.D., Analysis of high-efficiency silicon solar cells. *IEEE Trans. Electr. Dev.* ED-28:465-472 (1981)

14. Redfield, D., Mechanism of performance limitations in heavily doped silicon devices. *Appl. Phys. Lett.* 33:531-533 (1978)

15. Haug, A. and Schmid, W., Recombination mechanisms in heavily doped silicon. *Solid-State Electron.* 25:665-667 (1982)

16. Fossum, J.G., Mertens, R.P., Lee, D.S. and Nijis, J.F., Carrier recombination and lifetime in highly doped silicon. *Solid-State Electron.* 26:569-576 (1983)

17. Woerdman, J.P., Some optical and electrical properties of a laser-generated free-carrier plasma in silicon. *Phil Res. Rep.* Suppl. 7:1-81 (1971)

18. Svantesson, K.G. and Nilsson, N.G., The temperature dependence of the Auger recombination coefficient of undoped silicon. J. Phys. C: *Solid-State Phys.* 12:5111-5120 (1979)

19. Zimmermann, W., Experimental verification of the Shockley-Read-Hall recombination theory in silicon. *Electron. Lett.* 9:378-379 (1973); Cornu, J., Sittig, R. and Zimmermann, W., Analysis and measurement of carrier lifetimes in the various operating modes of power devices. *Solid-State Electron.* 17:1099-1106 (1974)

20. The data in Figure 7 were taken from References 8, 9, 10, 21, 22, 23.

21. Burtscher, J., Dannhaeuser, F. and Krausse, J., The recombination in thyristors and rectifiers in silicon: its influence on the forward-bias characteristic and the turn-off time (in German). *Solid-State Electron.* 18:35-63 (1975)

22. Passari, L. and Susi, E., Recombination mechanisms and doping density in silicon. *J. Appl. Phys.* 54:3935-3937 (1983)

23. Huber, D., Bachmeier, A., Wahlich, R. and Herzer, H., Minority carrier diffusion length and doping density in nondegenerate silicon, in: *Semiconductor Silicon/1986*, (Huff, H.R., Abe, T. and Kolbesen, B., eds.), pp. 1022-1032, Electrochem. Soc., Pennington, NJ (1986)

24. Bennett, H.S., Comparison of theoretical and empirical lifetimes for minority carriers in heavily doped silicon. *Solid-State Electr.* 27:893-897 (1984)

25. Possin, G.E., Adler, M.S. and Baliga, B.J., in: *Lifetime Factors in Silicon*, pp. 192-209 ASTM STP 712, Am. Soc. for Test. and Mat. (1980)

26. Roulston, D.J., Arora, N.D. and Chamberlain, S.G., Modeling and Measurement of minority-carrier lifetime versus doping in diffused layers of n^+-p silicon diodes. *IEEE Trans. Electr. Dev.* ED-29:284-291 (1982)

27. Wu, C.Y. and Chen, J.F., Doping and temperature dependences of minority carrier diffusion lengths and lifetime deduced from the spectral response measurements of pn junction solar cells. *Solid-State Electron.*, 25:679-682 (1982)

28. Fossum, J.G. and Lee, D.S., A physical model for the dependence of carrier lifetime on doping density in nondegenerate silicon. *Solid-State Electron.*, 25:741-747 (1982)

29. Landsberg, P.T. and Kousik, G.S., The connection between carrier lifetime and doping density in nondegenerate semiconductors. *J. Appl. Phys.* 56:1696-1700 (1984)

30. Hsu, F.C., Ko, P.K., Tang, S., Hu, C.M. and Muller, R.S., An analytical breakdown model for short-channel MOSFETs. *IEEE Trans. Electr. Dev.*, ED-29:1735-1740 (1982)

31. Schroder, D.K., The concept of generation lifetimes in semiconductors. *IEEE Trans. Electr. Dev.* ED-29:1336-1338 (1982)

32. Kishino, S., Matsushita, Y., Kanamori, M. and Iizuka, T., Thermally induced microdefects in Czochralski-grown silicon: nucleation and growth behavior. *Japan J. Appl. Phys.* 21:1-12 (1982)

33. Schroder, D.K., Hwang, J.M., Kang, J.S., Goodman, A.M. and Sopori, B.L., Lifetime and recombination concepts for oxygen-precipitated silicon, in *VLSI Science and Technology*/1985 (W.M. Bullis and S. Broydo, eds.), pp. 419-428, Electrochem Soc., Pennington, NJ (1985)

34. Schroder, D.K., Whitfield, J.D. and Varker, C.J., Recombination lifetime using the pulsed MOS capacitor. *IEEE Trans. Electr. Dev.* ED-31:462-467 (1984)

35. Neugroschel, A., Determination of lifetimes and recombination currents in pn junction solar cells, diodes, and transistors. *IEEE Trans. Electr. Dev.* ED-28:108-115 (1981)

36. Sah, C.T., Noyce, R.N. and Shockley, W., Carrier generation and recombination in pn junctions and pn junction characteristics. *Proc. IRE* 45:1128-1243 (1957)

37. Del Alamo, J., Swirhun, S. and Swanson, R.M., Measuring and modeling minority carrier transport in heavily doped silicon. *Solid-State Electron.* 28:47-54 (1985)

38. Schroder, D.K., Effective lifetimes in high quality silicon devices. *Solid-State Electron.* 27:247-251 (1984)

39. Chakravarti, S.N., Garbarino, P.L. and Murty, K., Oxygen precipitation effects on Si n^+p junction leakage behavior. *Appl. Phys. Lett.* 40:581-583 (1982)

40. Chappell, T.I., Chye, P.W. and Travel, M.A., Determination of the oxygen precipitate-free zone width in silicon wafers from surface photovoltage measurements. *Solid-State Electron.* 26:33-36 (1983)

41. Clemens, J.T., Mehta, D.A., Nelson, J.T., Pearce, C.W. and Sun, R.C.; *U.S. Patent* 4,216,489; Aug. 5, 1980; assigned to AT&T.

42. Slotboom, J.W., Theunissen, M.J.J. and deKock, A.J.R., Impact of silicon substrates on leakage currents. *IEEE Trans. Electron. Dev. Lett.* EDL-4:403-406 (1983)

43. Stevenson, D.T. and Keyes, R.J., Measurement of carrier lifetimes in germanium and silicon. *J. Appl. Phys.* 26:190-195 (1955)

44. Ryvkin, S.M., *Photoelectric Effects in Semiconductors*, New York: Consultants Bureau (1964)

45. Boulou, M. and Bois, D., Cathodoluminescence measurements of the minority-carrier lifetime in semiconductors. *J. Appl. Phys.* 48:4713-4721 (1977)

46. Eranen, S. and Blomberg, M., Simultaneous measurement of recombination lifetime and surface recombination velcoity. *J. Appl. Phys.* 56:2372-2374 (1984)

47. Gerhard, A.R. and Pearce, C.W., Measurement of minority carrier lifetime in silicon crystals by photoconductive decay technique, in *Lifetime Factors in Silicon*, pp. 161-170, Am. Soc. for Test. and Mat. (1980)

48. ANSI/ASTM Standard F28-75 (Reapproved 1981), Standard Test for measuring the minority carrier lifetime in bulk germanium and silicon. *1983 Annual Book for ASTM Stds.* Am. Soc. for Test. and Mat., pp. 40-45.

49. Ramsa, A.P., Jacobs, H. and Brand, F.A., Microwave techniques in measurement of lifetime in germanium. *J. Appl. Phys.* 30:1054-1060 (1959).

50. Larrabee, R.D., Measurement of semiconductor properties through microwave absorption. *RCA Rev.* 21:124-129 (1960)

51. Mada, Y., A nondestructive method for measuring the spatial distribution of minority carrier lifetime in Si wafer. *Japan. J. Appl. Phys.* 18:2171-2172 (1979)

52. Deri, R.J. and Spoonhower, J.P., Microwave photoconductivity lifetime measurements: experimental limitations. *Rev. Sci. Instrum.* 55:1343-1347 (1984)

53. Gossick, B.R., Post-injection barrier electromotive force of pn junctions. *Phys. Rev.* 91:1012-1013 (1953); On the transient behavior of semiconductor rectifiers. *J. Appl. Phys.* 26:1356-1365 (1955)

54. Choo, S.C. and Maxur, R.G., Open circuit voltage decay behavior of junction devices. *Solid-State Electron.* 13:553-564 (1970)

55. Lederhandler, S.R. and Giacoletto, L.J., Measurement of minority carrier lifetime and surface effects in junction devices. *Proc. IRE* 43:477-483 (1955)

56. Dhariwal, S.R. and Vasu, N.K., A generalized approach to lifetime measurement in pn junction solar cells. *Solid-State Electron.* 24:915-927 (1981)

57. Jain, S.C. and Muralidharan, R., Effect of emitter recombination on the open circuit voltage decay of a junction diode. *Solid-State Electron.* 24:1147-1154 (1981)

58. Berz. F. and Slatter, J.A.G., Effect of linear emitter recombination on OCVD determination of lifetime in pin diodes. *Solid-State Electron.* 25:963-967 (1982)

59. Basset, R.J., Fulop, W. and Hogarth, C.A., Determination of the bulk carrier lifetime in low-doped region of a silicon power rectifier by the method of open circuit voltage decay. *Int. J. Electron.* 35:177-192 (1973)

60. Mahan, J.E. and Barnes, D.L., Depletion layer effects on the open-circuit-voltage-decay lifetime measurement. *Solid-State Electron.* 24:989-994 (1981)

61. Green, M.A., Minority carrier lifetimes using compensated differential open circuit voltage decay. *Solid-State Electron.* 26:1117-1122 (1983)

62. Moore, A.R., Carrier lifetime in photovoltaic solar concentrator cells by the small signal open circuit decay method. *RCA Rev.* 40:549-562 (1980)

63. Rose, B.H. and Weaver, H.T., Determination of effective surface recombination velocity and minority-carrier lifetime in high-efficiency Si solar cells. *J. Appl. Phys.* 54:238-247 (1983), with corrections in *J. Appl. Phys.* 55:607 (1984)

64. Rose, B.H., Minority-carrier lifetime measurements on Si solar cells using I_{sc} and V_{oc} transient decay. *IEEE Trans. Electr. Dev.* ED-31:559-565 (1984)

65. Pell, E.M., Recombination rate in germanium by observation of pulsed reverse characteristic. *Phys. Rev.* 90:278-279 (1953)

66. Kingston, R.H., Switching time in junction diodes and junction transistors. *Proc. IRE* 42:829-834 (1954)

67. Lax, B. and Neustadter, S.F., Transient response of a pn junction. *J. Appl. Phys.* 25:1148-1154 (1954)

68. Kuno, H.J., Analysis and characterization of pn junction diode switching. *IEEE Trans. Electr. Dev.* ED-11:8-14 (1964)

69. Varker, C.J., Whitfield, J.D., Rao, K.V., and Demer, L.I., Minority carrier lifetime and reverse current inhomogeneities in pn diode arrays resulting from microdefects in Czochralski grown silicon crystals, in *Semiconductor Silicon* 1981, (H.R. Huff, R.J. Kriegler, Y. Takeishi, eds.) The Electrochemical Society, Pennington, N.J. (1981)

70. Moll, J.L., Ray, U.C. and Jain, S.C., Reverse recovery in pn junction diodes with built-in drift fields. *Solid-State Electron.* 26:1077-1081 (1983)

71. Dean, R.H. and Nuese, C.J., A refined step-recovery technique for measuring minority carrier lifetimes and related parameters in asymmetric pn junction diodes. *IEEE Trans. Electr. Dev.* ED-18:151-158 (1971)

72. Jain, S.C., Van Overstraeten, R., The influence of heavy doping effects on the reverse recovery storage time of a diode. *Solid-State Electron.* 26:473-481 (1983)

73. Kao, Y.C. and Davis, J.R., Correlation between reverse recovery time and lifetime of pn junction driven by a current ramp. *IEEE Trans. Electr. Dev.* ED-17:652-657 (1970)

74. Derdouri, M., Leturcq, P. and Munoz-Yague, A., A comparative study of methods of measuring carrier lifetime in pin diodes. *IEEE Trans. Electr. Dev.* ED-27:2097-2101 (1980)

75. Cooper, R.W., An investigation of recombination in gold-doped pin rectifiers. *Solid-State Electron.* 26:217-226 (1983)

76. Johnson, E.O., Measurement of the minority carrier lifetime with the surface photovoltage. *J. Appl. Phys.* 28:1349-1353 (1957)

77. Goodman, A.M., A method for the measurement of short minority carrier diffusion lengths in semiconductors. *J. Appl. Phys.* 32:2550-2552 (1961)

78. Chiang, C.L. and Wagner, S., On the theoretical basis of surface photovoltage technique. *IEEE Trans. Electr. Dev.* ED-32:1722-1726 (1985)

79. Goodman, A.M., Improvements in method and apparatus for determining minority carrier diffusion length. *IEE Trans. Electr. Dev. Meet.* Washington, DC, pp. 231-234 (1980)

80. ANSI/ASTM F391-78, Standard test method for minority carrier diffusion length in silicon by measurement of steady-state surface photovoltage. *1983 Annual Book of ASTM Stds.*, Am. Soc. for Test. and Mat., pp. 353-359

81. Kern, W., Cleaning solutions based on hydrogen peroxide for use in silicon semiconductor technology. *RCA Rev.* 31:187-206 (1970)

82. Nartowitz, E.S. and Goodman, A.M., Evaluation of Si optical absorption data for use in minority-carrier diffusion length measurements. *J. Electrochem. Soc.* 132:2992-2997 (1985)

83. Goodman, A.M., Goodman, L.A. and Grossenberger, H.F., Silicon-wafer process evaluation using minority-carrier diffusion length measurement by the SPV method. *RCA Rev.* 44:326-341 (1983)

84. Li, S.S., Determination of minority carrier diffusion length in InP by surface photovoltage measurement. *Appl. Phys. Lett.* 9:126-127 (1976)

85. Moore, A.R., Theory and experiment on the surface photovoltage diffusion length measurement as applied to amorphous silicon. *J. Appl. Phys.* 54:222-228 (1983)

86. Pankove, J.I. and Berkeyheiser, J.E., Mapping the quality of semiconductor wafers. *Rev. Sci. Instr.* 57:674-679 (1986)

87. Tomanek, P., Measuring the lifetime of minority carriers in MIS structures. *Solid-State Electron.* 12:301-303 (1969)

88. Muller, J. and Schiek, B., Transient responses of a pulsed MIS capacitor. *Solid-State Electron.* 13:1319-1332 (1970)

89. Wang, A.C. and Sah, C.T., New method for complete electrical characterization of recombination properties of traps in semiconductors. *J. Appl. Phys.* 57:4645-4656 (1985)

90. Soutschek, E., Muller, W. and Dorda, G., Determination of recombination lifetime in MOSFETs. *Appl. Phys. Lett.* 36:437-438 (1980)

91. Schroder, D.K., Whitfield, J.D. and Varker, C.J., Recombination lifetime using the pulsed MOS capacitor. *IEEE Trans. Electr. Dev.* ED-31:462-467 (1984)

92. Ross, B., Survey of literature on minority carrier lifetimes in silicon and related topics. in *Lifetime Factors in Silicon. ASTM STP712*, pp. 14-28 Am. Soc. for Test. and Mat. (1980)

93. Khan, A.A., Woollam, J.A. and Hermann, A.M., Minority carrier diffusion length measurements: a review and comparison of techniques. *Appl. Phys. Comm.* 2:17-56 (1982)

94. Zerbst, M., Relaxation effects at semiconductor-insulator interfaces (in German). *Z. Angew. Phys.* 22: 30-33 (1966)

95. van der Spiegel, J. and DeClerck, G.J., Theoretical and practical investigation of the thermal generation in gate controlled diodes. *Solid-State Electron.* 24:869-877 (1981)

96. Calzolari, P.U., Graffi, S. and Morandi, C., Field-enhanced carrier generation in MOS capacitors. *Solid-State Electron.* 17:1001-1011 (1974)

97. Kang, J.S. and Schroder, D.K., The pulsed MIS capacitor - a review. *Phys. Stat. Sol.* 89a:13-43 (1985)

98. Pierret, R.F., A linear sweep MOS-C technique for determining minority carrier lifetimes. *IEEE Trans. Electr. Dev.* ED-19:869-873 (1972)

99. Schroder, D.K. and Nathanson, H.C., On the separation of bulk and surface components of lifetime using the pulsed MOS capacitor. *Solid-State Electron.* 13:577-582 (1970)

100. Grove, A.S. and Fitzgerald, D.S., Surface effects on pn junctions: characteristics of surface space-charge regions under non-equilibrium conditions. *Solid-State Electron.* 9:783-806 (1966).

101. Schroder, D.K. and Guldberg, J., Interpretation of surface and bulk effects using the pulsed MIS capacitor. *Solid-State Electron.* 14:1285-1297 (1971)

102. Pierret, R.F., The gate-controlled diode s_o measurement and steady-state lateral current flow in deeply depleted MOS structures. *Solid-State Electron.* 17:1257-1269 (1974)

LIST OF SYMBOLS

A	diode area (cm^2)
A_G	gate area (cm^2)
A_J	junction area (cm^2)
B	radiative recombination coefficient (cm^3/s)
C	capacitance/unit area (F/cm^2)
C_F	final MOS-C capacitance/unit area (F/cm^2)
C_i	initial deep-depletion (DD) MOS-C capacitance/unit area (F/cm^2)
C_n	Auger recombination coefficient in n-type material (cm^6/s)
C_{ox}	oxide capacitance/unit area (F/cm^2)

C_p — Auger recombination coefficient in p-type material (cm^6/s)

D_n — electron diffusion coefficient (cm^2/s)

D_p — hole diffusion coefficient (cm^2/s)

ξ — electric field (V/cm)

E — energy (eV)

E_c — lowest conduction band energy (eV)

E_F — Fermi energy or Fermi level (eV)

E_G — band gap energy (eV)

E_i — intrinsic Fermi energy (eV)

E_T — G-R center energy (eV)

E_V — highest valence band energy (eV)

G — conductance (S)

G — Gibbs free energy (eV)

G — steady-state bulk generation rate ($cm^{-3}s^{-1}$)

G_s — steady-state bulk surface generation rate ($cm^{-2}s^{-1}$)

h — recombination parameter (cm^4/s)

I_D — drain current (A)

I_{diff} — diffusion current (A)

I_f — forward diode current (A)

I_{GIJ} — gate-induced junction current (A)

I_J — reverse diode current (A)

I_r — reverse diode current (A)

I_{rev} — diode leakage current (A)

I_s	surface current (A)
I_{scr}	diode space-charge region current (A)
k	Boltzmann's constant (8.617×10^{-5} eV/K)
K_{ox}	oxide dielectric constant (3.9 for SiO$_2$)
K_s	semiconductor dielectric constant (11.8 for Si)
L_n	electron diffusion length (cm)
L_n^*	modified electron diffusion length (cm)
L_p	hole diffusion length (cm)
L_p^*	modified hole diffusion length (cm)
n	electron concentration (cm^{-3})
n_1	defined in Equation 2(a)
n_i	intrinsic carrier concentration (cm^{-3})
n_o	equilibrium electron concentration (cm^{-3})
N_A	acceptor doping concentration (cm^{-3})
N_D	donor doping concentration (cm^{-3})
N_T	bulk G-R concentration (cm^{-3})
p	hole concentration (cm^{-3})
p_1	defined in Equation 2(b)
p_o	equilibrium hole concentration (cm^{-3})
q	magnitude of electron charge (1.602×10^{-19} coul)
Q_n	electron surface charge density (coul/cm^2)
Q_s	semiconductor charge (coul)
r	MOS-C gate radius (cm)

R	microwave reflectivity
R	optical reflectivity
s	surface generation velocity (cm/s)
s_c	surface recombination velocity at metal-semiconductor contact (cm/s)
s_o	surface generation velocity of a depleted surface (cm/s)
s_r	surface recombination velocity (cm/s)
t	time (s)
t_F	recovery time of a pulsed MOS-C (s)
t_p	pulse width (s)
t_S	reverse-recovery diode storage time(s)
T	temperature (K)
T_p	p-substrate thickness (cm)
v_n	thermal electron velocity (cm/s)
v_p	thermal hole velocity (cm/s)
V_{bi}	built-in pn junction potential (V)
V_d	diode voltage (V)
V_D	drain voltage (V)
V_G	gate voltage (V)
V_r	diode reverse voltage (V)
V_T	threshold voltage (V)
W	space-charge region (scr) width (cm)
W_F	final scr width for MOS-C in heavy inversion (cm)

W_{ox}	oxide thickness (cm)
α	optical absorption coefficient (cm^{-1})
Δn	excess electron concentration (cm^{-3})
Δp	excess hole concentration (cm^{-3})
ϵ_o	permittivity of free space (8.854x10^{-14}F/cm)
λ	wavelength (cm)
Φ	photon flux density (photons/s·cm^2)
$\emptyset_F$	Fermi potential (eV)
σ	conductivity (S/cm)
σ_n	bulk-state electron capture cross-section (cm^2)
σ_p	bulk-state hole capture cross-section (cm^2)
τ_{Auger}	Auger lifetime (s)
τ_g	generation lifetime (s)
τ_g'	effective generation lifetime (see Equation 101) (s)
$\tau_{g,eff}$	effective generation lifetime (see Equation 42) (s)
τ_n	electron lifetime (s)
τ_{no}	= $1/\delta_n v_n N_T$ (s)
τ_p	hole lifetime (s)
τ_{po}	= $1/\delta_p v_p N_T$ (s)
τ_r	recombination lifetime (s)
τ_{rad}	radiative lifetime (s)
τ_{scr}	space-charge region lifetime (s)
τ_{SRH}	Shockley-Read-Hall or multiphonon lifetime (s)
μ	mobility (cm^2/V·s)

9

Preparation and Properties of Polycrystalline-Silicon Films

Theodore D. Kamins

1.0 INTRODUCTION

Thin films of polycrystalline silicon have been used for a decade and a half as gate electrodes of metal-oxide-semiconductor (MOS) integrated circuits. The ability to self-align the gate electrodes of MOS field-effect transistors reduces capacitance and improves circuit speed. In addition to its use in conventional MOS integrated circuits, polysilicon is being used in new applications which demand very stringent control over its material properties; such advanced applications include high-value resistors in static RAMs, floating gates in electrically alterable ROMs, and emitters of bipolar transistors. One or even two layers of polysilicon are now used in virtually all advanced bipolar integrated circuits also, improving circuit performance. In addition to improving circuit speed, the compatibility of polycrystalline silicon with subsequent high-temperature processing allows its efficient integration into advanced integrated-circuit processing and permits fabrication of new device structures.

This excellent technological compatibility of polysilicon with integrated-circuit processing allows straightforward fabrication of convenient interconnections in VLSI circuits. Although a resistivity of less than 10^3 Ω-cm, eight orders of magnitude less than that used for static RAM load resistors, is routinely achieved, the lower bound on the resistivity of polysilicon can limit the performance of integrated circuits which use polysilicon interconnections to conduct signals long distances across a chip. Therefore, alternate materials are being incorporated into integrated-circuit interconnection structures to replace or augment polysilicon. In the future, interconnections will limit integrated-circuit performance even more severely.

Continued advances in the design of integrated circuits and the use of polysilicon in novel structures have brought into focus the need for a greater understanding of the detailed characteristics of polycrystalline-silicon films and their application in integrated circuits.

In this chapter we will consider the deposition technology of polysilicon, as well as its structural and electrical properties and their limitations. After examining the chemical vapor deposition (CVD) techniques used to form polysilicon, we will relate the deposition conditions to the physical and chemical properties of polysilicon and consider the effect of processing on these properties. We will examine the wide range of electrical behavior that can be achieved in CVD polysilicon films and study the close connection between the electrical properties and the structure, particularly the grain boundaries. The limitations on the amount of dopant that can be incorporated in substitutional sites in the polysilicon grains will be shown to be the basic limitation on the conductivity. Conduction through oxide grown on polysilicon will be related to the surface topology of the polysilicon. Present and potential applications of polysilicon in both MOS and bipolar integrated circuits will then be discussed.

2.0 DEPOSITION

Over the many years that polysilicon has been used in commercial integrated circuits, the equipment used for its deposition has changed greatly, evolving from low-capacity, silicon epitaxial reactors operating at atmospheric pressure to high-volume, low-pressure CVD systems. In all commercial applications, however, chemical vapor deposition has been used to deposit polysilicon. In early prototype investigations, physical vapor deposition by evaporation was studied, but the step coverage provided by this technique was inadequate to cover the irregular surface topology of the integrated circuit;

642 Handbook of Semiconductor Silicon Technology

by contrast, CVD techniques provided material with excellent conformal step coverage, leading to the rapid acceptance of this more complex formation technology.

At the time that chemical vapor deposition was first used for polysilicon deposition, the most common CVD system was the horizontal, atmospheric-pressure epitaxial reactor widely used in the late 1960s and early 1970s. This type of system (illustrated in Figure 1), allows operation over a wide temperature range, but its capacity is severely limited by the size of the susceptor, on which the wafers are placed in a single layer. The low-pressure CVD (LPCVD) reactor developed to overcome the limited capacity of the horizontal system, can form layers on 100-200 wafers simultaneously, although the range of conditions over which it can operate satisfactorily is severely limited. The basic elements of the LPCVD reactor are illustrated in Figure 2.

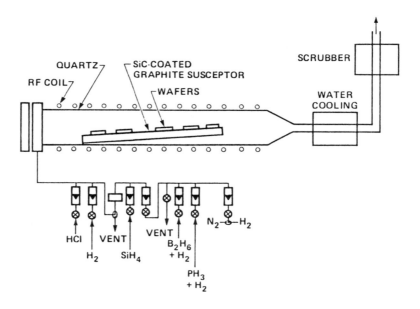

Figure 1. The versatile, horizontal, cold-wall, atmospheric-pressure reactor has the flexibility needed to develop many new CVD processes, but its limited wafer capacity makes it less desirable for routine manufacturing.

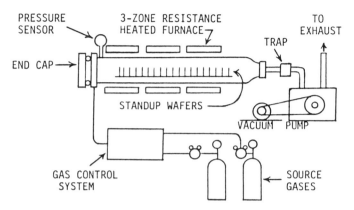

Figure 2. The hot-wall, low-pressure reactor is used for routine deposition of polysilicon because of its high wafer capacity and simplicity.

2.1 Gas Dynamics

In an open-flow CVD reactor, the reactant gas is continuously forced through the reactor; the silicon-containing gas is, in many cases, mixed with a carrier gas. Hydrogen is usually used as the carrier gas when the deposition occurs at higher temperatures ($\geq$ 800°C), where nitrogen may react with the silicon-containing gases; nitrogen can be used at lower temperatures, where hydrogen lowers the deposition rate because it is a reaction product.

Viscous forces exerted on the flowing gas by the susceptor and the walls of the deposition chamber slow the gas near these stationary surfaces, forming a *boundary layer*, which separates the *forced-convection* region from the wafer surfaces and the walls of the reaction chamber (Figure 3 (1),(2)). The silicon-containing molecules diffuse from the forced convection region through the boundary layer to the wafer surface. The rate of diffusion R can be written

$$R = D \frac{C_G}{\delta} \tag{1}$$

where C_G is the concentration of the silicon-containing species in the forced-convection region just above the boundary layer, D is the gas-phase diffusivity, and δ is the thickness of the boundary layer. Diffusion through the boundary layer is usually the most important of the gas-phase transport processes. Some reaction or decomposition of the

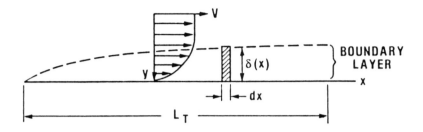

Figure 3. A *boundary layer* separates the rapidly moving gas in the forced-convection region from the stationary surfaces in the deposition chamber. (Courtesy of J. Andrews, AT&T Bell Laboratories.)

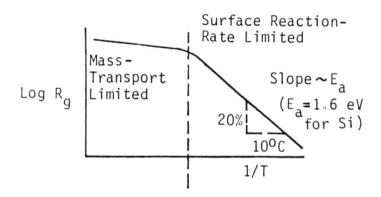

Figure 4. The deposition rate is a rapidly varying function of temperature in the surface-reaction-rate-limited regime of operation (low temperatures), while it changes only slowly with temperature in the mass-transport-limited regime (higher temperatures).

silicon-containing gas can occur within the boundary layer (*homogeneous reaction*), but the final reaction should occur on the surface itself (by *heterogeneous reaction*) so that a dense silicon film is formed. Once the silicon-containing gas (or its partially reacted products) reach the surface, it is adsorbed, and further chemical reactions take place to reduce it to silicon and reaction products, which are in turn desorbed. The overall reaction rate can be written as

$$R = Ck = Ck_0 \exp\left(-\frac{E_a}{kT}\right) \tag{2}$$

where the reaction-rate coefficient k is characterized by an apparent activation energy E_a.

Either diffusion through the boundary layer or reaction at the surface of the wafer may limit the overall deposition process. Gas-phase diffusion varies only slowly with temperature, increasing as $T^{1.5}$ or T^2, while the reaction rate increases rapidly with increasing temperature, varying as exp$(-E_a/kT)$. For silicon deposition, E_a is about 1.6 eV (38 kcal/mole). Therefore, at higher temperatures, the reaction proceeds rapidly, and diffusion of the gaseous silicon species through the boundary layer to the wafer surface limits the overall deposition process. In this *mass-transport-limited* regime, the deposition rate is only a weak function of temperature. As the temperature is reduced, the reaction rate decreases rapidly until it becomes the limiting step in the overall deposition process. In the *surface-reaction-limited* regime of operation, the deposition rate is a strong function of temperature (Figure 4). Near 700°C the deposition rate changes by about 25% for a 10°C temperature variation, so that excellent temperature control is needed in this operating regime to achieve the film thickness uniformity required for controllable integrated-circuit fabrication. [Note that over a considerable range of temperatures ($\geq$ 100°C), $C_G > C_S > 0$. In this range both mass transport and reaction influence the overall deposition process, and both must be considered.]

Proper process design considers the geometry of the particular reactor to be used for the deposition; the relative ease of controlling the gas flow and the temperature determine the choice of operating regime. If the gas flow is well controlled, but the temperature is difficult to control, the process should operate in the temperature-insensitive, mass-transport-limited regime. If the temperature is better controlled than the gas flow, operation in the surface-reaction-limited regime is preferred.

The importance of choosing the proper operating regime can be illustrated by contrasting the deposition conditions in horizontal, atmospheric-pressure reactors with those in the low-pressure reactors now widely used.

Horizontal, Atmospheric-Pressure Reactor: In the horizontal, atmospheric-pressure, reactor, a small quantity of the silicon-containing gas is mixed with a large amount of carrier gas, and this gas mixture is forced through the reactor. Only the wafers and their supporting susceptor are heated; the walls of the chamber remain relatively cool, and little deposit forms on them. The forced convection region is also relatively cool so that gases flowing in this region can travel long distances along the susceptor without reacting significantly. The majority of the temperature gradient occurs across the boundary layer. In this type of reactor, energy is often coupled into the susceptor by radio-frequency induction or by heating the opaque support plate with lamps without significantly heating the walls of the chamber. (The term "susceptor" is often used to describe the supporting plate on which the wafers sit even when heating methods other than rf-induction are used.) In either case, the temperature of the wafer can only be controlled within about 5 or 10°C, which would lead to unacceptable deposition rate and thickness variations if the reactor were operated in the reaction-limited regime. On the other hand, the gas flow, and especially diffusion through the well-defined boundary layer, is relatively well controlled, and operation of this reactor in the mass-transport-limited regime is preferred.

Although the horizontal reactor is very flexible and can operate over a wide temperature range because the forced-convection flow region remains cool, its wafer capacity is severely limited by the size of quartz envelopes available and the large area needed when wafers are placed in a single layer on the susceptor surface.

Low-Pressure Reactor: To overcome the capacity limitations of the horizontal reactor, the high-capacity, low-pressure reactor was developed. In this reactor the wafers are placed in a resistance-heated furnace similar to an oxidation or diffusion furnace and are closely spaced, generally perpendicular to the axis of the tube. Figure 5 show that, in this reactor, the gases flow first through the annular space between the chamber walls and the wafers. They then move between the closely spaced wafers to the wafer surfaces, where the reaction occurs. If mass transport limited the overall reaction, the deposition rate would be markedly higher near the readily accessible edges of the wafers. Therefore, in contrast to the horizontal reactor, this reactor must be operated in the reaction-limited regime. Fortunately, the temperature of the resistance-heated furnace used to heat this reactor can readily be controlled to a

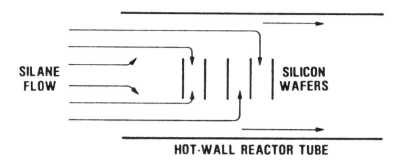

Figure 5. The narrow space between wafers in the low-pressure reactor makes gas diffusion to the centers of the wafers difficult. (Courtesy of J. Andrews, AT&T Bell Laboratories.)

fraction of a degree so that the rapid variation of deposition rate with temperature in this operating regime does not degrade control of the deposited film thickness.

Because of the long, narrow space through which the reactants must diffuse, obtaining reaction-limited operation requires considerable effort. Lowering the temperature alone is not adequate; the ease of mass transport must also be enhanced. As we have already seen, mass transport can be characterized by the ratio D/δ, where D is the gas-phase diffusivity and δ is the boundary-layer thickness or another characteristics dimension of the deposition system. The rapid variation of the diffusivity with pressure is the key to achieving surface-reaction-limited operation in this type of reactor. The diffusivity is inversely proportional to the total pressure; reducing the pressure by a factor of several thousand to a fraction of a torr (tens of pascals)[*] increases the ease of gas-phase diffusion by a similar amount. Making the mass transport easier by operating at reduced pressures thus moves the deposition process into the reaction-rate-limited regime, as desired. To maximize the deposition rate, the partial pressure of the silicon-containing gas must be

---

[*] 1 torr = 133 pascals (Pa); 1 atm = 1.013×10^5 Pa.

Table 1

Typical Polysilicon Deposition Conditions

	Atmospheric Pressure Reactors	Low Pressure Reactors	
Temperature (°C)	950	620	640
Silane partial pressure (torr)	0.3	0.2	0.2
Carrier Gas	H_2	None	N_2 or H_2
Total pressure (torr)	760	0.2	1.0
Deposition Rate (nm/min)	120	10	15
Wafer capacity	20	100	
Throughput (wafers/hour)	40	100-150	

comparable to the total pressure, and little or no carrier gas is used in this reactor. Typical deposition conditions are shown in Table 1.

One significant difference between the hot-wall LPCVD system and the cold wall reactor is the temperature of the gases in the forced-convection region. Because the gases are heated significantly in the forced-convection region of the hot-wall reactor, they may decompose or react in this region, leading to particles on the wafer surface or a porous film, as well as changing the deposition rate. Gas-phase decomposition is also promoted by the limited amount of carrier gas available to dilute the reactants and decrease the reaction probability. For example, a hydrogen carrier gas can suppress the thermal decomposition of silane because it is one of the products of the reaction

$$SiH_4(g) \rightarrow Si(s) + 2H_2(g) \qquad (3)$$

The presence of hot gases in the forced-convection region is even more deleterious when silicon compounds are to be deposited in the hot-wall reactor. One gas may tend to decompose before the desired reaction between different gases occurs, limiting the variety of gases that can be effectively used. For example silane decomposes more readily when it reacts with ammonia in the reaction

$$3SiH_4(g) + 4NH_3(g) \rightarrow Si_3N_4(s) + 12H_2(g) \qquad (4)$$

and the less easily decomposed, silicon-containing gas dichlorosilane (SiH_2Cl_2) is generally used in place of silane, leading to other problems.

2.2 Wafer-to-Wafer Uniformity

In either the horizontal reactor or the conventional low-pressure reactor, the gases are inserted from one end of the reaction chamber, and flow along the wafer load. For efficient utilization of the silicon-containing gas, its partial pressure must decrease along the wafer load. Unless compensated, this gas depletion causes the deposition rate to decrease along the wafer load, with unacceptable variations in film thickness. In each reactor a means related to the parameters limiting the deposition must be found to achieve a uniform thickness regardless of the position of the wafer.

As we have already discussed, the horizontal reactor is usually operated so that the overall deposition process is controlled by diffusion through the boundary layer. From Eq. 1 the diffusion rate of the silicon-containing species is just CD/δ, where C is the concentration of the silicon-containing gas in the forced-convection region. Because this gas is consumed as silicon is deposited, its concentration varies with distance x along the deposition chamber. The ratio $C(X)D/\delta$ can only be kept constant if one of the other variables can be made a compensating function of position. In practice, δ can be readily made to decrease with position to compensate for the decrease in C.

The boundary layer is formed by the viscous forces exerted on the gas in the forced-convection region by the susceptor and walls of the chamber. If the gas can be forced to travel faster, it resists the viscous forces more readily, and the boundary layer becomes thinner. Because the amount of gas blowing in the forced-convection region remains almost constant along the length of the deposition chamber, its velocity can be increased by decreasing the effective cross section of the chamber. By tilting the susceptor so that the cross section through which the gas must flow decreases along the direction of gas flow, the velocity increases, and the boundary layer thickness decreases. Consequently, the ratio $DC(x)/\delta(x)$ is less sensitive to gas depletion along the length of the deposition chamber. In practice, tilting the susceptor by 1 or 2° is adequate to compensate for several tens of percent gas depletion.

Different mechanisms limit the deposition process in the low-pressure reactor. In this reactor, the deposition rate depends strongly on temperature; consequently, a slight increase in temperature along the length of the deposition chamber can compensate for moderate gas depletion, and a uniform deposition rate can again be obtained along the length of the reaction chamber.

Severe gas depletion cannot be overcome, however, and proper process design involves a trade-off between deposition rate, efficient gas utilization, and uniformity. Figure 6 shows the deposition rate along the reaction chamber at several different temperatures (with the temperature uniform along the length of the chamber at each temperature). At the low temperature of 525°C, little of the silicon-containing gas is consumed, and the deposition rate is uniform along the length of the deposition chamber even without a temperature gradient. However, the very low deposition rate leads to an unacceptably low reactor throughput. The average deposition rate increases by about 10 times between 525 and 625°C, but at 625°C the rate varies moderately along the length of the

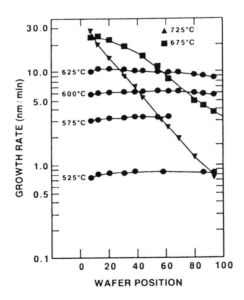

Figure 6. In the hot-wall, low-pressure reactor, the deposition rate decreases along the length of the reactor when a significant fraction of the reactant gas is consumed. This gas depletion is more important at higher deposition temperatures.

chamber. It is in this range that a temperature gradient can be used to improve uniformity. At still higher temperatures, the maximum deposition rate increases, as is desirable for higher wafer throughput, but severe gas depletion greatly decreases the deposition rate toward the downstream end of the deposition chamber. Because virtually all of the silicon-containing gas is being depleted before reaching this region, uniformity cannot be achieved by using a temperature gradient. Proper process design requires operating at a lower average temperature, at which the gas depletion is only moderate and can be compensated by using a temperature gradient. (The uniformity can also be improved somewhat by using a higher gas velocity so that the gas travels farther along the deposition chamber before entering into the reaction; however, this decreases the efficiency of gas use (3).)

Using a temperature gradient can have detrimental effects. Because the structure of polysilicon varies rapidly with deposition temperature, any device property which depends on the structure can vary with the position of the wafer in the reactor. As we will see when we discuss the oxidation of polysilicon, electrical conduction through oxide grown on polysilicon depends sensitively on the polysilicon structure; consequently, conduction through the oxide can vary with wafer position in the reactor during polysilicon deposition. Such variations cannot be tolerated in critical applications, such as in electrically erasable, programmable read-only memories (EEPROMs), which rely on controlled conduction through oxides grown on polysilicon. Therefore, for films which are to be used for critical applications, temperature gradients cannot be used, and modified LPCVD reactors must be employed.

Two different approaches can be taken to avoid the need for a temperature gradient. One approach merely modifies the standard, tube-type LPCVD reactor by injecting the silicon-containing gas at several locations along the length of the reaction chamber, rather than only at one end. The gases are still exhausted at one end of the tubular reactor chamber, as in the conventional reactor. The second approach uses a markedly different reactor geometry, in which the wafers are placed in a constant-temperature, bell-jar-shaped furnace, as shown in Figure 7. The gases are injected all along the wafer load; they flow through the narrow space between a single pair of wafers and then are immediately exhausted from the system. Both approaches provide good thickness uniformity without using a temperature gradient.

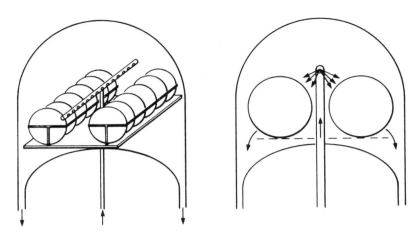

Figure 7. In the *vertical-flow* reactor the gases pass between a single pair of wafers and are then removed from the deposition chamber, decreasing the amount of gas depletion.

2.3 Silicon Gas Sources

Several silicon-containing gas sources are available for the chemical vapor deposition of silicon. The most commonly used species contain silicon plus varying amounts of hydrogen and chlorine. Although several of these gases are used to deposit silicon on silicon, nucleation on an insulating oxide or nitride surface may be difficult with the chlorine-containing species, and polysilicon is usually deposited by the thermal decomposition or *pyrolysis* of silane (SiH_4). The overall reaction can be written

$$SiH_4 \text{ (g)} \rightarrow Si \text{ (s)} + 2H_2 \text{ (g)} \tag{5}$$

When polysilicon is deposited on surfaces on which nucleation occurs more readily, other gases, such as dichlorosilane (SiH_2Cl_2), trichlorosilane ($SiHCl_3$) or silicon tetrachloride ($SiCl_4$) can be used.

All of the reactants in a CVD system enter into the reaction chamber as gases. This is most conveniently accomplished if the reactants are in the gas phase at room temperature so that they can be metered by conventional flowmeters and forced into the reaction chamber by the pressure difference between the gas source and the chamber. Silane is convenient in this respect because it is a gas at room temperature.

However, silane is highly explosive and must be used with great care; proper design of the storage and purging system is especially crucial. Although silane is *pyrophoric* and ignites spontaneously when exposed to oxygen, the reaction may not occur immediately if the silane and oxygen are not adequately mixed. A large quantity of silane may accumulate and then explode when contact between the silane and oxygen increases. To avoid silane accumulation if a leak occurs, any locations where stagnant gas can accumulate must be eliminated by forcing large quantities of air or nitrogen around the gas cylinder and other locations where leaks might occur. Extremely small leaks may not be a safety hazard, but the SiO_2 formed by the reaction of silane and oxygen from air can cause particles which degrade film quality. It can also coat the walls of the plumbing and mass flow controllers, changing the calibration of the latter.

In an atmospheric-pressure system, the gases are pushed through the reaction chamber by the higher pressure of the incoming gases. In a low-pressure system, a pump pulls the gases through the chamber. Mechanical pumps, sometimes augmented by a "Roots blower" are generally adequate to obtain the pressures used in typical LPCVD reactors. Because of the corrosive nature of the gases, the pumps and pump oil must be carefully selected. Particles resulting from reaction of the gas species being pumped must be frequently filtered from the pump oil to avoid damaging the pump.

After the gases leave the reaction chamber, they are cooled and flow through a "scrubber" which removes toxic or environmentally damaging species before being discharged to the atmosphere. This scrubber may be dedicated to a particular reactor, or one scrubber may serve an entire fabrication area. In the latter case, possible reactions between gases from various sources must be considered.

2.4 Doping During Deposition

In many cases polysilicon is deposited undoped, and dopant atoms are subsequently added by ion implantation or from a gas-phase source. In some cases, however, the process flow can be simplified by adding the desired dopant to the polysilicon during deposition. High concentrations of phosphorus are frequently added during the deposition when highly conducting gate electrodes and interconnections are required. Phosphorus is added by introducing phosphine gas (PH_3) into the deposition chamber along with the silane.

Because of the different reactivities of the two gases, however, the dopant concentration may vary along the wafer load. This variation is immediately visible as a variation in resistivity if the dopant concentration in the polysilicon grains is less than the solid solubility of the dopant in silicon. However, if the dopant concentration exceeds the solid solubility, excess dopant may be incorporated in the film without being electrically active. In this case, the resistivity appears uniform, but the excess dopant atoms can degrade the polysilicon, as will be discussed in Section 4.1. This degradation is especially important in the thinner polysilicon films are in advanced VLSI processes.

The dopant gases phosphine (PH_3), arsine (AsH_3), and diborane (B_2H_6) are convenient gas-phase dopant sources. However, adding large quantities of these dopant gases may itself alter the deposition process. The n-type dopant gases phosphine and arsine can severely depress the desposition rate when introduced in large quantities. The deposition rate may change by as much as a factor of ten, and the change is more severe for arsine than for phosphine (4). The effect is most visible in the surface-reaction-limited regime, in which the LPCVD reactor operates. Unlike phosphine and arsine, the p-type dopant gas diborane *increases* the deposition rate (5),(6).

Although the cause of these rate changes is not fully understood, it may be visualized in the following manner: Phosphine and arsine or their reaction products are strongly bound to surface adsorption sites, which are then unavailable to the incoming silicon-containing gas, decreasing the deposition rate. These adsorbed atoms on the surface can exert long-range repulsive forces so that the silicon-containing atoms are efficiently repelled by even moderate surface coverage. Phosphorus is known to efficiently repel other species from surfaces; in chemical kinetic studies, the walls of experimental chambers are often coated with phosphorus-containing solutions to impede the heterogeneous recombination of hydrogen atoms and other species at the walls (7). The efficiency of the repulsion reaction has been related to the electronegativity of the blocking species. The increase of deposition rate when boron is added is attributed to a catalytic effect of adsorbed boron atoms on the adsorption or decomposition of the silicon-containing species (8), thus providing an efficient parallel deposition path.

The severe decrease of the silicon deposition rate as n-type dopant gas is added during the deposition can be avoided by using other silicon-containing gases, such as disilane (Si_2H_6) (9),(10). Disilane decomposes into silane and silylene (SiH_2). The latter forms very strong bonds to silicon, and its adsorption is not readily blocked by phosphorus- or arsenic-containing species adsorbed on the depositing surface. While the deposition rate is reduced by a factor of ten for phosphorus concentrations in the $10^{20} cm^{-3}$ range when silane is used in the LPCVD reactor, the decrease is about twenty times less for disilane under the same conditions (9). However, achieving uniform deposition across a wafer may be more difficult with disilane because of its more reactive nature—the very reason that it is not blocked by adsorbed dopant atoms.

The dopant gases are, of course, highly toxic. Because large quantities are needed to dope the films during deposition, proper handling is required. In particular, decomposition of phosphine or the other gases at the exhaust end of the deposition system must be considered in system design. Additional apparatus is sometimes added to complete the dopant-gas reactions as the gases leave the main deposition chamber.

3.0 STRUCTURE

In the previous section we considered the deposition of polycrystalline-silicon films and saw that these films are now almost exclusively deposited in LPCVD reactors at a temperature near 625°C, although in the past they were often deposited at higher temperatures in horizontal, atmospheric-pressure systems. We saw that the operating regime is chosen to provide the greatest control over the deposition process in the particular reactor being used. In this section we will examine the structure and other properties of the films and consider the techniques needed to evaluate them. In our discussion we will emphasize the importance of surface migration of the adsorbed species during the deposition process in determining the structure of the deposited film.

3.1 Nucleation and Surface Processes

In our previous discussion we focused our attention on the processes occurring in the gas phase because these processes depend strongly on the reactor used for the deposition; we also considered the overall chemical reactions.

Now, we want to look at the processes occurring on the surface during and after nucleation because these processes influence the structure of the material formed.

As the silicon-containing species approaches the wafer surface, it may react in the gas phase, especially when the temperature in the boundary layer is high and the silicon-containing gas reacts at a low temperature. The reactions may form intermediate species, such as SiH_2, or a series of reactions may proceed to form elemental silicon. In the latter case, silicon atoms may agglomerate into small silicon grains in the gas phase. These grains then migrate to the wafer surface where they form an irregular, porous deposit, as well as loosely adherent particles. This gas-phase or *homogeneous* reaction does not form the dense, uniform films needed for integrated-circuit applications and is usually suppressed in favor of *heterogeneous* reactions, which occur on the surface. In the case of heterogeneous reaction, some initial portions of the overall reaction may occur in the gas phase, especially at atmospheric pressures and higher temperatures (11),(12),(13), but the final stage of the reaction forming elemental silicon occurs on the wafer surface. The adsorbed silicon species or its precursors diffuse on the surface, and the structure of the final film depends strongly on the amount of surface migration possible before the adsorbed silicon atoms are covered and immobilized by subsequently arriving silicon atoms.

The structure is, of course, also affected by the nature of the underlying substrate. If it is crystalline and atomically clean, the diffusing silicon atoms see an array of relatively low-energy sites at which it is favorable for them to locate. If they have enough thermal energy to migrate on the surface, they preferentially locate over these low-energy sites. If enough energy is available, the ordering is complete, and a single-crystal or *epitaxial* layer forms. At lower temperatures, the diffusing adsorbed surface atoms do not have enough energy to arrange themselves over the low energy sites, and the epitaxial relationship is lost. Random nuclei form on the wafer surface, and these nuclei join to form a polycrystalline deposit. Surface contamination can also hinder surface diffusion so that the diffusing atoms cannot reach low-energy sites, again preventing formation of a single-crystal layer. If the contamination is irregular, some very-low-energy sites may be available to attract diffusing atoms from the surrounding regions, and thick deposits may form at these locations, leading to defects in the films. Because the excess material near the defect is supplied by diffusion from the surrounding regions, the film may be thinner in the surrounding region than in the regions distant from the defect.

In many applications, however, the surface on which the polysilicon film is to be deposited is amorphous, rather than crystalline, and initial nucleation of the silicon film is quite different. Silicon atoms adsorbed on a perfectly clean, amorphous surface diffuse randomly and have a greater opportunity to re-evaporate (*desorb*) because they are less firmly bound to the surface. As the adsorbed atoms diffuse on the surface, they may encounter other diffusing atoms and form a pair. This atom pair is more stable than is an isolated atom and is less likely to desorb. As it diffuses on the surface, it may join other atoms, forming a larger, more stable cluster, until the cluster has a low probability of desorbing, and a *critical cluster* or stable nucleus is formed. Because the probability of diffusing atoms encountering each other depends strongly on the number of adsorbed atoms on the surface, it is a strong function of their arrival rate (through the partial pressure of the silicon-containing gas) and their desorption rate (through the temperature of the substrate and the binding energy of the diffusing atoms to the exposed surface). Because of the dependence on the binding energy, stable deposits may form on one surface while not forming on another surface under the same deposition conditions.

After stable nuclei form, additional adsorbed atoms diffusing on the surface can either initiate additional nuclei or join existing nuclei. When the existing nuclei are close enough together, additional atoms are more likely to join an existing nucleus, and the number of nuclei saturates and remains constant as the size of each nucleus grows (Figure 8), (4). The saturation number of nuclei depends on the substrate, the arrival rate of atoms, and the temperature. At higher temperatures, adsorbed atoms, can diffuse greater distances to join an already formed nucleus rather than initiating a new nucleus, and fewer, but larger, nuclei form (Fig. 9 (15)). Thus, more silicon atoms must be deposited before a continuous film forms as the nuclei impinge on each other. Because the individual nuclei are randomly oriented on the amorphous surface, discontinuities form when two differently oriented or differently aligned nuclei join. These discontinuities (*grain boundaries*) strongly influence the

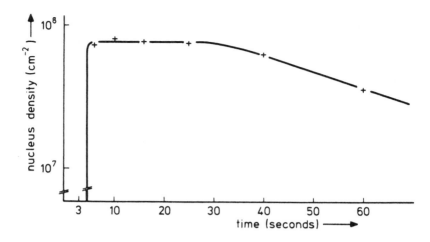

Figure 8. The density of stable nuclei as a function of exposure time to SiH_4, showing the incubation period, the rapid increase to a steady state value, and coalescence. From (14). Reprinted by permission of the publisher, The Electrochemical Society, Inc.

Figure 9. Replica micrographs of films 14 nm thick deposited at (a) 850°C and (b) 1025°C in an atmospheric-pressure reactor. At higher temperatures more surface diffusion can occur and fewer, but larger, nuclei are formed.

properties of the polycrystalline-silicon film, as we will see in some detail in the following sections.

Even after a continuous film forms, the structure is strongly influenced by the thermal energy available for surface migration. At very low temperatures, the adsorbed silicon atoms have little thermal energy and cannot diffuse significantly on the substrate surface before they are covered by subsequently arriving silicon atoms. Once they are covered, their random arrangement is locked into place, and an *amorphous* structure with no long-range order forms. (The amorphous silicon films of interest to us in this chapter are formed by CVD at a temperature slightly below the temperatures at which polysilicon is deposited; they are dense, contain little hydrogen, and differ markedly from the porous, amorphous films deposited at much lower temperatures by sputtering or glow-discharge techniques. We will discuss these higher-temperature CVD amorphous films further below.) At higher temperatures, adequate surface diffusion is possible to allow a crystalline structure to form. On an amorphous substrate, or even on a crystalline substrate at intermediate temperatures, the deposited film is polycrystalline. At high temperatures on a single-crystal substrate, the individual nuclei may each align themselves with the substrate crystal so that no grain boundaries form when the individual nuclei grow together, and an epitaxial layer forms.

Although temperature is the most important variable controlling the structure, pressure and deposition rate also affect the structure through their effect on surface diffusion, as discussed above. Just as surface migration decreases when less thermal energy is available at lower substrate surface temperatures, it also decreases when more of the available adsorption sites are filled by other (non-silicon) adsorbed atoms. As the arrival rate of silicon atoms increases, the amount of time during which the adsorbed atoms can diffuse on the surface before they are immobilized by subsequently arriving atoms decreases, and the amount of migration is again limited.

Each of these variables can modify the ordering of the growing grains, and they are consequently interrelated. For example, the temperature at which a polycrystalline structure, rather than an amorphous structure, forms during deposition is lower at lower total pressures because less carrier gas is adsorbed on the surface to impede rearrangement of the atoms diffusing on the surface. At atmospheric pressure, the transition between the deposition of amorphous and polycrystalline silicon occurs at about 680-700°C; in an LPCVD reactor at a typical pressure of about 0.2 torr, the transition temperature is about 580-600°C. Similarly, a given structure is formed at lower temperatures as the deposition rate decreases so that the adsorbed atoms can diffuse for a longer time on the surface before they are immobilized. (This effect is limited, however, by impurities. At very low silicon arrival rates, impurities can be the dominant species adsorbed on the surface, where they impede the surface migration of silicon atoms. Impurities are especially harmful in systems which are not sufficiently controlled for purity and vacuum integrity.) Within the temperature range where polycrystalline films are deposited, the structure of the polycrystalline film is also dominated by the amount of surface migration possible during deposition.

3.2 Evaluation Techniques

To learn more about the films, we must consider the techniques used to reveal their structure. Unfortunately, the most straightforward techniques, such as scanning-electron microscopy, provide little information about the structure, and more time-consuming techniques must be used.

Transmission Electron Microscopy: The most detailed information is obtained by transmission electron microscopy (TEM), in which a beam of electrons travels through a very thin section of the sample being studied and is then imaged. The specimen preparation needed for this technique is very time-consuming, and its use is generally limited to research and development, rather than being used for routine process monitoring.

Figure 10 (16) shows plan-view transmission electron micrographs of two films deposited by low-pressure CVD at a pressure of about 0.2 torr using SiH_4 without any additional carrier gas. The film shown in Figure 10a was deposited at 550°C and is basically amorphous with a few small crystalline inclusions. The corresponding transmission electron diffraction pattern shown in the inset exhibits the broad, diffuse rings characteristic of an amorphous film. Figure 10b shows a silicon film deposited at 625°C. The structure is polycrystalline, with an average grain size of about 70nm; the diffraction pattern shows the ring-and-dot pattern characteristic of a polycrystalline structure. Detailed examination of

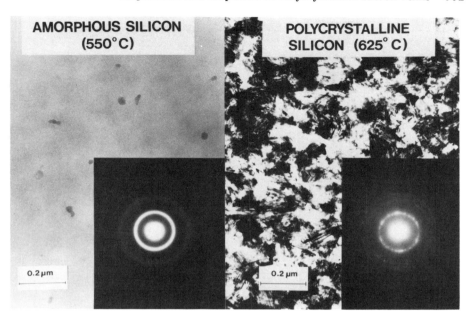

Figure 10. Transmission electron micrographs of 0.6 μm thick films deposited onto SiO_2 at 550°C and 625°C in a low-pressure reactor.

the diffraction patterns reveals that the second diffuse ring of the amorphous structure splits into two rings in the polycrystalline structure. This difference arises from the arrangement of the second-nearest atoms in the two cases. Although the nearest-neighbor spacing is not greatly different in amorphous and crystalline films, the second-nearest-neighbor spacing differs considerably. Because of bond-angle variations, a range of second-nearest-neighbor spacings is possible in the amorphous structure, and a diffuse band is seen in the diffraction pattern. In the crystalline material, only two different second-nearest-neighbor spacings are possible, leading to two well-defined rings in the diffraction pattern.

Detailed TEM examination of films deposited in the LPCVD reactor shows that the transition between an amorphous and a polycrystalline structure during deposition occurs near deposition temperatures of 580°C. Near this temperature, an anomalous, elongated structure sometimes appears. At 600°C, just above the transition temperature, a very fine-grain, but equi-axed, deposit forms. By 625°C, a well-defined, polycrystalline structure, which is approximately equi-axed in the plane of the film, is obtained. At higher temperatures within the very limited temperature range which can be studied in the LPCVD reactor, the grain size is only a weak function of temperature.

A significantly wider range of operating conditions can be explored in the horizontal, atmospheric-pressure reactor; the effect of temperature can be examined over the range from about 600°C to about 1200°C. Because adsorbed carrier gas impedes surface diffusion of the silicon-containing gas, the transition between an amorphous structure and a polycrystalline one occurs at a higher temperature (about 680°C) in this reactor than in the LPCVD reactor, as described above. An anomalous, elongated structure is again sometimes seen near the transition temperature; at higher temperatures, the more common, equi-axed structure forms. As the temperature increases, the grain size also increases, changing from about 50 nm at 700°C to about 300 nm at 1100°C. (The grain size would probably increase in the LPCVD reactor at higher temperatures also if a wider temperature range could be investigated; some evidence of this is seen even in the limited range which can be studied.) Transmission electron micrographs reveal that the grains within the polycrystalline films are often highly faulted and contain a high density of twin lamella and other, more serious defects.

When amorphous films are crystallized by annealing, the grains become even larger than the grains in initially polycrystalline films annealed at the same temperature. This difference in grain size affects the resistivity of the films when they are subsequently doped, as we will discuss further in a following section.

X-Ray Diffraction: Although transmission electron microscopy gives the most detailed information about the crystal structure, a quantitative comparison of films deposited under only slightly different conditions can be more readily obtained by X-ray diffraction. In the simplest arrangement of this technique, shown in Figure 11, X-rays are incident on the sample at an angle θ from the sample surface, and a detector is also located at an angle θ from the sample surface. As the angle θ of the sample is varied with respect to the fixed incident beam, the detector is simultaneously moved to a new position at an angle 2θ from the incident beam. (Because of the relation between the angles of the

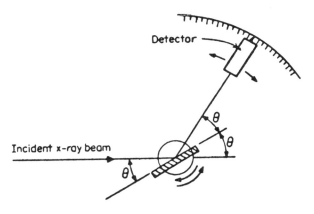

Figure 11. Geometrical arrangement of source, sample, and detector in the "θ-2θ" X-ray diffractometer.

sample and detector from the fixed incident beam this arrangement is often called a "θ-2θ" geometry.)

The X-rays are diffracted from crystal planes with a spacing d when Bragg's law

$$m\lambda = 2d \sin \theta \tag{6}$$

is satisfied. By varying the angle θ, a series of peaks in the X-ray signal appears as grains with different crystal planes parallel to their surface satisfy Bragg's law. Thus, the relative intensities of the various peaks give a quantitative indication of the prevalence of crystals with different orientations within the polycrystalline film. After correcting for the finite film thickness (17), comparison with standards (18) allows assessing the dominant crystal *texture* in a film and its variation with changing deposition conditions.

Figure 12 shows the normalized X-ray signal as a function of deposition temperature for polycrystalline silicon films about 0.5μm thick deposited in an LPCVD reactor using undiluted SiH_4 at a pressure of about 0.2 torr (19). The {110} texture dominates at most temperatures, with a strong peak near 625°C and a minimum near 675°C. Figure 13 shows similar X-ray data for atmospheric-pressure CVD (APCVD) films (20). Again a maximum and a minimum appear in the {110} texture; however, the temperatures at which these extrema occur are considerably higher for the APCVD films than for the LPCVD films. This difference can again be

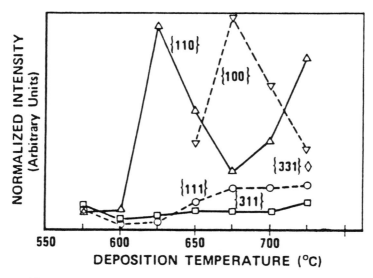

Figure 12. X-ray texture as a function of deposition temperature for 0.5 μm thick films deposited in a low-pressure CVD reactor.

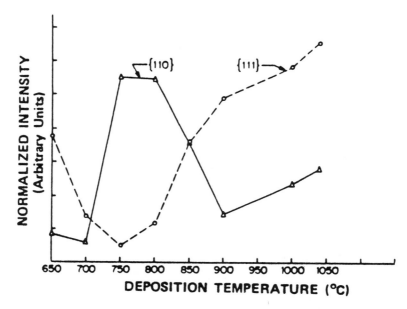

Figure 13. X-ray texture as a function of deposition temperature for 0.6 μm thick films deposited in an atmospheric-pressure CVD reactor.

attributed to the decreased surface diffusion possible in the atmospheric-pressure reactor because adsorbed carrier gas impedes surface diffusion of the adsorbed silicon atoms. Thus, a higher temperature is needed to provide the diffusion of the adsorbed silicon atoms required to form a given structure.

As the films become thicker, the texture continues developing. The relative importance of the {110} texture compared to other crystal orientations increases with increasing film thickness, indicating that this structure is favored during the continued deposition of the films, rather than only during the initial nucleation (*ie*, it is a *growth texture*, rather than a *nucleation texture*).

The dominant crystal orientation is important for several reasons. First, the surface structure appears to depend on the crystal orientation. As will be discussed later, the quality of oxide grown on the surface of a polysilicon film depends on the surface structure and, consequently, the grain orientation of the polycrystalline film. Second, both transmission electron microscopy of submicrometer films Figure 14a (21) and optical micrographs of thicker films (Figure 15) (22) show that the {110} texture is related to a columnar structure. Figure 15 shows that the columnar structure continues developing as the film becomes thicker, confirming X-ray measurements that indicate greater dominance of the {110} texture as the film thickness increases.

3.3 Dopant Diffusion

The columnar structure is especially important because it is related to very rapid diffusion of dopant atoms in polycrystalline silicon with {110} texture. In polycrystalline materials, grain boundaries provide disordered regions down which dopant atoms can readily diffuse. From the grain boundaries, the dopant atoms then diffuse into the surrounding crystalline grains. Thus, the "effective diffusivity" of dopant atoms in polycrystalline materials can be much greater than in single-crystals of the same material. The ease of dopant diffusion depends, of course, on the detailed structure of the grain boundaries and their direction. As indicated in Figure 16a, if the grains are randomly oriented, the dopant atoms must diffuse laterally, as well as vertically, to find easy diffusion paths along grain boundaries. Because dopant atoms diffuse much more readily along the disordered grain boundaries than within the grains, the effective diffusivity is

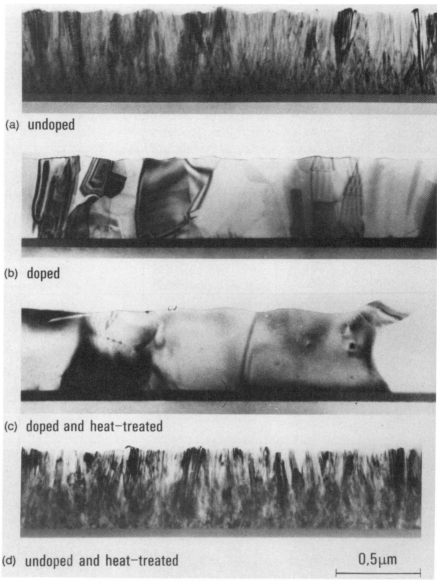

Figure 14. Cross section transmission electron micrographs of 0.6 μm-thick films deposited in a low-pressure reactor at 625°C. The initial columnar structure of undoped polysilicon films (a) changes little when the undoped films are annealed (d), but phosphorus doping increases the grain size (b), which is further enlarged by additional annealing of the doped films (c). From [21]. Reprinted by permission of the publisher, The Electrochemical Society, Inc.

Preparation and Properties of Polycrystalline-Silicon Films   667

Figure 15. Optical micrograph of polished and etched cross section of a thick polycrystalline-silicon film deposited onto SiO_2 at 1050°C, showing the development of the columnar structure as the film becomes thicker.

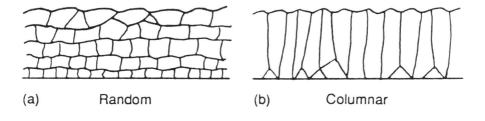

(a)   Random   (b)   Columnar

Figure 16. (a) The dopant atoms must diffuse both vertically and laterally to move through a random polycrystalline structure. (b) Diffusion is primarily vertical in a columnar structure.

greater in this polycrystalline structure than in single-crystal material, even though the path length of a diffusing dopant atom can be greater. However, if the structure is columnar (Figure 16b), the dopant atoms can diffuse vertically more easily, and the effective diffusivity is enhanced even more. Thus, a film with many columnar grains associated with the {110} texture allows very high diffusivity of dopant atoms, as shown in Figure 17.

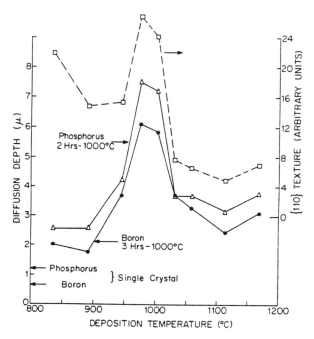

Figure 17. The diffusion depth in polycrystalline-silicon films depends on their structure and, therefore, the deposition conditions. These 15 μm thick films were deposited onto SiO_2 in an atmospheric-pressure reactor over the temperature range from 830 to 1170°C.

3.4 Grain Growth

Changes in the structure of the deposited silicon films during heat treatments after deposition, as well as the initial structure formed during deposition, can affect the processing and device behavior. The amorphous films deposited at lower temperatures are not stable when heated to higher temperatures, as occurs during subsequent processing of an integrated circuit; annealing, even at 800°C, crystallizes the films

(23). On the other hand, the initially polycrystalline films are much more stable on annealing, provided that they are not highly doped, and must be heated to about 1100°C before significant changes occur in their structure (23).

However, if polycrystalline-silicon films are heavily doped with phosphorus or arsenic, grain growth is greatly enhanced. The transmission electron micrograph of Figure 14d (21) shows that annealing an undoped film at 1000°C does not change the columnar structure obtained during deposition. When the film is heavily doped with phosphorus, however, significant grain growth occurs during doping and annealing (Figure 14b and c (21)). This dopant-enhanced grain growth is a strong function of the dopant concentration. Little enhanced grain growth occurs below a phosphorus concentration of 1×10^{20} cm^{-3}, while the grain growth is greatly accelerated in the high-10^{20} cm^{-3} range (24). When the grains grow, they first achieve a diameter about equal to the film thickness (by *primary* or *normal recrystallization*), with most of the grains being about the same size. When the grain size increases further (by *secondary recrystallization*) some grains grow at the expense of neighboring grains, and the grain size becomes less uniform. The driving force for grain growth is the reduction in grain-boundary energy as grain boundaries are eliminated by grain growth. When grain-boundary reduction drives grain growth, the grain size increases as the square root of time (24). Minimizing surface energy can also provide the driving force for grain growth because the surface energy of different orientations differ (25).

Like phosphorus, arsenic accelerates grain growth at intermediate concentrations; however, for arsenic the grain-growth enhancement diminishes at very high dopant concentrations (26), probably because arsenic clusters impede grain-boundary motion. Boron appears to have little effect on grain growth by itself. However, when boron, a p-type dopant is added to a heavily doped n-type film, it decreases the enhanced grain growth (27). This behavior indicates the importance of the Fermi-level location for grain growth. It has been postulated that vacancies with different charge states (which depend on the location of the Fermi level in the band gap) accelerate grain-boundary migration. Lowering the Fermi level by introducing a p-type dopant into an n-type sample reduces the concentration of charged vacancies and, therefore, the rate of grain growth.

3.5 Optical Reflection

Some information about the structure of deposited silicon films can be obtained by measuring the reflectance of the films deposited on an oxidized wafer over a suitable wavelength range. This measurement is especially useful as a rapid technique for distinguishing between amorphous- and polycrystalline-silicon films. The output of a typical spectrophotometer, which is related to the reflectance of the silicon-oxide-silicon structure being measured, is shown in Figure 18 for an amorphous-silicon film deposited on an oxidized wafer in an LPCVD reactor at 550°C and for polycrystalline-silicon films similarly deposited at 625 and 700°C.

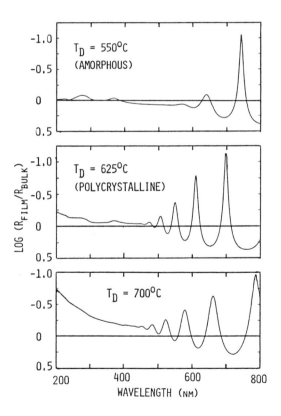

Figure 18. Reflectance as a function of wavelength for silicon films deposited at 550, 625, and 700°C using a single-crystal reference wafer.

At short wavelengths ($\lambda < 400$ nm), the light cannot penetrate the silicon films, and the observed signal is primarily an indication of the surface roughness of the film. The low reflectance of the amorphous film indicates a very smooth surface. (The two small features at 280 and 370 nm arise from the difference between the amorphous structure of the film and the crystalline structure of the reference wafer; the structure present in the single-crystal silicon at these wavelengths is absent in the amorphous films.) In contrast, the significant signal from the polysilicon films in the short-wavelength range indicates that they are markedly rougher and that the surface roughness is greater for the film deposited at the higher deposition temperature. The rms value of surface roughness of the polycrystalline films can be calculated from their reflectance in the short-wavelength region by the formula (28)

$$\sigma_0 = \frac{0.12\lambda}{\cos\phi}\sqrt{\log_{10}\left(\frac{R_0}{R}\right)} \tag{7}$$

where λ is the wavelength, ϕ is the angle of incidence (measured from the normal to the sample surface), R is the reflectance of the sample, and R_0 is the reflectance of an ideally smooth surface of the same material.

Because of the limited surface diffusion during deposition, amorphous silicon films are generally much smoother than are polycrystalline films. The smooth surface is retained even after an amorphous film is crystallized by annealing, probably because the surface is stabilized by the thin native oxide that forms on the surface when the film is exposed to air. The smoother surfaces of initially amorphous films increases the quality of oxides grown on crystallized films deposited in the amorphous form and may have important consequences for pattern definition during fabrication of fine-geometry circuits.

Near a wavelength of 500 nm, light can travel through the thickness of a polycrystalline film so that constructive and destructive interference occur between light reflected from the top surface of the silicon film and that reflected from the underlying interfaces. The amorphous film is still highly absorbing at this wavelength, and little interference is seen. It is only for wavelengths well above 600 nm that significant light can penetrate the amorphous film. After annealing, the crystallized, initially amorphous film becomes more transparent, and stronger interference peaks are seen in the reflectance spectra.

Once the structure of the film (*ie*, amorphous or polycrystalline) is known, the corresponding optical properties and the reflectance spectrum can be used to determine the silicon film thickness if the thickness of the underlying dielectric layer is known. Incident light is reflected from the top surface of the silicon film, from the underlying silicon-dielectric interface, and from the dielectric-substrate interface. These reflected beams combine to produce constructive and destructive interference as the wavelength varies. The locations of the extrema are then related to the thickness of the silicon film (29). The data-reduction algorithm is contained in commercially available measurement equipment, which calculates the thickness assuming a fixed underlying silicon-dioxide thickness (usually 100 nm).

The index of refraction of polycrystalline silicon is only slightly different from that of single-crystal silicon. However, the index of refraction of amorphous silicon films is considerably higher (19), so the same values cannot be used in the data-reduction algorithm. After annealing to convert the amorphous structure to a polycrystalline one, the index of refraction decreases to approximately that of crystalline silicon. The peaks in the reflected signal consequently occur at different locations. (Note that the thickness of these dense amorphous films does not change significantly when they are annealed to convert them to polysilicon.)

3.6 Etch Rate

The chemical properties of amorphous silicon films also differ from those of polycrystalline silicon. Amorphous silicon is attacked by hydrofluoric acid while polycrystalline silicon is not. Again, once the amorphous film is crystallized by annealing, its properties become similar to those of polycrystalline silicon, and it is no longer etched by HF-based solutions (30).

3.7 Summary

In this section we have examined the relation between the deposition conditions and the structure of the film formed. We saw that lower deposition temperatures, higher deposition rates, and the presence of foreign atoms on the surface impede the development of the structure in the depositing film. The {110} texture is related to a columnar structure which allows rapid dopant diffusion down grain boundaries. The optical properties of amorphous silicon differ significantly from those of crystalline silicon, and nondestructive optical measurements can be used to obtain some information about the structure and thickness of the deposited films.

4.0 OXIDATION

In most integrated circuits, the polysilicon is electrically isolated from overlying conductors (either metal or additional layers of polysilicon) by silicon dioxide, which may be formed by thermal oxidation of the polysilicon or by chemical vapor deposition. In most integrated circuits these oxide layers must simply be highly insulating. However, specialized devices, such as the electrically erasable, programmable read-only memories (EEPROMs) used with increasing frequency in VLSI circuits require a thin oxide with well-controlled conductivity above the polysilicon.

Numerous studies have shown that the oxidation rate of polysilicon can differ substantially from that of single-crystal silicon and also that the electrical properties of the oxide grown on polysilicon are different from those of a similar thickness of oxide grown on single-crystal silicon. In this section, we will first examine the differences in the oxidation rates of polysilicon and single-crystal silicon and relate these differences to the structure of the polysilicon. We will see that in some applications the differences in the oxidation rate can be used constructively, while in other applications, they complicate integrated-circuit fabrication. Then, we will consider conduction through oxide grown on polysilicon, and show that the surface irregularities on the polysilicon surface lead to high conduction through oxides thermally grown on polysilicon.

4.1 Oxide Growth on Polysilicon

Undoped Films: Substantial differences can exist between the thickness of oxide grown on polysilicon and that of oxide simultaneously grown on single-crystal silicon. For lightly doped films one of the dominant factors leading to this difference is the presence of grains with different orientations in the polysilicon. To demonstrate the importance of differently oriented grains, in one study (31) thick layers of polysilicon were polished so that the resulting smooth surface contained regions with different crystal orientations. When oxidized, each differently oriented grain oxidizes at a rate characteristic of that particular orientation of crystalline silicon. Grains with a given orientation exhibit the same oxide color as seen on the similarly oriented, single-crystal control wafers. These differences are accentuated under surface-reaction-limited oxidation conditions and reduced when the oxidation is performed under diffusion-limited conditions. Although polishing the samples removed the faceted surface structure which causes the exposed crystal planes to differ from the grain orientation, the study did show that the macroscopically measured oxide thickness can be expected to be a suitable average of the oxide thicknesses grown on differently oriented grains.

A similar trend is seen in oxides grown on the fine-grain polysilicon typically used in integrated circuits. The oxide thicknesses grown on thin polysilicon films deposited either in an atmospheric-pressure reactor at 960°C or in a low-pressure reactor at 625°C are between the oxide thicknesses grown on rapidly oxidizing (111)-oriented, single-crystal silicon and slowly oxidizing (100)-oriented silicon (32). Because the oxide thickness grown depends on the dominant crystal orientations in the film being oxidized, it can differ significantly for polysilicon deposited under different conditions, necessitating process modification when the polysilicon deposition conditions change significantly. We saw in Section 11 that the crystalline texture of polysilicon films depends strongly on the deposition temperature; the oxide thickness should vary correspondingly.

Heavily Doped Films: Although oxidation of lightly doped polysilicon is similar to that of lightly doped single-crystal silicon, marked differences are seen between the oxidation of heavily doped polysilicon and similarly doped single-crystal silicon. Heavily doped single-crystal silicon oxidizes much more rapidly than does lightly doped single-crystal silicon because of the excess point defects present in the heavily doped material. As the influence of the added point defects begins to dominate the oxidation process, the differences between the oxidation rates of the different orientations of silicon decrease significantly. However, the oxidation-rate enhancement on polysilicon is usually much less than that on single-crystal silicon doped at the same time (32).

Figure 19 (32) shows the thickness of oxide grown on two different orientations of single-crystal silicon and on polysilicon deposited under two different deposition conditions. Varying amounts of phosphorus were added by gaseous diffusion from a POCl$_3$ source to sets of samples each containing all four types of material, and then all samples were simultaneously oxidized under surface-reaction-limited conditions. The resulting oxide thicknesses are shown as functions of the sheet resistance measured on the (100)-oriented single-crystal silicon contained in each set of samples. As the dopant concentration increases (decreasing sheet resistance), the oxide thickness grown on the polysilicon becomes a smaller fraction of that grown on single-crystal silicon, even though the amount of dopant added to the polysilicon is expected to be at least as great as that added to the single-crystal silicon.

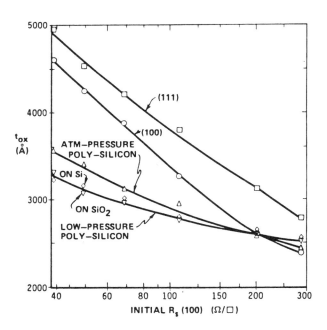

Figure 19. Oxide thickness grown on phosphorus-doped single-crystal and polycrystalline silicon during a 150 min, 850°C, pyrogenic steam oxidation as a function of the sheet resistance R_S measured on the (100)-oriented, single-crystal silicon wafer in each set.

This apparently anomalous behavior is best understood by considering the interaction of diffusion and oxidation. In Section 3.3 we saw that impurities diffuse much more rapidly in polysilicon than in single-crystal silicon. During doping and oxidation, therefore, the phosphorus added near the surface can diffuse toward the back of the silicon film more readily in polysilicon than in single-crystal silicon. The surface concentration is lower; and, consequently, a thinner oxide is grown. The differences between the oxide thickness grown on the different types of polysilicon depends on the ease of diffusion in each sample, which is governed by the detailed grain structure.

An alternate explanation suggests that the difference between the oxide thicknesses grown on polysilicon and on single-crystal silicon is dominated by the different electrical activity of the dopant in each type of material. Because the dopant is less active in polysilicon, the Fermi level is closer to the intrinsic Fermi level during oxidation; the charged point defects which enhance the oxidation are less numerous; and the oxide grown is thinner. The different behavior of the point defects in single-crystal silicon and in polysilicon may also affect the oxidation rate. Grain boundaries and other structural defects in polysilicon can act as recombination sites for point defects, increasing their concentration gradients near the surface. If the silicon interstitials injected by the oxidation process can diffuse away from the oxidizing surface more readily, the oxidation rate should increase, rather than decrease.

Differences in the dopant-diffusion rates between polysilicon and single-crystal silicon can also explain the *more rapid* oxidation of heavily doped polysilicon than single-crystal silicon sometimes seen. We can understand this behavior by considering the finite thickness of the polysilicon. Table 2 (32) shows that the oxide thickness grown on heavily doped polysilicon also depends on the thickness of the polysilicon film and is greater on thinner films than on thicker films. In thick films, the dopant can readily diffuse away from the polysilicon surface, reducing the surface dopant concentration and the oxidation rate, as discussed above. In thinner polysilicon films, however, the finite thickness confines the dopant atoms; as dopant atoms approach the back surface, the increasing concentration of dopant there decreases the concentration gradient driving the diffusion. The surface concentration remains high, and the oxide grown is, consequently, thicker.

Table 2

Oxide Thicknesses Grown on n^+ Polysilicon Films of Different Thicknesses During a 75 min, 850°C, Pyrogenic Steam Oxidation

Polysilicon thickness (μm)	Oxide thickness (nm)	
	LP	AP
0.5	393	378
1.0	280	301
1.5	276	310
n^+ (100) single crystal	515	

The increase in oxide thickness with increasing dopant concentration does not continue indefinitely, however. It saturates at a value approximately corresponding to the solid solubility of phosphorus in silicon at the oxidation temperature. As shown in Figure 20 (33), when oxidation occurs at 750°C, the oxide growth rate remains constant for phosphorus chemical concentrations greater than about 10^{21} cm-3, which is close to the solid solubility of phosphorus in silicon at this temperature, especially if segregation of the dopant at grain boundaries is considered. Thus, the oxide thickness depends on the phosphorus concentration from the concentration at which the Fermi level at the oxidation temperature departs from the intrinsic Fermi level up to the concentration corresponding to the solid solubility of the dopant in silicon.

The considerably greater oxide thickness grown on heavily doped polysilicon than on lightly doped, single-crystal silicon can be used advantageously in the fabrication of many integrated circuits. In a typical silicon-gate integrated circuit, the heavily doped, n-type polysilicon gate is adjacent to the lightly doped single-crystal silicon regions which subsequently form the source and drain of an MOS transistor. After definition of the polysilicon, however, these single-crystal regions are still lightly doped. Oxidation under suitable conditions produces a thick oxide on the polysilicon, while only a thin oxide is grown on the single-crystal silicon. The

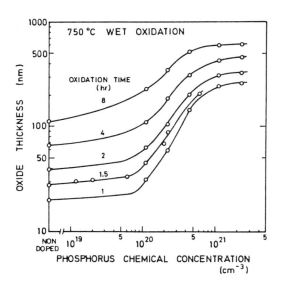

Figure 20. Oxide thickness grown on heavily phosphorus-doped polysilicon during a 750°C, wet-oxygen oxidation. From (33). Reprinted by permission of the publisher, The Electrochemical Society, Inc.

thick oxide on the polysilicon can serve as an implant mask during processing, or it can be used to reduce capacitance in the finished circuit. The differential oxidation rate is also useful in reducing critical mask alignment; the thin oxide on the single-crystal silicon can be removed while not exposing the heavily doped polysilicon, which is covered by a much thicker oxide.

Grain-Boundary Oxidation: The disordered structure near the grain boundaries might be expected to oxidize more rapidly than the crystalline structure near the center of the grains. In the study of polished samples of undoped, large-grain polysilicon, deposited at a high temperature, careful examination of the grain-boundary regions did not reveal any enhanced oxidation at the grain boundaries. Because grain boundaries occupy a larger fraction of fine-grain polysilicon, grain-boundary effects might be expected to play a more important role in the oxidation of fine-grain material, especially in films deposited at lower temperatures, which might be expected to be less ordered. If the grain boundaries oxidize rapidly, the oxide formed there should cause considerable compressive stress in the oxidized film. However, no tendency toward stress is caused by oxidation of *undoped* films (34). In fact, a slight increase of tensile stress is found, suggesting that the heat treatment orders the structure near the grain boundaries.

On the other hand, in phosphorus-doped films oxidation causes compressive stress (34). In addition, examination of the local film thickness near the grain boundaries as oxidation proceeds shows that the the entire thickness of the polysilicon film is consumed first at the grain boundaries while unoxidized silicon still remains near the centers of the grains (34). Although differences in the thickness of the polysilicon film near the center of the grains and near the grain boundaries could also cause the polysilicon to be completely consumed by oxidation near the grain boundaries first, these observations suggest that the region near the grain boundaries oxidizes more rapidly than does the silicon away from the grain boundaries. Because the grain boundaries are the lowest portions of the polysilicon surface before oxidation and they oxidize most rapidly, the surface roughness of the polysilicon is expected to increase as oxidation proceeds.

Other studies have confirmed the more rapid oxidation near grain boundaries. High resolution, cross-section transmission electron microscopy shows that Si-P precipitates can form at grain boundaries when the phosphorus concentration exceeds its solid solubility at the oxidation temperature (35). Because this phase oxidizes more rapidly than does silicon, the oxide near the grain boundaries can be considerably thicker than that over the centers of the grains. Consider a polysilicon film doped to solid solubility at an intermediate temperature. When it is oxidized at a higher temperature, the phosphorus concentration is below solid solubility at the oxidation temperature, and no enhanced oxidation is expected at the grain boundaries. At lower oxidation temperatures the phosphorus concentration is greater than solid solubility; excess phosphorus concentrates at the grain boundaries; and the grain-boundary regions oxidize more rapidly than do the centers of the grains. In addition to forming more rapidly, the oxide grown over grain-boundary precipitates appears to contain a high phosphorus concentration and, therefore, etches more rapidly than does SiO_2. As the oxide grown on polysilicon is etched, narrow grooves can be left in the polysilicon surface. In the extreme case, the indentation can extend through the entire thickness of the polysilicon film so that the underlying oxide (*e.g.*, the gate oxide) can be attacked.

4.2 Oxide-Thickness Evaluation

Although measuring the oxide thickness on single-crystal silicon nondestructively by optical techniques is straight forward using an ellipsometer or a spectrophotometer, measuring the oxide thickness grown on polysilicon is more complex because of the multilayer structure on which the oxide is grown. At the wavelengths typically used for oxide-thickness measurements (*eg*, λ = 628 nm for ellipsometry or λ = 400-800 nm for spectrophotometry), polysilicon is transparent, and the reflected signal being analyzed is influenced by reflections from the underlying interfaces, as well as from the top and bottom interfaces of the oxide layer grown on the polysilicon. In theory, the reflectance of the total multilayer structure can be analyzed, and the oxide thickness of the polysilicon can be extracted. However, in most cases variations in the thicknesses of the underlying layers make the indicated thickness of the oxide grown on the polysilicon layer too uncertain for practical use.

The optical techniques can, however, be adapted to the ultraviolet wavelength range, in which polysilicon is opaque. In this case, the reflected signal is dominated by interference at the top and bottom of the oxide layer grown on the polysilicon. The wavelength range from 200 to 400 nm is suitable for this type of measurement. However, near the lower end of this wavelength range, the surface roughness of the polysilicon also affects the reflected signal, and at λ=280 and 370 nm, structural bands of crystalline silicon can influence the reflected signal. The interference of light reflected from the two surfaces of the top oxide is strong, however, and the technique is a useful, nondestructive, method of measuring the oxide thickness grown on polysilicon. Figure 21 shows the interference signal obtained over the wavelength range from 200 to 500 nm, with interference in the top oxide below about 400 nm and interference in the multilayer structure at higher wavelengths. Ellipsometry using ultraviolet light can also be employed to determine the oxide thickness, again taking advantage of the fact that silicon is opaque to ultraviolet light.

Of course, the oxide thickness can be determined destructively by etching a step and measuring the step height with a surface profilometer. Although this technique is the most straightforward, it can be more time consuming than the optical techniques, its resolution is limited by the surface roughness of the polysilicon, and it cannot readily be used on device wafers.

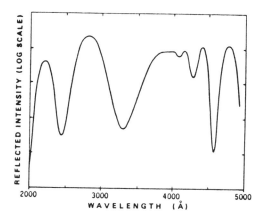

Figure 21. The reflectance in the wavelength range from 200 to 400 nm arises from interference in the oxide above the polysilicon, while the signal at longer wavelengths is influenced by reflection at interfaces beneath the polysilicon as well.

4.3 Conduction Through Oxide Grown on Polysilicon

When oxide is thermally grown on polysilicon, the current flow through it is typically much greater than that through a similar thickness of oxide grown on single-crystal silicon (36). This higher conduction is caused by the rough surface of the polysilicon (36). When a voltage is applied, surface "asperities" (Figure 22) and other irregularities increase the *local* electric field so that it is markedly greater than the *average* electric field calculated from the applied voltage and the oxide thickness. Because the conduction in SiO_2 increases rapidly with increasing electric field, the macroscopically observed current corresponding to this high local electric field is much greater than expected.

In applications requiring controlled conduction through the oxide, the rough surface of polysilicon is utilized for proper device performance, and the structure of the polysilicon must be carefully controlled during its deposition and subsequent processing. In particular, conduction through the oxide depends strongly on the magnitude of the surface roughness, which is related to the dominant crystal orientations of the film. As we have seen, both the surface roughness and the dominant crystal orientations are strong functions of the deposition temperature and also depend on

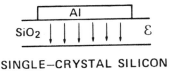

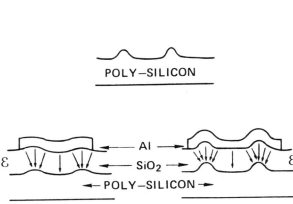

Figure 22. The local electric field is equal to the average electric field for oxide grown on the smooth surface of single-crystal silicon, but the local field can be much greater than the average field for oxide grown on polysilicon because of field enhancement near the irregularities on the polysilicon surface.

other deposition parameters (19),(20). In particular, the surface roughness is correlated with the {110} texture (37), which is a very sensitive function of deposition temperature near the temperature of 625°C normally used for polysilicon deposition in a low-pressure CVD reactor (16).

The temperature gradient used along the length of conventional, low-pressure CVD reactors to compensate for gas depletion (Section 2.2) leads to wafer-to-wafer variations in the structure of the films deposited and, consequently, in conduction through oxide subsequently thermally grown on the films. The surface roughness, and hence the oxide conduction for which a particular device is designed, may only be obtained over a limited portion of the deposition chamber, and the yield may be very low unless the temperature gradient is eliminated and other methods are used to obtain a uniform deposition rate over all the wafers in the reactor.

In addition to the deposition conditions, the surface

roughness, and consequently the oxide conduction and the breakdown voltage, depend on the doping in the polysilicon. At high concentrations the oxide grown on the phosphorus-containing precipitates, which may form at grain boundaries, may be inferior to those of stoichiometric SiO_2. In addition to degradation of oxide properties arising from surface roughness and phosphorus-containing precipitates, the oxide grown on the sharp edges and corners of device features can lead to high local electric fields, again causing enhanced conduction through the oxide.

In the majority of integrated circuits, the oxide above the polysilicon is simply used to isolate the polysilicon from overlying conductors, and the lowest possible conduction is desired. In these cases the enhanced conduction can be reduced by depositing silicon films with smoother surfaces. At lower deposition temperatures, for example, amorphous silicon, rather than polycrystalline silicon, is deposited; the surface of the amorphous material is quite smooth (38), as discussed in Section 3.5, but this desired attribute is obtained at the expense of a lower deposition rate. Although the amorphous silicon crystallizes to become polysilicon during subsequent processing, its surface remains smooth, probably because a stabilizing native oxide forms during exposure to air. Its surface remains smooth even after diffusion doping and annealing (37). Smoothing the surface by laser melting the top of the polysilicon has also been demonstrated, with the resulting improvement in the oxide leakage current and breakdown voltage (39).

Not only does the basic surface roughness of the polysilicon modify the conduction through the subsequently grown oxide, but the oxidation process itself can markedly change the surface topology. The asperities can be accentuated during oxidation to produce sharp "protuberances" extending far above the polysilicon surface; in the extreme case, small regions of silicon may detach from the polysilicon surface to become isolated silicon inclusions within the oxide (40). Geometrical effects can be especially severe when polysilicon is defined (*eg*, to form gate electrodes) before the polysilicon is oxidized. At the top edges of polysilicon patterns, oxidation can form sharp points or "horns" at which the local electric field is greatly increased (41), especially if nearby silicon inclusions within the oxide reduce the effective oxide thickness. In addition, an oxide wedge can form under the polysilicon near its edge by lateral oxidation, distorting the shape of the gate electrode (42), and lateral oxidation of the side of a polysilicon line can cause an overhang at the edge of the gate electrode.

Conduction through oxide grown on polysilicon and changes in the device geometry during oxidation depend sensitively on the oxidation conditions employed (43). In general, high-temperature, dry oxidation produces a less conductive oxide than does low-temperature, wet oxidation. Consequently, enhanced conduction through oxides grown on polysilicon is becoming more serious as lower oxidation temperatures find increasingly widespread use in advanced device processing.

Unlike thermal oxides, chemically vapor deposited oxides do not accentuate the irregularities of the polysilicon surface. They tend to be conformal or to smooth the surface irregularities, as well as reducing the deleterious effects of oxidizing a heavily doped surface. Therefore, a CVD oxide is often used above a thermal oxide when low conductance is needed. Use of this oxide is also often more compatible with the remainder of the integrated-circuit fabrication process than is thermal oxidation.

5.0 CONDUCTION

Polycrystalline-silicon (polysilicon) films formed by chemical vapor deposition are used in a wide range of different VLSI applications requiring very different electrical properties. Very high-value load resistors for static random-access-memory (RAM) cells utilize the high resistance of lightly doped polysilicon to provide a technologically convenient and stable resistor that limits the current flowing in the cell. For the gate electrode and interconnections of MOS integrated circuits, the lowest resistance possible is needed, and a resistivity less than 10^{-3} Ω-cm—eight orders of magnitude less than for static RAM loads—is routinely obtained for this application. However, this lower bound on the resistivity of polysilicon has become a limitation on the circuit performance of silicon-gate integrated circuits which use polysilicon interconnections to conduct signals long distances across a chip (44). As feature sizes become smaller and intrinsic device delays decrease on chips of increasing overall dimensions, the resistance of polysilicon interconnections is becoming a more serious limitation on integrated-circuit performance.

In this section we will examine the wide range of

electrical properties that can be achieved in CVD polysilicon films and show that the high resistance of lightly and moderately doped polysilicon is related to the grain boundaries. Methods of modifying the electrical properties by controlling the number of grain boundaries or their electrical activity will be examined. We will also see that the resistivity of highly doped polysilicon films is limited by the amount of dopant that can be incorporated into substitutional sites in the polysilicon grains.

5.1 Lightly and Moderately Doped Films

For static RAMs and other devices in which high resistivities are needed, the absolute value of the resistivity and its control are important. At low dopant concentrations the resistivity of polysilicon can be as much as six orders of magnitude greater than that of similarly doped single-crystal silicon, as shown in Figure 23. In this range the resistivity approaches 10^6 Ω-cm and depends only weakly on the average dopant concentration in the film, while the resistivity decreases rapidly at intermediate dopant concentrations to approach that of single-crystal silicon. Control of the value of a high-value resistor is, therefore, best in very lightly doped material, while achieving accurate values of resistance in the intermediate dopant concentration range is difficult.

The large resistivity difference between polysilicon and single-crystal silicon has been observed in material deposited over a wide range of conditions, with deposition temperatures varying from less than 600°C to more than 1100°C. The dopant has been added during deposition in some cases while it has been introduced by ion implantation or diffusion in other experiments, with similar results. Two mechanisms, both involving grain boundaries, cause the electrical behavior of polysilicon to depart from that of single-crystal silicon at low and moderate dopant concentrations:

(1) Dopant atoms physically segregate at grain boundaries because of their lower energy at the disordered structure there.

(2) Carriers contributed by dopant atoms substitutionally incorporated into the grains are trapped in deep levels at the grain boundaries.

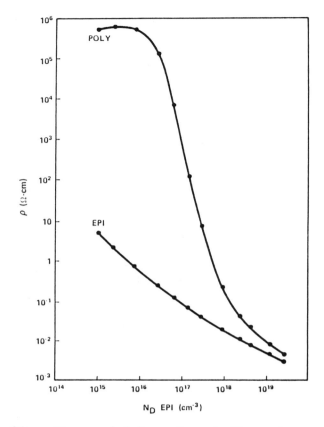

Figure 23. The resistivity of polysilicon is much greater than that of similarly doped, single-crystal, epitaxial silicon. It changes slowly at low dopant concentrations, but decreases rapidly at intermediate dopant concentrations and approaches the resistivity of single-crystal silicon at high dopant concentrations.

Dopant Segregation at Grain Boundaries: At a given temperature the dopant atoms are distributed between the low-energy positions at grain boundaries, where they are electrically inactive, and substitutional positions within the grains, where they can contribute to the conduction process. This grain-boundary segregation is especially important for the n-type dopants phosphorus and arsenic (45),(46), with some evidence for the segregation of antimony (47). Grain-boundary segregation does not appear to be important for boron.

The dopant atoms can move reversibly between the grains and the grain boundaries during successive heat cycles at different temperatures, with consequent reversible changes in resistivity (38). For those dopant species that do segregate to grain boundaries, a greater fraction of the dopant atoms segregate at lower temperatures if the process is allowed to reach equilibrium. As device processing temperatures decrease to the 800-900°C temperature range, especially in the latter stages of the device-fabrication process, grain-boundary segregation and the consequent increase in resistivity becomes of more concern. The use of rapid thermal processing, which employs high temperatures for limited times, may reverse the trend toward more severe dopant segregation at grain boundaries.

The amount of dopant segregation, and consequently the film resistivity, is determined not only by the final processing temperature, but also by the time at this temperature. At lower annealing temperatures (below about 700°C) the time required to reach equilibrium can be much greater than typical processing times, and dopant loss to grain boundaries during common low-temperature annealing cycles is less severe than expected solely from equilibrium considerations.

Although much of the evidence for grain-boundary segregation is inferred from electrical measurements, advanced analytical techniques have recently allowed direct observation of dopant accumulation at grain boundaries. Higher concentrations of arsenic are seen at grain boundaries than within the grains (46),(49), providing direct evidence for dopant segregation there. Phosphorus is also found to segregate to grain boundaries, with more segregation occurring after lower-temperature heat cycles (50), as was previously inferred from electrical measurements. In addition to dopant motion to the grain boundaries, more subtle interactions have also been proposed; for example, it has been suggested that arsenic at the grain boundaries enhances scattering there and possibly causes grain-boundary reconstruction as it segregates (46). The latter effect would not be reversible, but its influence on the electrical properties has not been conclusively observed. In addition to the dopants, other impurities have directly been observed to segregate to grain boundaries in polysilicon (51).

Carrier Trapping at Grain Boundaries: To predict the number of free carriers in polysilicon films, the substitutional dopant concentration must first be found by subtracting from the total dopant concentration the number of dopant atoms segregated at the grain boundaries. Even after this loss is considered, however, the conductivity of polysilicon is still much less than that of single-crystal silicon containing the same concentration of substitutional dopant atoms. Free carriers (either electrons or holes) are contributed to the conduction or valence band by substitutional dopant atoms located within the grains. However, many of these free carriers are quickly immobilized or trapped at low-energy positions at the grain boundaries and, consequently, cannot contribute to conduction (52),(53).

The charge trapped at the grain boundaries is compensated by oppositely charged depletion regions surrounding the grain boundaries. From Poisson's equation, the charge in the depletion regions causes curvature in the energy bands, leading to an energy barrier of height qV_B (as shown in Figure 24), which impedes the movement of any remaining free carriers from one grain to another (52),(53). The barrier height depends strongly on the substitutional dopant concentration and the trap density and is similar for n-type and p-type dopants. At low dopant concentrations virtually all the carriers contributed to the conduction or valence band by the substitutional dopant atoms are trapped at grain boundaries; very few are free to contribute to conduction; and the resistivity approaches that of intrinsic silicon. Because of the low dopant concentration, the compensating depletion regions surrounding the grain boundaries extend completely through the grains (Figure 24a), and no neutral regions remain. Because of the low dopant concentration, the energy bands within the grains have little curvature; the barriers to conduction are small; and any free carriers present can readily pass from one grain to another.

As the dopant concentration increases, more carriers are trapped at the grain boundaries, and the curvature of the energy bands increases (Figure 24b), raising the height of the potential barriers and making carrier transport from one grain to another more difficult. In the simplest formulation of the carrier-trapping model, the grain-boundary traps are completely filled when the dopant concentration exceeds a critical value. Neutral regions form within the grains as further dopant atoms are added, and the barriers to carrier transport between grains decrease with increasing dopant concentration.

Carriers often travel from one grain to another primarily by thermionic emission over the potential barriers. In some ranges of dopant concentration, other conduction

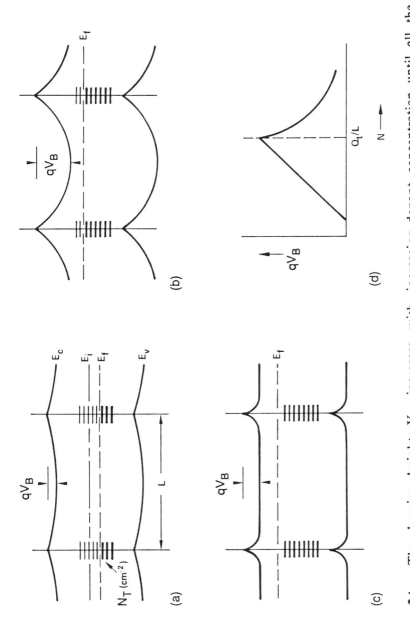

Figure 24. The barrier height V_B increases with increasing dopant concentration until all the traps are filled. Above the critical dopant concentration $N^* = N_T/L$, it decreases as neutral regions form within the grains.

mechanisms, such as tunneling, can also contribute substantially to the current flow. When thermionic emission dominates, the conduction process can be understood by considering the effect of an applied voltage on the barrier height. An applied voltage decreases the barrier to carrier transport in one direction, while it increases the barrier in the opposite direction. The total thermionic emission current can be written as (54)

$$J = qnv_c \exp\left(-\frac{qV_B}{kT}\right)\left[\exp\left(\frac{qV_G}{2kT}\right) - \exp\left(-\frac{qV_G}{2kT}\right)\right] \quad (8)$$

or

$$J = 2qnv_c \exp\left(-\frac{qV_B}{kT}\right)\sinh\left(\frac{qV_G}{2kT}\right) \quad (10.9) \quad (9)$$

where n is the free-carrier concentration, v_c is the carrier velocity, V_B is the barrier height, and V_G is the voltage applied across one grain boundary. Current flow in polysilicon films has been found to fit this expression well, although the grain size used to find the voltage applied across one grain boundary is often treated as a variable parameter.

For fine-grain polysilicon at low applied voltages, the voltage drop across each grain boundary is small, and the general expression can be simplified to obtain a linear relation between current and applied voltage; the conductivity of the film can then be written as

$$\sigma = \frac{q^2 nv_c L}{kT}\exp\left(-\frac{qV_B}{kT}\right) \quad (10)$$

The conductivity is a strong function of temperature with an apparent activation energy $E_a = qV_B$, which depends on the dopant concentration and the grain size.

Based on this simple model we can define a critical dopant concentration N^* at which all the grain-boundary traps are filled with carriers from the substitutional dopant atoms (55). This concentration divides the region with high resistivity from that with rapidly decreasing resistivity. It occurs approximately when the substitutional dopant concentration (after subtracting the dopant atoms segregated at grain boundaries) times the grain size equals the trap density Q_T per unit area of grain boundary. Thus, the critical dopant concentration varies approximately inversely with grain size L: $N^* = Q_T/L$. The inverse proportionality between critical dopant concentration and grain size applies over a wide range of grain sizes, varying from 20 nm to over 20 μm (55).

Below the critical dopant concentration the free carrier concentration is low, and the resistivity is high. As the critical dopant concentration is approached, the barrier height increases, the ease of carrier motion from one grain to another (the "effective mobility") decreases, and the resistivity remains high, even though the carrier concentration increases somewhat. The barrier height is maximum at the critical dopant concentration, and the "effective mobility: is minimum. Above the critical dopant concentration the carrier concentration increases and the barrier height decreases; in this range the resistivity decreases rapidly with increasing dopant concentration until it approaches the resistivity of single-crystal silicon.

The basic model presented above describes the current flow in terms of thermionic emission of carriers over the entire height of the depletion-region barrier. However, the barriers are narrower at higher potentials, and tunneling through the top portion of the barriers is likely. Although thermionic emission dominates at higher temperatures, the combination of thermal excitation and tunneling (*thermionic field emission*) can become important at lower temperatures (56). To consider the effect of the grain boundary itself on the carrier movement, a high, but narrow, rectangular potential barrier is added to the curved barrier arising from carrier depletion near the grain boundary, as shown in Figure 25 (57). Thermionic field emission across this type of composite barrier agrees well with experimental data. The narrow, rectangular barrier becomes especially important at high dopant concentrations when the curved, depletion-region barrier diminishes, as we will see in Section 10.5.2. The conduction models show that the conduction depends strongly on the height and thickness of the grain-boundary potential barrier. Detailed measurements on various types of grain boundaries suggest that the potential barrier is larger for high-angle grain boundaries than for low-angle grain boundaries (58). However, because of the detailed structure of the grain boundaries, the potential barrier can vary from one grain boundary to another in the same sample and even along a single grain boundary.

Either type of majority carrier can be trapped at the grain boundaries. The similar behavior of electrons and holes implies that the grain-boundary traps are located near

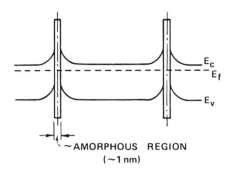

Figure 25. The impedance of the grain boundary is modeled by considering a high, narrow barrier at the grain boundary, in addition to the parabolic, depletion-region barrier.

mid-gap or are distributed symmetrically around mid-gap. More detailed observation shows that both the resistivity and the activation energy of the resistivity decrease monotonically with increasing dopant concentration when boron is added, as shown in Figure 26 (15). However, for n-type dopants, the resistivity and its activation energy first increase somewhat as dopant is added, before reaching a maximum and decreasing. This asymmetry suggests that the dominant traps are located somewhat below mid-gap, pinning the Fermi level there in lightly doped polysilicon. Other studies also indicate that the trap density is highest near mid-gap (59) and that the Fermi level at the grain boundaries appears to be pinned about 0.62 eV below the conduction band edge (60), consistent with the observation of an electronic state related to dangling silicon bonds about 0.65 eV below the conduction band edge (61). (Dangling bonds are related to trivalent silicon, which has been found at grain boundaries in polysilicon (62).

The simplest carrier-trapping model suggests that a discrete level exists at the grain boundary and that the Fermi level is pinned close to this level until all the carrier traps there are filled as more dopant is added. These assumptions lead to a well-defined maximum in the barrier height as the dopant concentration increases. Because of the varying properties of grain boundaries, however, the traps can be distributed in energy within the bandgap, allowing the number of carriers trapped to continue increasing as the Fermi level moves through the distribution. In this case, the dopant concentration at which the traps are completely filled

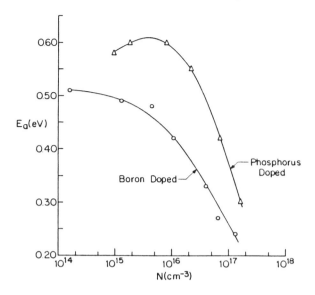

Figure 26. Dopant-concentration dependence of the activation energy of the conductivity in thick polysilicon films deposited in an atmospheric-pressure reactor.

need not coincide with the formation of a neutral region within the interior of the grain (63). A neutral region may exist within a grain even though the traps are not completely filled, and there may be a range of dopant concentrations over which the barrier height varies only weakly with dopant concentration.

The barrier properties not only depend on the physical nature of the grain boundary; they can also be modulated by the electrical bias applied and the current flowing. If minority carriers traveling across the grain boundary fall into grain-boundary traps, they can compensate some of the trapping levels, reducing the barrier height, with a subsequent increase in the conductivity (64). Excess majority carriers can also reduce the barrier height. They first fill the traps; further majority carriers remain in the grains to compensate some of the depletion-region charge and reduce the barrier height. In the extreme case, carrier injection can lead to switching action as the current increases, and a bistable device may be obtained (65),(66),(67). The dynamics of charge trapping and emission can also lead to an anomalous capacitive current (68).

While we have considered the dopant-segregation and

carrier-trapping models independently, interactions may occur. For example, dopant atoms segregating to the grain boundaries may change the density of carrier traps there; the conduction decreases because dopant atoms are lost from substitutional sites within the grains, but increase because of reduced carrier trapping at the grain boundaries (69). For example, phosphorus segregation appears to decrease the trap density and lowers the barrier height (70). However, dopant segregation usually dominates, and modification of the carrier traps is secondary. In addition to barriers at the grain boundaries, the conduction is limited by defects within the grains themselves. As the polysilicon is annealed, the grain size increases, and the concentration of defects within the grains decreases; both effects lower the resistivity.

In this section we have seen that three interrelated effects lead to the marked differences between the electrical behavior of polysilicon and single-crystal silicon: First, some dopant species physically segregate to the grain boundaries, where they do not contribute to the conduction. Second, some carriers provided by substitutional dopant atoms within the grains are immobilized by carrier traps at the grain boundaries and are lost to the conduction process. Third, potential barriers at the grain boundaries impede the movement of the remaining free carriers from one grain to another.

Conductivity Control and Grain-Boundary Modification: As we have seen, above the critical dopant concentration N^*, the resistivity of polysilicon decreases rapidly with increasing dopant concentration, and precise resistivity control is difficult. Polysilicon doped in this intermediate range is widely used for high-value load resistors in static RAMs. Although static RAMs can function with a wide range of load resistances, as device feature size decreases, noise and radiation-induced charge limit the highest value of resistance which can be tolerated, and the increasing number of devices on a chip places a lower limit on the resistance which can be used without excessive power dissipation. Therefore, for better control, reduced sensitivity of the resistivity is needed at intermediate dopant concentrations.

Because grain boundaries strongly influence the electrical properties of polysilicon, attempts have been made to modify the grain boundaries themselves or to alter their effect on the electrical behavior. As will be discussed further in Section 6.2, one approach requires the melting and subsequent solidification of fine-grain polysilicon into material with few or no grain boundaries (71). Because most of the grain boundaries are removed, the resistivity depends much less strongly on the dopant concentration, and the value of resistors is less sensitive to slight variations in the number of dopant atoms. For example, for a resistivity in the intermediate range, melting with a laser can reduce the resistivity variation from a factor of ten to a factor of two for a 30% change in the implanted dose (72). The resistivity of lightly doped polysilicon films also changes rapidly with temperature, varying exponentially with reciprocal temperature. This sensitivity can be reduced by compensating the dopant in phosphorus-doped polysilicon with boron to reduce the activation energy E_a (73).

Another approach to improving the electrical properties of polysilicon films attempts to remove the deleterious effects of the grain boundaries, rather than removing the grain boundaries themselves. Grain boundaries in polysilicon arise from misorientation of the adjacent grains during deposition. Dangling bonds associated with the lattice discontinuities at the grain boundaries lead to the trapping states there. Active hydrogen atoms are often used to terminate dangling silicon bonds at Si-SiO_2 interfaces and can be used similarly at grain boundaries in polysilicon films to complete many of the unsatisfied bonds which cause the grain-boundary traps. Deep-level transient spectroscopy (DLTS) on passivated and unpassivated grain boundaries confirms that this hydrogen *passivation* of the grain boundaries reduces the number of effective trapping states (74). As the number of trapped carriers decreases, the potential barrier associated with the grain boundary also decreases, as indicated in Figure 27 (75). The resistivity of moderately doped polycrystalline silicon is markedly reduced by introducing hydrogen (76),(77).

Hydrogen can be added to polysilicon by several different techniques. Although the active, atomic hydrogen needed to passivate the dangling bonds at the grain boundaries cannot be readily obtained by annealing in molecular hydrogen, a hydrogen plasma dissociates molecular hydrogen, creating atomic hydrogen, which then diffuses into the grain boundaries and passivates the grain-boundary traps (76). Hydrogen can also be added effectively by diffusing it from a hydrogen-containing, plasma-deposited nitride layer above the polysilicon (78). The hydrogen in these layers moves readily and can affect the threshold voltage of standard MOS transistors in bulk silicon integrated circuits, in which it is often used for final device passivation. In polysilicon

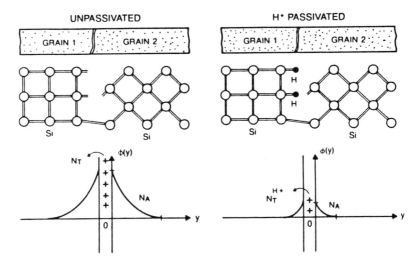

Figure 27. As the dangling bonds at the grain boundary are "passivated" with hydrogen, the depletion barrier surrounding the boundary decreases. From (75). Copyright 1985 IEEE. Reprinted with permission.

transistors, however, this active, mobile hydrogen can be used advantageously to passivate the grain-boundary defects and improve the electrical characteristics of polysilicon. Active hydrogen can also be provided by implanting hydrogen into the polysilicon film and annealing at a low temperature to move it to the grain boundaries (77). The hydrogen appears to be relatively stable at room temperature, but evolves from the structure in the 400-500°C temperature range. Therefore, it must be added after all high-temperature heat treatments are complete. In addition to elements such as hydrogen, lithium can also passivate grain boundaries (79) and neutralize recombination centers (80).

5.2 Heavily Doped Films

The major use of polysilicon films in VLSI circuits is as the gate electrode and one level of interconnections in MOS integrated circuits. The material used for interconnections must have extremely low resistivity to increase circuit speed and avoid voltage drops along long lines; obtaining the lowest possible resistivity in polysilicon is critical for circuit operation. However, as more dopant atoms are added to polysilicon, the resistivity decreases to a limiting value and then remains constant (81). The limiting resistivities are about 400 $\mu\Omega$-cm in phosphorus-doped polysilicon and about 2000 $\mu\Omega$-cm in arsenic- or boron-doped polysilicon.

As in single-crystal silicon, the amount of dopant that can be actively incorporated into polysilicon grains appears to be limited by the solid solubility of the dopant species (82), with the same solid solubility as in single-crystal silicon (83). Variations between different dopant species are consistent with this interpretation (82); the resistivity of both single-crystal silicon and polysilicon is greater when the material is doped with arsenic than when it is doped with phosphorus (82), corresponding to the higher solid solubility of phosphorus in silicon (84). The resistivity of polysilicon appears to be about a factor of two higher than that of correspondingly doped single-crystal silicon in both cases. Consistent with this observation, independent Hall-mobility measurements show that the mobility in very highly phosphorus-doped polycrystalline silicon is about half that in single-crystal silicon (about 30 cm^2/V-s compared to 60 cm^2/V-s) (85).

At high dopant concentrations, the barrier to conduction created by the depletion regions surrounding the grain boundaries decreases, as expected from Poisson's equation, and other mechanisms limit the mobility. The high, narrow, rectangular grain-boundary barrier (56),(57) discussed in Section 5.1 can account for this reduced mobility. The need for carriers to tunnel through this additional barrier limits the mobility and conductivity of heavily doped polysilicon. A barrier thickness of about 1 nm and a barrier height corresponding to that of amorphous silicon are consistent with the experimental data (57),(56). Differences in the electrical properties are also seen for different n-type dopant species. For example, the mobility is lower for arsenic than for phosphorus (82),(87), because of the greater ionized-impurity scattering and grain-boundary scattering for arsenic (84).

Although dopant atoms can be physically added to polysilicon at concentrations above solid solubility, these additional atoms are not electrically active. For at least phosphorus, the excess dopant atoms can precipitate at grain boundaries and may seriously degrade the integrated circuit, as discussed in Section 4.1.

After doping, further processing can change the resistivity because of the increased grain-boundary segregation and the lower solubility of the common dopants in crystalline silicon at lower temperatures. Thus, the trend toward lower processing temperatures for VLSI fabrication limits the conductivity that can be obtained in polysilicon. Using short, high-temperature heat cycles near the end of the device-processing sequence can minimize these effects and improve the conductivity (88). Because grain boundaries and other defects limit the mobility even in heavily doped polysilicon, the mobility can be improved somewhat by reducing the number of grain boundaries, but the improvement in mobility is limited to approximately a factor of two.

Although the limited conductivity of polysilicon is becoming an increasingly severe constraint on the performance of small-geometry, high-performance integrated circuits, polysilicon will most likely be augmented by more highly conducting layers, such as the refractory metal silicides or the refractory metals themselves. Retaining the superior electrical properties and excellent stability of transistors with polysilicon directly over the gate insulator is strong motivation for augmenting, rather than replacing, polysilicon in the near future.

5.3 Summary

Grain-boundary effects cause the electrical properties of polycrystalline silicon to differ from those of single-crystal silicon. At low and moderate dopant concentrations, dopant segregation to and carrier trapping at grain boundaries reduces the conductivity of polysilicon markedly compared to that of similarly doped single-crystal silicon. Because the properties of moderately doped polysilicon are limited by grain boundaries, modifying the carrier traps at the grain boundaries by introducing hydrogen to saturate dangling bonds improves the conductivity of polysilicon. Removing the grain boundaries by melting and recrystallization improves the properties further. When polysilicon is used as an interconnecting layer in integrated circuits, its limited conductivity can degrade circuit performance. At high dopant concentrations, the active carrier concentration is limited by the solid solubility of the dopant species in crystalline silicon.

6.0 APPLICATIONS

The development of polysilicon technology was driven by the use of polysilicon as a gate electrode for integrated circuits. However, once it was developed, polysilicon technology found use in an increasing number of applications, not only in MOS integrated circuits, but in bipolar technology as well. In this section we will look at some of the present and potential applications of polysilicon in integrated-circuit technology.

6.1 Silicon-Gate MOS Technology

In the mid-1960s most gate electrodes were made from aluminum, which was deposited after the source and drain regions were doped. The aluminum had to overlap onto these regions to insure that a continuous channel from source to drain was formed when the gate was biased to turn on the transistor. The required alignment tolerance caused a significant Miller feedback capacitance between the gate and the drain, which slowed circuit speed.

By using a material which allows the source and drain regions to be self-aligned to the gate electrode, as shown in Figure 28, this capacitance can be markedly reduced (Figure 28). Providing self-aligned interconnections not only reduces the Miller capacitance, but it also makes this capacitance more uniform from one device to another; this device matching is often critical for improved circuit performance. The material used for such a self-aligned structure needs to be compatible with the high temperatures required for the source and drain diffusions and other high-temperature processing steps which follow gate-electrode formation.

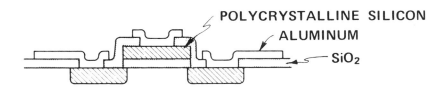

Figure 28. In a silicon-gate transistor, the gate electrode can be used to define the location of the source and drain dopant atoms, thereby eliminating the need for a large overlap of the gate electrode. This self-aligned structure reduces the overlap capacitance.

Although other high-temperature materials, such as tungsten and molybdenum were investigated, the compatibility of polycrystalline silicon with integrated-circuit fabrication led to its rapid adoption (89). In the initial attempts to use thin films of silicon for gate electrodes, the silicon layers were deposited by evaporation. However, these evaporated films could not adequately cover the steps on the integrated-circuit surface between the gate oxide and the field oxide. At the time when polysilicon technology was being developed, formation of epitaxial silicon by chemical vapor deposition was widely used in bipolar integrated circuits, and it was recognized that polysilicon could be deposited by similar techniques if the substrate was an oxide-coated surface, rather than single-crystal silicon. Because of the readily available technology and the conformal step coverage of films produced by chemical vapor deposition, this method of forming polysilicon layers was quickly adopted.

One of the terms entering into the threshold-voltage expression of an MOS transistor is the difference between the work functions of the gate and the silicon (*ie*, the *"metal"-semiconductor work-function difference* Φ_{MS}). When polysilicon is used as the gate electrode, the metal-semiconductor work-function difference depends on the dopant concentration in the polysilicon, as well as that in the silicon substrate. For maximum conductivity, the polysilicon is very heavily doped, and the Fermi level is generally assumed to be very near the edge of the conduction or valence band for *n*-type or *p*-type polysilicon, respectively. Thus, the threshold voltage changes by the band-gap of silicon (1.1 eV) when the dopant type in the polysilicon gate is changed. For the highest conductivity, the polysilicon is usually *n*-type; however, in CMOS circuits, it may be more convenient to dope the gate of the *p*-channel transistor *p*-type, especially if the dopant is added during the source-drain implantation. (In this case, a silicide or other low-resistance conducting material is usually placed over the polysilicon gate to lower the interconnection resistance and short the p^+n^+ diode that would otherwise form between the gates of *p*-channel and *n*-channel transistors.) Choosing the correct dopant type for the gate electrode can also produce the desired threshold voltage without using complex dopant profiles in the single-crystal silicon, which can degrade transistor operation (*eg*, by reducing the carrier mobility).

When the gate electrode is doped during the source-drain implantation only, the dopant concentration may be lower than when a separate doping cycle is used, and the Fermi level may not be at the band edge. The flat-band voltage is usually comparable to that in devices with the polysilicon doped from the vapor phase (90), but more detailed investigation does reveal some differences (91),(2).

Not only does using polysilicon gates allow self-aligning the source and drain regions to the gate electrode, but the polysilicon can also be used as an additional partial level of interconnections. In addition, placing the polysilicon interconnections over thick field oxide reduces the capacitance below that of diffused interconnections. The additional level of interconnections provided by polysilicon cannot be used as effectively as other levels, however, because the processing sequence of a typical silicon-gate integrated circuit prevents forming a heavily doped region of single-crystal silicon under the polysilicon line. However, even with this limitation, polysilicon interconnections provide additional flexibility in integrated-circuit layout. As integrated-circuits complexity increased, interconnecting devices became more critical, and the additional flexibility provided by polysilicon interconnections was of increasing benefit.

The ability of polysilicon to withstand subsequent high-temperature processing also allowed development of more reliable integrated-circuit fabrication processes. Coverage of surface steps by evaporated aluminum interconnections has always limited yield and reliability. When polysilicon gates with their ability to withstand high temperatures are used, a phosphorus-containing, silicon-dioxide layer can be deposited over the polysilicon and heated until it becomes viscous and "flows," reducing the abruptness of the steps which the aluminum must cover. This *phosphosilicate glass (PSG)* can be flowed before the contact windows are opened through the oxide to the underlying heavily doped source and drain regions. It can also be flowed ("reflowed") after the windows are opened to provide a more easily covered step at the edges of contact regions.

In addition to its applications in normal silicon-gate structures, the high-temperature capability of polysilicon allowed development of more complex structures, which provided denser integrated circuits even with the same minimum feature size. For example, the size of the one-transistor, dynamic random-access-memory cell can be reduced significantly by using two different layers of polysilicon—one as the gate of the access transistor and one as the counterelectrode of the storage capacitor. Figure 29a

(93) shows that the single-layer-polysilicon, dynamic RAM cell requires space between the gate and capacitor counterelectrodes. The space between these two polysilicon features is filled with a diffused region to prevent a barrier between the inversion layers under the electrodes and to allow charge transfer to the storage region when the access transistor is turned on. When two levels of polysilicon are used (Figure 29b), the access transistor and the storage region are separated only by the thickness of the oxide grown on the side of the first level of polysilicon, rather than by the minimum horizontal feature size. Thus, the cell size is reduced appreciably, but at the expense of more complex processing. Although the electrodes are separated by the finite thickness of the oxide grown on the side of the first layer of polysilicon, fringing fields reduce the barrier between the two electrodes when both are biased to induce inversion layers in the substrate. Charge-coupled devices (CCDs) also use closely spaced, independent polysilicon electrodes above single-crystal silicon to allow charge transfer between adjacent inversion layers in a compact device structure.

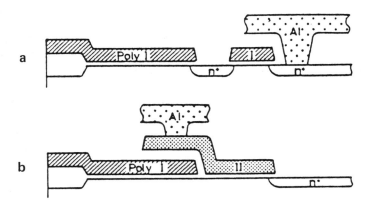

Figure 29. Dynamic RAM cells achieved with (a) a single layer of polysilicon and (b) two layers of polysilicon. Using two layers eliminates the need for the lateral separation between the gate and storage electrodes and allows a more compact cell. From (93). Copyright 1985 IEEE. Reprinted with permission.

Fabrication of device structures with two levels of polysilicon often takes advantage of the difference in the oxide thickness grown on lightly doped single-crystal silicon and on heavily doped polysilicon, as discussed in Section 4.1. The oxide under the second level of polysilicon serves as the capacitor dielectric for the storage cell and should be as thin as possible for maximum charge storage. On the other hand, the oxide between the two levels of polysilicon should be as thick as possible to reduce unwanted parasitic capacitance between these two electrodes and increase the breakdown voltage. The more rapid oxidation rate of heavily doped polysilicon compared to lightly doped single-crystal silicon provides the desired difference in oxide thickness. Oxidation conditions can be chosen to maximum this difference, consistent with other limitations on processing conditions.

Gettering: In some processes, the high-temperature capability of polysilicon also allows efficient gettering of heavy metals away from critical device regions. The single-crystal silicon on the back of the wafer can be exposed by removing oxide layers from the back before the polysilicon on the front of the wafer is heavily doped with phosphorus from the gas-phase so that phosphorus is also added to the back of the single-crystal wafer. The phosphorus-containing region on the back of the wafer can attract heavy metals during all subsequent high-temperature device processing. This gettering can be important in providing highly reliable integrated circuits in high-volume production. For example, removing the metals from the region under the storage capacitor of a dynamic RAM reduces the generation-recombination centers there, and increases the time allowed between refresh cycles. However, backside gettering is becoming less important today as integrated-circuit fabrication temperatures are reduced and thicker wafers are used. Alternate gettering techniques are being developed which do not rely on diffusion of heavy metals through the entire thickness of the silicon wafer.

EEPROMs: Using two layers of polysilicon can also be valuable in other devices. For example, a substantial amount of electrically erasable, programmable read-only memory (EEPROM) is frequently included on a microprocessor or other computer chip to allow storage of code specific to a particular application and other information that must be infrequently modified. This type of nonvolatile memory element requires one level of polysilicon above another in the gate region of the device. The first level is not electrically connected and "floats". The charge stored in this *floating gate* determines the condition of the underlying single-crystal silicon between the source and drain regions. If it contains excess electrons, it attracts holes and turns off an *n*-channel device. The opposite charge creates an inversion layer in the silicon and allows current to flow between source and drain. Application of a voltage of the appropriate potential allows charge to be transferred through the gate oxide to charge the floating region and "write" a cell of this programmable read-only memory. The charge may be generated by avalanching the drain-substrate junction to create free carriers, which are then attracted to or repelled from the floating gate by the potential on the overlying *control gate*. The charge can be removed from the floating gate by tunneling through the oxide separating the two layers of polysilicon. If this method of erasing the cell is used, the surface texture of the lower layer of polysilicon is critical. We have already seen that the conduction through oxide grown on polysilicon depends strongly on the asperities on the polysilicon surface. Control of the deposition conditions and the resulting structure of the polysilicon is more critical for this application than probably for any other.

High-Value Resistors: After polysilicon technology was developed for silicon-gate structures, it quickly found use in other applications, such as static RAMs. A typical static RAM circuit uses a cross-coupled structure that allows current to flow in one or other other of two parallel current paths. This static RAM cell can be formed by using two transistors in each leg of the circuit and two access transistors, one connected to the intermediate note of each leg of the circuit, forming a 6-transistor cell. To conserve area, the two load transistors can be replaced by high-value resistors to obtain a 4-transistor, 2-resistor cell (94). The requirements on the resistance are not severe. It must be low enough to provide adequate current to retain charge on the intermediate node of each leg of the circuit and prevent noise from causing the cross-coupled cell to change state; it must be high enough to limit the current flowing through the driver transistor that is turned on so that the power dissipation remains acceptable. Resistors in the range of 10^9 Ω are generally satisfactory.

Using lightly doped polysilicon for this application is attractive; it is compatible with standard integrated-circuit processing and can even be placed above an active element on the circuit to reduce the cell size. It only needs to be shielded from dopant introduction during doping of the polysilicon gate or the source and drain regions. As we saw in Figure 23, undoped polysilicon has a resistivity in the mid-10^5 Ω-cm range. With a typical thickness of a few hundred nanometers, a sheet resistance of about 10^{10} $\Omega/\square$ is easily obtained. In some cases, this value is satisfactory; more frequently, however, a somewhat lower resistance is needed to insure that adequate current is available to avoid stray charge from switching the state of the cell. A small amount of dopant can readily be added by ion implantation. Although control of the resistivity in the lightly doped region is difficult (Figure 23), the wide range of resistance allowed in this circuit makes lightly doped polysilicon acceptable.

Because hydrogen can change the resistance of high-value resistors by saturating dangling bonds at grain boundaries, the effect of hydrogen during device fabrication and operation is of concern. Moderately doped films are most sensitive to the barrier heights at grain boundaries, and are affected most strongly by the addition of hydrogen. Because the high-valued polysilicon resistors used in static RAMs are often doped in this range, the value of the resistors can be inadvertently changed by hydrogen introduced during the later stages of the fabrication cycle or during operation. Hydrogen from a plasma nitride passivation layer is of special concern; as we have seen, hydrogen from this source effectively passivates dangling bonds in polysilicon, and it can move during device operation, changing the value of the resistor. If circuit margins are not adequate, these changes pose a severe, long-term reliability problem.

Although polysilicon load resistors are adequate for many static RAM cells, improved circuit performance is obtained if the element connected to the transistor being turned off can conduct enough current to charge the intermediate node rapidly, and a voltage-variable load element is advantageous. If the perpendicular electric field provided by a gate electrode can modulate the conductivity of a polysilicon resistor to even a moderate degree, the circuit performance of the static RAM can be improved. To obtain this load element and simultaneously increase circuit density, the polysilicon layer can be placed above a thin oxide on the gate electrode of the bulk MOS transistor so that one gate electrode can control both the bulk transistor and the voltage-variable polysilicon resistor. As an extension of this concept, the possibility of constructing an MOS transistor with its *channel* in a layer of polysilicon has been explored for this and other applications.

6.2 MOS Transistors in Polysilicon

Operation of conventional integrated circuits requires compromises to avoid interactions between nearby elements. CMOS devices suffer from latch-up between neighboring n-channel and p-channel transistors; capacitive coupling to the high-relative-permittivity silicon can degrade circuit performance. Charge generation in the underlying silicon can reduce the radiation hardness of circuits in bulk silicon; charge injected at one node may travel to another node through the substrate, changing the charge state of dynamic memory cells. More conceptually, restricting integrated circuits to a single plane of devices at the surface of a bulk silicon wafer limits the type of devices and circuits possible. The ability to use the third dimension would allow fabricating novel device structures and should also reduce the interconnection length between devices.

The concept of placing the active layer of an MOS transistor within a thin layer of silicon has been explored for a number of years. MOS transistors with their conducting channels in polysilicon were fabricated early in the development of MOS integrated circuits (95),(96). Figure 30 shows a p-channel transistor with its channel in a layer of polysilicon and the corresponding drain characteristics. Although the shape of the characteristics is similar to that of bulk MOS transistors, a large gate voltage must be applied before significant drain current flows, and the transistor has a very low transconductance. This inefficiency of the gate voltage in inducing a conducting channel can be attributed to the high concentration of allowed trapping states within the forbidden gap of polysilicon films, especially at the grain boundaries. As a gate voltage is applied, the energy bands must be bent to induce a conducting inversion layer near the silicon surface. Before a conducting channel can be induced in a material containing a high concentration of traps, the energy levels corresponding to many of the traps must be moved through the Fermi level (96); that is, the charge state of many of the traps must be changed. Much of the applied gate voltage is used to charge or discharge trapping levels, rather than inducing conducting free carriers in an inversion layer, and excessively high gate voltages must be applied before conduction can occur between source and drain. Thus, the characteristics of transistors built in moderately doped polysilicon films are dominated by the behavior of the traps, as was the resistivity of polysilicon.

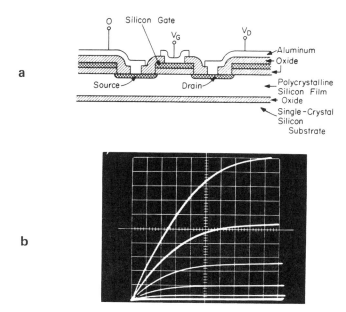

Figure 30. (a) Cross section of an MOS transistor with its *channel* within a polysilicon film and second polysilicon layer forming the gate electrode. (b) Corresponding drain characteristics of a *p-channel transistor:* I_d = -50 µA/div, V_d = -2V/div; Vg = -2.5V/step, 8 steps.

Because most of the traps are located near the grain boundaries, the transistor can be visualized as several transistors in series. Comparatively high-quality grains are separated from each other by highly defective grain boundaries. The regions above the central portions of the grains have a threshold voltage determined by the normal transistor equations for single-crystal silicon; the regions above the grain boundaries have a much higher threshold voltage. A moderate gate voltage bends the bands and induces conducting channels over the central regions of the grains, but barriers to conduction still exist at the surface near the grain boundaries, and no continuous conducting channel is formed. Conduction is not obtained until a considerably higher gate voltage is applied to change the charge state of the traps at the grain boundaries near the surface of the polysilicon and induce conducting channels where the grain boundaries intersect the surface. Consequently, the observed threshold voltage is much higher than expected from the average dopant concentration.

Even when a continuous conducting channel has been induced, the transistor characteristics are still influenced by the number of trapping states near the surface of the silicon. If the number of states that must be moved through the Fermi level to change the surface potential slightly and increase the drain current is small, the current can increase rapidly because the effective channel length of the current-limiting region is very short (only above the grain boundaries; the remaining regions are heavily inverted already). On the other hand, if the trap density is high, the charge state of many traps must still be changed as the gate voltage is increased, and the *effective mobility* remains low. The latter case applies in the fine-grain polysilicon normally deposited in integrated-circuit fabrication processes.

More useful transistors can be obtained by employing *accumulation-mode* transistors in place of the more common *inversion-mode* transistors normally used in silicon integrated circuits. In bulk silicon transistors, useful devices can only be obtained by modulating an inversion layer near the surface of the silicon; the conduction of the thick wafer under the gate cannot readily be modulated by changing the gate voltage. In addition to confining the conducting layer, the use of an inversion-mode transistor electrically separates the conducting channel from the common substrate by a depletion region and provides the needed isolation between transistors.

In a thin-film transistor, however, an oxide layer lies immediately beneath the silicon film, and parallel conduction in the bulk of the film can be small if the film is lightly doped. An accumulation layer can be induced near the surface of the film to provide the transistor action, while little current flows in the absence of this accumulation layer (especially if the bulk of the film is depleted of free carriers at zero gate voltage). Generating an accumulation layer requires less bending of the energy bands than does forming an inversion layer, so the applied gate voltage does not need to change the charge state of as many trapping levels within the polysilicon film as in an inversion-mode transistor. Conduction modulation is easier, and the magnitude of the transistor threshold voltage is lower. The characteristics of an accumulation-mode transistor depend on the thickness of the polysilicon film; a thin film can be completely depleted so that little drain current flows at zero gate voltage, while the neutral region in a thicker film can lead to significant "leakage" current.

Even in accumulation-mode transistors in fine-grain

polysilicon, however, the threshold voltage remains too high to be readily compatible with the voltages commonly used in integrated circuits. Although these transistors may be useful in specialized applications, the characteristics of the polysilicon must be modified before transistors in polysilicon can be used in more standard integrated circuits.

Two different approaches can be taken to reduce the effective trap density in polysilicon to levels which allow more useful transistors to be fabricated. Because these trapping states are predominately located at grain boundaries, we must reduce the effectiveness of the trapping states at the grain boundaries or modify the grain boundaries themselves. In the first approach, the defects are *passivated* (*eg*, by adding active hydrogen), as discussed in Section 5.1, so that they do not provide active levels within the forbidden gap to act as traps. A smaller fraction of the applied gate voltage is then used to change the charge state of the traps, and it is more effective in inducing and modulating a conducting channel; the threshold voltage decreases, and the mobility increases (97),(8). Although their properties do not approach those of transistors in bulk wafers, hydrogen-passivated transistors may be quite satisfactory for selected applications, such as the load elements of static RAMs or the peripheral circuitry for active-matrix-addressed displays.

However, even when the grain boundaries are passivated with hydrogen to the maximum extent practical, the number of active grain-boundary states is still large, and the transistor characteristics are markedly inferior to those of transistors in single-crystal silicon. To achieve transistors of much higher quality, the grain boundaries themselves must be removed or at least reduced in number. This can be accomplished by melting the entire thickness of the finegrain polysilicon film with a scanned heat source, such as a cw laser, an electron beam, a graphite rod, or a high-intensity incoherent light source. When the silicon solidifies, only a relatively small number of grains nucleate, and these grains then grow into the rapidly cooling molten silicon to produce large-grain material containing few grain boundaries, and consequently, few grain-boundary traps. The entire thickness of the fine-grain silicon must be melted so that the solidifying silicon does not regrow on the remaining small grains to again produce a fine-grain structure. A moving heat source is generally used so that the lateral temperature gradient promotes growth of the few nucleating grains; alternatively, the desired lateral thermal gradient can be obtained by locally varying the amount of power absorbed to melt the silicon (*ie*, by locally varying the reflectivity and using a narrow-band optical source) or by locally varying the thermal conductance.

Transistors subsequently fabricated in this recrystallized silicon have properties approaching those of transistors in single-crystal silicon, and are much superior to transistors fabricated in fine-grain polysilicon. Field-effect mobilities are typically about 1/2 to 2/3 those of transistors in bulk silicon wafers (99). These transistors are quite useful in a number of applications, but the few remaining grain boundaries impede their use in VLSI applications. If a grain boundary is parallel to the current-flow direction and intersects the source and drain regions, the source-drain dopant can readily diffuse along the grain boundary during device fabrication (100) to reduce the effective channel length or even cause a short circuit in a small-geometry transistor. If the grain boundary is perpendicular to the current-flow direction, it presents an added impedance to current flow. Large-geometry transistors, contain many grain boundaries, and the transistor characteristics of one device do not differ markedly from those of another. In small-geometry transistors, however, some transistors will contain grain boundaries and others do not. The resulting variation in transistor characteristics cannot be tolerated in high-performance circuits, and the few remaining grain boundaries must be removed.

Two techniques can be used to produce material free of grain boundaries in the active transistor regions. In one approach, windows (*seeding regions*) are periodically opened in the oxide above the single-crystal substrate before the fine-grain polysilicon is deposited. As the molten silicon solidifies during recrystallization, the crystal structure of the substrate propagates into the solidifying material, first vertically, then laterally over the oxide. In this manner, grain-boundary-free, single-crystal silicon is obtained over the insulating oxide. However, the distance that the grain-boundary-free material can propagate laterally is limited, making the technique difficult to control. Alternatively, regions of transparent material of varying thicknesses can be placed above the silicon before recrystallization so that the amount of power absorbed from a narrow-band optical source varies from one region to another. The regions which absorb the least power are coolest and solidify first so that any grain boundaries which subsequently form are located away from these regions. In this manner, the location of defects can be controlled photolithographically, and they can be placed outside of the active device regions (i.e., in the field regions). This technique can be coupled with the "seeding technique" discussed above to control the orientation of the grain-boundary-free material. Transistors subsequently fabricated in these single-crystal films have properties very similar to those of transistors in bulk silicon wafers, although the floating neutral region beneath the channel leads to the so-called *kink-effect* (101) and other, more subtle differences (102),(103).

6.3 Bipolar-Circuit Applications

Although polycrystalline silicon is most frequently used in MOS integrated circuits, it is becoming crucial in high-performance, bipolar integrated circuits also. Its use decreases parasitic resistance and capacitance and also reduces reverse carrier injection from the base into the emitter.

In modern bipolar transistors, the minority-carrier transit time through the thin base region is very short, and the operating speed of the transistor is limited by parasitic elements. The resistance of the *extrinsic base region* (ie, the region between the *intrinsic* base under the emitter and the external base contact) can degrade device performance significantly, as can the capacitance associated with the depletion region between the extrinsic base and the collector. In typical bipolar transistors, the extrinsic base region is heavily doped to reduce its resistance (while increasing the capacitance). However, in conventional structures the extrinsic base region must be separated from the emitter region by an alignment tolerance, and the decrease in base resistance provided by the heavily doped, extrinsic base region is limited; it is also variable because of varying mask alignment. To reduce the need for a trade-off between low base resistance, low capacitance, and ease of mask alignment, a polysilicon layer can be used to provide most of the extrinsic base. The size of the extrinsic-base region in the single-crystal silicon is reduced, and most of the extrinsic-base region is placed in a layer of polysilicon above an oxide, as shown in Figure 31a. The extrinsic-base—collector capacitance is reduced in this structure because a thick layer of low-permittivity oxide separates the two regions over most of the area. In addition, the polysilicon of the extrinsic

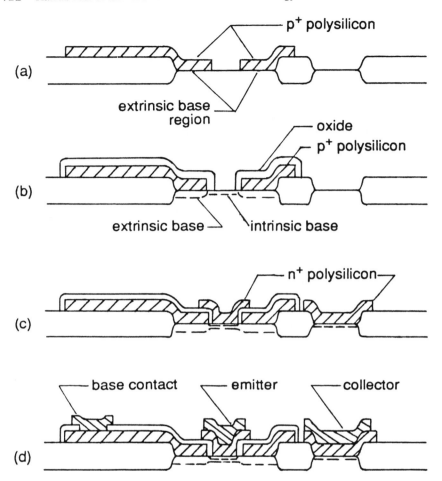

Figure 31. Two layers of polysilicon can be used in the polysilicon-emitter transistor, with an oxide on the sidewall of the first layer separating the two. (a) The extrinsic-base region is formed by the first polysilicon layer which extends over the field oxide. (b) Boron is diffused from the polysilicon into the single-crystal silicon to make contact to the intrinsic base, which can be added by implantation. The polysilicon is oxidized, and the oxide is removed from the single-crystal region which is to become the emitter. (c) A second layer of polysilicon is deposited and defined either before or after adding the emitter doping to the single-crystal silicon. (d) Contact to the base region is formed over the oxide, reducing the area of the single-crystal extrinsic - base region.

base can be used to self-align the emitter region to the extrinsic base so that the two are closely spaced and independent of alignment variations. Thus, using a polysilicon extrinsic-base region greatly improves the performance of bipolar transistors.

Polysilicon can also be advantageously applied to the emitter region of bipolar transistors (Figure 31c). In the most straightforward application, the polysilicon serves as a source for the dopant atoms diffused into the single-crystal silicon to form the emitter. When high concentrations of dopant are added directly to single-crystal silicon by either gas-phase diffusion or by ion implantation, defects can form in the silicon. Because the heat cycles which can be used after emitter doping must be very limited to avoid excessive dopant diffusion, removing this damage adequately is often difficult. If the dopant is added to a polysilicon layer above the single-crystal silicon and then diffused from the polysilicon into the single-crystal silicon, the single-crystal region is not damaged, and a high quality emitter can be formed. Because the dopant diffusivity is much lower in single-crystal silicon than in polysilicon, an emitter formed in this manner can be very shallow.

Using a polysilicon region above the emitter offers other advantages in addition to providing a convenient source of dopant atoms. Such *polysilicon-emitter* bipolar transistors can have much higher gains than do similar transistors in which metal makes direct contact to the single-crystal emitter. One of the factors degrading transistor performance is the injection of holes from the p-type base region into the n-type emitter. This *reverse injection* adds to the base current a component which reduces the *emitter injection efficiency* and, therefore, the gain of the transistor. Placing a layer of polysilicon above the emitter reduces the reverse injection and increases the overall transistor gain. The detailed mechanism for this reduction in reverse hole injection is not fully understood, and several explanations have been proposed.

First, the role of the interfacial oxide between the polysilicon and the single-crystal silicon appears to be critical (104). A thin, residual oxide layer between the polysilicon and the single-crystal silicon may impede hole transfer from the base into the emitter. Alternatively, holes may be injected into the polysilicon to diffuse toward the emitter contact, where they would recombine. Because the polysilicon layer is typically thicker than the single-crystal emitter region, the hole gradient is reduced, and less injection occurs (105). The lower lifetime of minority carriers in polysilicon, however, limits the benefit of which can be gained by this mechanism, suggesting that the role of the polysilicon-single-crystal interface is crucial in obtaining the highest transistor gain. More detailed observations show that the gain enhancement decreases when the thin oxide between the single-crystal silicon and the polysilicon becomes discontinuous. Although the maximum gain enhancement can be obtained with a thin, continuous interfacial oxide, practical use of such a structure is questionable because control of this interfacial oxide is difficult: if it is too thin, the gain is reduced; if it is too thick, the emitter resistance is excessive. However, even without the gain related to the interfacial oxide, using polysilicon in the emitter region offers substantial advantages, and most high-performance bipolar processes use polysilicon-emitter transistors.

An alternate method of reducing minority carrier injection from the base into the emitter is to use a material with a wider bandgap which presents a large barrier to reverse minority-carrier injection. Oxygen-doped polycrystalline silicon (_semi-insulating polycrystalline silicon_ or _SIPOS_) has been used for this purpose (106). Because of the alignment between the bandgaps of single-crystal silicon and SIPOS, electron injection across the heterojunction is not significantly impeded, but the unwanted reverse injection of holes into the emitter is greatly reduced, improving the injection efficiency.

6.4 Polysilicon Diodes

In addition to transistors, other devices can be fabricated in polysilicon films. The simplest device is a lateral p-n junction diode (107),(108). A structure in which both the p-type and the n-type regions are heavily doped can readily be obtained in a CMOS process which employs p^+ and n^+ polysilicon gates for the p-channel and n-channel transistors, respectively. (As mentioned above, these diodes can be undesired parasitic elements between the two types of transistors unless a silicide is formed over the polysilicon or the two gate regions are connected together by other means. However, in other cases, the diode may be used advantageously.) The diode quality factor is typically about 1.8-2 (107), indicating that recombination in the space-charge regions dominates; the breakdown voltage is low because of the heavy doping on both sides of the junction. The breakdown voltage can be increased by reducing the dopant concentration on one side of the junction. However, because of the rapid increase in resistivity of polysilicon with decreasing dopant concentration, the series resistance in the polysilicon diode can rapidly limit the device performance. The low lifetime in polysilicon does, however, reduce minority-carrier storage, and polysilicon diodes are expected to switch rapidly.

6.5 Polysilicon Pressure Sensors

Polycrystalline silicon is also used in pressure sensors. In this application, a thin layer of polysilicon is placed over an oxide, and resistors are formed in the polysilicon. A hole is etched through the substrate from the back, leaving a thin membrane of oxide and polysilicon. When pressure is applied to the membrane, the polysilicon distorts, and the values of the resistors change. In a typical circuit, unstrained resistors are located nearby over the thick silicon wafer for comparison so that small changes in resistance can readily be measured. Because of the grain boundaries, the sensitivity of these polysilicon pressure transducers is only about half that of single-crystal silicon pressure sensors, but their ease of fabrication and isolation from the substrate make polysilicon pressure transducers useful devices.

6.6 Device Isolation

In addition to forming part of the active device structure, polysilicon is also used for isolation between adjacent devices. The oldest application of polysilicon that is still in current production is its use as the supporting layer for the single-crystal islands of dielectrically isolated integrated circuits (109). To fabricate a circuit using this technology, a single-crystal substrate of the appropriate conductivity type (eg, n-type to serve as the collector of a bipolar transistor) is first oxidized, and the oxide is patterned. Isolation moats are etched into the substrate to a depth greater than the thickness of the final silicon pockets. If (100)-oriented silicon is used, the etching terminates when the etched grooves reach the apex of a V formed by the intersection of {111} planes; the depth of the grooves is determined by the width of the oxide removed from these regions. After a heavily doped region, which serves as the buried layer or subcollector, is added and the surface is oxidized, a thick layer of polysilicon is deposited.

Because this layer of polysilicon serves as a mechanical support, it must be approximately as thick as a starting silicon wafer, and high deposition rates are needed. Therefore, the deposition is usually carried out at a high temperature using a chlorinated silicon compound, such as dichlorosilane or silicon tetrachloride. Using a high temperature also reduces wafer deformation (warpage) during deposition by increasing the desorption of gaseous impurities and increasing surface migration. After polysilicon deposition, most of the original single-crystal wafer is removed by mechanically lapping and polishing the wafer until the previously formed grooves appear. The structure is then inverted, and transistors are fabricated in the exposed single-crystal silicon pockets. Because of wafer distortion during the process, the pockets are often displaced substantially from their original location. Registration of devices to the pockets is difficult, and this technique can only be used for low-density circuitry. However, it does provide the excellent isolation required for very-high-voltage or radiation-tolerant integrated circuits. Some of the limitations can be removed by using self-terminating etching techniques (110),(111),(112) so that grooves do not have to be formed before polysilicon deposition. The registration limitation is then removed, but controlling the polysilicon-deposition and silicon-removal processes remains difficult, restricting the technique to specialized applications.

In another isolation technique that also uses polysilicon formed at high temperatures, polysilicon and epitaxial silicon are simultaneously deposited. The substrate is first oxidized, and the oxide is defined to leave it in the regions where polysilicon is to be formed. During subsequent silicon deposition, the material grown on the exposed silicon substrate is epitaxial and that on the oxide is polycrystalline. Using this mixed structure takes advantage of the rapid diffusion of dopant atoms in polysilicon. Dopant can readily diffuse through the entire thickness of the polysilicon regions while only penetrating a small distance into single-crystal silicon. This phenomenon can be used to form p-n junction isolation regions in a bipolar integrated circuit without excessively long diffusion times. The lateral diffusion into the single-crystal regions is also significantly reduced. To further lower the diffusion times required, doped oxides can be used to nucleate the polysilicon, so that diffusion occurs into the polysilicon regions both from the top and from the bottom. A heavily doped *collector sink* can also be formed by this mixed epitaxial-polycrystalline deposition to lower the parasitic collector resistance. Because rapid diffusion is essential for this technique to be useful, the structure of the polysilicon must be carefully controlled to provide columnar grains with the well-defined vertical paths needed for rapid diffusion, as discussed in Section 3.3. In addition, of course, the deposition conditions must be compatible with the simultaneous formation of high-quality epitaxial silicon.

An isolation structure discussed more recently uses the familiar fine-grain polysilicon. In this *trench-isolation* technique (113), a groove is etched in the single-crystal silicon where isolation regions are needed. This groove is formed by anisotropic dry etching so that a narrow, deep "trench" with vertical sidewalls is obtained. The walls of the groove are then oxidized, and polysilicon is deposited to fill the groove completely. The deposition conditions are chosen to be well within the surface-reaction-limited regime so that the silicon-containing gas can readily travel down the entire length of the narrow groove. Otherwise, filling can be incomplete, leaving a void which produces a mechanically unstable structure. If the deposition on the walls of the groove is well controlled, the groove is completely filled once a layer half as thick as the width of the groove is deposited. A polysilicon layer of about the same thickness is also deposited on the oxide-covered top surface of the wafer, and this polysilicon is then removed without using a mask by anisotropic dry etching until the underlying oxide surface is exposed. After the oxide is removed from the single-crystal regions and a protective oxide is formed over the tops of the polysilicon-filled grooves, transistors are fabricated in the single-crystal regions. This technique can be used for either bipolar or MOS integrated circuits. It is very useful in dense, small-geometry CMOS integrated circuits for suppressing latchup by increasing the effective path length between adjacent devices. It can also be used to form the storage capacitor of a one-transistor, dynamic RAM cell so that a large effective storage area is obtained without occupying much of the wafer surface (114). In a further refinement, both the access transistor and storage capacitor can be placed on the sidewalls of an etched trench (115).

As a modification of the trench-isolation technique, a selective deposition process can be employed in which silicon is deposited in the grooves, but not on the top surface of the wafer (116). In this case, the walls of the groove are covered with oxide, but the oxide is removed from the bottom of the groove; an isolation diffusion can be added to the exposed silicon at the bottom of the groove, if desired. Silicon is then deposited selectively on the exposed silicon at the bottom of the groove, and the groove is filled from the bottom without having silicon nucleate on the top surface of the wafer. Control of the etching and deposition processes is critical for this technique to be successful.

6.7 Summary

In this section we have examined some of the more important uses of polysilicon films in integrated circuits. Although the dominant use of polysilicon is for gate electrodes and interconnections in MOS integrated circuits, virtually all new, high-performance bipolar technologies employ one, or even two, layers of polysilicon to improve transistor performance and increase packing density. The availability of polysilicon has also led to its use in floating-gate memory elements and in high-value load resistors.

Devices with their active elements within polysilicon layers are also being investigated. Polysilicon diodes are attractive because of their compatibility with CMOS integrated-circuit processing, and polysilicon transistors can be employed both in silicon integrated-circuit applications and for large-area circuits, such as the active matrix and peripheral circuits of displays.

As integrated-circuit technology continues developing, the uses of polysilicon will undoubtedly expand, even as its use for interconnecting elements in MOS integrated circuits becomes restricted by its limited conductivity. Even for MOS integrated circuits, however, the compatibility of polysilicon with integrated-circuit processing makes its continued use attractive, and it will probably be augmented by other materials, rather than being replaced in the foreseeable future. The widespread use of polysilicon in present applications and its expansion into new applications make it a material of continuing importance in silicon integrated-circuit technology and require continued and expanded understanding of its deposition and properties.

REFERENCES

Reviews

Kamins, T.I., *Polycrystalline Silicon for Integrated-Circuit Applications*, Kluwer Academic Publishing, Norwell MA (1988)

Kazmerski, L.L., *Polycrystalline and Amorphous Thin Films and Devices*, Academic Press, New York (1980)

deGraaff, H.C., in *Polycrystalline Semiconductors*, ed. G. Harbeke, p. 170, Springer-Verlag, Berlin, New York (1985)

Deposition

1. Eversteyn, F.C., Severin, P.J.W., van den Brekel, C.H.J. and Peek, H.L., A stagnant layer model for the epitaxial growth of silicon from silane in a horizontal reactor, *J. Electrochem. Soc.* 117, 925-931 (1970)

2. Grove, A.S., *Physics and Technology of Semiconductor Devices*, p. 18, Wiley, New York (1967)

3. Rosler, R.S., Low pressure CVD production processes for poly, nitride, and oxide, *Solid State Technology* pp. 63-70 (1977)

4. Eversteyn, C., and Put, B.H., Influence of AsH_3, PH_3, and B_2H_6 on the growth rate and resistivity of polycrystalline silicon films deposited from a SiH_4-H_2 mixture, *J. Electrochem. Soc.* 120, 106-109 (1973)

5. Hall, L.H., and Koliwad, K.M., Low temperature chemical vapor deposition of boron doped silicon films, *J. Electrochem. Soc.* 120, 1438-1440 (1973)

6. Nakayama, S., Kawashima, I., and Murota, J., Boron doping effect on silicon film deposition in the Si_2H_6-B_2H_6-He gas system, *J. Electrochem. Soc.* 133, 1721-1724 (1986)

7. Goodman, D.W. and Rye, R.R., in *Tungsten and Other Refractory Metals for VLSI Applications*, Ed. R.S. Blewer, Materials Research Society, Pittsburgh, PA (1986)

8. Yasuda, Y., and Moriya, T., Marked effects of borondoping on the growth and properties of polycrystalline silicon films, *Semiconductor Silicon 1973*, pp. 271-284, Electrochemical Society, Pennington, NJ (1973)

9. Meakin, D.B. and Ahmed, W., LPCVD of *in-situ* phosphorus doped polysilicon from PH_3/Si_2H_6 mixtures, Fall 1986 *Electrochemical Society Meeting*, abstract 267, pp. 398-399 (1986)

10. Ahmed, W. and Meakin, D.B., Phosphorus-doped silicon films prepared by low pressure chemical vapour deposition of disilane and phosphine, *Thin Solid Films 148*, L63-L65 (1987)

Structure

11. Claassen, A.P., Bloem, J., Valkenburg, W.G.J.N. and van den Brekel, C.H.J., The deposition of silicon from silane in a low-pressure hot-wall system, *J. Crystal Growth 57*, 259-266 (1982)

12. Coltrin, M.E., Kee, R.J. and Miller, J.A., A mathematical model of the coupled fluid mechanics and chemical kinetics in a chemical vapor deposition reactor, *J. Electrochem. Soc.* 131, 425-434 (1984)

13. Meyerson, B.S. and Jasinski, J.M., Silane pyrolysis rates for the modeling of chemical vapor deposition, *J. Appl. Phys.* 61, 785-787 (1987)

14. Claassen, W.A.P. and Bloem, J., The nucleation of CVD silicon on SiO_2 and Si_3N_4 substrates. I. The SiH_4-HCl-H_2 system at high temperatures, *J. Electrochem. Soc.* 127, 194-202 (1980)

15. Kamins, T.I., Chemically vapor deposited polycrystalline-silicon films, *IEEE Trans. Parts, Hybrids, and Packaging*, PHP-10, 221-229 (1974)

16. Brown, W.A. and Kamins, T.I., An analysis of LPCVD system parameters for polysilicon, silicon nitride and silicon dioxide deposition, *Solid State Technology* 84, pp. 51-57 (1979)

17. Cullity, B.D., *Elements of X-Ray Diffraction*, Addison-Wesley, Reading MA (1956)

18. Smith, J.V., Editor, *X-Ray Powder Data File* American Society for Testing and Materials, Philadelphia, PA (1960)

19. Kamins, T.I., Structure and properties of LPCVD silicon films, *J. Electrochem. Soc.* 127, 686-690 (1980)

20. Kamins, T.I. and Cass, T.R., Structure of chemically deposited polycrystalline-silicon films, *Thin Solid Films*, 16, 147-165 (1973)

21. Falckenberg, R., Doering, E. and Oppolzer, H., Surface roughness and grain growth of thin P-doped polycrystalline Si-films, Fall 1979 *Electrochem. Soc. Meeting*, abstract 570, pp. 1429-1432 (1979)

22. Kamins, T.I., Manoliu, J. and Tucker, R.N., Diffusion of impurities in polycrystalline silicon, *J. Appl. Phys.* 43, 83-91 (1972)

23. Kamins, T.I., Mandurah, M.M. and Saraswat, K.C., Structure and stability of low pressure chemically vapor-deposited silicon films, *J. Electrochem. Soc.* 125, 927-932 (1978)

24. Wada, Y. and Nishimatsu, S., Grain growth mechanism of heavily phosphorus-implanted polycrystalline silicon, *J. Electrochem. Soc.* 125, 1499-1504 (1978)

25. Thompson, C.V. and Smith, H.I., Surface-energy-driven secondary grain growth in ultrathin (< 100 nm) films of silicon, *Appl. Phys. Lett.* 44, 603-605 (15 1984)

26. Mei, L., Rivier, M., Kwark, Y. and Dutton, R.W., Grain-growth mechanisms in polysilicon, *J. Electrochem. Soc.* 129, 1791-1795 (1982)

27. Kim, H-J. and Thompson, C.V., Compensation of grain growth enhancement in doped silicon films, *Appl. Phys. Lett.* 48, 399-401 (1986)

28. Chiang, K.L., Dell'Oca, C.J. and Schwettmann, F.N., Optical evaluation of polycrystalline silicon surface roughness, *J. Electrochem. Soc.* 126, 2267-2269 (1979)

29. Dell'Oca, C.J., Nondestructive thickness determination of polycrystalline silicon deposited on oxidized silicon, *J. Electrochem. Soc.* 119, 108-111 (1972)

30. Boxall, B.A., A change of etch rate associated with the amorphous to crystalline transition in CVD layers of silicon, *Solid-State Electron.* 20, 873-874 (1977)

Oxidation

31. Kamins, T.I. and MacKenna, E.L., Thermal oxidation of polycrystalline silicon films, *Metallurgical Transactions of AIME*, 2, 2292-2294 (1971)

32. Kamins, T.I., Oxidation of phosphorus-doped low pressure and atmospheric pressure CVD polycrystalline- silicon films, *J. Electrochem. Soc.* 126, 838-844 (1979)

33. Sunami, H., Thermal oxidation of phosphorus-doped polycrystalline silicon in wet oxygen, *J. Electrochem. Soc.* 125, 892-897 (1978)

34. Irene, E.A., Tierney, E. and Dong, D.W., Silicon oxidation studies: Morphological aspects of the oxidation of polycrystalline silicon, *J. Electrochem. Soc.* 127, 705-713 (1980)

35. Bravman, J.C. and Sinclair, R., Transmission electron microscopy studies of the polycrystalline silicon-SiO_2 interface, *Thin Solid Films*, 104, 153-161 (1983)

36. DiMaria, D.J. and Kerr, D.R., Interface effects and high conductivity in oxides grown from polycrystalline silicon, *Appl. Phys. Lett.* 27, 505-507 (1975)

37. Duffy, M.T., McGinn, J.T., Shaw, J.M., Smith, R.T., Soltis, R.A. and Harbeke, G., LPCVD polycrystalline silicon: Growth and physical properties of diffusion-doped, ion-implanted, and undoped films, *RCA Review* 44, 313-325 (1983)

38. Sternheim, M., Kinsbron, E., Alspector, J. and Heimann, P.A., Properties of thermal oxides grown on phosphorus in-situ doped polysilicon, *J. Electrochem Soc.* 130, 1735-1740 (1983)

39. Yaron, G., Hess, L.D., and Kokorowski, S., Application of laser processing for improved oxides grown from polysilicon, *IEEE Trans. Electron Devices*, ED-27, 964-969 (1980)

40. Marcus, R.B., Sheng, T.T. and Lin, P., Polysilicon/SiO_2 interface microtexture and dielectric breakdown, *J. Electrochem. Soc.* 129, 1282-1289 (1982).

41. Ham, W.E., Abrahams, M.S. and Buiocchi, C.J., A note on the ability of CVD polysilicon to deposit nearly inaccessible areas of IC topology, *J. Electrochem. Soc.* 128, 1623-1624 (1981)

42. Sunami, H., Koyanagi, M. and Hashimoto, N., Intermediate oxide formation in double-polysilicon gate MOS structure, *J. Electrochem. Soc.* 127, 2499-2506 (1980)

43. Anderson, R.M. and Kerr, D.R., Evidence for surface asperity mechanism of conductivity in oxide grown on polycrystalline silicon, *J. Appl. Phys.* 48, 4834-4836 (1977)

Electrical Properties

44. Saraswat, K. and Mohammadi, F., Effect of scaling of interconnections on the time delay of VLSI circuits, *IEEE Trans. Electron Devices*, ED-29, 645-650 (1982)

45. Mandurah, M.M., Saraswat, K.C., Helms, C.R. and Kamins, T.I., Dopant segregation in polycrystalline silicon, *J. Appl. Phys.* 51, 5755-5763 (1980)

46. Wong, C.Y., Grovenor, C.R.M., Batson, P.E. and Smith, D.A., Effect of arsenic segregation on the electrical properties of grain boundaries in polycrystalline silicon, *J. Appl. Phys.* 57, 438-442 (1985)

47. Tandon, J.L., Harrison, H.B., Neoh, C.L., Short, K. T. and Williams, J.S., The annealing behavior of antimony implanted polycrystalline silicon, *Appl. Phys. Lett.* 40, 228-230 (1982)

48. Mandurah, M.M., Saraswat, K.C. and Kamins, T.I. Arsenic segregation in polycrystaline silicon, *Appl. Phys. Lett.* 36, 683-685 (1980)

49. Grovenor, C.R.M., Batson, P.E., Smith, D.A. and Wong, C., As segregation to grain boundaries in Si, *Phil. Mag.* A 50, 409-423 (1984)

50. Rose, J.H. and Gronsky, R., Scanning transmission electron microscope microanalytical study of phosphorus segregation at grain boundaries in thin-film silicon, *Appl. Phys. Lett.* 41, 993-995 (1982)

51. Kazmerski, L.L., Ireland, P.J. and Ciszek, T.F., Evidence for the segregation of impurities to grain boundaries in multigrained silicon using Auger electron spectroscopy and secondary ion mass spectroscopy, *Appl. Phys. Lett.* 36, 323-325 (1980)

52. Kamins, T.I., Hall mobility in chemically deposited polycrystalline silicon, *J. Appl. Phys.* 42, 4357-4365 (1971)

53. Seto, J.Y.W., The electrical properties of polycrystalline silicon films, *J. Appl. Phys.* 46, 5247-5254 (1975)

54. Korsh, G.J. and Muller, R.S., Conduction properties of lightly doped, polycrystalline silicon, *Solid-State Electron.* 21, 1045-1051 (1978)

55. Ghosh, A.K., Fishman, C. and Feng, T., Theory of the electrical and photovoltaic properties of polycrystalline silicon, *J. Appl. Phys.* 51, 446-454 (1980)

56. Lu, N. C-C, Gerzberg, L., Lu, C-Y and Meindl, J.D., A conduction model for semiconductor-grain-boundary-semiconductor barriers in polycrystalline-silicon films, *IEEE Trans. Electron Devices*, ED-30, 137-149 (1983)

57. Mandurah, M.M., Saraswat, K.C. and Kamins, T.I., A model for conduction in polycrystalline silicon—Part I: Theory, *IEEE Trans. Electron Devices*, ED-28, 1163-1171 (1981)

58. Martinez, J., Criado, A. and Piqueras, J., Grain boundary potential determination in polycrystalline silicon by the scanning light spot technique, *J. Appl. Phys.* 52, 1301-1305 (1981)

59. Seager, C.H. and Pike, G.E., Grain boundary states and varistor behavior in silicon bicrystals, *Appl. Phys. Lett.* 35, 709-711 (1979)

60. Seager, C.H., Grain boundary recombination: Theory and experiment in silicon, *J. Appl. Phys.* 52, 3960-3968 (1981)

61. Jackson, W.B., Johnson, N.M. and Biegelsen, D.K., Density of gap states of silicon grain boundaries determined by optical absorption, *Appl. Phys. Lett.* 43, 195-197 (1983)

62. Schubert, W.K. and Lenahan, P.M., Spin dependent trapping in a polycrystalline silicon integrated circuit resistor, *Appl. Phys. Lett.* 43, 497-499 (1983)

63. Baccarani, G., Ricco, B. and Spadini, G., Transport properties of polycrystalline silicon films, *J. Appl. Phys.* 49, 5565-5570 (1978)

64. Landsberg, P.T. and Abrahams, M.S., Effects of surface states and of excitation on barrier heights in a simple model of a grain boundary or a surface, *J. Appl. Phys.* 55, 4284-4293 (1984)

65. Kenyon, P.K. and Dressel, H., Negative resistance switching in near-perfect crystalline silicon film resistors, *J. Vac. Sci. Technol.* A 2, 1486-1490 (1984)

66. Lu, C-Y, Lu, N.C-C and Shih, C-C , Resistance switching characteristics in polycrystalline silicon film resistors, *J. Electrochem. Soc.* 132, 1193-1196 (1985)

67. Mahan, J.E., Threshold and memory switching in polycrystalline silicon, *Appl. Phys. Lett.* 41, 479-481 (1982)

68. Seager, C.H. and Pike, G.E., Anomalous low-frequency grain-boundary capacitance in silicon, *Appl. Phys. Lett.* 37, 747-749 (1980)

69. Taniguchi, M., Hirose, M., Osaka, Y., Hasegawa, S. and Shimizu, T., Current transport in doped polycrystalline silicon, *Japan. J. Appl. Phys.* 19, 665-673 (1980)

70. Loh, E., Interpretation of dc characteristics of phosphorus-doped polycrystalline silicon films: Conduction across low-barrier grain boundaries, *J. Appl. Phys.* 54, 4463-4466 (1983)

71. Gat, A., Gerzberg, L., Gibbons, J.F., Magee, T.J., Peng, J. and Hong, J.D., cw laser anneal of polycrystalline silicon: Crystalline structure, electrical properties, *Appl. Phys. Lett.* 33, 775-778 (1978)

72. Yaron, G., Hess, L.D. and Olsen, G.L., Electrical characteristics of laser-annealed polysilicon resistors for device applications, *Proc. Materials Research Society Symposium*, eds. C. W. White and P. S. Peercy, Academic Press, New York (1980)

73. Lee, M-K, Lu, C-Y, Chang, K-Z, and Shih, C., On the semi-insulating polycrystalline silicon resistor, *Solid-State Electron.* 27, 995-1001 (1984)

74. Srivastava, P.C., Bourgoin, J.C., Rabajo, F. and Arroyo, J.M., Transient capacitance spectroscopy in polycrystalline silicon, *J. Appl. Phys.* 53, 8633-8638 (1982)

75. Malhi, S.D.S., Shichijo, H., Banerjee, S.K., Sundaresan, R., Elahy, M., Pollack, G.P., Richardson, W.F., Shah, A.H., Hite, L.R., Womack, R.H., Chatterjee, P.K., and Lam, H.W., Characteristics and three-dimensional integration of MOSFETs in small-grain LPCVD polycrystalline silicon, *IEEE Trans. Electron Devices*, ED-32, 258-281 (1985)

76. Makino, T. and Nakamura, H., The influence of plasma annealing on electrical properties of polycrystalline Si, *Appl. Phys. Lett.* 35, 551-552 (1979)

77. Chen, D.L., Greve, D.W. and Guzman, A.M. Influence of hydrogen implantation on the resistivity of polycrystalline silicon, *J. Appl. Phys.* 57, 1408-1410 (1985)

78. Pollack, G.P., Richardson, W.F., Malhi, S.D.S., Bonifield, T., Shichijo, H., Banerjee, S., Elahy, M., Shah, A.H., Womack, R. and Chatterjee, P.K., Hydrogen passivation of polysilicon MOSFETs from a plasma nitride source, *IEEE Electron Device Lett.* EDL-5, 468-470 (1984)

79. Young, R.T., Lu, M.C., Westbrook, R.D. and Jellison, G.E. Jr, Effect of lithium on the electrical properties of grain boundaries in silicon, *Appl. Phys. Lett.* 38, 628-630 (1981)

80. Miller, G.L. and Orr, W.A., Lithium doping of polycrystalline silicon, *Appl. Phys. Lett.* 37, 1100-1101 (1980)

81. Wada, Y. and Nishimatsu, S., Resistivity lowering limitations of heavily doped polycrystalline silicon, *Denki Kagaku*, 47, 118-123 (1979)

82. Murota, J. and Sawai, T., Electrical characteristics of heavily arsenic and phosphorus doped polycrystalline silicon, *J. Appl. Phys.* 53, 3702-3708 (1982)

83. Makino, T. and Nakamura, H., Resistivity changes of heavily-boron-doped CVD-prepared polycrystalline silicon caused by thermal annealing, *Solid-State Electron.* 24, 49-55 (1981)

84. Solmi, S., Severi, M., Angelucci, R., Baldi, L. and Bilenchi, R., Electrical properties of thermally and laser annealed polycrystalline silicon films heavily doped with arsenic and phosphorus, *J. Electrochem. Soc.* 129, 1811-1818 (1982)

85. Mandurah, M.M., Saraswat, K.C. and Kamins, T.I., Phosphorus doping of low pressure chemically vapor-deposited silicon films, *J. Electrochem. Soc.* 126, 1019-1023 (1979)

86. Joshi, D.P., and Srivastava, R.S., A model of electrical conduction in polycrystalline silicon, *IEEE Trans. Electron Devices*, ED-31, 920-927 (1984)

87. Lifschitz, N., Solubility of implanted dopants in polysilicon: Phosphorus and arsenic, *J. Electrochem. Soc.* 130, 2464-2467 (1983)

88. Chow, R. and Powell, R.A., Activation and redistribution of implants in polysi by RTP, *Semiconductor International*, pp. 108-113 (1985)

Applications

89. Faggin, F. and Klein, T., Silicon gate technology, *Solid-State Electron.* 13, 1125-1144 (1970)

90. Peters, D., Implanted-silicided polysilicon gates for VLSI transistors, *IEEE Trans. Electron Devices*, ED-33, 1391-1393 (1986)

91. Lifshitz, N., Dependence of the work-function difference between the polysilicon gate and silicon substrate on the doping level in polysilicon, *IEEE Trans. Electron Devices*, ED-32, 617-621 (1985)

92. Lifshitz, N., Luryi, S. and Sheng, T.T., Influence of the grain structure on the Fermi level in polycrystalline silicon: A quantum size effect? *Appl. Phys. Lett.* 51, 1824-1826 (1987)

93. Sunami, H., Cell structure for future DRAMs, *Tech. Digest* International Electron Devices Meeting (Washington DC), paper 29.1, pp. 694-697 (1985)

94. *eg*, McKenny, V.G., A 5 V-only 4-K static RAM, *Digest* 1977 IEEE International Solid-State Circuits Conference, paper WAM-1.3, pp. 16-17 (1977)

95. Fa, C.H. and Jew, T.T., The poly-silicon insulated-gate field-effect transistor, *IEEE Trans. Electron Devices*, ED-13, 290-291 (1966)

96. Kamins, T.I., Field-effects in polycrystalline-silicon films, *Solid-State Electron.* 15, 789-799 (1972)

97. Kamins, T.I., and Marcoux, P.J., Hydrogenation of transistors fabricated in polycrystalline-silicon films, *IEEE Electron Device Lett.* EDL-1, 159-161 (1980)

98. Malhi, S.D.S., Shah, R.R., Shichijo, H., Pinizzotto, R.F., Chen, C.E., Chatterjee, P.K. and Lam, H.W., Effects of grain boundary passivation on the characteristics of p-channel MOSFETs in LPCVD polysilicon, *Electronics Lett.* 19, 993-994 (1983)

99. Lee, K.F., Gibbons, J.F., Saraswat, K.C. and Kamins, T.I., Thin film MOSFET's fabricated in laser-annealed polycrystalline silicon, *Appl. Phys. Lett.* 35, 173-175 (1979)

100. Johnson, N.M., Biegelsen, D.K. and Moyer, M.D., Grain boundaries in p-n junction diodes fabricated in laser-recrystallized silicon thin films, *Appl. Phys. Lett.* 38, 900-902 (1981)

101. Tihanyi, J. and Schlotterer, H., Properties of ESFI MOS transistors due to the floating substrate and the finite volume, *IEEE Trans. Electron Devices*, ED-22, 1017-1023 (1975)

102. Sasaki, N., Charge pumping in SOS-MOS transistors, *IEEE Trans. Electron Devices*, ED-28, 48-52 (1981)

103. Lim, H.K., and Fossum, J.G., Transient drain current and propagation delay in SOI CMOS, *IEEE Trans. Electron Devices*, ED-31, 1251-1258 (1984)

104. Ashburn, P. and Soerowirdjo, B., Comparison of experimental and theoretical results on polysilicon emitter bipolar transistors, *IEEE Trans. Electron Devices*, ED-31, 853-860 (1984)

105. Ning, T.H. and Isaac, R.D., Effect of emitter contact on current gain of silicon bipolar devices, *IEEE Trans. Electron Devices*, ED-27, 2051-2055 (1980)

106. Matsushita, T., Oh-uchi, N., Hayashi, H. and Yamoto, H., A silicon heterojunction transistor, *Appl. Phys. Lett.* 35, 549-550 (1979)

107. Manoliu, J. and Kamins, T.I., p-n junctions in polycrystalline-silicon films, *Solid-State Electron.* 15, 1103-1106 (1972)

108. Dutoit, M. and Sollberger, F., Lateral polysilicon p-n diodes, *J. Electrochem. Soc.* 125, 1648-1651 (1978)

109. Davidsohn, U.S. and Lee, F., Dielectric isolated integrated circuit substrate processes, *Proc. IEEE*, 57, 1532-1537 (1969)

110. van Dijk, H.J.A. and de Jonge, J., Preparation of thin silicon crystals by electrochemical thinning of epitaxially grown structures, *J. Electrochem. Soc.* 117, 553-554 (1970)

111. Meek, R.L., Electrochemically thinned N/N^+ epitaxial silicon—Method and application, *J. Electrochem. Soc.* 118, 1240-1246 (1971)

112. Kamins, T.I., A new dielectric isolation technique for bipolar integrated circuits using thin single-crystal silicon films, *Proc. IEEE*, 60, 915-916 (1972)

113. Rung, R.D., Momose, H. and Nagakubo, Y., Deep trench isolated CMOS device, *Tech. Digest* 1982 International Electron Devices Meeting, paper 9.6, pp. 237-240 (1982)

114. Sunami, H., Kure, T., Hashimoto, N., Itoh, K., Toyabe, T. and Asai, S., A corrugated capacitor cell (CCC) *IEEE Trans. Electron Devices*, ED-31, 746-753 (1984)

115. Richardson, W.F., Bordelon, D.M., Pollack, G.P., Shah, A.H., Malhi, S.D.S., Shichijo, H., Banerjee, S.K., Elahy, M., Womack, R.H., Wang, C.P., Gallia, J., Davis, H.E. and Chatterjee, P.K., A trench transistor cross-point dRAM cell, *Tech. Digest* 1985 IEEE International Electron Devices Meeting, paper 29.6, pp. 714-717 (1985)

116. Silvestri, V.J., Si selective epitaxial trench refill, Fall 1986 *Electrochemical Society Meeting*, abstract 269, pp. 402-403 (1986)

10

Silicon Phase Diagrams

Richard A. Seilheimer

1.0 INTRODUCTION

The phase diagram represents the relationship between different phases in solids and liquids, and their solutions. A simple example will show some of the ideas. We will look at a solution of two metals, tin (Sn) and lead (Pb), which most people know as solder. The melting point of pure tin is 232°C and that of lead is 327.5°C. A solution of these two elements melts at a lower temperature (183°C). The phase diagram for this system is shown in Figure 1. This figure shows the stable phases as a function of temperature and composition at one atmosphere of pressure. Some important information is given on this chart. On the left axis the melting point of lead is shown and on the right side is the melting point of tin. When tin and lead are melted they are completely soluble in each other. This fact is shown as a single field labelled L, standing for liquid. A mixture of tin and lead melts at a temperature less than either of the pure elements. The field on the left hand side of the chart is a solution of tin dissolved in lead, labelled (Pb). On the right hand side is a solution of lead dissolved in tin, labelled beta-Sn. In the lower central region below 183°C, is a mixture of (Pb) and beta-Sn. While in the liquid state complete solubility exists, and in the solid state only limited solubility exists. This results in the above mixture region. How a diagram such as this is determined and and how it is used is the subject of this chapter. For a high temperature solder we may select an alloy containing 95% Pb and 5% Sn which melts at slightly over 300°C. For a low temperature melting solder of high fluidity, we would select 61.9% Sn and 38.1% Pb, the eutectic composition. For an alloy that may not need as much fluidity we would pick a composition in the two phase region of liquid and (Pb), a composition of 13% Sn and 87% Pb.

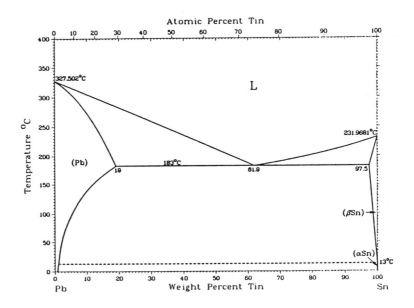

Figure 1: The Pb-Sn Phase Diagram (9).

2.0 PHASE DIAGRAMS

Phase diagrams are used to represent the state relationships between materials as a function of temperature. The states may be solid, liquid or gases, and the materials may take a variety of forms depending on interactions between them. The most common form is used to show the interactions between two materials and is called a binary phase diagram. When three materials are considered the diagram is a ternary phase diagram. In some complex alloys or ceramics, six, seven, eight, or more elements may be represented (10).

In this chapter the concepts and principles of binary phase diagrams of silicon will be given. The chapter will also give the latest information available on silicon binary phase diagrams and presents several examples of the use which may be made of this information.

2.1 The Phase Rule

A phase diagram is the graphical representation of the equilibrium states of an alloy for various temperature. The phase diagram is constructed from experimental data using the laws of thermodynamics. In this chapter we will also discuss the methods used to determine these diagrams, the thermodynamic rules and principles involved, and explain some of the important features and the uses of them. The information available on silicon alloy phase diagrams is presented later in figures 25-72. The basis of the diagram is the phase rule of Willard Gibbs (1). This diagram, called either a phase or equilibrium diagram, presents the relationship between the phases as a function of temperature and composition. It is based on equilibrium and gives no information about the kinetics of reaching equilibrium. The phase rule states, the sum of the number of degrees of freedom of a system must equal the sum of the components plus two when in equilibrium, and is given by Equation (1).

$$P + F = C + 2 \qquad (1)$$

The degrees of freedom are the number of intensive variables that can be altered without bringing about the disappearance of a phase or the formation of a new phase. Intensive variables are normally temperature, pressure or composition. A phase is any portion of a system which is physically homogeneous within itself and separated by a boundary from any other portion of the system. A simple example of a system would be a single phase solid pure metal. If the temperature is raised so that a portion of the system melts, we have a two phase system consisting of a liquid phase and solid phase. The phase rule as applied to the melting pure metal system explains the constant melting temperature. The number of phases are two, solid and liquid. There is only one component, the pure element, thus $2 + F = 1 + 2$, and F, the number of degrees of freedom is one. Since there are two thermodynamic variables, temperature and pressure, they may not be varied independently. If the pressure is held constant there is one temperature at which the melting takes place. If the pressure is changed the temperature will also change or one of the phases will disappear. Systems in which the pressure variable is held constant are called condensed systems. The reduced phase rule for such systems may be represented by Equation (2) in which P refers to only solid and liquid phases.

$$P + F = C + 1 \qquad (2)$$

In binary systems in which the pressure is constant three coexisting phases produce an invariant condition, two phases produce a univariant condition, and one phase a bivariant condition.

2.2 Free Energy Of Alloy Systems

The free energy concept can be used to explain the phase behavior of an alloy system. According to the second law of thermodynamics, in a system in which the temperature and pressure are fixed, the thermodynamic potential or Gibbs Free Energy, G, must be a minimum. The thermodynamic potential is given by:

$$G = U - TS + pV \qquad (3)$$

where U is the internal energy, T the temperature, S the entropy, p the pressure, and V the volume. In alloy systems where we are concerned with solids and liquids the pV term is small and may be neglected without great error. The condition of equilibrium is that the free energy $F = U - TS$ should be minimized. This quantity, F, is known as the Helmholtz Free Energy. The internal energy and the pV term may be taken together and are called the enthalpy. In this expression the enthalpy is largely determined by the energies of interaction of the two kinds of atoms, say A and B in the liquid solution, and the entropy is affected primarily by its change due to mixing of the two kinds of atoms. If is it assumed that the two metals form an ideal solution, the enthalpy varies linearly with composition,as shown in Figure 2. The entropy of mixing goes through a minimum with composition being higher for the pure elements and lower for the mixture. The free energy curve may be drawn as in Figure 2. If this same logic is applied to a solid solution the entropy term still produces a minimum in the free energy. However the enthalpy term become very large as large amounts of element B are forced into a solid solution with metal A. The curves for the free energy of the two solutions A in B and B in A as shown in Figure 3. Since the region in the center between the two minimums would be lower in free energy for a mixture of A in B and B in A this is the form this system will assume. It is shown as a dotted line a-b in Figure 3.

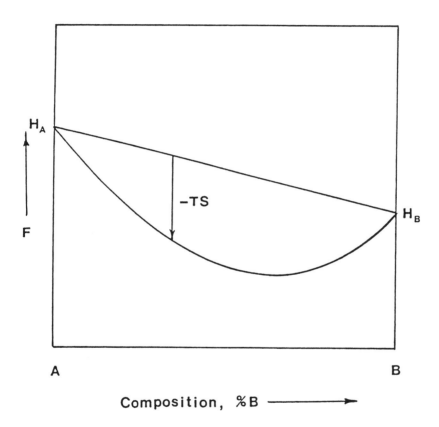

Figure 2: The Free Energy Curve for an Ideal Solution of Two Metals.

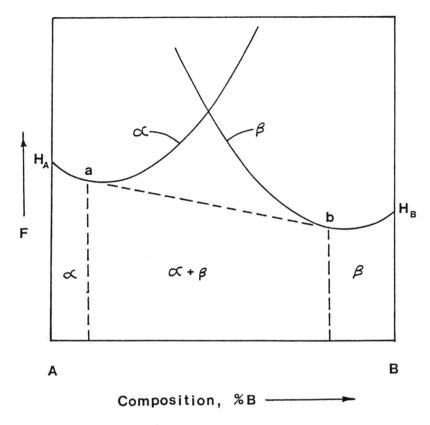

Figure 3: Free Energy Curve for Two Solid Solutions.

2.2.1 Complete Solubility: For a system in which the solid is completely soluble in all proportions, the free energy curves for the liquid and solid phases are as shown in Figure 4, for a temperature above melting for all compositions. In Figure 5 are shown the free energy curves at a lower temperature at which both liquid and solid phases are stable. The free energy curves now intersect. Because the lowest free energy state is the stable state we see that, the liquid phase is stable for some compositions and the solid is stable phase for other compositions. If a line is drawn tangent to the liquid and solid curves it is seen to have a lower energy than either liquid or solid in this region. The tangent line represents a linear combination of liquid and solid phases. Thus between compositions p and q in Figure 5 the equili-

Silicon Phase Diagrams 737

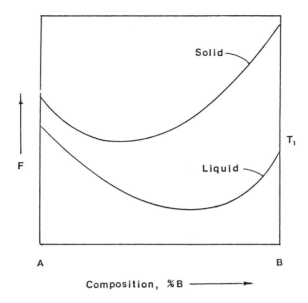

Figure 4: Free Energy Curve for Liquid and Solid Solutions, Above Melting Point.

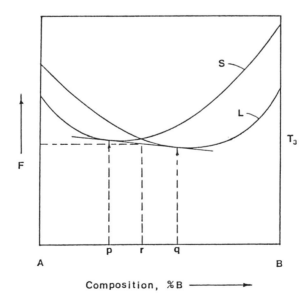

Figure 5: Free Energy Curve for Liquid and Solid Phases, Liquid plus Solid Region.

738 Handbook of Semiconductor Silicon Technology

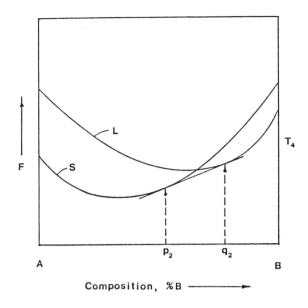

Figure 6: Free Energy Curve for Liquid and Solid Phases, Less Liquid

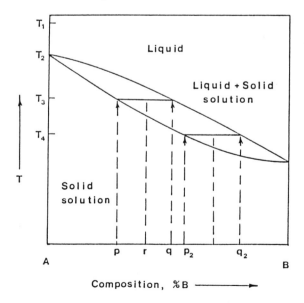

Figure 7: Phase Diagram Corresponding to Above Free Energy Curve.

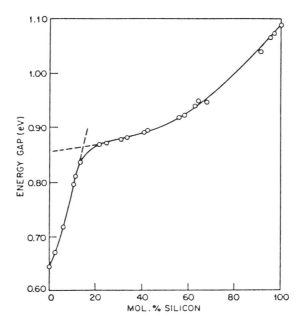

Figure 8: Indirect Energy Levels in Si-Ge System, (18).

brium state consists of a combination of both liquid and solid. In Figure 6 the mixture is still stable but at a different composition. Figure 7 shows the phase diagram corresponding to these free energy curves. This type phase diagram is referred to as an isomorphous type, and is found in only one alloy system involving silicon, that with germanium. While these two metals are completely soluble in both the liquid and solid states not all properties vary smoothly across the phase diagram. The indirect energy gap, as revealed by optical gap for indirect absorption, clearly is not continuous as shown by Figure 8 which shows a break at 0.15 mole percent Silicon (3).

2.2.2 Eutectic And Peritectic Systems: The case of limited solubility in the solid state and complete solubility in the liquid state a different type phase diagram is produced. For this discussion refer to Figure 9a and b. At temperature T1, the liquid state has the lowest free energy and is the stable state. When the temperature is lowered to T2, for the set of free energy diagrams Figure 10a, the minimum energy states are, from left to right, the alpha phase solid, a mixture of alpha and liquid, the liquid phase, a mixture of beta and liquid, and the beta phase solid. At temperature T2 in the set of curves Figure 10b there are a different set of stable phases, alpha, alpha plus liquid, and liquid. The free energy of the beta phase at this temperature is still higher than that of the liquid. At the temperature where the three minimums in the free energy states line up in a straight line, a distinct state is formed. In case (a) the composition of the minimum free energy of the liquid state lies between the compositions of the solid phases. The point is called the eutectic point. In the case (b) the composition is not between the solid phase composition minimums but lies outside of this range. The point, given by the composition and temperature of the minimum free energy of the liquid phase is called the peritectic point. At a temperature slightly below the peritectic temperature the free energy curves are as shown in Figure 12b. The stable states may be read from left to right as, solid alpha, alpha plus beta, solid beta, solid beta plus liquid, and liquid. Note the two mixed phase regions. At a much lower temperature where the liquid state is not stable, shown in Figure 13a and 13b. The phases in both type systems are found to be the same, reading from left to right, alpha, alpha plus beta, and beta, all of which are solid phases. Case (a) is called a eutectic system and (b) is called a peritectic

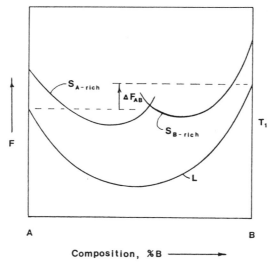

Figure 9a: Free Energy Curve for Eutectic System, Liquid Stable.

Silicon Phase Diagrams   741

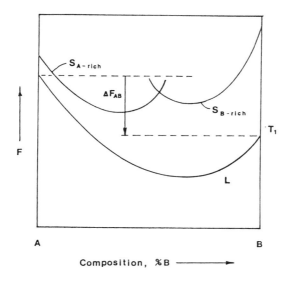

Figure 9b: Free Energy Curve for Peritectic System, Liquid Stable.

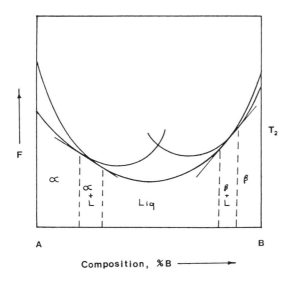

Figure 10a: Free Energy Curve for Eutectic System, Alpha,-Alpha+Liq,Liq,Liq+Beta,Beta.

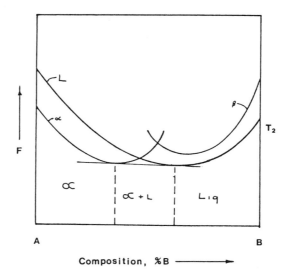

Figure 10b: Free Energy Curve for Peritectic System, Alpha,Alpha+Liq,Liq.

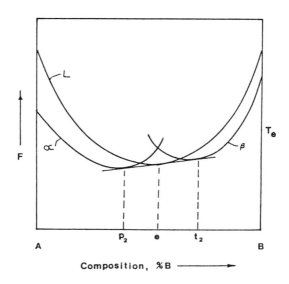

Figure 11a: Free Energy Curve for Eutectic System, Eutectic Temperature.

Silicon Phase Diagrams 743

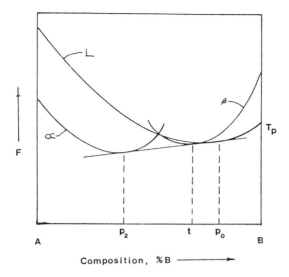

Figure 11b: Free Energy Curve for Peritectic System, Peritectic Temperature.

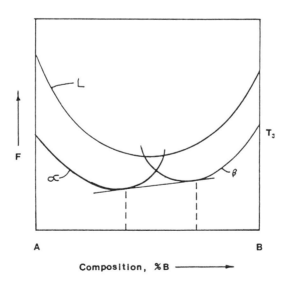

Figure 12a: Free Energy Curve for Eutectic System, Alpha, Alpha+Beta, Beta.

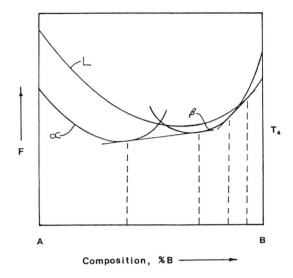

Figure 12b: Free Energy Curve for Peritectic System, Alpha, Alpha+Beta, Beta, Beta+Liq.

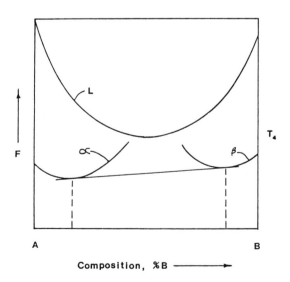

Figure 13a: Free Energy Curve for Eutectic System, Alpha, Alpha+Beta, Beta.

Silicon Phase Diagrams   745

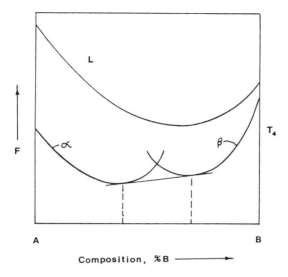

Figure 13b: Free Energy Curve for Peritectic System, Alpha, Alpha+Beta, Beta.

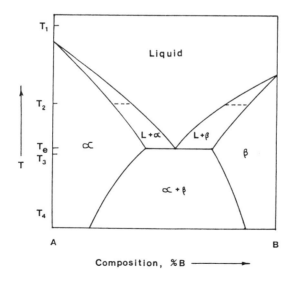

Figure 14a: Phase Diagram For Eutectic System.

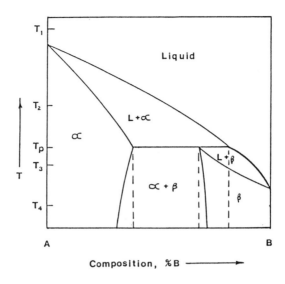

Figure 14b: Phase Diagram for Peritectic System.

system. The main difference noted in the behavior above is caused by the differences in the free energy of the liquid phase for the two extremes of composition. This difference in the pure component liquid is indicated in Figure 9a and 9b as delta F. The difference in the free energy of the pure liquids A and B is larger in case 9b. This results in entirely different looking phase diagrams. Typical phase diagrams for these systems are shown in Figures 14a and 14b.

2.2.3 Intermediate Phases: Often it is found that an intermediate phase forms of a fixed ratio of atoms, say AB. For the system in which complete solubility exists in the liquid phase with limited solubility in the solid phase, and an intermediate composition phase is also formed, a discussion follows. Figure 15 illustrates a high temperature where the free energy of the liquid phase is the lowest. As the temperature is lowered to T_M, the free energy of the liquid and solid phase change as shown by the dotted line. At this temperature, the first solid phase appears at composition M. Further cooling gives rise to the solid phases indicated in Figures 16 and 17. The corresponding phase diagram is shown in Figure 18 for intermediate phase formation. The points of tangency in the free energy diagrams that form the two phase mixed region are indicated in the phase diagram.

2.3 Solid Solutions

Solid solutions usually exist over a limited range of compositions. They may be either substitutional, in which the atoms exist in the same crystal structure, or they may be interstitial, in which one atom, the smaller in size, occupy the spaces between the atoms of the larger size. The range of composition over which a solid solution exists is called the range of solubility. The amount usually increases with temperature, and may form an ordered solution at lower temperatures. Atomic proportions at which ordered solutions exist are found at 1:1 and 1:3 atom ratios (2).

2.3.1 Limitations of Solubility:
The features of atomic structure which govern the extent to which one metal may dissolve another were given by Hume-Rothery (4). These factors include 1) the relative size factor, 2) the chemical factor, 3) the relative valency factor, and 4) the lattice type factor. Each of these factors will be discussed in detail below.

2.3.2 Relative Size Factor:
A metal will dissolve substitutionally only those elements that have a comparable size. The size of an atom in a pure metal is ordinarily determined as the distance from the center of one atom to the center of its nearest neighbor. The size factor is the difference in atom diameters divided by the weighted average diameter, expressed as a percentage. The size factor must be within 10% of the solute atoms size for appreciable solubility to exist. If the size factor is greater than 15%, solid solution formation tends to be very small, much less than one-percent. The size factor for Ge in Si is 4% and these two metals form a continuous solid solution. For most other metals the solubility in Si is very limited. Silicon is roughly four times as large as copper and copper has a very small solubility in silicon (.0000007%). For interstitial compounds this size factor is related to the open spaces between the atoms, called the interstices.

2.3.3 The Chemical Factor:
Rather than forming a solid solution, two metals will form an intermediate phase if they differ chemically. The intermediate phases shown in Figures 15 to 18 are examples of this behavior. The chemical differences are measured by their electronegativity, which is a measure of the tendency to attract electrons in a chemical

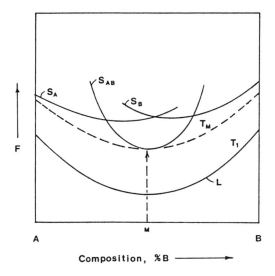

Figure 15: Free Energy Curve for System with Intermediate Composition Compound Formation, Liquid Stable.

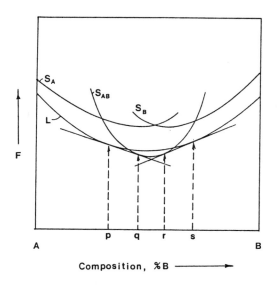

Figure 16: Free Energy Curves Illustrating Equilibrium Between Intermediate Phase Compound and the Liquid.

Silicon Phase Diagrams 749

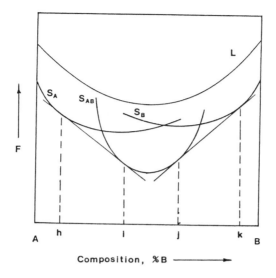

Figure 17: Free Energy Curves Illustrating Equilibrium Between an Intermediate Phase and Two Primary Solid Solutions.

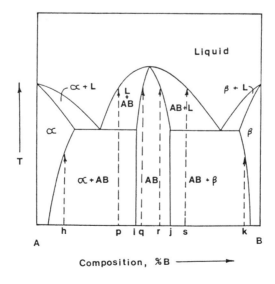

Figure 18: Phase Diagram Corresponding to Figs. 15-17.

Table I

Electronegativities of the Elements

Element	X	Element	X	Element	X	Element	X
H	2.2	Mn	1.5	Tc	1.9	Os	2.2
Li	1.0	Fe(II)	1.8	Ru	2.2	Ir	2.2
Be	1.5	Fe(III)	1.9	Rh	2.2	Pt	2.2
B	2.0	Co	1.8	Pd	2.2	Au	2.4
C	2.0	Ni	1.8	Ag	1.9	Hg	1.9
N	3.05	Cu(I)	1.9	Cd	1.7	Tl	1.8
O	3.5	Cu(II)	2.0	In	1.7	Pb	1.8
F	3.9	Zn	1.6	Sn(II)	1.8	Bi	1.9
Na	0.9	Ga	1.6	Sn(IV)	1.9	Po	2.0
Mg	1.2	Ge	1.9	Sb	2.05	At	2.2
Al	1.5	As	2.0	Te	2.3	Fr	0.65
Si	1.9	Se	2.45	I	2.65	Ra	0.9
Cl	3.15	Br	2.85	Cs	0.7	Ac	1.1
K	0.8	Rb	0.8	Ba	0.9	Th	1.3
Ca	1.0	Sr	1.0	La-Lu	1.1-1.2	Pa	1.5
Sc	1.3	Y	1.3	Hf	1.3	U	1.7
Ti	1.5	Zr	1.6	Ta	1.3	Np-No	1.3
Cr	1.6	Mo	1.8	Re	1.9		

Values taken from *Lange's Handbook of Chemistry*, 13th edition, J. Dean, ed., McGraw Hill, NY.

bond. Table I lists the electronegativity of the metallic elements (5). When one element is strongly electropositive and the other strongly electronegative, the two form true chemical compounds in which the general chemistry valence laws are obeyed. Most intermediate phases are not of this nature but can be considered as structures in which the atoms are held together with an essentially metallic bond. The difference in electronegativity reduces solubility when a solution is formed.

2.3.4 **Relative Valency:** Other things being equal, a metal of lower valence will be more likely to dissolve one of higher valance, than vice versa. The relative valency effect is well demonstrated in univalent metals such as copper, silver and gold, when dissolving high valence metals. It is easily understood by first considering a metal such as silicon which crystallizes according to the (8 - N) rule. Each silicon atom has four neighbors with which it shares four electrons completing the octet to form homopolar bonds. If an atom of silicon is replaced by one of copper, there is a deficiency of electrons to form the covalent bonds. This restricts the solubility of copper in silicon (.0000007% Cu maximum). When silicon is dissolved in copper, appreciable solubility exists (11.25% Si in Cu). Metals that crystallize by the (8 - N) rule also tend to dissolve large amounts of other metals of the same valence within the same group of the periodic chart. Silicon and germanium have complete mutual solubility for this reason. This effect also explains the solubility of phosphorous and arsenic used as dopants to increase the conductivity of silicon by providing extra electrons not used in bonding. The high solubility of silicon for phosphorous and arsenic is a direct result of the fact that both have a higher valence than silicon (+5 vs +4).

2.3.5 **Lattice Type Factor:** The lattice type of both metals usually must be the same in order to form an alloy of appreciable solubility. Only metals of the same crystal structure can form complete solid solutions in each other. Silicon and germanium are both diamond type crystals, and form complete solid solutions. Very few other materials have this crystal structure, and no other element forms a complete solid solution with silicon. All other metals of the diamond crystal structure have an unfavorable size factor for the formation of solid solutions.

3.0 PHASE CHANGES

When a material changes state, such as from liquid to solid, the change is similar to a chemical reaction and similar equations are written to describe it. The reactions of interest in binary phase diagrams may be divided into five categories, which are listed below with the corresponding reactions.

1. Isomorphous or continuous solid solution 1 liquid phase ---> 1 solid phase

2. Eutectic   1 liquid phase ---> 2 solid phases

3. Eutectoid   1 solid phase ---> 2 solid phases

4. Peritectic   2 solid phases ---> 1 solid phase

5. Others (e.g., spinoidal)

The isomorphous or continuous solid solution diagram is shown for the Silicon-Germanium system, in Figure 19.

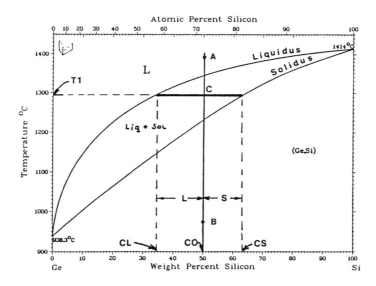

Figure 19: The Lever Law Illustration, Ge-Si Phase Diagram.

3.1 The Lever Rule

The lever rule allows one to calculate the relative amounts of each phase in a two phase regions. The vertical line at 50% Si in Figure 19 represents a composition of interest. If this composition is cooled from the liquid state at point A to the solid state at point B, it traverses the two phase field labelled liquid + alpha to the single phase alpha field. If a horizontal line is drawn through point C at T1 = 1295°C in the two phase field, it will intersect the solidus line on the left and the liquidus line on the right. The intersection of this line with the solidus is the composition of the solid in equilibrium with the liquid, given by the intersection with the liquidus line. The lever rule may be used the calculate the relative amounts of liquid and solid in any two phase fields. If a composition of the original liquid is given as C_O and the temperature is lowered to the two phase region, as shown by the line at T_1, the composition of the liquid is C_L and the solid that has formed by C_S. The amount of silicon in the liquid, is W_L x C_L, and the amount of silicon in the solid is W_S x C_S. The total amount of silicon is C_0 x $(W_L + W_S)$ = $(C_L$ x $W_L)$ + $(C_S$ x $W_S)$. Solving for the ratio of the weights of the solid and liquid we get,

$$\frac{W_L}{W_S} = \frac{C_L - C_0}{C_0 - C_S} = \frac{L}{S} \tag{4}$$

where L and S are the lengths shown in Figure 19. This may be used to calculate the amounts of each phase in a two phase field at a given temperature. this may be used to examine the effect of temperature changes by calculating the amounts of each phase at different temperatures and comparing them.

3.2 Intermediate Phases

When the solubility of the solid solution is exceeded a new phase is formed. The new phases formed may be grouped into three main classes, (a) electrochemical compounds, (b) size factor compounds, and (c) electron compounds. This grouping is not distinct, but relates to the cause of the effect creating the compound. The word "compound" is commonly used to describe all types although class (a) is the only type relating to the valancy laws of chemistry.

The phase Mg_2Si is an example of the electrochemical type compound. These types of compounds are formed when one element is strongly electropositive and the other strongly electronegative. These compounds are true chemical compounds with either covalent or ionic bonding. They usually are insulators or conduct electrically by means of defect structures if at all. With the exception of this group of compounds most intermediate phases should be regarded as structures in which metallic bonding takes place.

Size factor compounds usually form in systems where there is a large difference in the size of the atoms. The size difference is in the range of 20 to 30 percent and is the result of a lowering of the energy when the atoms are packed together in a regular ordered array. The structures are of high coordination often higher than the close packed cubic structures. The Laves phases with compositions given by AB_2 have size differences of 22.5 per cent (7). In this phase each A atom has 12 B neighbors and 4 A neighbors giving a coordination of 16. Each B atom has 12 A neighbors giving an average coordination number of 13.3. These high coordination numbers are due to the metallic bonding characteristic and not covalent or ionic bonding. The electronic compounds are those in which the phase has a definite ratio of valance electrons to atoms. These ratios are frequently found to be 3:2, 21:13, 7:4, or multiple of these ratios. Silicon and copper form several of these compounds.

In an equilibrium diagram for binary alloys at atmospheric pressure, there will be regions of single phase and two phase regions alternating across the diagram. Each time a phase boundary is crossed the number of phases changes by one, either increasing or decreasing. Consider cooling an alloy of silicon and germanium, Figure 19, from the liquid to the solid state. In the liquid region there is one phase, liquid. On cooling to the liquid plus solid region, the number of phases has increased by one. Further cooling to the solid region reduces the number of phases by one.

4.0 TECHNIQUES FOR DETERMINATION OF PHASE DIAGRAMS

The most precise method of locating the phase boundaries is the use of X-ray analysis to determine the lattice spacing. Within a single phase region, such as a solid solution, the lattice spacing will vary smoothly with composition. This variation is often linear, starting with the spacing of the pure metal for zero amount of the second metal and changing in direct proportion to the amount of second element. Within a two phase region the lattice spacings of the two phases are invariant with changes in composition at constant temperature. Thus by measuring the lattice constants for a series of alloys of two metals and plotting the lattice constant vs. composition, the phase boundaries will be located at the intersections of the straight lines.

The lattice constants may be measured at elevated temperature and the high temperature phases may be determined. The X-ray analysis technique also provides information about the type of crystal structure in high temperature phases. Other properties may be used to determine the phase boundaries. One of the most useful is the thermal analysis technique. In this technique an alloy is heated or cooled at a uniform rate and the temperature of the sample is observed. If the sample either absorbs heat or gives off heat it will be cooler or hotter than the furnace. These temperature differences occur when a new phase is formed or disappears. The simplest example of this is the thermal arrest that occurs on cooling a molten material as it freezes and liberates the latent heat of fusion. Differences in the thermal coefficient of expansion also give information when a sample is heated or cooled. Additional techniques involving electrical, optical and mechanical properties are often used in combination to produce a phase diagram (8).

4.1 Determination of Ternary Phase Diagrams

If information is desired about a three component system, a ternary, it is first necessary to determine the binary phase diagrams for each combination pair, and then investigate the three elements together. Often a ternary diagram will represent several hundred or more alloy compositions with analysis at many temperatures. Efforts are under way at this time to develop a larger number of ternary diagrams of industrial interest. Recent development of higher temperature superconductors, all of which are ternary compounds will spur this effort along.

4.2 Silicon Phase Diagrams

The American Society for Metals and the National Bureau of Standards Data Program for Alloy Phase Diagrams is a recent effort to evaluate all available information on metal alloys and publish the results in a series of monographs (9). The binary phase diagrams presented later are some of the results of this effort. Presented here are the binary alloy phase diagrams for silicon with other elements (10). Another source of information about ceramic materials is published by the American Ceramics Society (11).

5.0 SEGREGATION COEFFICIENT AND ZONE REFINING

In many alloy systems which exist as a mixture of two phases at low temperature, as the temperature rises the limit of solubility increases. The alloy exists as a mixture because the free energy of the mixture is lower than that of a homogeneous phase of the same composition. As the temperature is increased the entropy term of the free energy relation F + E - TS increases and becomes more important. Thus with increasing temperatures, disordered homogeneous solutions become preferred over phase mixtures. For the case of very low solubility, such as most impurities of electronic grade silicon, the variation of solubility with temperature (12) is given by

$$c = \exp\left(\frac{-zV}{kT}\right) \qquad (5)$$

where c is the concentration of the impurity, z is the co-ordination number of the structure, V is the bond energy, k the Boltzmann constant, and T is the absolute temperature.

Semiconductor applications of silicon require very high purity material. One purification method for metals is zone refining. The phase diagram is used to explain the zone refining below. A phase diagram of silicon with a small amount of impurity or alloying elements as shown in Figure 20. Consider an alloy of composition A as it is cooled from the liquid state to the solid state. At temperature 1, the liquid is of uniform composition A_0. As the temperature is lowered to point 2, the liquid begins to solidify.

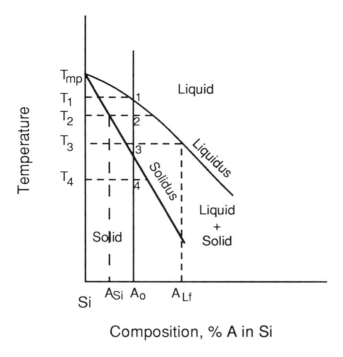

Figure 20: Portion of Phase Diagram for Segregation Coefficient.

The composition of the solid formed is given by the intersection of a horizontal constant temperature line with the solidus line, A_{Si}. Note that the solid contains less impurity than the liquid from which it is forming ($A_{Si} < A_o$). As heat is removed and more solid material is formed, the composition of both the liquid and the solid changes. Since the first solid formed has less of the alloying element the liquid must have more of the alloying element. At an temperature in the two-phase region the composition of the liquid at any point is read from the phase diagram at the intersection of the constant temperature (horizontal) line with the liquidus line. The solid composition is given by the intersection with the solidus line as discussed before. When the material is all solid, at temperature 4 on the Figure 20, the solid equilibrium composition is A_o, the composition of the liquid we started with. The last liquid to solidify has composition A_{Lf}. For thermal equilibrium to exist, it is necessary for diffusion to take place in the solid material to remove the differences that exist in the first formed solid and the last liquid to solidify. These differences in composition result in "coring" in cast metal parts and are the basis of the zone-refining method of purification.

In zone refining a melt region is formed at one end of a bar and moved along the length of the bar. In the molten zone shown in Figure 21, the solidifying material, on the left of the molten zone, will have a concentration lower in the impurity than the liquid. For the phase diagram of Figure 20, segregation during normal freezing of a binary alloy will concentrate the solute at the surface first solidified. This concentration in the solid can be expressed by the relationship

$$C = kC_0(1 - g)^{k-1} \qquad (6)$$

where C_0 is the initial solute concentration in the melt, k is the segregation coefficient, and g is the length solidified. This equation assumes that diffusion in solid is zero, and that there is no concentration gradient in the liquid. The segregation coefficient is also assumed constant throughout the freezing process.

In zone refining, a bar of uniform composition, C_0, is heated to melting at one end, and a narrow molten zone traverses its length. At the beginning of the bar the concentration is lowered by a factor of k, the distribution coefficient. The distribution coefficient is defined as the ratio of the concentration in the solid to that of the liquid at equilibrium (13). The liquid is enriched in the impurity as the molten zone moves down the length of the bar. The process is quite effective in moving the impurities from one end of the bar to the other. Referring to figure 20, when the molten zone moves away from the end, the material that first solidifies has composition A_{Si}. Assuming that k < 1, which is normal for impurities in most materials, moving the molten zone down the bar enriches the liquid until the composition A_{Lf} is reached. At this point the concentration remains constant until the end of the bar is reached. At the end of the bar, the composition is given by equation 6, with the concentration increasing rapidly. The composition along the bar is shown in Figure 21. The concentration C at any point along the bar, except for the end of the bar, is given by

$$\frac{C}{C_o} = 1 - (1-k)\exp\left(\frac{-kL}{M}\right) \qquad (7)$$

where M = width of the molten zone. Thus it can be seen that the solute or impurity can be moved from one end to the other. For silicon production, this method can be used to improve the purity of the ingot. Multiple passes can be made to reach very high purity. In the case of silicon, the rod is vertical, and the molten zone "floats" as it moves from one end to the other. For this reason, purification and crystal growth using this technique are termed Float Zone Refining and Crystal Growth, respectively.

From Figure 21 it can be seen that for the central section of the length, concentration of impurity is very uniform. This can be used to produce very uniform solute distribution along the length of the rod by a process called zone leveling. In this manner, dopants can be added in a controlled and uniform manner to silicon. For details and a derivation of the equations, the reader is referred to the original paper by Pfann (14).

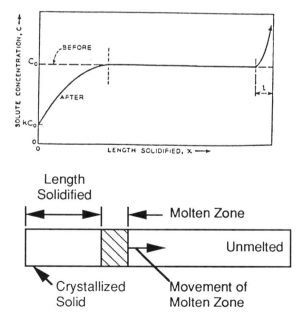

Figure 21: Zone Refining (14).

The distribution coefficient may be obtained directly from the phase diagram, however at very low concentrations the accuracy may be poor. It is the ratio of the slope of the liquidus to that of the solidus line on the phase diagram. Previously electrical measurements were used for determination of the distribution coefficient. Another method is to determine the change in impurity content when a portion of a doped crystal is melted and recrystallized. For more accurate data on solubility and distribution, electron paramagnetic resonance combined with neutron activation analysis (15) is now used. The original work on the distribution coefficients is still in use today (16). Figure 22 shows the solid solubilities of impurities in silicon. An exception to the above behavior is boron. Because the distribution coefficient for boron is very close to one, the recrystallization by zone refining is difficult. Because of the low solubility of these elements, a logarithmic scale has been used in Figure 22 for the concentration.

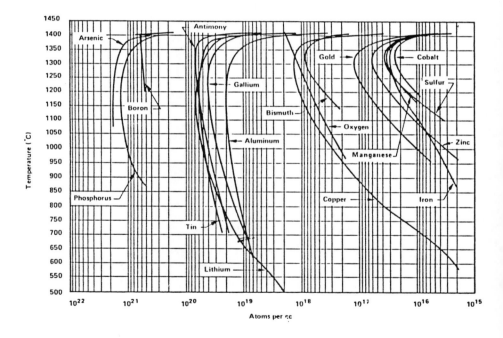

Figure 22: Solubilities of Impurities in Silicon (16).

6.0 RETROGRADE SOLUBILITY

Consider a phase diagram for silicon and an element which forms a complete solution in the liquid state and is insoluble, or nearly insoluble in the solid state. If the eutectic temperature is low in relation to the melting point of silicon, the solubility of the element will exhibit retrograde solubility. For retrograde solubility, maximum solubility exists at a temperature higher than the eutectic temperature. For all solid materials the solubility increases, at low temperatures, according to the relationship, equation 5. This increase is due to the higher entropy of the disordered state forming defects such as vacancies.

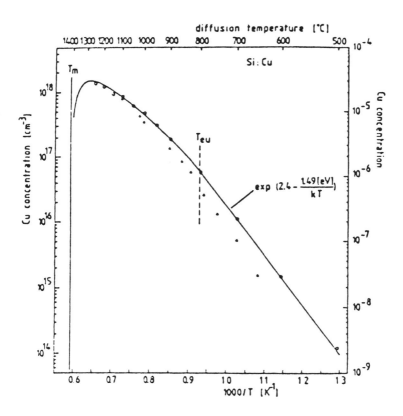

Figure 23:   Retrograde Solubility of Cu in Si (15).

Since we form a complete solution in the liquid state, the amount in solid solution must go to zero at the temperature increase up to the melting point. The result of the solubility increase in the solid phase and the reduction in the amount of solid phase as the melting point of the pure element is approached, results in a maximum in the solubility. For this to occur, there must be a large difference in the melting point of the pure metal and the alloy phase The retrograde solubility of copper in silicon is shown in Figure 23 (15). Fischler (17) found an empirical relationship, Figure 17, between the maximum solubility and the distribution coefficient. This has been shown to be true from thermodynamic considerations (18).

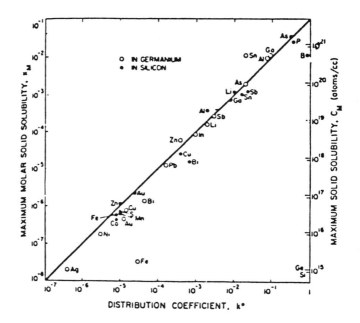

Figure 24: Correlation Between Maximum Solubility and Segregation Coefficient in Silicon (18).

The relationship is

$$x_m = 0.1\, k_o \tag{8}$$

Where x_m is the maximum solubility and k_o is the distribution coefficient. Thus by knowing the maximum solubility the distribution coefficient may be estimated.

SILICON PHASE DIAGRAMS

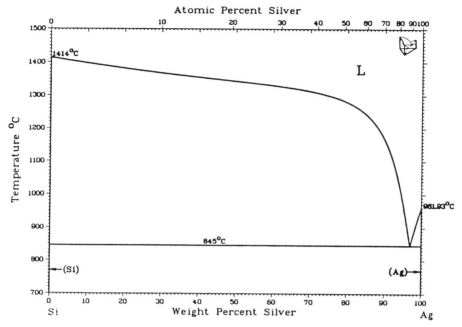

Figure 25: Silver - Silicon Phase Diagram (9).

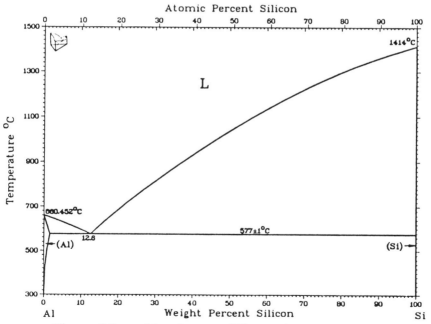

Figure 26: Aluminum - Silicon Phase Diagram (9).

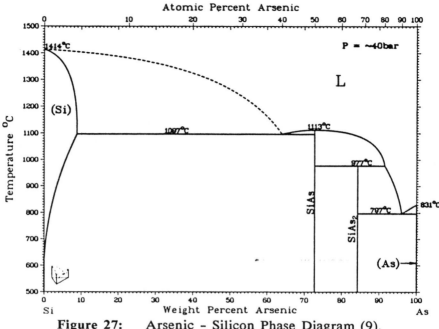

Figure 27: Arsenic - Silicon Phase Diagram (9).

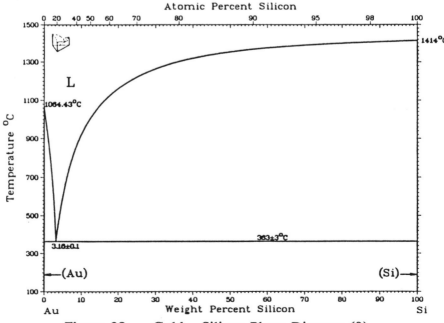

Figure 28: Gold - Silicon Phase Diagram (9).

Silicon Phase Diagrams 765

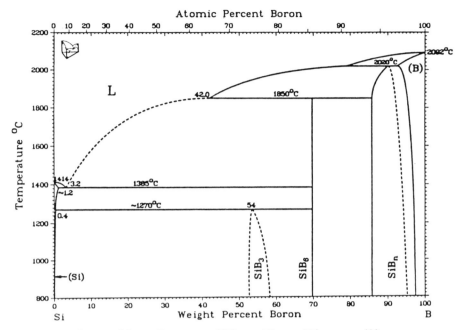

Figure 29: Boron - Silicon Phase Diagram (9).

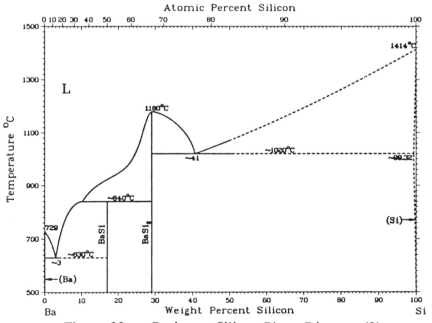

Figure 30: Barium - Silicon Phase Diagram (9).

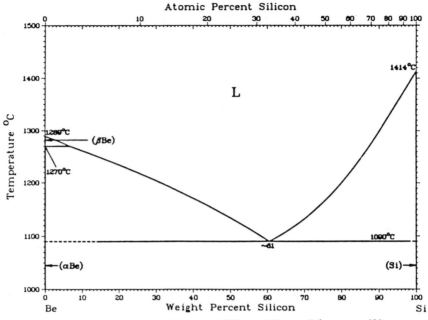

Figure 31: Beryllium - Silicon Phase Diagram (9).

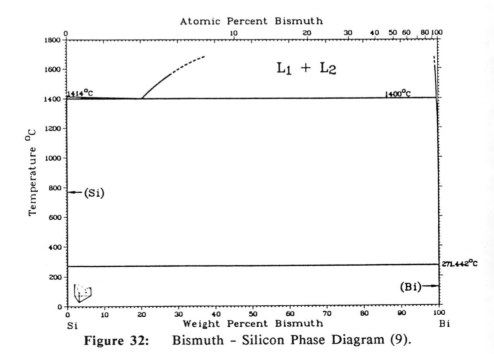

Figure 32: Bismuth - Silicon Phase Diagram (9).

Silicon Phase Diagrams 767

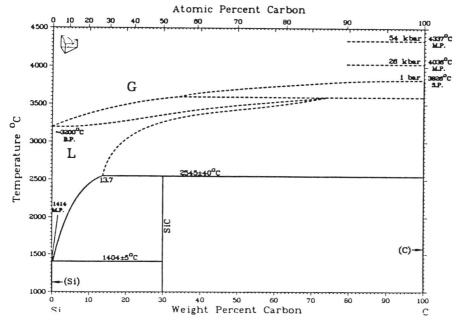

Figure 33: Carbon - Silicon Phase Diagram (9).

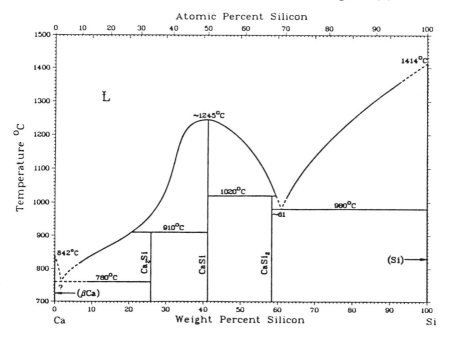

Figure 34: Calcium - Silicon Phase Diagram (9).

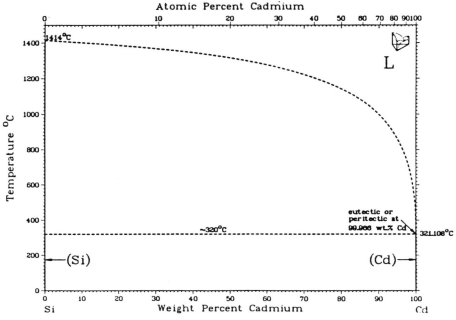

Figure 35: Cadmium - Silicon Phase Diagram (9).

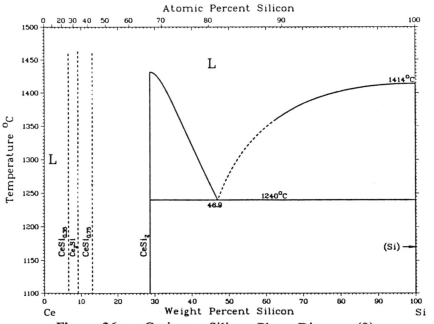

Figure 36: Cerium - Silicon Phase Diagram (9).

Silicon Phase Diagrams 769

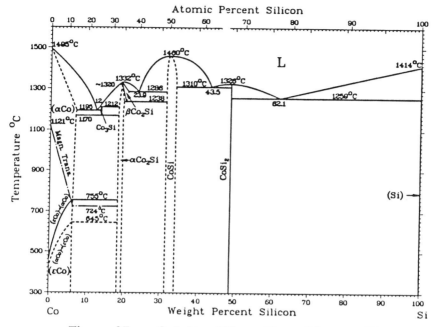

Figure 37: Cobalt - Silicon Phase Diagram (9).

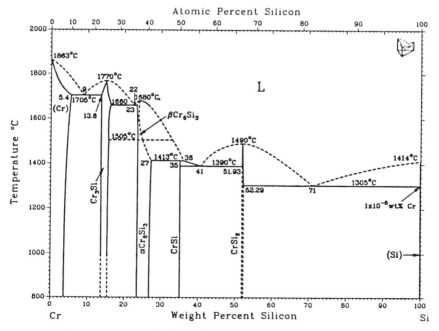

Figure 38: Chromium - Silicon Phase Diagram (9).

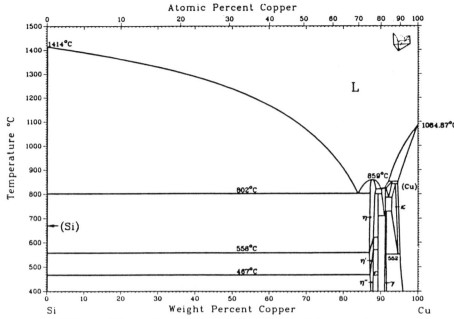

Figure 39: Copper - Silicon Phase Diagram (9).

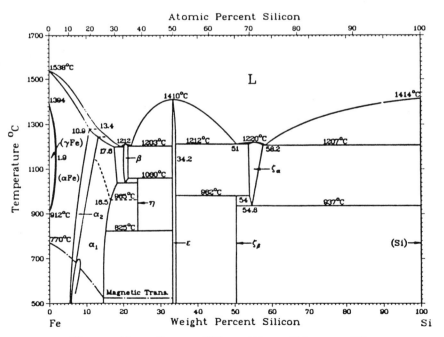

Figure 40: Iron - Silicon Phase Diagram (9).

Silicon Phase Diagrams 771

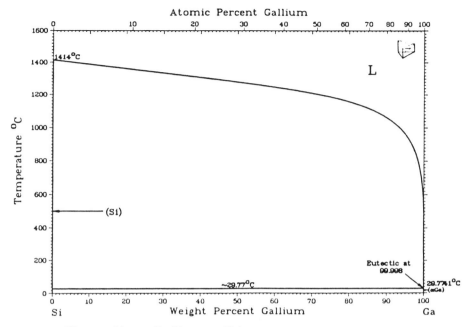

Figure 41: Gallium – Silicon Phase Diagram (9).

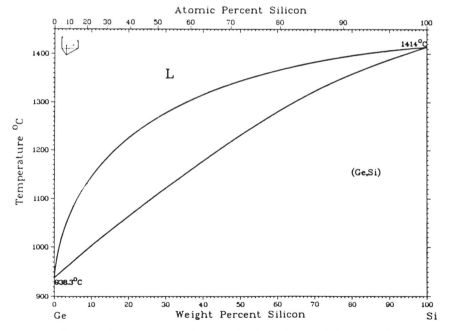

Figure 42: Germanium – Silicon Phase Diagram (9).

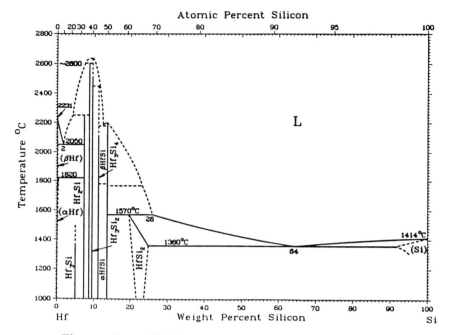

Figure 43: Hafnium - Silicon Phase Diagram (9).

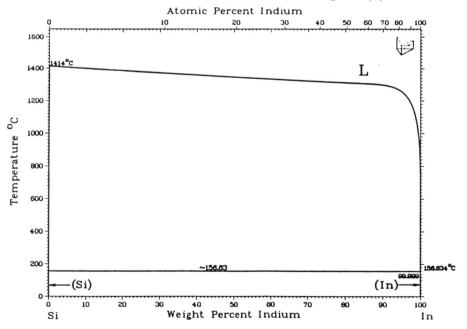

Figure 44: Indium - Silicon Phase Diagram (9).

Silicon Phase Diagrams 773

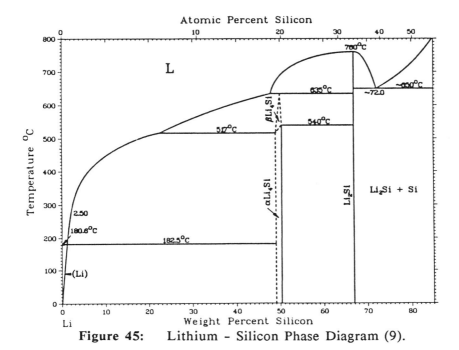

Figure 45: Lithium - Silicon Phase Diagram (9).

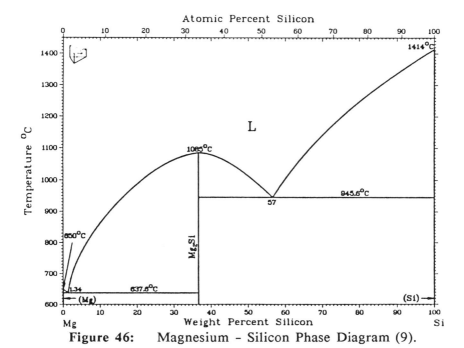

Figure 46: Magnesium - Silicon Phase Diagram (9).

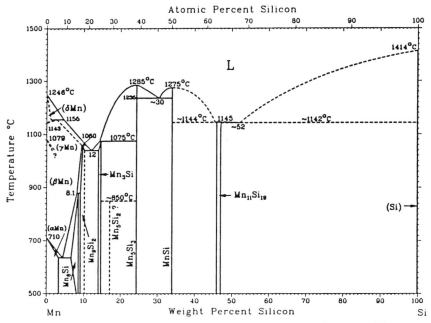

Figure 47: Manganese - Silicon Phase Diagram (9).

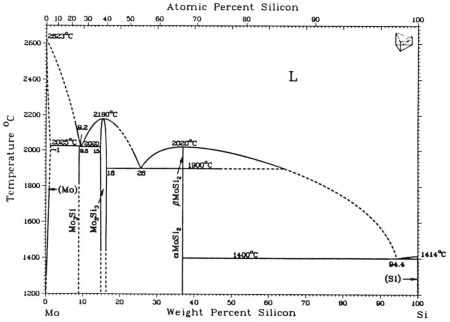

Figure 48: Molybdenum - Silicon Phase Diagram (9).

Silicon Phase Diagrams 775

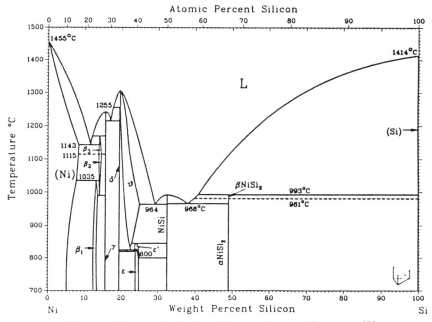

Figure 49: Nickel - Silicon Phase Diagram (9).

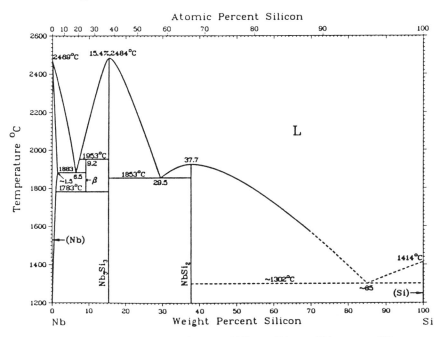

Figure 50: Niobium - Silicon Phase Diagram (9).

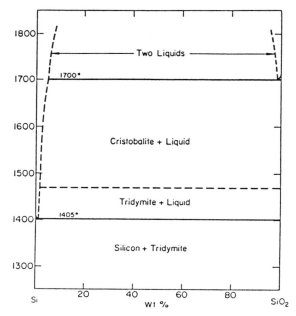

Figure 51: Oxygen - Silicon Phase Diagram (9).

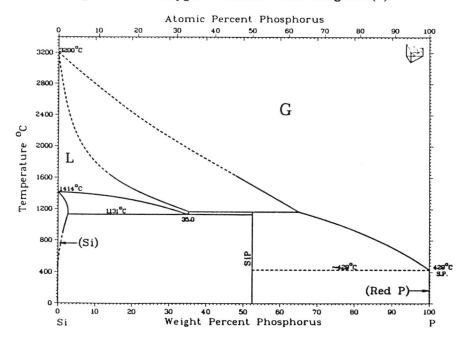

Figure 52: Phosphorus - Silicon Phase Diagram (9).

Silicon Phase Diagrams 777

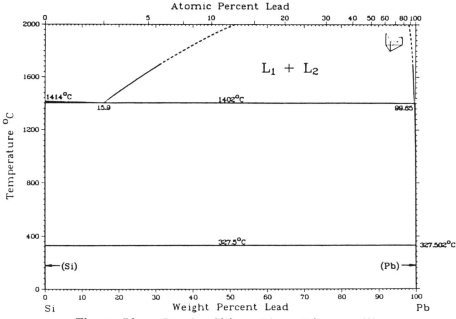

Figure 53: Lead - Silicon Phase Diagram (9).

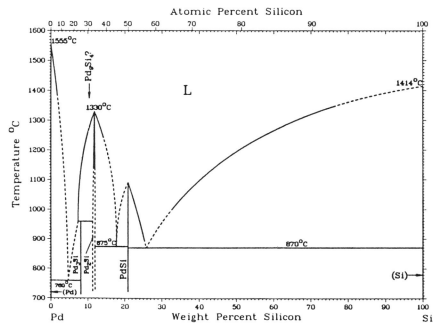

Figure 54: Palladium - Silicon Phase Diagram (9).

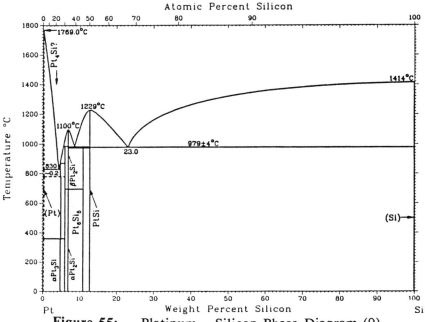

Figure 55: Platinum - Silicon Phase Diagram (9).

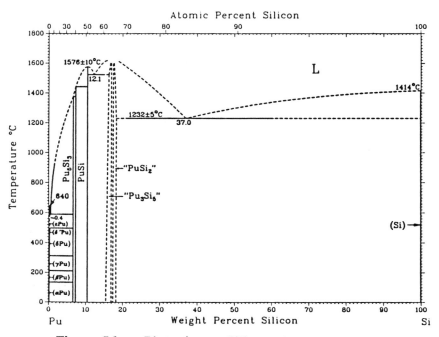

Figure 56: Plutonium - Silicon Phase Diagram (9).

Silicon Phase Diagrams 779

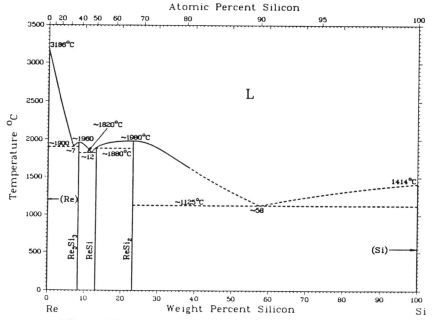

Figure 57: Rhenium - Silicon Phase Diagram (9).

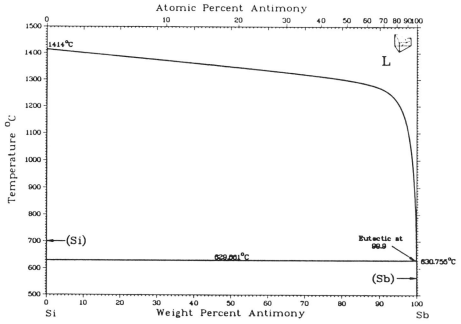

Figure 58: Antimony - Silicon Phase Diagram (9).

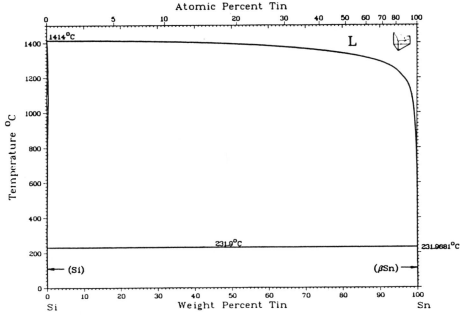

Figure 59: Tin − Silicon Phase Diagram (9).

Si-Rich Region of the Si-Sn Phase Diagram

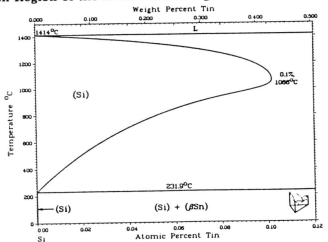

Figure 60: Detail of Tin − Silicon Phase Diagram (9).

Silicon Phase Diagrams 781

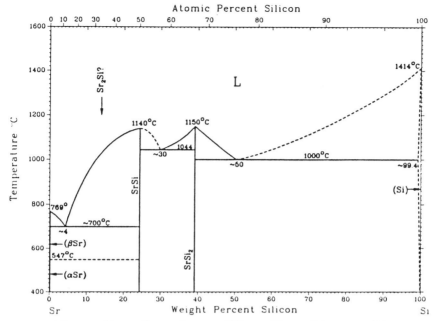

Figure 61: Strontium - Silicon Phase Diagram (9).

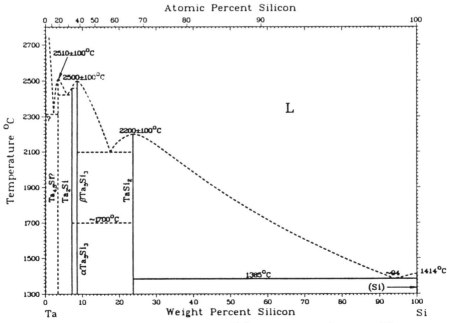

Figure 62: Tantalum - Silicon Phase Diagram (9).

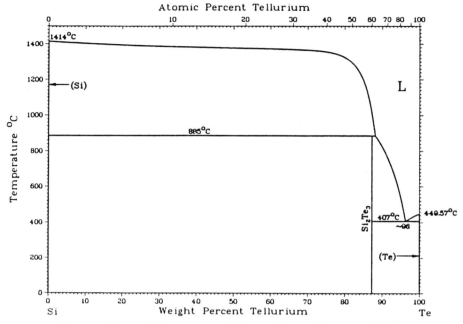

Figure 63: Tellurium – Silicon Phase Diagram (9).

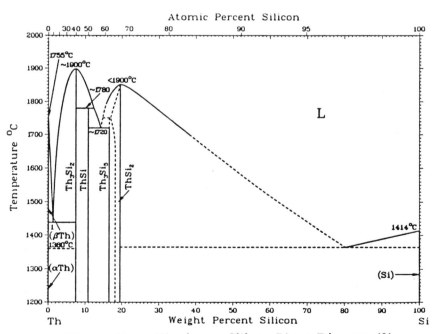

Figure 64: Thorium – Silicon Phase Diagram (9).

Silicon Phase Diagrams 783

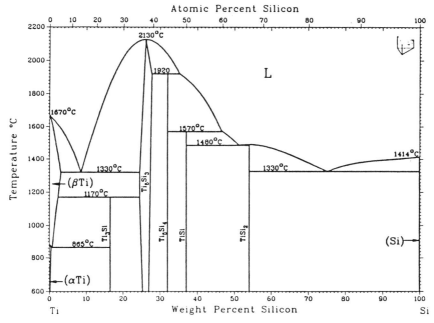

Figure 65: Titanium - Silicon Phase Diagram (9).

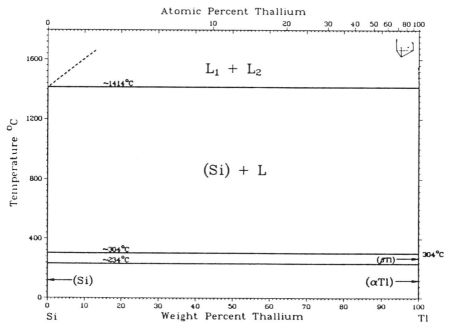

Figure 66: Thallium - Silicon Phase Diagram (9).

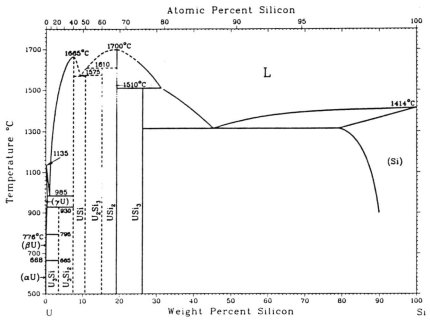

Figure 67: Uranium - Silicon Phase Diagram (9).

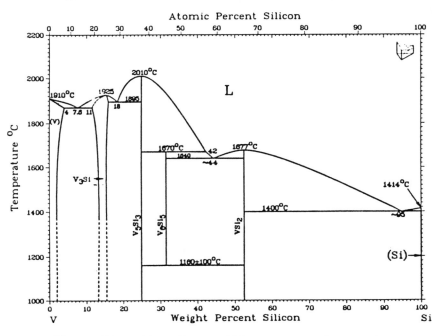

Figure 68: Vanadium - Silicon Phase Diagram (9).

Silicon Phase Diagrams 785

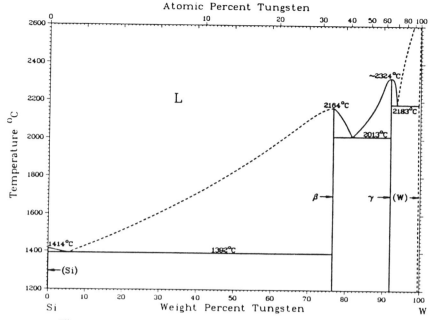

Figure 69: Tungsten - Silicon Phase Diagram (9).

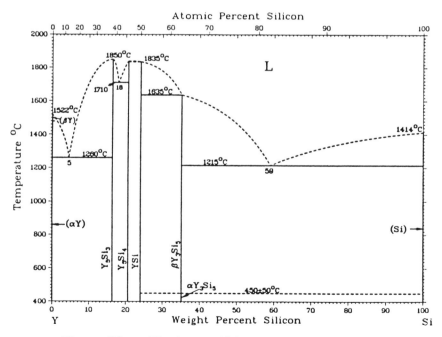

Figure 70: Yttrium - Silicon Phase Diagram (9).

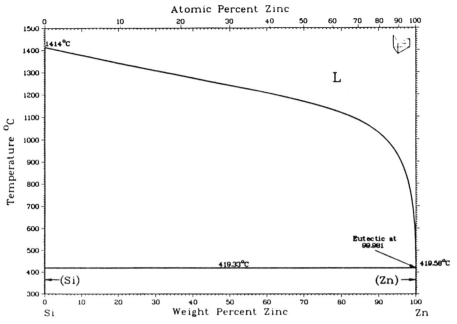

Figure 71: Zinc - Silicon Phase Diagram (9).

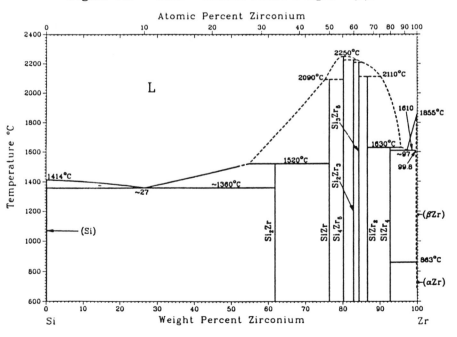

Figure 72: Zirconium - Silicon Phase Diagram (9).

REFERENCES

1. Gibbs, J.W., *The Collected Works of J. Willard Gibbs*, Dover, N.Y. (1961)

2. Cottrell, A.H., *Theoretical Structural Metallurgy*, second ed., p. 126, St. Martins Press, New York (1960)

3. Phillips, J.C., *Bands and Bonds in Semiconductors*, Academic Press, New York, (1973)

4. Hume-Rothery, W. and Raynor, G.V., *The Structure of Metals and Alloys*, 4th. rev. ed., p. 100, Inst. of Metals, London, (1962)

5. Dean, J.A. (ed.), *Lange's Handbook of Chemistry*, 13th. ed., McGraw-Hill, New York, (1985)

6. Guy, A.G., *Elements of Physical Metallurgy*, second ed., pp. 171-173 Addison-Wesley, Reading, MA, (1959)

7. Hume-Rothery, op. cit., p. 228

8. Barrett, C.S., *Structure of Metals*, second ed., p. 196-212, McGraw-Hill, New York (1952)

9. *Bulletin of Alloy Phase Diagrams*, American Society for Metals, Metals Park, OH

10. Massalski, T.B., (ed.), *Binary Alloy Phase Diagrams*, Vol. I and II, ASM, Metals Park, OH (1986)

11. Levin, E.M., et. al., (eds), Phase Diagrams for Ceramists, Vols I to V, *Am. Cer. Soc.*, Columbus, OH (1969-1985)

12. Hall, R.N., Segregation of Impurities During the Growth of Germanium and Silicon Crystals, *J. Phys. Chem.* 57, 827 (1953)

13. Cottrell, A.H., op. cit., p. 157

14. Pfann, W.G., Principles of Zone-Melting, *Trans. AIME* 194, 747-753 (1952)

15. Weber, E.R., Transition Metals in Silicon, *Appl. Phys.* A30, 1-22 (1983)

16. Trumbore, F., *Bell Sys. Tech. J.* 39, 205 (1960)

17. Fischler, S., Correlation Between Maximum Solubility and Distribution Coefficients for Impurities in Ge and Si, *J. Appl. Phys.* 33, 1685 (1962)

18. Statz, H., Ltr. to Ed., *J. Appl. Phys.* 34, 700 (1963)

Index

A Center, 506
Abrasive, 197
Activation Analysis, 459, 488, 530, 534, 760
Adatom, 293
Alcohol Dryer, 276
Aluminum, 2, 39, 161, 265
Amorphous, 273
Antimony, 139, 265, 307, 321, 487, 494
Antireflection Coating, 410
Arc Furnace, 2, 36
Area Defect, 553
Arsenic, 139, 321, 669
Arsine, 302, 305, 654
Autodoping, 305, 312, 335
Avalanche Multiplication, 567
Backseal, 310
Backside Damage, 240
Bake, 283
Band Structure, 383
Bell Jar
  deposit, 45
  metal, 46, 52, 64
  quartz, 73
  reactor, 42, 49
Blade
  dressing, 213
  tension, 209
Boiling Point, 5, 10, 292
Boron, 39, 139
Bow, 213, 247, 521
Brillouin Zone, 384
Bulk Recombination, 572
Burgers Vector, 370
Buried Layer, 260, 322
C-V Profile, 311
Carbon, 39, 150, 177, 283, 365, 417, 451, 489, 521
Carbon in Silicon
  complexes, 530
  properties, 528
  solubility, 527

Carrier Density, 394
Carrier Lifetime, 550
Chlorosilanes
  accentric factor, 10
  autoignition temperature, 10
  boiling point, 5, 10, 292
  critical constants, 9
  enthalpy of vaporization, 19
  equilibrium partial pressure, 16, 297, 303
  flash point, 10
  free energy, 14, 16
  heat capacity, 13
  liquid density, 10
  materials of construction, 26
  melting point, 10, 292
  molecular weight, 10
  physical properties, 9
  standard enthalpy of formation, 14
  standard entropy of formation, 14
  surface tension, 20
  thermal conductivity, 22, 24
  vapor pressure, 9, 11, 292
  viscosity, 21
Cleaning, 235
Coefficient
  absorption, 414
  distribution, 107, 109
  effective distribution, 110, 121
Compression Modulus, 424
Conduction, 394
Conduction Band, 372
Conductivity, 385
Convection, 115
Conversion Factors, 28
Creep, 318
Critical Constants, 9
Cropping, 202
Crucible
  quartz, 125, 147, 364
  rotation, 118
  silicon nitride, 128, 154
Crystal

habit, 352
mounting, 211
shaping, 201
Crystal Growth
continuous, 95
Czochralski, 79, 84, 94
dendritic web, 167
dislocation free, 128, 174
double crucible, 146
EFG, 167
float zone, 85
future, 177
heat balance, 99
mechanisms, 105
pedestal, 96
pullers, 125
stacking faults, 476
theory, 97
Crystallography, 323, 349
CVD, 262, 641
Czochralski
continuous, 162
crystal habit, 352
deformation, 428
diameter control, 136, 201
hexagonal habit, 135
interface shape, 102
magnetic, 165
pull rate, 100
semicontinuous, 159
square habit, 134, 166
temperature gradient, 100
Dangling Bonds, 692
Dark Resistance, 583
Defect
bulk, 483
concentration, 694
crystal, 362
formation, 98
Frenkel, 362
lifetime, 551
Schottky, 362
swirl, 176
Degrees of Freedom, 733
Dember Voltage, 591
Dendrite, 168, 371
Density of States, 376
Denuded Zone, 148, 366, 479, 481, 525, 570
Device
bipolar, 259, 526, 711
CCD, 527
CMOS, 262, 310, 487, 524, 580, 624
DRAM, 262, 486, 526

EEPROM, 651, 673
FET, 272
forward-biased diode, 571
JFET, 262
MOS, 262, 486, 576, 579, 604, 677, 684
MOSFET, 567, 624
NMOS, 524
photoconductor, 566
photodiode, 566
processing, 524
RAM, 684
reverse-biased diode, 574
SRAM, 526
Diamond Saw, 202, 207
Diborane, 302, 654
Dichlorosilane
epitaxy, 290
hazards, 25
kinetics, 295
manufacture, 7, 37
precursor, 1
pyrolysis, 288
rating, 66, 649, 652, 716
Dichroism, 502
Die, 171
Diffusion
barrier, 348
current, 574
length, 579, 598, 614
pipe, 320
rate, 643
Disproportionation, 287
Disilane, 655
Dislocation, 98, 314, 368, 467, 521
Dislocation Density, 268
Disposal, 199
Donor
activation energy, 382
density, 386
thermal, 148, 238, 365
Dopant, 301, 364, 394
Doping
antimony, 178
nitrogen, 95
oxygen, 149, 714
phosphorus, 175
precipitate, 697
stacking faults, 477
techniques, 138
transmutation, 175, 653
Edge Contour, 218, 221, 248
Elastic Constants, 422
Electrical Characterization, 390
Electron Density, 386

Electronic Conduction, 385
Ellipsometry, 680
Emissivity, 420
Energy Structure, 379
Enthalpy of Vaporization, 19
Environment, 201, 653
Epitaxy
  denuding, 486
  dopants, 302
  equipment, 325
  future, 335
  growth rate, 271, 289
  kinetics, 293
  liquid phase, 270
  mechanism, 293
  molecular beam, 263
  reduced pressure, 312
  silicon, 258
  slip free, 330
Equilibrium
  partial pressure, 16, 297, 303
  silicon yield, 18
Etch, 198, 206, 284, 672
Etch Pit, 357, 370
Etchant
  acidic, 199
  basic, 201
  caustic, 357
  Secco, 475, 531
  Sirtl, 229
  Wright, 320, 531
Eutectic, 732
Exhaust Gas Recovery, 52
Explosion Hazard, 24
Feedstock Ratings, 61
Fermi Function, 375
Fermi Level, 669
Ferrosilicon, 36
Fiber Optics, 52
Fire Hazard, 24
Flash Point, 10
Flatness, 246
Flats, 205, 244
Float Zone, 173, 365, 428, 459, 533, 565, 756
Fluorosilicic Acid, 8
Forbidden Energy Gap, 399
Fracture, 198, 422, 430
Free Carrier Absorption, 419, 493
Free Energy, 14, 16, 97, 107, 128, 299, 461, 552, 756
Fumed Silica, 6
Future
  crystal growth, 177

  epitaxy, 335
  polysilicon, 86
  wafers, 249
GaAs, 559
Gallium, 265
Gas Phase Clean, 277
Gate Electrode, 696
Gate Voltage, 612
Generation Lifetime, 566, 568
GeO_4, 518
Getter, 148, 239, 365, 484, 623, 703
Gold, 39, 368, 401
Grain
  boundary, 657, 665
  growth, 669
  size, 690
Graphite, 127, 280
Grinding, 196, 203, 226
Hall Effect, 397
Hardness, 433
Haze, 233, 283, 320
Health Hazard, 24
Heat of Vaporization, 19
Heat Treatment, 238
Heating
  induction, 328, 332
  radiant, 334
  resistance, 332
Heavy Metals, 39
Heteroepitaxy, 259
High-Level Injection, 561
Hillock, 316
Hole Density, 386
Homoepitaxy, 259
Hydrofluoric Acid, 275, 672
Hydrogen, 366
Impurities
  absorption, 417
  acceptor, 374
  atomic, 363
  concentration, 107
  deep level, 399
  donor, 374
  dopant, 419
  mass transport, 106
  retrograde solubility, 761
  segregation coefficient, 756
  solubility, 140, 363, 760
Index of Refraction, 410
Integrated Circuit, 260
Interfacial Breakdown, 124
Intrinsic Carrier Density, 378
Inversion Layer, 611
Iron, 320

Kerf Loss, 217
Komatsu Process, 64
Lapping, 197, 222, 225
Lattice
  adsorption, 416
  damage, 220
  scattering, 386
Levitation, 175
Lifetime
  electron, 565
  generation, 550
  generation-recombination centers, 553
  hole, 565
  metallic impurities, 554
  photoconductive decay, 582
  recombination, 550
Lifetime Measurement
  comparison, 601
  diode reverse recovery, 595
  gate controlled diode, 617
  open circuit voltage decay, 588
  photoconductive decay, 582
  pulsed MOS capacitor, 604, 614
  surface photovoltage, 598
  surface treatment, 602
Line Defect, 553
Low-Level Injection, 560
LPCVD, 642, 651
Magnesium, 519
Magnetic Field, 112, 119, 155, 365, 397
Majority Carrier Concentration, 575
Mass Transport, 647
Materials of Construction, 26
Megasonic Clean, 276, 314
Melt Back, 129
Melt Level, 136
Melting Point, 10, 292
Metal
  chlorides, 286
  in silicon, 367
Microcracks, 211
Microprecipitate, 320
Microwave, 587
Miller Indices, 351
Minority Carrier Concentration, 575, 593
Minority Carrier Lifetime, 413, 585
Molecular Weight, 10
Monochlorosilane, 7
Morphology, 313
Neck-In, 129
Nickel, 321
Nitrogen, 154, 366, 417, 451, 533
Nucleation, 299
Nuclei, 290

Optical Properties, 409
Orientation, 245, 313, 355, 433
Out Diffusion, 480
Oxidation Precipitation
  effect of doping, 473
  lifetime, 570
  morphology, 506
  retardation, 472
Oxide, 277, 665
Oxygen, 39, 153, 156, 364, 417, 451, 570
Oxygen Cluster, 459, 463, 505, 522
Oxygen in Silicon
  A center, 507
  diffusion coefficient, 456
  diffusivity, 456
  effects on thermal donor, 508
  heat of solution, 455
  interstitial, 496, 504, 514
  IR Absorption, 492
  measurement, 487
  out diffusion, 479
  phase diagram, 460
  properties, 452, 500
  segregation coefficient, 458
  solubility, 453, 505
  substitutional, 472, 499, 504
  thermal donor, 516
Oxygen Precipitate
  critical radius, 467
  morphology, 467
  nucleation, 463
  nucleation rate, 465
  nuclei density, 470
  thermal history, 469
Packaging, 241
Passivation, 695
Pattern Shift, 324
Peltier Effect, 409
Peritectic, 739
Phase Diagram
  determination, 754
  example, 731
  free Energy, 734
  intermediate phase, 746, 753
  lever rule, 753
  Si-Ge, 739, 108
Phase Rule, 733
Phosphine, 302, 653
Phosphorus, 139, 669
Phosphosilicate Glass, 701
Photoconductivity, 372, 413
Photoluminescence, 419
Photoresist, 220
Photoresistance, 583

Index 793

Photovoltaic, 34, 54, 159, 604
Piezoresistivity, 425
Pitting, 286
Plasma Resonance, 420
POCl$_3$, 675
Point Defect, 98, 239, 362, 553
Poisson's Ratio, 425
Polishing, 196, 226, 231, 233
Polysilicon
  backseal, 310
  carbon, 531
  conversion efficiency, 55
  counterelectrode, 702
  demand, 34
  deposition efficiency, 51
  deposition rate, 50, 51, 645
  emitter, 713
  epitaxy, 288
  evaluation, 84
  feedstock comparison, 77
  feedstocks, 37, 60
  film, 640
  future, 86
  gas depletion, 649
  gate, 700, 704
  grains, 697
  impurities, 39
  index of refraction, 672
  interconnections, 701
  layer, 273
  manufacture, 33
  manufacturers, 35
  oxygen doped, 714
  pop corn, 49
  precursors, 1, 4, 35, 652
  pressure sensor, 715
  reactors, 41
  recrystallized, 710
  rod, 6, 43, 47, 48, 159, 173
Polysilicon Film
  evaluation, 660
  grain size, 660
  structure, 662
  structure-temperature, 659
  texture, 663
Polysilicon Evaluation
  carbon, 85
  heavy metals, 85
  minority carrier life time, 85
  oxygen, 85
  photoluminescence, 84
  resistivity, 84
Polysilicon Film
  accumulation layer, 708
  amorphous, 659, 683
  boundary layer, 643
  carrier trapping, 687
  CCD, 702
  collector sink, 717
  columnar, 665
  conducting channel, 706
  conduction, 684
  critical cluster, 657
  deposition, 648
  device isolation, 715
  diode, 714
  dopant segregation, 686
  EEPROM, 703
  energy barrier, 688
  gas dynamics, 643
  grain orientation, 673, 681
  heterogeneous reaction, 656
  homogeneous reaction, 656
  hydrogen, 705
  laser melt, 695
  MOS, 699, 706
  oxidation rate, 673
  oxide thickness, 673
  RAM, 703
  resistivity, 685
  resistor, 704
  structure, 651, 670
  surface diffusion, 665
  thermionic emission, 690
  thickness, 647
  trap density, 708
  trench isolation, 717
Polysilicon Oxide
  conduction, 681
  dopant effect, 675
  grain boundary, 678
  point defects, 674
  precipitates, 679, 683
  thickness, 677, 680
Polysilicon Plant
  capital cost, 53, 55, 59
  operating cost, 59
Precipitate, 371
Precipitation, 430
Preclean, 275
Pyramid, 316
Quality Control, 241
Quartz, 347
RCA Clean, 275
Reactor
  cylinder, 330
  epitaxial, 641
  flow models, 326

fluidized bed, 81
free space, 78
geometry, 645
horizontal, 327, 645
low-pressure, 646
pressure, 334
quartz, 331, 334
vertical, 329
Recombination, 585
Recombination Lifetime, 557
Rectification, 405
Reflectance, 670
Regrowth, 273
Resistivity, 81, 140, 152, 385, 390, 394, 654, 694
S-pit, 320
Safety, 24, 43, 201, 276, 653, 655
Saturation Current, 574
Screw Dislocation, 105
Seebeck Effect, 407
Segregation Coefficients, 756
Self-Interstitial, 551
Shear Modulus, 425, 425
Si_2O, 495
$SiCl_2$, 6, 287, 295
Side Facets, 130
Siemens Process, 34, 42
SiI_2, 287
Silane
adsorption, 294
epitaxy, 279
hazards, 25
manufacture, 7
precursor, 1
pyrolysis, 288
rating, 61
reaction rate, 294
Si_3N_4, 533, 648, 652
Silane Properties (see Chlorosilanes)
Silanes, 38, 290
Silica, 2, 35, 52, 431
Silica Sol, 230
Silicide, 27, 365, 451, 698
Silicon
absorption, 412
deposition, 18
deposition rate, 303
electrical properties, 371
etching, 18
extrinsic, 383
intrinsic, 383
lattice constant, 349
material properties, 347
mechanical properties, 422

mechanical strength, 427
metallurgical grade, 2, 4, 36, 95
metals in, 320
moduli, 424
phase diagrams, 731, 763
physical constants, 433
plastic deformation, 427
precursors, 1
thermal Properties, 422
thermodynamic constants, 433
uncompensated, 390, 394
Silicon Carbide, 6, 157, 451, 529
Silicon Monoxide, 3, 39, 126, 148, 451, 454, 488
Silicon Nitride, 274, 310, 431, 451, 533, 648
Silicon Tetrabromide, 96
Silicon Tetrachloride
by-product, 4
conversion efficiency, 57
decomposition rate, 293
DOE, 34
epitaxy, 291
kinetics, 296
manufacture, 5
markets, 6
precursor, 1
rating, 75
residence time, 57, 652, 652, 716
Silicon Tetrafluoride, 9
Silicone
industry, 2, 291
oils, 50, 52, 54
Silylene, 294, 655, 656
SiO_3, 512
SiO_4, 509
SiO_x, 493, 506, 521, 530
Slicing, 208
Slim Rod, 46
Slip, 198, 317, 332, 368, 422, 428, 522
Slurry Sawing, 217
Sodium, 39
Solar Cell, 95, 161, 168, 173, 410, 555, 560, 592, 624
Solar Energy, 413
Solid Solution, 747
Space-Charge Region, 571, 591
Spectroscopy
deep-level transient, 401
FTIR, 85, 530
ICP, 86
IR, 489, 536
SIMS, 489, 577
TEM, 660

Spreading Resistance, 302, 483, 576
Stacking Fault, 269, 313, 368, 463, 474, 478
Stagnant Layer, 110
Standard Enthalpy of Formation, 14
Standard Entropy of Formation, 14
Statistics, 375
Stress-Strain Curve, 429
Striation, 114, 141
Strip Heater, 273
Sulfur, 519
Supercooling, 121, 143, 161
Supersaturation, 299, 455
Surface
  generation Velocity, 619
  inspection, 248
  photovoltage, 599
  recombination, 572, 592
  roughness, 671
  treatment, 602
Swarf, 211
Swirl, 472, 504, 533
Thermal Conductivity
  silanes, 22, 24
  silicon, 432
Thermal Donor, 463, 474, 507
Thermal Excitation, 568
Thermal Expansion, 317, 431
Thermal Shock, 133
Thermoelectric Effect, 407
Thickness, 245
Thomson Effect, 409
Tin, 270
Titanium, 320
Traps, 587
Tribromosilane, 34
Trichlorosilane
  conversion efficiency, 50
  decomposition, 5
  dissociation, 42, 52
  epitaxy, 291
  feedstock, 41
  hazards, 25
  kinetics, 296
  manufacture, 3, 37
  precursor, 1
  purification, 4
  rating, 70, 527
Twin Planes, 131
Unit Cell, 350
Vacancy, 362, 551, 669
Valence Band, 372
Vapor Pressure, 9, 11, 292
Viscosity, 21

Volume Defect, 553
Wafer
  diameter, 244
  flatness, 224
  marking, 240
  measurement, 244
  mounting, 232
  preparation, 192
  shaping, 196
  specifications, 242
  strength, 523
  thickness, 224
Wafering, 207, 216, 249
Warp, 149, 179, 247, 422, 430, 523, 716
Water Adsorption, 280
Wave Guides, 6
Wires
  Rogers/Heitz, 47
X-Ray Diffraction, 356, 662, 754
Young's Modulus, 424
Zerbst Plot, 615